TJ
267
K57
2011

Principles of Turbomachinery

Principles of Turbomachinery

Seppo A. Korpela
The Ohio State University

WILEY

A JOHN WILEY & SONS, INC., PUBLICATION

Copyright © 2011 by John Wiley & Sons, Inc. All rights reserved.

Published by John Wiley & Sons, Inc., Hoboken, New Jersey.
Published simultaneously in Canada.

No part of this publication may be reproduced, stored in a retrieval system or transmitted in any form or by any means, electronic, mechanical, photocopying, recording, scanning or otherwise, except as permitted under Section 107 or 108 of the 1976 United States Copyright Act, without either the prior written permission of the Publisher, or authorization through payment of the appropriate per-copy fee to the Copyright Clearance Center, Inc., 222 Rosewood Drive, Danvers, MA 01923, (978) 750-8400, fax (978) 750-4470, or on the web at www.copyright.com. Requests to the Publisher for permission should be addressed to the Permissions Department, John Wiley & Sons, Inc., 111 River Street, Hoboken, NJ 07030, (201) 748-6011, fax (201) 748-6008, or online at http://www.wiley.com/go/permission.

Limit of Liability/Disclaimer of Warranty: While the publisher and author have used their best efforts in preparing this book, they make no representation or warranties with respect to the accuracy or completeness of the contents of this book and specifically disclaim any implied warranties of merchantability or fitness for a particular purpose. No warranty may be created or extended by sales representatives or written sales materials. The advice and strategies contained herein may not be suitable for your situation. You should consult with a professional where appropriate. Neither the publisher nor author shall be liable for any loss of profit or any other commercial damages, including but not limited to special, incidental, consequential, or other damages.

For general information on our other products and services please contact our Customer Care Department within the United States at (800) 762-2974, outside the United States at (317) 572-3993 or fax (317) 572-4002.

Wiley also publishes its books in a variety of electronic formats. Some content that appears in print, however, may not be available in electronic formats. For more information about Wiley products, visit our web site at www.wiley.com.

Library of Congress Cataloging-in-Publication Data:

Korpela, S. A.
  Principles of turbomachinery / Seppo A. Korpela. — 1st ed.
      p. cm.
  Includes index.
  ISBN 978-0-470-53672-8 (hardback)
  1. Turbomachines. I. Title.
  TJ267.K57 2011
  621.406—dc23   2011026170

Printed in the United States of America.

10 9 8 7 6 5 4 3 2 1

*To my wife Terttu,
to our daughter Liisa,
and to the memory of our
daughter Katja*

CONTENTS

Foreword   xiii

Acknowledgments   xv

1 Introduction   1
- 1.1 Energy and fluid machines   1
  - 1.1.1 Energy conversion of fossil fuels   1
  - 1.1.2 Steam turbines   2
  - 1.1.3 Gas turbines   3
  - 1.1.4 Hydraulic turbines   4
  - 1.1.5 Wind turbines   5
  - 1.1.6 Compressors   5
  - 1.1.7 Pumps and blowers   5
  - 1.1.8 Other uses and issues   6
- 1.2 Historical survey   7
  - 1.2.1 Water power   7
  - 1.2.2 Wind turbines   8
  - 1.2.3 Steam turbines   9
  - 1.2.4 Jet propulsion   10
  - 1.2.5 Industrial turbines   11
  - 1.2.6 Note on units   12

2 Principles of Thermodynamics and Fluid Flow

- 2.1 Mass conservation principle
- 2.2 First law of thermodynamics
- 2.3 Second law of thermodynamics
  - 2.3.1 $T\,ds$ equations
- 2.4 Equations of state
  - 2.4.1 Properties of steam
  - 2.4.2 Ideal gases
  - 2.4.3 Air tables and isentropic relations
  - 2.4.4 Ideal gas mixtures
  - 2.4.5 Incompressibility
  - 2.4.6 Stagnation state
- 2.5 Efficiency
  - 2.5.1 Efficiency measures
  - 2.5.2 Thermodynamic losses
  - 2.5.3 Incompressible fluid
  - 2.5.4 Compressible flows
- 2.6 Momentum balance
- Exercises

3 Compressible Flow through Nozzles

- 3.1 Mach number and the speed of sound
  - 3.1.1 Mach number relations
- 3.2 Isentropic flow with area change
  - 3.2.1 Converging nozzle
  - 3.2.2 Converging–diverging nozzle
- 3.3 Normal shocks
  - 3.3.1 Rankine–Hugoniot relations
- 3.4 Influence of friction in flow through straight nozzles
  - 3.4.1 Polytropic efficiency
  - 3.4.2 Loss coefficients
  - 3.4.3 Nozzle efficiency
  - 3.4.4 Combined Fanno flow and area change
- 3.5 Supersaturation
- 3.6 Prandtl–Meyer expansion
  - 3.6.1 Mach waves
  - 3.6.2 Prandtl–Meyer theory
- 3.7 Flow leaving a turbine nozzle
- Exercises

4 Principles of Turbomachine Analysis

4.1	Velocity triangles		106
4.2	Moment of momentum balance		108
4.3	Energy transfer in turbomachines		109
	4.3.1	Trothalpy and specific work in terms of velocities	113
	4.3.2	Degree of reaction	116
4.4	Utilization		117
4.5	Scaling and similitude		124
	4.5.1	Similitude	124
	4.5.2	Incompressible flow	125
	4.5.3	Shape parameter or specific speed	128
	4.5.4	Compressible flow analysis	128
4.6	Performance characteristics		130
	4.6.1	Compressor performance map	131
	4.6.2	Turbine performance map	131
	Exercises		132

5 Steam Turbines — 135

5.1	Introduction		135
5.2	Impulse turbines		137
	5.2.1	Single-stage impulse turbine	137
	5.2.2	Pressure compounding	146
	5.2.3	Blade shapes	150
	5.2.4	Velocity compounding	152
5.3	Stage with zero reaction		158
5.4	Loss coefficients		160
	Exercises		162

6 Axial Turbines — 165

6.1	Introduction		165
6.2	Turbine stage analysis		167
6.3	Flow and loading coefficients and reaction ratio		171
	6.3.1	Fifty percent (50%) stage	176
	6.3.2	Zero percent (0%) reaction stage	178
	6.3.3	Off-design operation	179
6.4	Three-dimensional flow		181
6.5	Radial equilibrium		181
	6.5.1	Free vortex flow	183
	6.5.2	Fixed blade angle	186
6.6	Constant mass flux		187
6.7	Turbine efficiency and losses		190
	6.7.1	Soderberg loss coefficients	190

CONTENTS

		6.7.2	Stage efficiency	191
		6.7.3	Stagnation pressure losses	192
		6.7.4	Performance charts	198
		6.7.5	Zweifel correlation	203
		6.7.6	Further discussion of losses	204
		6.7.7	Ainley–Mathieson correlation	205
		6.7.8	Secondary loss	209
	6.8	Multistage turbine		214
		6.8.1	Reheat factor in a multistage turbine	214
		6.8.2	Polytropic or small-stage efficiency	216
		Exercises		217

7 Axial Compressors — 22

7.1	Compressor stage analysis	22
	7.1.1 Stage temperature and pressure rise	22
	7.1.2 Analysis of a repeating stage	22
7.2	Design deflection	230
	7.2.1 Compressor performance map	23
7.3	Radial equilibrium	23
	7.3.1 Modified free vortex velocity distribution	23
	7.3.2 Velocity distribution with zero-power exponent	23
	7.3.3 Velocity distribution with first-power exponent	24
7.4	Diffusion factor	24
	7.4.1 Momentum thickness of a boundary layer	24
7.5	Efficiency and losses	24
	7.5.1 Efficiency	24
	7.5.2 Parametric calculations	25
7.6	Cascade aerodynamics	25
	7.6.1 Blade shapes and terms	25
	7.6.2 Blade forces	25
	7.6.3 Other losses	25
	7.6.4 Diffuser performance	25
	7.6.5 Flow deviation and incidence	25
	7.6.6 Multistage compressor	25
	7.6.7 Compressibility effects	26
	Exercises	26

8 Centrifugal Compressors and Pumps — 26

8.1	Compressor analysis	26
	8.1.1 Slip factor	26
	8.1.2 Pressure ratio	26

	8.2	Inlet design	274
		8.2.1 Choking of the inducer	278
	8.3	Exit design	281
		8.3.1 Performance characteristics	281
		8.3.2 Diffusion ratio	283
		8.3.3 Blade height	284
	8.4	Vaneless diffuser	285
	8.5	Centrifugal pumps	290
		8.5.1 Specific speed and specific diameter	294
	8.6	Fans	302
	8.7	Cavitation	302
	8.8	Diffuser and volute design	305
		8.8.1 Vaneless diffuser	305
		8.8.2 Volute design	306
		Exercises	309
9	**Radial Inflow Turbines**		**313**
	9.1	Turbine analysis	314
	9.2	Efficiency	319
	9.3	Specific speed and specific diameter	323
	9.4	Stator flow	329
		9.4.1 Loss coefficients for stator flow	333
	9.5	Design of the inlet of a radial inflow turbine	337
		9.5.1 Minimum inlet Mach number	338
		9.5.2 Blade stagnation Mach number	343
		9.5.3 Inlet relative Mach number	345
	9.6	Design of the Exit	346
		9.6.1 Minimum exit Mach number	346
		9.6.2 Radius ratio r_{3s}/r_2	348
		9.6.3 Blade height-to-radius ratio b_2/r_2	350
		9.6.4 Optimum incidence angle and the number of blades	351
		Exercises	356
10	**Hydraulic Turbines**		**359**
	10.1	Hydroelectric Power Plants	359
	10.2	Hydraulic turbines and their specific speed	361
	10.3	Pelton wheel	363
	10.4	Francis turbine	370
	10.5	Kaplan turbine	377
	10.6	Cavitation	380
		Exercises	382

11 Hydraulic Transmission of Power — 385

    11.1 Fluid couplings — 385
        11.1.1 Fundamental relations — 386
        11.1.2 Flow rate and hydrodynamic losses — 388
        11.1.3 Partially filled coupling — 390
    11.2 Torque converters — 391
        11.2.1 Fundamental relations — 392
        11.2.2 Performance — 394
    Exercises — 398

12 Wind turbines — 401

    12.1 Horizontal-axis wind turbine — 402
    12.2 Momentum and blade element theory of wind turbines — 403
        12.2.1 Momentum Theory — 403
        12.2.2 Ducted wind turbine — 407
        12.2.3 Blade element theory and wake rotation — 409
        12.2.4 Irrotational wake — 412
    12.3 Blade Forces — 415
        12.3.1 Nonrotating wake — 415
        12.3.2 Wake with rotation — 419
        12.3.3 Ideal wind turbine — 424
        12.3.4 Prandtl's tip correction — 425
    12.4 Turbomachinery and future prospects for energy — 429
    Exercises — 430

Appendix A: Streamline curvature and radial equilibrium — 431
    A.1 Streamline curvature method — 431
        A.1.1 Fundamental equations — 431
        A.1.2 Formal solution — 435

Appendix B: Thermodynamic Tables — 437

References — 449

Index — 453

Foreword

Turbomachinery is a subject of considerable importance in a modern industrial civilization. Steam turbines are at the heart of central station power plants, whether fueled by coal or uranium. Gas turbines and axial compressors are the key components of jet engines. Aeroderivative gas turbines are also used to generate electricity with natural gas as fuel. Same technology is used to drive centrifugal compressors for transmitting this natural gas across continents. Blowers and fans are used for mine and industrial ventilation. Large pumps are often driven with steam turbines to provide feedwater to boilers. They are used in sanitation plants for wastewater cleanup. Hydraulic turbines generate electricity from water stored in reservoirs, and wind turbines do the same from the flowing wind.

This book is on the principles of turbomachines. It aims for a unified treatment of the subject matter, with consistent notation and concepts. In order to provide a ready reference to the reader, some of the developments have been repeated in more than one chapter. This also makes possible the omission of some chapters from a course of study. The subject matter becomes somewhat more general in three of the later chapters.

Acknowledgments

The subject of turbomachinery occupied a central place in mechanical engineering curriculum some half a century ago. In the early textbooks fluid mechanics was taught as a part of a course on turbomachinery, and many of the pioneers of fluid dynamics worked out the many technical issues related to these machines. The field still draws substantial interest. Today the situation has been turned around, and books on fluid dynamics introduce turbomachines in one or two chapters. The same relationship existed with thermodynamics and steam power plants, but today an introduction to steam power plants is usually found in a single chapter in an introductory textbook on thermodynamics.

The British tradition on turbomachinery is long and illustrious. There W. J. Kearton established a center at the University of Liverpool nearly a century ago. His book *Steam Turbine Theory and Practice* became a standard reference source. After his retirement J. H. Horlock occupied the Harrison Chair of Mechanical Engineering there for a decade. His book *Axial Flow Compressors* appeared in 1958 and its complement, *Axial Flow Turbines*, in 1966. Whereas Horlock's books are best suited for advanced workers in the field, at University of Liverpool, S. L. Dixon's textbook *Fluid Mechanics and Thermodynamics of Turbomachinery* appeared in 1966, and its later editions continue in print. It is well suited for undergraduates. Another textbook in the British tradition is the *Gas Turbine Theory* by H. Cohen and G. F. C. Rogers. It was first published in 1951 and in later editions still today. At a more advanced level are R. I. Lewis's *Turbomachinery Performance Analysis* from 1996, N. A. Cumpsty's *Compressor Aerodynamics* published in 1989, and the *Design of Radial Turbomachines* by A. Whitfield and N. C. Baines in 1990.

More than a generation of American students learned this subject from D. G. Sheppard's *Principles of Turbomachinery* and later from the short *Turbomachinery—Basic Theory and Applications* by E. Logan, Jr. The venerable A. Stodola's *Steam and Gas Turbines* has been translated to English, but many others classic works, such as W. Traupel's *Thermische Turbomaschinen* and the seventh edition of *Strömungsmachinen*, by Pfleiderer and Petermann, require a good reading knowledge of German.

I am indebted to all the above mentioned authors for their fine efforts to make the study of this subject enjoyable.

My introduction to the field of turbomachinery came thanks to my longtime colleague the late Richard H. Zimmerman. After working on other areas of mechanical engineering for many years, I returned to this subject after Reza Abhari invited me to spend a summer at ETH in Zurich. There I also met Anestis Kalfas, now also at the Aristotle University of Thessaloniki. I am grateful to both of them for sharing their lecture notes, which showed me how the subject was taught at the institutions of learning where they had completed their studies and how they have developed it further. I am grateful to my former student and friend, V. Babu, a professor of Mechanical Engineering of the Indian Institute of Technology Madras, for reading the manuscript and making many helpful suggestions for improving it. Undoubtedly some errors have remained, and I will be thankful for readers who take the time to point them out by e-mail to me at the address: korpela.1@osu.edu.

I am grateful for permission to use graphs and figures from various published works and wish to acknowledge the generosity of the various organization for granting the permission to use them. These include Figures 1.1 and 1.2 from Siemens press photo, Siemens AG Figure 1.3, from Schmalenberger Strömungstechnologie AG; Figures 1.6 and 7.1 are by courtesy of MAN Diesel & Turbo SE, and Figures 4.12 and 4.11 are published by permission of BorgWarner Turbo Systems. Figure 3.7 is courtesy of Professor D. Papamaschou. Figures 10.3 and 10.11 are published under the GNU Free Documentation licenses with original courtesy of Voith Siemens Hydro. The Figure 10.9 is reproduced under the Gnu Free Documentation licence, with the original photo by Audrius Meskauskas. Figure 1.5 is also published under Gnu Free Documentation licence, and so is Figure 1.4 and by permission from Aermotor. The Institution of Mechanical Engineers has granted permission to reproduce Figures 3.14, 6.16, 7.5, 7.6, and 7.16. Figure 4.10 is published under agreement with NASA. The *Journal of the Royal Aeronautical Society* granted permission to publish Figure 6.11. Figures 6.19 and 6.20 are published under the Crown Stationary Office' Open Government Licence of UK. Figure 3.11 has been adapted from J. H. Keenan *Thermodynamics*, MIT Press and Figure 9.6 from O. E. Balje, *Turbo machines A guide to Selection and Theory*. Permission to use Figures 12.13 and 12.15 from *Wind Turbin Handbook* by T. Burton, N. Jenkins, D. Sharpe, and E. Bossanyi has been granted by John Wiley & Sons.

I have been lucky to have Terttu as a wife and a companion in my life. She has been and continues to be very supportive of all my efforts.

S. A. K

CHAPTER 1

INTRODUCTION

1.1 ENERGY AND FLUID MACHINES

The rapid development of modern industrial societies was made possible by the large-scale extraction of fossil fuels buried in the earth's crust. Today oil makes up 37% of world's energy mix, coal's share is 27%, and that of natural gas is 23%, for a total of 87%. Hydropower and nuclear energy contribute each about 6% which increases the total from these sources to 99%. The final 1% is supplied by wind, geothermal energy, waste products, and solar energy. Biomass is excluded from these, for it is used largely locally, and thus its contribution is difficult to calculate. The best estimates put its use at 10% of the total, in which case the other percentages need to be adjusted downward appropriately [54].

1.1.1 Energy conversion of fossil fuels

Over the the last two centuries engineers invented methods to convert the chemical energy stored in fossil fuels into usable forms. Foremost among them are methods for converting this energy into electricity. This is done in steam power plants, in which combustion of coal is used to vaporize steam and the thermal energy of the steam is then converted to shaft work in a steam turbine. The shaft turns a generator that produces electricity. Nuclear power plants work on the same principle, with uranium, and in rare cases thorium, as the fuel.

Principles of Turbomachinery. By Seppo A. Korpela
Copyright © 2011 John Wiley & Sons, Inc.

Oil is used sparingly this way, and it is mainly refined to gasoline and diesel fuel. The refinery stream also yields residual heating oil, which goes to industry and to winter heating of houses. Gasoline and diesel oil are used in internal-combustion engines for transportation needs, mainly in automobiles and trucks, but also in trains. Ships are powered by diesel fuel and aircraft, by jet fuel.

Natural gas is largely methane, and in addition to its importance in the generation of electricity, it is also used in some parts of the world as a transportation fuel. A good fraction of natural gas goes to winter heating of residential and commercial buildings, and to chemical process industries as raw material.

Renewable energy sources include the potential energy of water behind a dam in a river and the kinetic energy of blowing winds. Both are used for generating electricity. Water waves and ocean currents also fall into the category of renewable energy sources, but their contributions are negligible today.

In all the methods mentioned above, conversion of energy to usable forms takes place in a *fluid machine*, and in these instances they are *power-producing* machines. There are also *power-absorbing* machines, such as pumps, in which energy is transferred into a fluid stream.

In both power-producing and power-absorbing machines energy transfer takes place between a fluid and a moving machine part. In *positive-displacement machines* the interaction is between a fluid at high pressure and a reciprocating piston. Spark ignition and diesel engines are well-known machines of this class. Others include piston pumps, reciprocating and screw compressors, and vane pumps.

In *turbomachines* energy transfer takes place between a *continuously flowing fluid stream and a set of blades rotating about a fixed axis*. The blades in a pump are part of an *impeller* that is fixed to a shaft. In an axial compressor they are attached to a compressor *wheel*. In steam and gas turbines the blades are fastened to a disk, which is fixed to a shaft, and the assembly is called a turbine *rotor*. Fluid is guided into the rotor by *stator vanes* that are fixed to the *casing* of the machine. The inlet stator vanes are also called *nozzles*, or *inlet guidevanes*.

Examples of power-producing turbomachines are steam and gas turbines, and water and wind turbines. The power-absorbing turbomachines include pumps, for which the working fluid is a liquid, and fans, blowers, and compressors, which transfer energy to gases.

Methods derived from the principles of thermodynamics and fluid dynamics have been developed to analyze the design and operation of these machines. These subjects, and heat transfer, are the foundation of *energy engineering*, a discipline central to modern industry.

1.1.2 Steam turbines

Central station power plants, fueled either by coal or uranium, employ steam turbines to convert the thermal energy of steam to shaft power to run electric generators. Coal provides 50% and nuclear fuels 20% of electricity production in the United States. For the world the corresponding numbers are 40% and 15%, respectively. It is clear from these figures that steam turbine manufacture and service are major industries in both the United States and the world.

Figure 1.1 shows a 100-MW steam turbine manufactured by Siemens AG of Germany. Steam enters the turbine through the nozzles near the center of the machine, which direct the flow to a rotating set of blades. On leaving the first stage, steam flows (in the sketch) toward the top right corner) through the rest of the 12 stages of the high-pressure section in this turbine. Each stage consists of a set rotor blades, preceded by a set of stator vanes.

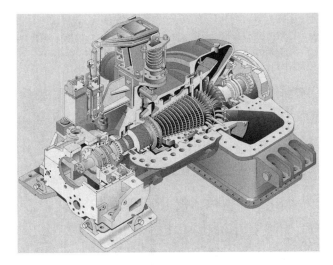

Figure 1.1 The Siemens SST-600 industrial steam turbine with a capacity of up to 100-MW. (Courtesy Siemens press picture, Siemens AG.)

The stators, fixed to the casing (of which one-quarter is removed in the illustration), are not clearly visible in this figure. After leaving the high-pressure section, steam flows into a two-stage low-pressure turbine, and from there it leaves the machine and enters a condenser located on the floor below the turbine bay. Temperature of the entering steam is up to $540°C$ and its pressure is up to 140 bar. Angular speed of the shaft is generally in the range 3500–15,000 rpm (rev/min). In this turbine there are five bleed locations for the steam. The steam extracted from the bleeds enters feedwater heaters, before it flows back to a boiler. The large regulator valve in the inlet section controls the steam flow rate through the machine.

In order to increase the plant efficiency, new designs operate at supercritical pressures. In an ultrasupercritical plant, the boiler pressure can reach 600 bar and turbine inlet temperature, $620°C$. Critical pressure for steam is 220.9 bar, and its critical temperature is $373.14°C$.

1.1.3 Gas turbines

Major manufacturers of gas turbines produce both jet engines and industrial turbines. Since the 1980s, gas turbines, with clean-burning natural gas as a fuel, have also made inroads into electricity production. Their use in combined cycle power plants has increased the plant overall thermal efficiency to just under 60%. They have also been employed for stand-alone power generation. In fact, most of the power plants in the United States since 1998 have been fueled by natural gas. Unfortunately, production from the old natural gas-fields of North America is strained, even if new resources have been developed from shale deposits. How long they will last is still unclear, for the technology of gas extraction from shale deposits is new and thus a long operating experience is lacking.

Figure 1.2 shows a gas turbine manufactured also by Siemens AG. The flow is from the back toward the front. The rotor is equipped with advanced single-crystal turbine blades, with a thermal barrier coating and film cooling. Flow enters a three-stage turbine from an annular combustion chamber which has 24 burners and walls made from ceramic tiles. These turbines power the 15 axial compressor stages that feed compressed air to the combustor. The fourth turbine stage, called a *power turbine*, drives an electric generator in a combined cycle power plant for which this turbine has been designed. The plant delivers a power output of 292-MW.

Figure 1.2 An open rotor and combustion chamber of an SGT5-4000F gas turbine. (Courtesy Siemens press picture, Siemens AG.)

1.1.4 Hydraulic turbines

In those areas of the world with large rivers, water turbines are used to generate electrical power. At the turn of the millennium hydropower represented 17% of the total electrical energy generated in the world. The installed capacity at the end of year 2007 was 940,000 MW, but generation was 330,000 MW, so their ratio, called a *capacity factor*, comes to 0.35.

With the completion of the 22,500-MW Three Gorges Dam, China has now the world's largest installed capacity of 145,000 MW, which can be estimated to give 50,000 MW of power. Canada, owing to its expansive landmass, is the world's second largest producer of hydroelectric power, with generation at 41,000 MW from installed capacity of 89,000 MW. Hydropower accounts for 58% of Canada's electricity needs. The sources of this power are the great rivers of British Columbia and Quebec. The next largest producer is Brazil, which obtains 38,000 MW from an installed capacity of 69,000 MW. Over 80% of Brazil's energy is obtained by water power. The Itaipu plant on the Paraná River, which borders Brazil and Paraguay, generates 12,600 MW of power at full capacity. Of nearly the same size is Venezuela's Guri dam power plant with a rated capacity of 10,200 MW, based on 20 generators.

The two largest power stations in the United States are the Grand Coulee station in the Columbia River and the Hoover Dam station in the Colorado River. The capacity of the Grand Coulee is 6480 MW, and that of Hoover is 2000 MW. Tennessee Valley Authority operates a network of dams and power stations in the Southeastern parts of the country. Many small hydroelectric power plants can also be found in New England. Hydroelectric power in the United States today provides 289 billion kilowatthours (kwh) a year, or 33,000 MW, but this represents only 6% of the total energy used in the United States. Fossil fuels still account for 86% of the US energy needs.

Next on the list of largest producers of hydroelectricity are Russia and Norway. With its small and thrifty population, Norway ships its extra generation to the other Scandinavian countries, and now with completion of a high-voltage powerline under the North Sea, also to western Europe. Norway and Iceland both obtain nearly all their electricity from hydropower.

1.1.5 Wind turbines

The Netherlands has been identified historically as a country of windmills. She and Denmark have seen a rebirth of wind energy generation since 1985 or so. These countries are relatively small in land area and both are buffeted by winds from the North Sea. Since the 1990s Germany has embarked on a quest to harness its winds. By 2007 it had installed wind turbines on most of its best sites with 22,600 MW of installed capacity. The installed capacity in the United States was 16,600 MW in the year 2007. It was followed by Spain, with an installed capacity of 15,400 MW. After that came India and Denmark.

The capacity factor for wind power is about 0.20, thus even lower than for hydropower. For this reason wind power generated in the United States constitutes only 0.5% of the country's total energy needs. Still, it is the fastest-growing of the renewable energy systems. The windy plains of North and South Dakota and of West and North Texas offer great potential for wind power generation.

1.1.6 Compressors

Compressors find many applications in industry. An important use is in the transmission of natural gas across continents. Natural-gas production in the United States is centered in Texas and Louisiana as well as offshore in the Gulf of Mexico. The main users are the midwestern cities, in which natural gas is used in industry and for winter heating. Pipelines also cross the Canadian border with gas supplied to the west-coast and to the northern states from Alberta. In fact, half of Canada's natural-gas production is sold to the United States.

Russia has 38% of world's natural-gas reserves, and much of its gas is transported to Europe through the Ukraine. China has constructed a natural-gas pipeline to transmit the gas produced in the western provinces to the eastern cities. Extensions to Turkmenistan and Iran are in the planning stage, as both countries have large natural-gas resources.

1.1.7 Pumps and blowers

Pumps are used to increase pressure of liquids. Compressors, blowers, and fans do the same for gases. In steam power plants condensate pumps return water to feedwater heaters, from which the water is pumped to boilers. Pumps are also used for cooling water flows in these power plants.

Figure 1.3 shows a centrifugal pump manufactured by Schmalenberger Strömungstechnologie GmbH. Flow enters through the eye of an impeller and leaves through a spiral volute. This pump is designed to handle a flow rate of $100\,\mathrm{m^3/h}$, with a 20 m increase in its head.

In the mining industry, blowers circulate fresh air into mines and exhaust stale, contaminated air from them. In oil, chemical, and process industries, there is a need for large blowers and pumps. Pumps are also used in great numbers in agricultural irrigation and municipal sanitary facilities.

Figure 1.3 A centrifugal pump. (Courtesy Schmalenberger GmbH.)

Offices, hospitals, schools and other public buildings have heating, ventilating, and air conditioning (HVAC) systems, in which conditioned air is moved by large fans. Pumps provide chilled water to cool the air and for other needs.

1.1.8 Other uses and issues

Small turbomachines are present in all households. In fact, it is safe to say that in most homes, only electric motors are more common than turbomachines. A pump is needed in a dishwasher, a washing machine, and the sump. Fans are used in the heating system and as window and ceiling fans. Exhaust fans are installed in kitchens and bathrooms. Both an airconditioner and a refrigerator is equipped with a compressor, although it may be a screw compressor (which is not a turbomachine) in an air-conditioner. In a vacuum cleaner a fan creates suction. In a car there is a water pump, a fan, and in some models a turbocharger. All are turbomachines.

In addition to understanding the fluid dynamical principles of turbomachinery, it is important for a turbomachinery design engineer to learn other allied fields. The main ones are material selection, shaft and disk vibration, stress analysis of disks and blades, and topics covering bearings and seals. Finally, understanding control theory is important for optimum use of any machine.

In more recent years, the world has awoken to the fact that fossil fuels are finite and that renewable energy sources will not be sufficient to provide for the entire world the material conditions that Western countries now enjoy. Hence, it is important that the machines that make use of these resources be well designed so that the remaining fuels are used with consideration, recognizing their finiteness and their value in providing for some of the vital needs of humanity.

1.2 HISTORICAL SURVEY

This section gives a short historical review of turbomachines. Turbines are power-producing machines and include water and wind turbines from early history. Gas and steam turbines date from the beginning of the last century. Rotary pumps have been in use for nearly 200 years. Compressors developed as advances were made in aircraft propulsion during the last century.

1.2.1 Water power

It is only logical that the origin of turbomachinery can be traced to the use of flowing water as a source of energy. Indeed, waterwheels, lowered into a river, were already known to the Greeks. The early design moved to the rest of Europe and became known as the *norse mill* because the archeological evidence first surfaced in northern Europe. This machine consists of a set of radial paddles fixed to a shaft. As the shaft was vertical, or somewhat inclined, its efficiency of energy extraction could be increased by directing the flow of water against the blades with the aid of a *mill race and a chute*. Such a waterwheel could provide only about one-half horsepower (0.5 hp), but owing to the simplicity of its construction, it survived in use until 1500 and can still be found in some primitive parts of the world.

By placing the axis horizontally and lowering the waterwheel into a river, a better design is obtained. In this *undershot waterwheel*, dating from Roman times, water flows through the lower part of the wheel. Such a wheel was first described by the Roman architect and engineer Marcus Vitruvius Pollio during the first century B.C.

Overshot waterwheel came into use in the hilly regions of Rome during the second century A.D. By directing water from a chute above the wheel into the blades increases the power delivered because now, in addition to the kinetic energy of the water, also part of the potential energy can be converted to mechanical energy. Power of overshot waterwheels increased from 3 hp to about 50 hp during the Middle Ages. These improved overshot waterwheels were partly responsible for the technical revolution in the twelfth–thirteenth century. In the William the Conquerer's *Domesday Book* of 1086, the number of watermills in England is said to have been 5684. In 1700 about 100,000 mills were powered by flowing water in France [12].

The genius of Leonardo da Vinci (1452–1519) is well recorded in history, and his notebooks show him to have been an exceptional observer of nature and technology around him. Although he is best known for his artistic achievements, most of his life was spent in the art of engineering. Illustrations of fluid machinery are found in da Vinci's notebooks, in *De Re Metallica*, published in 1556 by Agricola [3], and in a tome by Ramelli published in 1588. From these a good understanding of the construction methods can be gained and of the scale of the technology then in use. In Ramelli's book there is an illustration of a mill in which a grinding wheel, located upstairs, is connected to a shaft, the lower end of which has an enclosed impact wheel that is powered by water. There are also illustrations that show windmills to have been in wide use for grinding grain.

Important progress to improve waterwheels came in the hands of the Frenchman Jean Victor Poncelet (1788–1867), who curved the blades of the undershot waterwheel, so that water would enter tangentially to the blades. This improved its efficiency. In 1826 he came up with a design for a horizontal wheel with radial inward flow. A water turbine of this design was built a few years later in New York by Samuel B. Howd and then improved by James Bicheno Francis (1815–1892). Improved versions of Francis turbines are in common use today.

About the same time in France an outward flow turbine was designed by Claude Burdin (1788–1878) and his student Benoît Fourneyron (1802–1867). They benefited greatly from the work of Jean-Charles de Borda (1733–1799) on hydraulics. Their machine had a set of guidevanes to direct the flow tangentially to the blades of the turbine wheel. Fourneyron in 1835 designed a turbine that operated from a head of 108 m with a flow rate of 20 liters per second (L/s), rotating at 2300 rpm, delivering 40 hp as output power at 80% efficiency.

In the 1880s in the California gold fields an impact wheel, known as a *Pelton wheel*, after Lester Allen Pelton (1829–1918) of Vermillion, Ohio, came into wide use.

An axial-flow turbine was developed by Carl Anton Henschel (1780–1861) in 1837 and by Feu Jonval in 1843. Modern turbines are improvements of Henschel's and Jonval's designs. A propeller type of turbine was developed by the Austrian engineer Victor Kaplan (1876–1934) in 1913. In 1926 a 11,000-hp Kaplan turbine was placed into service in Sweden. It weighed 62.5 tons, had a rotor diameter of 5.8 m, and operated at 62.5 rpm with a water head of 6.5 m. Modern water turbines in large hydroelectric power plants are either of the Kaplan type or variations of this design.

1.2.2 Wind turbines

Humans have drawn energy from wind and water since ancient times. The first recorded account of a windmill is from the Persian-Afghan border region in 644 A.D., where these vertical axis windmills were still in use in more recent times [32]. They operate on the principle of drag in the same way as square sails do when ships sail downwind.

In Europe windmills were in use by the twelfth century, and historical research suggests that they originated from waterwheels, for their axis was horizontal and the masters of the late Middle Ages had already developed gog-and-ring gears to transfer energy from a horizontal shaft into a vertical one. This then turned a wheel to grind grain [68]. An early improvement was to turn the entire windmill toward the wind. This was done by centering a round platform on a large-diameter *vertical post* and securing the structure of the windmill on this platform. The platform was free to rotate, but the force needed to turn the entire mill limited the size of the early *postmills*. This restriction was removed in a *towermill* found on the next page, in which only the platform, affixed to the top of the mill, was free to rotate. The blades were connected to a windshaft, which leaned about 15° from the horizontal so that the blades would clear the structure. The shaft was supported by a wooden main bearing at the *blade end* and a thrust bearing at the *tail end*. A band brake was used to limit the rotational speed at high wind speeds. The power dissipated by frictional forces in the brake rendered the arrangement susceptible to fire.

Over the next 500 years, to the beginning of the industrial revolution, progress was made in windmill technology, particularly in Great Britain. By accumulated experience, designers learned to move the position the spar supporting a blade from midcord to quarter-chord position, and to introduce a nonlinear twist and leading edge camber to the blade [68]. The blades were positioned at a steep angles to the wind and made use of the lift force, rather than drag. It is hard not to speculate that the use of lift had not been learned from sailing vessels using *lanteen sails* to tack.

A towermill is shown in Figure 1.4a. It is seen to be many meters tall, and each of the four quarter-chord blades is about one meter in width. The blades of such mills were covered with either fabric or wooden slats. By an arrangement such as is found in window shutters today, the angle of attack of the blades could be changed at will, providing also a braking action at high winds.

(a) (b)

Figure 1.4   A traditional windmill (a) and an American farm windmill (b) for pumping water.

The American windmill is shown in Figure 1.4b. It is a small multibladed wind turbine with a vertical vane to keep it oriented toward the wind. Some models had downwind orientation and did not need to be controlled in this way. The first commercially successful wind turbine was introduced by Halladay in 1859 to pump water for irrigation in the Plains States. It was about 5 m in diameter and generated about one kilowatt (1 kW) at windspeed of 7 m/s [68]. The windmill shown in the figure is a 18-steel-bladed model by Aermotor Company of Chicago, a company whose marketing and manufacturing success made it the prime supplier of this technology during the 1900–1925.

New wind turbines with a vertical axis were invented during the 1920s in France by G. Darrieus and in Finland by S. Savonius [66]. They offer the advantage of working without regard to wind direction, but their disadvantages include fluctuating torque over each revolution and difficulty of starting. For these reasons they have have not achieved wide use.

1.2.3 Steam turbines

Although the history of steam to produce rotation of a wheel can be traced to Hero of Alexandria in the year 100 A.D., his invention is only a curiosity, for it did not arise out of a historical necessity, such as was imposed by the world's increasing population at the beginning of the industrial revolution. Another minor use to rotate a roasting spit was suggested in 1629 Giovanni de Branca. The technology to make shafts and overcome friction was too primitive at this time to put his ideas to more important uses. The age of steam began with the steam engine, which ushered in the industrial revolution in Great Britain. During the eighteenth century steam engines gained in efficiency, particularly when James Watt in 1765 reasoned that better performance could be achieved if the boiler and the condenser were separate units. Steam engines are, of course, positive-displacement machines.

Sir Charles Parsons (1854–1931) is credited with the development of the first *steam turbine* in 1884. His design used multiple turbine wheels, about 8 cm in diameter each to drop the pressure in *stages* and this way to reduce the angular velocities. The first of Parson's turbines generated 7.5 kW using steam at inlet pressure of 550 kPa and rotating at 17,000 rpm. It took some 15 years before Parsons' efforts received their proper recognition.

An impulse turbine was developed in 1883 by the Swedish engineer Carl Gustav Patrik de Laval (1845–1913) for use in a cream separator. To generate the large steam velocities he also invented the supersonic nozzle and exhibited it in 1894 at the Columbian World's Fair in Chicago. From such humble beginnings arose rocketry and supersonic flight. Laval's turbines rotated at 26,000 rpm, and the largest of the rotors had a tip speed of 400 m/s. He used flexible shafts to alleviate vibration problems in the machinery.

In addition to the efforts in Great Britain and Sweden, the Swiss Federal Institute of Technology in Zurich [Eidgenössische Technische Hochschule, (ETH)] had become an important center of research in early steam turbine theory through the efforts of Aurel Stodola (1859–1942). His textbook *Steam and Gas Turbines* became the standard reference on the subject for the first half of last century [75]. A similar effort was led by William J. Kearton (1893–?) at the University of Liverpool in Great Britain.

1.2.4 Jet propulsion

The first patent for gas turbine development was issued to John Barber (1734–c.1800) in England in 1791, but again technology was not yet sufficiently advanced to build a machine on the basis of the proposed design. Eighty years later in 1872 Franz Stolze (1836–1910) received a patent for a design of a gas turbine power plant consisting of a multistage axial flow compressor and turbine on the same shaft, together with a combustion chamber and a heat exchanger. The first U.S. patent was issued to Charles Gordon Curtis (1860–1953) in 1895.

Starting in 1935, Hans J. P. von Ohain (1911–1998) directed efforts to design gas turbine power plants for the Heinkel aircraft in Germany. The model He178 was a fully operational jet aircraft, and in August 1939 it was first such aircraft to fly successfully.

During the same timeframe Sir Frank Whittle (1907–1996) in Great Britain was developing gas turbine power plants for aircraft based on a centrifugal compressor and a turbojet design. In 1930 he filed for a patent for a single-shaft engine with a two-stage axial compressor followed by a radial compressor from which the compressed air flowed into a straight-through burner. The burned gases then flowed through a two-stage axial turbine on a single disk. This design became the basis for the development of jet engines in Great Britain and later in the United States.

Others, such as Alan Arnold Griffith (1893–1963) and Hayne Constant (1904–1968), worked in 1931 on the design and testing of axial-flow compressors for use in gas turbine power plants. Already in 1926 Griffith had developed an aerodynamic theory of turbine design based on flow past airfoils.

In Figure 1.5 shows the De Havilland Goblin engine designed by Frank Halford in 1941. The design was based on the original work of Sir Frank Whittle. It is a turbojet engine with single-stage centrifugal compressor, and with can combustors exhausting the burned combustion gases into a turbine that drives the compressor. The remaining kinetic energy leaving the turbine goes to propulsive thrust.

Since the 1950s there has been continuous progress in the development of gas turbine technology for aircraft power plants. Rolls Royce in Great Britain brought to the market its Olympus twin-spool engine, its Dart single-spool engine for low-speed aircraft, and in

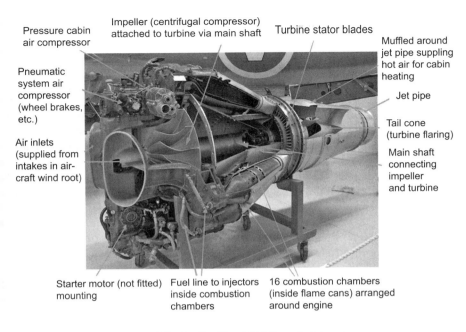

Figure 1.5 De Havilland Goblin turbojet engine.

1967 the Trent, which was the first three-shaft turbofan engine. The Olympus was also used in stationary power plants and in marine propulsion.

General Electric in the United States has also a long history in gas turbine development. Its I-14, I-16, I-20, and I-40 models were developed in the 1940s. The I-14 and I-16 powered the Bell P-59A aircraft, which was the first American turbojet. It had a single centrifugal compressor and a single-stage axial turbine. Allison Engines, then a division of General Motors, took over the manufacture and improvement of model I-40. Allison also began the manufacture of General Electric's TG series of engines.

Many new engines were developed during the latter half of the twentieth century, not only by Rolls Royce and General Electric but also by Pratt and Whitney in the United States and Canada, Rateau in France, and by companies in Soviet Union, Sweden, Belgium, Australia, and Argentina. The modern engines that power the flight of today's large commercial aircraft by Boeing and by Airbus are based on the Trent design of Rolls Royce, or on General Electric's GE90 [7].

1.2.5 Industrial turbines

Brown Boveri in Switzerland developed a 4000-kW turbine power plant in 1939 to Neuchatel for standby operation for electric power production. On the basis of this design, an oil-burning closed cycle gas turbine plant with a rating of 2 MW was built the following year.

Industrial turbine production at Ruston and Hornsby Ltd. of Great Britain began by establishment of a design group in 1946. The first unit produced by them was sold to Kuwait Oil Company in 1952 to power pumps in oil fields. It was still operational in

Pumps and compressors

The centrifugal pump was invented by Denis Papin (1647–1710) in 1698 in France. To be sure, a suggestion to use centrifugal force to effect pumping action had also been made by Leonardo da Vinci, but neither his nor Papin's invention could be built, owing to the lack of sufficiently advanced shop methods. Leonhard Euler (1707–1783) gave a mathematical theory of the operation of a pump in 1751. This date coincides with the beginning of the industrial revolution and the advances made in manufacturing during the ensuing 100 years brought centrifugal pumps to wide use by 1850. The Massachusetts pump, built in 1818, was the first practical centrifugal pump manufactured. W. D. Andrews improved its performance in 1846 by introducing double-shrouding. At the same time in Great Britain engineers such as John Appold (1800–1865) and Henry Bessemer (1813–1898) were working on improved designs. Appold's pump operated at 788 rpm with an efficiency of 68% and delivered 78 L/s and a head of 5.9 m.

The same companies that in 1900 built steam turbines in Europe also built centrifugal blowers and compressors. The first applications were for providing ventilation in mines and for the steel industry. Since 1916 compressors have been used in chemical industries, since 1930 in the petrochemical industries, and since 1947 in the transmission of natural gas. The period 1945–1950 saw a large increase in the use of centrifugal compressors in American industry. Since 1956 they have been integrated into gas turbine power plants and have replaced reciprocating compressors in other applications.

The efficiencies of single stage centrifugal compressors increased from 70% to over 80% over the period 1935–1960 as a result of work done in companies such as Rateau, Moss-GE, Birmann-DeLaval, and Whittle in Europe and General Electric and Pratt & Whitney in the United States. The pressure ratios increased from 1.2 : 1 to 7 : 1. This development owes much to the progress that had been made in gas turbine design [26].

For large flow rates multistage axial compressors are used. Figure 1.6 shows such a compressor, manufactured by Man Diesel & Turbo SE in Germany. It has 14 axial stages followed by a centrifugal compressor stage. The rotor blades are seen in the exposed rotor. The stator blades are fixed to the casing, the lower half of which is shown. The flow is from right to left. The flow area decreases toward the exit, for in order to keep the axial velocity constant, as is commonly done, the increase in density on compression is accommodated by a decrease in the flow area.

1.2.6 Note on units

The Système International (d'Unités) (SI) system of units is used in this text. But it is still customary in some industries English Engineering system of units and if other reference books are consulted one finds that many still use this system. In this set of units mass is expressed as pound (lbm) and foot is the unit of length. The British gravitational system of units has *slug* as the unit of mass and the unit of force is pound force (lbf), obtained from Newton's law, as it represents a force needed to accelerate a mass of one slug at the rate of one foot per second squared. The use of slug for mass makes the traditional British gravitational system of units analogous to the SI units. When pound (lbm) is used for mass it ought to be first converted to slugs (1 slug = 32.174 lbm), for then calculations follow smoothly as in the SI units. The unit of temperature is Fahrenheit or Rankine. Thermal

Figure 1.6 Multistage compressor. (Courtesy MAN Diesel & Turbo SE.)

energy in this set of units is reported in British thermal units or Btu's for short. As it is a unit for energy, it can be converted to one encountered in mechanics by remembering that 1 Btu = 778.17 ft lbf. The conversion factor to SI units is 1 Btu = 1055 J. Power is still often reported in horsepower, and 1 hp = 0.7457 kW. The flow rate in pumps is often given in gallons per minute (gpm). The conversion to standard units is carried out by noting recalling that 1 gal = 231 in^3. World energy consumption is often given in quads. The conversion to SI units is 1 quad =1.055 EJ, where EJ is exajoule equal to 10^{18} J.

CHAPTER 2

PRINCIPLES OF THERMODYNAMICS AND FLUID FLOW

This chapter begins with a review of the conservation principle for mass for steady uniform flow, after which follows the first and second laws of thermodynamics, also for steady uniform flow. Next, thermodynamic properties of gases and liquids are discussed. These principles enable the discussion of turbine and compressor efficiencies, which are described in relation to thermodynamic losses. The final section is on the Newton's second law for steady and uniform flow.

2.1 MASS CONSERVATION PRINCIPLE

Mass flow rate $\dot{m}$ in a *uniform flow* is related to density ρ and velocity V of the fluid, and the cross-sectional area of the flow channel A by

$$\dot{m} = \rho V_n A$$

When this equation is used in the analysis of steam flows, specific volume, which is the reciprocal of density, is commonly used. The subscript n denotes the direction *normal* to the flow area. The product $V_n A$ arises from the scalar product $\mathbf{V} \cdot \mathbf{n} = V \cos\theta$, in which $\mathbf{n}$ is a unit normal vector on the surface A and θ is the angle between the normal and the direction of the velocity vector. Consequently, the scalar product can be written in the two alternative forms

$$\mathbf{V} \cdot \mathbf{n} A = V A \cos\theta = V_n A = V A_n$$

Principles of Turbomachinery. By Seppo A. Korpela
Copyright © 2011 John Wiley & Sons, Inc.

in which A_n is the area normal to the flow. The principle of conservation of mass for a uniform steady flow through a control volume with one inlet and one exit takes the form

$$\rho_1 V_1 A_{n1} = \rho_2 V_2 A_{n2}$$

Turbomachinery flows are steady only in a time-averaged sense; that is, the flow is periodic with a period equal to the time taken for a blade to move a distance equal to the spacing between adjacent blades. Despite the unsteadiness, in elementary analysis all variables are assumed to have steady values.

If the flow has more than one inlet and exit, then, in steady uniform flow, conservation of mass requires that

$$\sum_i \rho_i V_i A_{ni} = \sum_e \rho_e V_e A_{ne} \qquad (2.1)$$

in which the sums are over all the inlets and exits.

■ **EXAMPLE 2.1**

Steam flows at the rate $\dot{m} = 0.20\,\text{kg/s}$ through each nozzle in the bank of nozzles shown in Figure 2.1. Steam conditions are such that at the inlet specific volume is $0.80\,\text{m}^3/\text{kg}$ and at the outlet it is $1.00\,\text{m}^3/\text{kg}$. Spacing of the nozzles is $s = 5.0\,\text{cm}$, wall thickness at the inlet is $t_1 = 2.5\,\text{mm}$, and at the outlet it is $t_2 = 2.0\,\text{mm}$. Blade height is $b = 3.0\,\text{cm}$. Nozzle angle is $\alpha_2 = 70°$. Find the steam velocity at the inlet and at the outlet.

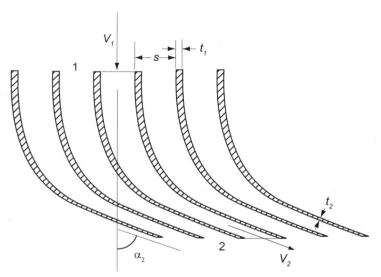

Figure 2.1 Turning of flow by steam nozzles.

Solution: The area at the inlet is

$$A_1 = b(s - t_1) = 3(5 - 0.25) = 14.25\,\text{cm}^2$$

Velocity at the inlet is solved from the mass balance

$$\dot{m} = \rho_1 V_1 A_1 = \frac{V_1 A_1}{v_1}$$

which gives
$$V_1 = \frac{\dot{m}v_1}{A_1} = \frac{0.20 \cdot 0.80 \cdot 100^2}{14.25} = 112.3\,\text{m/s}$$

At the exit the flow area is
$$A_2 = b(s\cos\alpha_2 - t_2) = 3[5\cos(70°) - 0.20] = 4.53\,\text{cm}^2$$

hence the velocity is
$$V_2 = \frac{\dot{m}v_2}{A_2} = \frac{0.2 \cdot 1.00 \cdot 100^2}{4.53} = 441.5\,\text{m/s}$$
∎

2.2 FIRST LAW OF THERMODYNAMICS

For a uniform steady flow in a channel, the first law of thermodynamics has the form

$$\dot{m}\left(u_1 + p_1v_1 + \frac{1}{2}V_1^2 + gz_1\right) + \dot{Q} = \dot{m}\left(u_2 + p_2v_2 + \frac{1}{2}V_2^2 + gz_2\right) + \dot{W} \quad (2.2)$$

The sum of specific internal energy u, kinetic energy $V^2/2$, and potential energy gz is the specific energy $e = u + \frac{1}{2}V^2 + gz$ of the fluid. In the potential energy term g is the acceleration of gravity and z is a height. The term p_1v_1, in which p is the pressure, represents the work done by the fluid in the flow channel just upstream of the inlet to move the fluid ahead of it into the control volume, and it thus represents energy flow into the control volume. This work is called *flow work*. Similarly, p_2v_2 is the flow work done by the fluid inside the control volume to move the fluid ahead of it out of the control volume. It represents energy transfer as work leaving the control volume. The sum of internal energy and flow work is defined as enthalpy $h = u + pv$. The heat transfer rate into the control volume is denoted as $\dot{Q}$ and the rate at which work is delivered is $\dot{W}$. Equation (2.2) can be extended to multiple inlets and outlets in the same manner as was done in Eq. (2.1).

Dividing both sides by $\dot{m}$ gives the first law of thermodynamics the form

$$h_1 + \frac{1}{2}V_1^2 + gz_1 + q = h_2 + \frac{1}{2}V_2^2 + gz_2 + w$$

in which $q = \dot{Q}/\dot{m}$ and $w = \dot{W}/\dot{m}$ denote the heat transfer and work done per unit mass. By convention, heat transfer *into* the thermodynamic system is taken to be a positive quantity, as is *work done by the system* on the surroundings.

The sum of enthalpy, kinetic energy, and potential energy is called the *stagnation enthalpy*

$$h_0 = h + \frac{1}{2}V^2 + gz$$

and the first law can also be written as

$$h_{01} + q = h_{02} + w$$

In the flow of gases the potential energy terms are small and can be neglected. Similarly, for pumps, the changes in elevation are small and potential energy difference is negligible.

Only for some water turbines is there a need to retain the potential energy terms. When the change in potential energy is neglected, the first law reduces to

$$h_1 + \frac{1}{2}V_1^2 + q = h_2 + \frac{1}{2}V_2^2 + w$$

In addition, even if velocity is large, the difference in kinetic energy between the inlet and exit may be small. In such a case first law is simply

$$h_1 + q = h_2 + w$$

Turbomachinery flows are nearly adiabatic, so q can be dropped. Then work delivered by a turbine is given as

$$w = h_{01} - h_{02}$$

and the work done on the fluid in a compressor is

$$w = h_{02} - h_{01}$$

The compressor work has been written in a form that gives the work done a positive value. Hence the convention of thermodynamics of denoting work out from a system as positive and work in as negative is ignored, and the equations are written in a form that gives a positive value for work, for both a turbine and a compressor.

■ **EXAMPLE 2.2**

Steam flows adiabatically at a rate $\dot{m} = 0.01$ kg/s through a diffuser, shown in Figure 2.2, with inlet diameter $D_1 = 1.0$ cm. Specific volume at the inlet $v_1 = 2.40 \text{ m}^3/\text{kg}$. Exit diameter is $D_2 = 2.5$ cm, with specific volume at the outlet $v_2 = 3.80 \text{ m}^3/\text{kg}$. Find the change in enthalpy neglecting any change in the potential energy.

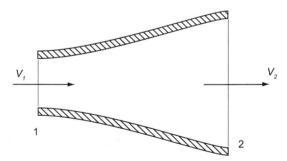

Figure 2.2 Flow through a diffuser.

Solution: The areas at the inlet and outlet are

$$A_1 = \frac{\pi D_1^2}{4} = \frac{\pi \, 0.01^2}{4} = 7.85 \cdot 10^{-5} \text{ m}^2$$

$$A_2 = \frac{\pi D_2^2}{4} = \frac{\pi \, 0.025^2}{4} = 4.91 \cdot 10^{-4} \text{ m}^2$$

The velocity at the inlet is

$$V_1 = \frac{\dot{m} v_1}{A_1} = \frac{0.01 \cdot 2.4}{7.85 \cdot 10^{-5}} = 305.6 \,\text{m/s}$$

and at the outlet it is

$$V_2 = \frac{\dot{m} v_2}{A_2} = \frac{0.01 \cdot 3.8}{4.91 \cdot 10^{-4}} = 77.4 \,\text{m/s}$$

Since no work is done and the flow is adiabatic, the stagnation enthalpy remains constant $h_{01} = h_{02}$. With negligible change in potential energy, this equation reduces to

$$h_2 - h_1 = \frac{1}{2} V_1^2 - \frac{1}{2} V_2^2 = \frac{1}{2}(305.6^2 - 77.4^2) = 43.7 \,\text{kJ/kg}$$

■

2.3 SECOND LAW OF THERMODYNAMICS

For a uniform steady flow in a channel the second law of thermodynamics takes the form

$$\dot{m}(s_2 - s_1) = \int_{\ell_1}^{\ell_2} \frac{\dot{Q}'}{T} d\ell + \int_{\ell_1}^{\ell_2} \dot{s}'_p \, d\ell \qquad (2.3)$$

in which s is the entropy. On the right-hand side (RHS) $\dot{Q}'$ is the rate at which heat is transferred from the walls of the flow channel into the fluid per unit length of the channel. The incremental length of the channel is $d\ell$, and the channel extends from location ℓ_1 to ℓ_2. The absolute temperature T in this expression may vary along the channel. In the second term on the RHS, $\dot{s}'_p$ is the rate of entropy production per unit length of the flow channel. If the heat transfer is internally reversible, entropy production is the result of internal friction and mixing in the flow. In order for the heat transfer to be reversible, the temperature difference between the walls and the fluid has to be small. In addition, the temperature gradient in the flow direction must be small. This requires the flow to move rapidly so that energy transfer by bulk motion far exceeds the transfer by conduction and radiation in the flow direction.

As Eq. (2.3) shows, when heat is transferred into the fluid, its contribution is to increase the entropy in the downstream direction. If, on the other hand, heat is transferred from the fluid to the surroundings, its contribution is to reduce the entropy. Entropy production $\dot{s}'_p$ is caused by irreversibilities in the flow and is always positive, and its contribution is to increase the entropy in the flow direction. For the ideal case of an *internally reversible process* entropy production vanishes.

2.3.1 *T ds* equations

The first law of thermodynamics for a *closed system* relates the work and heat interactions to a change in internal energy U. For infinitesimal work and heat interactions the first law can be written as

$$dU = \delta Q - \delta W$$

For a simple compressible substance, defined to be one for which the only relevant work is compression or expansion, reversible work is given by

$$\delta W_s = p\, dV$$

This expression shows that when a fluid is compressed so that its volume decreases, work is negative, meaning that work is done on the system. For an internally reversible process the second law of thermodynamics relates heat transfer to a change in entropy by

$$\delta Q_s = T\, dS$$

in which it must be remembered that T is the absolute temperature. Hence, for an internally reversible process, the first law takes the differential form

$$dU = T\, dS - p\, dV$$

Dividing by the mass of the system converts this to an expression

$$du = T\, ds - p\, dv$$

between specific properties. Although derived for reversible processes, this is a *relationship between intensive properties*, and for this reason it is valid for all processes; reversible, or irreversible. It is usually written as

$$T\, ds = du + p\, dv \qquad (2.4)$$

and is called the first *Gibbs equation*.

Writing $u = h - pv$ and differentiating gives $du = dh - p\, dv - v\, dp$. Substituting this into the first Gibbs equation gives

$$T\, ds = dh - v\, dp \qquad (2.5)$$

which is the second *Gibbs equation*.

2.4 EQUATIONS OF STATE

The *state principle* of thermodynamics guarantees that a thermodynamic state for a simple compressible substance is completely determined by specifying two independent thermodynamic properties. Other properties are then functions of these independent properties. Such functional relations are called *equations of state*.

In this section the equations of state for steam and those of ideal gases are reviewed. In addition, ideal gas mixtures are considered as they arise in combustion of hydrocarbon fuels. Combustion gases flow through the gas turbines of a jet engine and through industrial turbines burning natural gas. Preliminary calculations can be carried out using properties of air since air is 78% of nitrogen by volume, which, although contributing to formation of nitric oxides, is otherwise largely inert during combustion. Later in the chapter a better model for combustion gases is discussed, but for accurate calculations the actual composition is to be taken into account. Also in many applications, such as in oil and gas production, mixtures rich in complex molecules flow through compressors and expanders. Their equations of state may be very complicated, particularly at high pressures.

2.4.1 Properties of steam

It has been found that a useful way to present properties of steam is to construct a chart, such as is shown in Figure 2.3, with entropy on the abscissa and temperature on the ordinate. On the heavy line water exists as a saturated liquid on the descending part on the left and as saturated vapor on the right. Away from this vapor dome, on the right water is superheated vapor, that is to say *steam*; and to the left, water exists as a compressed liquid. The state at the top of the vapor dome is called a *critical state*, with pressure $p_c = 220.9$ bar and temperature $T_c = 374.14°$C. At this condition entropy is $s_c = 4.4298$ kJ/(kg · K) and enthalpy is $h_c = 2099.6$ kJ/kg. Below the vapor dome water exists as a two-phase mixture of saturated vapor and saturated liquid. Such a state may exist in the last stages of a steam turbine where the saturated steam is laden with water droplets.

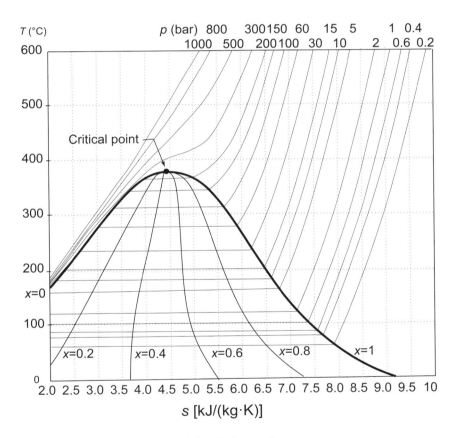

Figure 2.3 *Ts*-diagram for water.

The lines of constant pressure are also shown in Figure 2.3. As they intersect the vapor dome, their slopes become horizontal across the two-phase region. Thus they are parallel to lines of constant temperature, with the consequence that temperature and pressure are *not independent* properties in the two-phase region. To specify the thermodynamic state in this region, a quality denoted by x is used. It is defined as the mass of vapor divided by the mass of the mixture. In terms of quality, thermodynamic properties of a two-phase mixture are calculated as a weighted average of the saturation properties. Thus, for example

$$h = (1-x)h_f + xh_g$$

or

$$h = h_f + xh_{fg}$$

in which h_f denotes the enthalpy of saturated liquid, h_g that of saturated vapor, and the difference is denoted by $h_{fg} = h_g - h_f$. Similarly, entropy of the two-phase mixture is

$$s = s_f + xs_{fg}$$

and its specific volume is

$$v = v_f + xv_{fg}$$

Integrating the second Gibbs equation $Tds = dh - vdp$ between the saturated vapor and liquid states at constant pressure gives

$$h_{fg} = Ts_{fg}$$

The first law of thermodynamics shows that the amount of heat transferred to a fluid flowing at constant pressure, as it is evaporated from its saturated liquid state to saturated vapor state, is

$$q = h_g - h_f = h_{fg}$$

and this is therefore also

$$q = T(s_g - s_f) = Ts_{fg}$$

States with pressure above the critical pressure have the peculiar property that if water at such pressures is heated at constant pressure, it converts from a liquid state to a vapor state without ever forming a two-phase mixture. Thus, neither liquid droplets nor vapor bubbles can be discerned in the water during the transformation. This region is of interest because in a typical supercritical steam power plant built today water is heated at supercritical pressure of 262 bar to temperature 566°C, and in ultrasupercritical power plants steam generator pressures of 600 bar are in use. Steam at these pressures and temperatures then enters high-pressure (HP) steam turbine, which must be designed with these conditions in mind.

Steam tables, starting with those prepared by H. L. Callendar in 1900, and Keenan and Kays in 1936, although still in use, are being replaced by computer programs today. Steam tables, found in Appendix B, were generated by the software EES, a product of the company F-chart Software, in Madison, Wisconsin. It was also used to prepare Figures 2. and 2.4. Its use is demonstrated in the following example.

■ **EXAMPLE 2.3**

Steam at $p_1 = 6000\,\text{kPa}$ and $T_1 = 400°C$ expands reversibly and adiabatically through a steam turbine to pressure $p_2 = 60\,\text{kPa}$. (a) Find the exit quality and (b) the work delivered if the change in kinetic energy is neglected.

Solution: (a) The thermodynamic properties at the inlet to the turbine are first found from the steam tables, or calculated using computer software. Either way shows that $h_1 = 3177.0\,\text{kJ/kg}$ and $s_1 = 6.5404\,\text{kJ/(kg} \cdot \text{K)}$. Since the process is reversible and adiabatic, it takes place at constant entropy and $s_2 = s_1$. The exit state is in the two-phase region, and steam quality is calculated from

$$x_2 = \frac{s_2 - s_\mathrm{f}}{s_\mathrm{g} - s_\mathrm{f}} = \frac{6.5404 - 1.1451}{7.5314 - 1.1451} = 0.8448$$

in which $s_\mathrm{f} = 1.1451\,\mathrm{kJ/(kg \cdot K)}$ and $s_\mathrm{g} = 7.5314\,\mathrm{kJ/(kg \cdot K)}$ are the values of entropy for saturated liquid and saturated vapor at $p_2 = 60\,\mathrm{kPa}$. Exit enthalpy is then obtained from

$$h_2 = h_\mathrm{f} + x_2 h_\mathrm{fg} = 359.79 + 0.8448 \cdot 2293.1 = 2297.0\,\mathrm{kJ/kg}$$

(b) Work delivered is

$$w_\mathrm{s} = h_1 - h_2 = 3177.0 - 2297.0 = 880\,\mathrm{kJ/kg}$$

The calculations have been carried out using the EES script shown below.

```
"State 1"
    T1=400 [C]
    p1=6000 [kPa]
    h1=ENTHALPY(Steam,P=p1,T=T1)
    s1=ENTROPY(Steam, P=p1,T=T1)
"State 2"
    p2=60 [kPa]
    s2=s1
    sf2=ENTROPY(Steam,P=p2,X=0)
    sg2=ENTROPY(Steam,P=p2,X=1)
    x2=(s2-sf2)/(sg2-sf2)
    hf2=ENTHALPY(Steam,P=p2,X=0)
    hg2=ENTHALPY(Steam,P=p2,X=1)
    h2=(1-x2)*hf2+x2*hg2
"Performance Calculations"
    wt=h1-h2
```

The results are:

h1=3177 [kJ/kg]   h2=2297 [kJ/kg]
hf2=359.8 [kJ/kg]   hg2=2653 [kJ/kg]
p1=6000 [kPa]   p2=60 [kPa]
s1=6.54 [kJ/kg-K]   s2=6.54 [kJ/kg-K]
T1=400 [C]   x2=0.8448   wt=879.9 [kJ/kg]

Calculation of enthalpy and steam quality at state 2 could have been shortened by simply writing

```
p2=60 [kPa]
h2=ENTHALPY(Steam, P=p2, S=s1)
x2=QUALITY(Steam, P=p2, S=s1)
```

∎

The Ts diagram is a convenient representation of the properties of steam, for lines of constant temperature on this chart are horizontal in the two-phase region, as are the lines of constant pressure. Isentropic processes pass through points along vertical lines. Adiabatic irreversible processes veer to the right of vertical lines, as entropy must increase. These make various processes easy to visualize. An even more useful representation is one in which entropy is on the abscissa and enthalpy is on the ordinate. A diagram of this kind was developed by R. Mollier in 1906. A Mollier diagram, with accurate steam properties calculated using EES, is shown in Figure 2.4.

The enthalpy drop used in the calculation of the work delivered by a steam turbine is now represented as a vertical distance between the end states. If the exit state is inside the vapor dome, there is a practical limit beyond which exit steam quality cannot be reduced. In a *condensing steam turbine* quality at the exit is generally kept above the line $x = 0.955$. Below this value droplets form, and, owing to their higher density, they do not turn as readily as vapor does, and thus on their impact on blades, they cause damage. A complicating factor in the analysis is the lack of thermodynamic equilibrium as steam crosses into the vapor dome. Droplets take a finite time to form, and if the water is clean and free of nucleation sites, their formation is delayed. Also, if the quality is not too low, by the time droplets form, steam may have left the turbine.

The line below which droplet formation is likely to occur is called the *Wilson line*. It is about 115 kJ/kg below the saturated vapor line, with a steam quality 0.96 at low pressure of about 0.1 bar. The quality decreases to 0.95 along the Wilson line as pressure increases to 14 bar. Steam inside the vapor dome is *supersaturated* above the Wilson line, a term that arises from water existing as vapor at conditions at which condensation should be taking place.

■ **EXAMPLE 2.4**

Steam from a steam chest of a single-stage turbine at $p_1 = 3$ bar and $T_1 = 440°C$ expands reversibly and adiabatically through a nozzle to pressure of $p = 1$ bar. Find the velocity of the steam at the exit.

Solution: Since the process is isentropic, the states move down along a vertical line on the Mollier chart. From the chart, steam tables — or using EES, enthalpy of steam in the reservoir — is determined to be $h_1 = 3358.7$ kJ/kg, and its entropy is $s_1 = 8.1536$ kJ/(kg · K). For an isentropic process, the exit state is determined by $p_2 = 1$ bar and $s_2 = 8.1536$ kJ/(kg · K). Enthalpy, obtained by interpolating in the tables, is $h_2 = 3039.2$ kJ/kg.

Assuming that the velocity in the steam chest is negligible, the exit velocity is obtained from

$$h_1 = h_2 + \frac{1}{2}V_2^2$$

or

$$V_2 = \sqrt{2(h_1 - h_2)} = \sqrt{2(3358.7 - 3039.4)\,1000} = 799.1\,\text{m/s}$$

An EES script used to solve this example is shown below. Conversion between kilojoules and joules is carried out by the statement convert(kJ,J):

```
"State 1"
    p1=3 [bar]
    T1=440 [C]
    h1=ENTHALPY(Steam, P=p1, T=T1)
    s1=ENTROPY(Steam, P=p1, T=T1)
"State 2"
```

EQUATIONS OF STATE 25

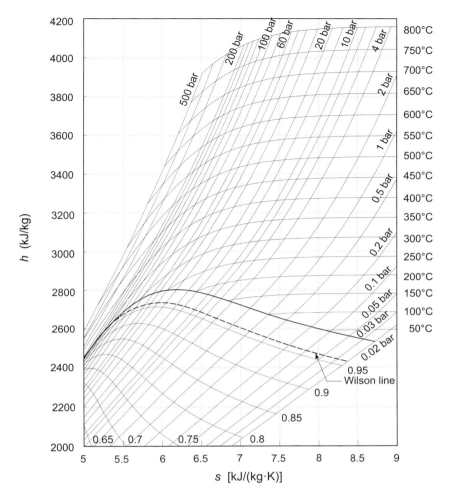

Figure 2.4 Mollier diagram for steam.

```
p2=1 [bar]
s2=s1
h2=ENTHALPY(Steam, P=p2, S=s2)
V2=sqrt(2*(h1-h2)*convert(kJ,J))
```

The results are:

```
h1=3359 [kJ/kg]      h2=3039 [kJ/kg]
s1=8.154 [kJ/kg-K]   s2=8.154 [kJ/kg-K]
p1=3 [bar]    p2=1 [bar]
T1=440 [C]    V2=799.3 [m/s]
```

∎

To the left of the saturated liquid line water exists as a *compressed liquid*. Since specific volume and internal energy do not change appreciably as a result of water being compressed,

their values may be approximated as
$$v(T,p) \approx v_f(T)$$
$$u(T,p) \approx u_f(T)$$

Enthalpy can then be obtained from
$$h(p,T) \approx u_f(T) + pv_f(T)$$

which can also be written as
$$h(p,T) = u_f(T) + p_f(T)v_f(T) + (p - p_f(T))v_f(T)$$

or as
$$h = h_f + v_f(p - p_f) \tag{2.6}$$

in which explicit dependence on temperature has been dropped and it is understood that all the properties are given at the saturation temperature.

Consider next the calculation of a change in enthalpy along an isentropic path from the saturated liquid state to a compressed liquid state at higher pressure. Integration of
$$T ds = dh - v\, dp$$

along an isentropic path, assuming v to be constant, gives
$$h = h_f + v_f(p - p_f) \tag{2.7}$$

This equation is identical to Eq. (2.6). Both approximations use the value of specific volume at the saturation state.

■ EXAMPLE 2.5

Water as saturated liquid at $p_1 = 6\,\text{kPa}$ is pumped to pressure $p_2 = 3400\,\text{kPa}$. Find the specific work done by assuming the process to be reversible and adiabatic, assuming that the difference in kinetic energy between inlet and exit is small and can be neglected. Also calculate the enthalpy of water at the state with temperature $T_2 = 36.17°\text{C}$ and pressure $p_2 = 3400\,\text{kPa}$.

Solution: Since at the inlet to the pump water exists as saturated liquid, its temperature is $T_1 = 36.17°\text{C}$, specific volume is $v_1 = v_f = 0.0010065\,\text{m}^3/\text{kg}$, and entropy is $s_1 = s_f = 0.5208\,\text{kJ}/(\text{kg}\cdot\text{K})$. At this state its enthalpy $h_1 = h_f = 151.473\,\text{kJ/kg}$.

Along the isentropic path from state 1 to state 2s, Eq. (2.7), gives the value of enthalpy $h_{2sa} = 154.889\,\text{kJ/kg}$. On the other hand, the value using EES at $p_{2s} = 3400\,\text{kPa}$ and $s_{2s} = 0.5208\,\text{kJ}/(\text{kg}\cdot\text{K})$ is $h_{2s} = 154.886\,\text{kJ/kg}$, which for practical purposes is the same as the approximate value. Hence the work done is
$$w_s = h_{2s} - h_1 = 154.89 - 151.47 = 3.42\,\text{kJ/kg}$$

From Eq. (2.6) at pressure $3400\,\text{kPa}$ an approximate value for enthalpy becomes
$$h_{2ta} = 151.473 + (3400 - 6) \cdot 0.0010065 = 154.889\,\text{kJ/kg}$$

whereas an accurate value obtained by EES for compressed liquid is $154.509\,\text{kJ/kg}$. These values are shown at points 1 and 2t in Figure 2.5.

■

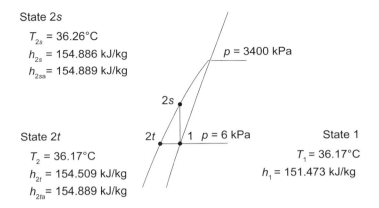

Figure 2.5 An illustration of how to obtain an approximate value for the enthalpy of compressed liquid.

2.4.2 Ideal gases

An ideal gas model assumes that internal energy is only a function of temperature $u = u(T)$ and the *equation of state* relates pressure and specific volume to temperature by

$$pv = RT \quad \text{or} \quad p = \rho RT \tag{2.8}$$

in which R is an ideal gas constant. It is equal to the universal gas constant, $\bar{R} = 8.314 \, \text{kJ}/(\text{kmol} \cdot \text{K})$, divided by the molecular mass M of the gas, so that it is calculated according to $R = \bar{R}/M$. The ideal gas model has been shown to be valid for various gases at low pressures. From Eq. (2.8) it follows that enthalpy for an ideal gas can be written in the form $h = u + RT$, and this shows that enthalpy is also a function of temperature only.

Specific heats for an ideal gas at constant volume and constant pressure simplify to

$$c_v(T) = \left(\frac{\partial u}{\partial T}\right)_v = \frac{du}{dT} \quad \text{so that} \quad du = c_v(T)dT$$

and

$$c_p(T) = \left(\frac{\partial h}{\partial T}\right)_p = \frac{dh}{dT} \quad \text{so that} \quad dh = c_p(T)dT$$

Differentiating next, $h = u + RT$ gives

$$dh = du + R\,dT \quad \text{or} \quad c_p(T)dT = c_v(T)dT + R\,dT$$

from which it follows that

$$c_p(T) = c_v(T) + R$$

Thus even if specific heats depend on temperature, their difference does not. Henceforth the explicit dependence on temperature is not displayed. With $\gamma = c_p/c_v$ denoting the ratio of specific heats, the relations

$$c_v = \frac{R}{\gamma - 1} \qquad c_p = \frac{\gamma R}{\gamma - 1} \tag{2.9}$$

follow directly. The values of c_v, c_p, and γ are shown for air in Figure 2.6.

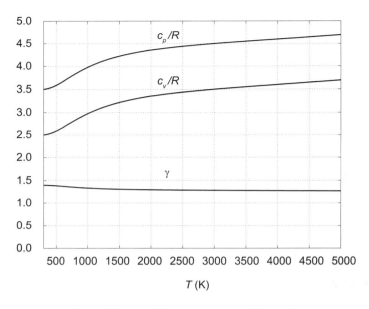

Figure 2.6 Specific heats for air and their ratio.

An approximate value for the ratio of specific heats is obtained from the *equipartition of energy principle* of kinetic theory of gases. It states that each degree of freedom of molecule contributes $\frac{1}{2}R$ to the specific heat at constant volume. For a monatomic gas there are three translational degrees of freedom: one for each of the three orthogonal coordinate directions. This means that for monatomic gases

$$c_v = \frac{3}{2}R \qquad c_p = \frac{5}{2}R \qquad \gamma = \frac{5}{3} = 1.67$$

If a molecule of a diatomic gas is regarded as a dumbbell, the rotational degrees of freedom about the two axes giving the largest moments of inertia contribute each one degree of freedom and the third is neglected. The vibrational degrees of freedom are not excited at relatively low temperatures. Hence, for diatomic gases the specific heats are

$$c_v = \frac{5}{2}R \qquad c_p = \frac{7}{2}R \qquad \gamma = \frac{7}{5} = 1.40$$

Since air is made up mainly of the diatomic N_2 and O_2, Figure 2.6 shows that the equipartition principle explains the low-temperature behavior of specific heats very well. Activation of greater number of vibrational modes takes place as temperature is increased.

Products of combustion flowing through a gas turbine consist of complex molecules, and the reasoning above suggests that the ratio of specific heats for them is closer to unity than for diatomic molecules, for all three rotational and low-level vibrational modes are excited. For combustion gases the value $\gamma = 1.333$ is appropriate. For superheated steam at low pressures the value $\gamma = 1.3$ is acceptable, and for steam that is just below the saturated vapor line Zeuner's empirical equation $\gamma = 1.035 + 0.1x$ is often used, with x as the steam quality. At saturation condition $x = 1$, and this gives $\gamma = 1.135$ for saturated steam.

2.4.3 Air tables and isentropic relations

In this section the influence of temperature variation of specific heats on the thermodynamic properties of air are considered. Entropy for ideal gases can be determined by first writing

$$T\,ds = dh - v\,dp$$

in the form

$$ds = c_p \frac{dT}{T} - R\frac{dp}{p}$$

and integrating. This gives

$$s(T_2, p_2) - s(T_1, p_1) = s^0(T_2) - s^0(T_1) - R \ln \frac{p_2}{p_1} \qquad (2.10)$$

in which s^0 is defined as

$$s^0(T) = \int_{T_{\text{ref}}}^{T} c_p(T) \frac{dT}{T}$$

Entropy is assigned the value zero at the reference state, $T_{\text{ref}} = 0\,\text{K}$ and $p_{\text{ref}} = 1\,\text{atm}$. The value of entropy at temperature T and pressure p is then calculated from

$$s(T, p) = s^0(T) - R \ln \frac{p}{p_{\text{ref}}}$$

For a reversible process $s_2 = s_1$, and Eq. (2.10) shows that

$$\frac{p_2}{p_1} = \exp\left[\frac{s^0(T_2) - s^0(T_1)}{R}\right]$$

which can be also be written as

$$\frac{p_2}{p_1} = \frac{\exp[s_2^0(T_2)/R]}{\exp[s_1^0(T_1)/R]}$$

Defining a reduced pressure as

$$p_{\text{r}}(T) = \exp \frac{s^0(T)}{R}$$

it is seen that p_{r} is only a function of temperature. The ratio of pressures at the endpoints of a reversible process can now be expressed as

$$\frac{p_2}{p_2} = \frac{p_{\text{r}2}}{p_{\text{r}1}}$$

Specific volume ratio can be obtained from the pressure ratio by using the ideal gas law $pv = RT$ to recast the pressure ratio into the form

$$\frac{p_2}{p_1} = \frac{RT_2}{v_2} \frac{v_1}{RT_1} = \frac{p_{\text{r}2}}{p_{\text{r}1}}$$

Solving for the specific volume ratio yields

$$\frac{v_2}{v_1} = \left[\frac{RT_2}{p_{\text{r}}(T_2)}\right] \left[\frac{p_{\text{r}}(T_1)}{RT_1}\right]$$

Now defining $v_{\text{r}}(T) = RT/p_{\text{r}}(T)$ allows the specific volume ratio to be written as

$$\frac{v_2}{v_1} = \frac{v_{\text{r}2}}{v_{\text{r}1}}$$

The values of $s^0(T), p_{\text{r}}(T)$ and $v_{\text{r}}(T)$ are listed in the air Table (B.4) in Appendix B.

EXAMPLE 2.6

Air enters a compressor at $p_1 = 100\,\text{kPa}$ and $T_1 = 300\,\text{K}$. It is compressed isentropically to $p_2 = 1200\,\text{kPa}$. Assuming that there is no change in the kinetic energy between the inlet and the exit, find the work done by the compressor using the air tables.

Solution: Reversible work done is

$$w_c = h_{2s} - h_1$$

At $T_1 = 300\,\text{K}$, $p_{r1} = 1.386$ and $h_1 = 300.19\,\text{kJ/kg}$. For an isentropic process

$$p_{r2} = \frac{p_2}{p_1} p_{r1} = \frac{1200}{100} 1.386 = 16.632$$

Temperature corresponding to this value of p_{r2} is $T_{2s} = 603.5\,\text{K}$ and $h_{2s} = 610.64\,\text{kJ/kg}$. Hence

$$w_c = h_{2s} - h_1 = 610.64 - 300.19 = 310.45\,\text{kJ/kg}$$

■

When specific heats are assumed to be constant, integrating the $T ds$ equations gives

$$s_2 - s_1 = c_v \ln \frac{T_2}{T_1} + R \ln \frac{v_2}{v_1}$$

or

$$s_2 - s_1 = c_p \ln \frac{T_2}{T_1} - R \ln \frac{p_2}{p_1}$$

For an isentropic process, the first of these gives

$$\frac{v_2}{v_1} = \left(\frac{T_2}{T_1}\right)^{-1/(\gamma-1)}$$

and the second one can be written as

$$\frac{p_2}{p_1} = \left(\frac{T_2}{T_1}\right)^{\gamma/(\gamma-1)}$$

Eliminating the temperature ratios gives

$$\frac{p_2}{p_1} = \left(\frac{v_1}{v_2}\right)^{\gamma} = \left(\frac{\rho_2}{\rho_1}\right)^{\gamma}$$

The next example illustrates the use of these equations for the same conditions as in the previous example.

EXAMPLE 2.7

Air enters a compressor at $p_1 = 100$ kPa and $T_1 = 300$ K. It is compressed isentropically to $p_2 = 1200$ kPa. Find the work done by the compressor assuming constant specific heats with $\gamma = 1.4$ and using variable specific heats and EES.

Solution: Work done is
$$w_c = c_p(T_{2s} - T_1)$$

Temperature T_{2s} is found from
$$T_{2s} = T_1 \left(\frac{p_2}{p_1}\right)^{(\gamma-1)/\gamma} = 300 \cdot 12^{0.4/1.4} = 610.18 \text{ K}$$

Hence
$$w_c = c_p(T_{2s} - T_1) = 1.0045\,(610.18 - 300) = 311.58 \text{ kJ/kg}$$

Carrying out the calculations with EES gives

```
"State 1"
    p1=100 [kPa]
    T1=300 [K]
    s1=ENTROPY(Air,P=p1,T=T1)
    h1=ENTHALPY(Air,T=T1)
"State 2"
    p2=1200 [kPa]
    s2s=s1
    h2s=ENTHALPY(Air,P=p2, S=s2s)
    T2s=TEMPERATURE(Air, H=h2s)
"Work"
    wc=h2s-h1
```

The results are:

h1=300.4 [kJ/kg]   h2s=611.2 [kJ/kg]
p1=100 [kPa]   p2=1200 [kPa]
s1=5.705 [kJ/kg-K]   s2s=5.705 [kJ/kg-K]
T1=300 [K]   T2s=603.7 [K]   wc=310.8 [kJ/kg]

Owing to the relatively small temperature range, the error made in assuming constant specific heats is quite small. The difference between the computer calculation and using the air tables arises from interpolation and it is insignificant.

2.4.4 Ideal gas mixtures

Kinetic theory of ideal gas mixtures originates from the intuitive notion that the pressure on the walls of a vessel containing a gas is caused by the momentum of colliding molecules. This suggests that at relatively low densities each molecular species may be assumed to act independently.

Dalton's model is based on such a consideration, and it states that the mixture pressure is equal to the sum of the component pressures p_i, which each of the molecular species in the mixture would exert if it were to exist alone at the mixture temperature and volume. Expressed algebraically, this is

$$p = p_1 + p_2 + \cdots + p_n$$

When ideal gas behavior can be assumed, the component pressure p_i can be represented by

$$p_i = \frac{N_i \bar{R} T}{V}$$

in which N_i is the number of moles of the ith component, T is the mixture temperature and V is the mixture volume. The value of the universal gas constant in SI units is $\bar{R} = 8.314 \, \text{kJ}/(\text{kmol} \cdot \text{K})$.

For the mixture the ideal gas law is

$$p = \frac{N \bar{R} T}{V}$$

and N is the number of moles in the mixture. Dividing the last two equations by each other gives

$$p_i = \frac{N_i}{N} p = y_i p$$

and here y_i is the mole fraction and p_i is the *partial pressure* of the ith component. It is equal to the component pressure only for ideal gases.

Other properties of an ideal gas mixture can be obtained by a generalization of Dalton's rule, called the *Gibbs–Dalton* rule. Thus internal energy of a mixture is given by

$$U = U_1 + U_2 + \cdots + U_n$$

Since the internal energy of the ith component can be written as

$$U_i = N_i \bar{u}_i$$

in which $\bar{u}_i$ is the internal energy per mole of the ith species, the internal energy of the mixture can be expressed as

$$U = N_1 \bar{u}_1 + N_2 \bar{u}_2 + \cdots + N_n \bar{u}_n$$

Dividing this by the total number of moles gives

$$\bar{u} = y_i \bar{u}_1 + y_2 \bar{u}_2 + \cdots + y_n \bar{u}_n$$

On a mass basis internal energy can be written as

$$U = m_1 u_1 + m_2 u_2 + \cdots + m_n u_n$$

in which m_i is the mass of the ith component. Dividing this by the mass of the mixture gives

$$u = x_1 u_1 + x_2 u_2 + \cdots + x_n u_n$$

and here $x_i = m_i/m$ is the mass fraction of the ith component. Similar equations hold for enthalpy:

$$\bar{h} = y_i \bar{h}_1 + y_2 \bar{h}_2 + \cdots + y_n \bar{h}_n$$

$$h = x_i h_1 + x_2 h_2 + \cdots + x_n h_n$$

Using the Gibbs–Dalton rule, entropy of the ith component in an ideal gas mixture behaves as if it existed alone at the mixture temperature and its own partial pressure. Thus

$$S = N_1 \bar{s}_1(T, p_1) + N_2 \bar{s}_2(T, p_2) + \cdots + N_n \bar{s}_n(T, p_n)$$

or

$$S = m_1 s_1(T, p_1) + m_2 s_2(T, p_2) + \cdots + m_n s_n(T, p_n)$$

On a molar basis the specific entropy is

$$\bar{s} = y_1 \bar{s}_1(T, v) + y_2 \bar{s}_2(T, v) + \cdots + y_n \bar{s}_n(T, v)$$

and on a mass basis it is

$$s = x_1 s_1(T, p_1) + x_2 s_2(T, p_2) + \cdots + x_n s_n(T, p_n)$$

Gibbs equation for the ith component can be written as

$$T ds_i = dh_i - v_i \, dp_i$$

in which $v_i = V/m_i$ is the specific volume of the ith component. Using the ideal gas law $p_i v_i = R_i T$ puts the Gibbs equation into the form

$$ds_i = c_{p,i} \frac{dT}{T} - R_i \frac{dp_i}{p_i}$$

By assuming the specific heat to be constant, integrating this gives

$$\Delta s_i = c_{p,i} \ln \frac{T_2}{T_1} - R_i \ln \frac{p_{i,2}}{p_{i,1}}$$

or on a molar basis

$$\Delta \bar{s}_i = \bar{c}_{p,i} \ln \frac{T_2}{T_1} - \bar{R} \ln \frac{p_{i,2}}{p_{i,1}}$$

The pressure term on the right is called the *entropy of mixing*. In combustion reactions, once the combustion is complete, the mixture of combustion products may be considered a pure substance, just as is done for atmospheric air. Expansion through a turbine then takes place at a constant mixture composition and the entropy of mixing vanishes. If the specific heats are assumed to be constant, then, in order to carry out the calculations, it only remains to determine the specific heat and molecular mass of the mixture.

The molecular mass of the mixture is obtained from

$$\mathcal{M} = \frac{m_1 + m_2 + \cdots + m_n}{N} = \frac{N_1 \mathcal{M}_1 + N_2 \mathcal{M}_2 + \cdots + N_n \mathcal{M}_n}{N}$$
$$= y_1 \mathcal{M}_1 + y_2 \mathcal{M}_2 + \cdots + y_n \mathcal{M}_n$$

or

$$\mathcal{M} = \sum_{i=1}^{n} y_i \mathcal{M}_i$$

The mixture specific heat is

$$\bar{c}_p = \sum_{i=1}^{n} y_i \bar{c}_{pi} \quad \text{and} \quad c_p = \frac{\bar{c}_p}{\mathcal{M}}$$

From earlier studies of combustion it may be recalled that combustion of methane with stoichiometric amount of theoretical air leads to the chemical equation

$$CH_4 + 2(O_2 + 3.76N_2) \rightarrow CO_2 + 2H_2O + 7.52N_2$$

Assuming that the water in the products remains as vapor, the total number of moles in the gaseous products is 10.52. If the amount of theoretical air is 125% of the stoichiometric amount, then the previous chemical equation becomes

$$CH_4 + 2.5(O_2 + 3.76N_2) \rightarrow CO_2 + 2H_2O + 0.5O_2 + 9.40N_2$$

and the number of moles of gaseous products is 12.90. The next example illustrates the calculation of the mixture specific heat.

■ EXAMPLE 2.8

Consider the combustion of methane with 125% of theoretical air. Find the molecular mass of the mixture and the specific heat at constant pressure.

Solution: The number of moles of each species has been calculated above and are as follows: $N_{CO_2} = 1$, $N_{H_2O} = 2$, $N_{O_2} = 0.5$, and $N_{N_2} = 9.4$. Hence the total number of moles is $N = 12.9$ and the mole fractions are $y_{CO_2} = 0.0775$, $y_{H_2O} = 0.1550$, $y_{O_2} = 0.0388$, and $y_{N_2} = 0.7287$. The molecular masses and specific heats of common gases are listed in the Appendix B. Using them, the molecular mass of the mixture is given by

$$\begin{aligned} M &= y_{CO_2} M_{CO_2} + y_{H_2O} M_{H_2O} + y_{O_2} M_{O_2} + y_{N_2} M_{N_2} \\ &= 0.0775 \cdot 44.0 + 0.1550 \cdot 18.0 + 0.0388 \cdot 32.0 + 0.7287 \cdot 28.0 \\ &= 27.845 \text{ kg/kmol} \end{aligned}$$

The molar specific heat at constant pressure is then

$$\begin{aligned} \bar{c}_p &= y_{CO_2} \bar{c}_{p\,CO_2} + y_{H_2O} \bar{c}_{p\,H_2O} + y_{O_2} \bar{c}_{p\,O_2} + y_{N_2} \bar{c}_{p\,N_2} \\ &= 0.0775 \cdot 37.3292 + 0.1550 \cdot 33.5702 + 0.0388 \cdot 29.3683 + 0.7287 \cdot 29.1533 \\ &= 30.480 \text{ kJ/(kmol} \cdot \text{K)} \end{aligned}$$

The mixture specific heat is

$$c_p = \frac{\bar{c}_p}{M} = 1.0946 \text{ kJ/(kg} \cdot \text{K)}$$

■

As pointed out by Cohen et al. [15], it has been found that for combustion products of jet fuel it is sufficiently accurate to use the values

$$c_p = 1148 \text{ J/(kg} \cdot \text{K)} \qquad R = 287 \text{ J/(kg} \cdot \text{K)} \qquad \gamma = \frac{4}{3}$$

As inspection of Figure 2.6 shows that the value of γ decreases and that of c_p increases as temperature increases. Hence, if the actual mean temperature during a process is lower than that for which these values apply, then the value of γ is too large in the calculation in which it is used to determine the temperature change, and therefore this leads to a excessively large change in the temperature. But then the value of c_p is too low and the product $c_p \Delta T$ to determine the enthalpy change during the process is nearly correct, as it involves compensating errors. By a similar argument the constant values

$$c_p = 1004.5 \text{ J/(kg} \cdot \text{K)} \qquad R = 287 \text{ J/(kg} \cdot \text{K)} \qquad \gamma = 1.4$$

can be used for air.

2.4.5 Incompressibility

The important distinction between an *incompressible fluid* and *incompressible flow* is introduced next. Incompressibility may, on one hand, mean that specific volume does not change with pressure, but it is allowed to change with temperature. A stricter model is to have the specific volume remain an absolute constant. In liquid water even large changes in pressure lead to only small changes in the specific volume, and by this definition it is nearly incompressible, even if its specific volume changes appreciably with temperature. In the flow of gases at low speeds pressure changes are mild and the *flow is considered incompressible*, even if the fluid is clearly compressible.

With these distinctions in mind, consider a strictly incompressible fluid. With v constant, the first Gibbs equation reduces to

$$du = T ds$$

This shows that internal energy changes only if the entropy changes. If the flow is adiabatic, entropy increases only as a result of irreversibilities, and hence this can be the only cause of an increase in internal energy. Similarly, if the flow is reversible and adiabatic, then internal energy must remain constant. As a consequence, the first law of thermodynamics in such a flow takes the form

$$\frac{p_1}{\rho} + \frac{1}{2}V_1^2 + gz_1 = \frac{p_2}{\rho} + \frac{1}{2}V_2^2 + gz_2 + w_s \qquad (2.11)$$

Thermal energy terms are completely absent, and this equation involves only mechanical energy. When no work is done, it reduces to

$$\frac{p_1}{\rho} + \frac{1}{2}V_1^2 + gz_1 = \frac{p_2}{\rho} + \frac{1}{2}V_2^2 + gz_2 \qquad (2.12)$$

which is the familiar Bernoulli equation. Its usual development shows that for inviscid flows

$$p + \frac{1}{2}\rho V^2 + \rho g z = p_0$$

is constant along a streamline, with the constant p_0 called the *Bernoulli constant*.

2.4.6 Stagnation state

Stagnation state is defined by the equations

$$h_0 = h + \frac{1}{2}V^2 + gz \qquad (2.13)$$

$$s_0 = s$$

It is a *reference state* that may not correspond to any actual state in the flow. As was pointed out earlier, enthalpy h_0 is called the *stagnation enthalpy* and h is now called the *static enthalpy*. Other properties, such as pressure, temperature, specific volume, or density are designated similarly. This definition fixes to each static state in the flow a corresponding unique stagnation state. The stagnation state is arrived at by a *thought experiment* in which the flow is decelerated isentropically to zero velocity while it descends or ascends to a reference elevation.

From the definition of a stagnation state, integrating $Tds = dh - v\,dp$ from a static state to its stagnation state gives the following equation, since $ds = 0$:

$$h_0 - h = \int_p^{p_0} v\,dp = \int_p^{p_0} \frac{dp}{\rho}$$

For an *incompressible fluid* this reduces to

$$h_0 - h = \frac{p_0}{\rho} - \frac{p}{\rho}$$

Substituting for h_0 from Eq. (2.13) into this gives

$$p_0 = p + \frac{1}{2}\rho V^2 + \rho g z \qquad (2.14)$$

This is the same equation that defines the Bernoulli constant, which is now seen to define the stagnation pressure for an incompressible fluid. This expression can also be used in low-speed compressible flow as an approximation to the true stagnation pressure.

2.5 EFFICIENCY

In this section various measures of efficiency for turbomachinery flows and their relationship to thermodynamic losses are discussed.

2.5.1 Efficiency measures

Work delivered by a turbine is given as the difference between inlet and exit stagnation enthalpy. A greater amount of work would be delivered along a reversible path to the same exit pressure. With w the actual work and w_s the isentropic work, their ratio

$$\eta_{tt} = \frac{w}{w_s} = \frac{h_{01} - h_{03}}{h_{01} - h_{03s}} \qquad (2.15)$$

is called a *total-to-total* efficiency. In the analysis of a turbine stage inlet to a stator (nozzle) is given label 1 and 3 is the exit state from the rotor. Label 2 is reserved to identify a state between the stator and the rotor. The process line for an adiabatic expansion between static states h_1 and h_3 is shown in Figure 2.7, which also shows the process line between the stagnation states h_{01} and h_{03}. In addition to the constant pressure lines corresponding to these states a line of constant stagnation pressure p_{03i} is drawn. This stagnation pressure corresponds to an end state along a reversible path with the *same amount* of work as in the actual process. As will be shown below, the loss of stagnation pressure $\Delta p_0 = p_{03i} - p_0$ is a measure of irreversibility in the flow. However, a stagnation pressure loss calculated in this way is only an estimate, and for a stage the losses across a stator and rotor need to be calculated separately. This is discussed in Chapter 5 and Chapter 6.

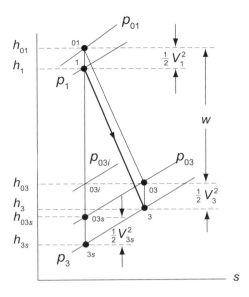

Figure 2.7 Thermodynamic states used to define a turbine efficiency.

If no attempt is made to diffuse the flow to low velocity, the exit kinetic energy, for example, a single-stage turbine, is wasted. For such a turbine a *total-to-static* efficiency is used as a measure of the efficiency. By this definition efficiency is given as

$$\eta_{ts} = \frac{h_{01} - h_{03}}{h_{01} - h_{3s}} \tag{2.16}$$

and the larger value of the denominator, caused by the wasted kinetic energy, reduces the efficiency.

The total-to-total efficiency is clearly also

$$\eta_{tt} = \frac{h_1 + \frac{1}{2}V_1^2 - h_3 - \frac{1}{2}V_3^2}{h_1 + \frac{1}{2}V_1^2 - h_{3s} - \frac{1}{2}V_{3s}^2} \tag{2.17}$$

The flow expands between the static states with enthalpy h_1 and h_3, with states 01 and 03 as the corresponding stagnation reference states. For an isentropic expansion to pressure p_3, the static enthalpy at the exit is h_{3s}. To find its corresponding stagnation pressure, the exit velocity V_{3s} would have to be known. A consistent theory can be developed if it is assumed that the state 03s lies on the constant-pressure line p_{03}. Then integrating the Gibbs equation along the constant-pressure p_3 line and also along the constant p_{03} line gives the two equations

$$s_3 - s_1 = c_p \ln \frac{T_3}{T_{3s}} \qquad s_3 - s_1 = \ln \frac{T_{03}}{T_{03s}}$$

from which

$$\frac{T_3}{T_{3s}} = \frac{T_{03}}{T_{03s}} \quad \text{or} \quad \frac{T_{03}}{T_3} = \frac{T_{03s}}{T_{3s}}$$

From the definition of a stagnation state the following two equations are obtained

$$\frac{T_{03}}{T_3} = 1 + \frac{V_3^2}{2c_p T_3} \qquad \frac{T_{03s}}{T_{3s}} = 1 + \frac{V_{3s}^2}{2c_p T_{3s}}$$

and the equality of the temperature ratios on the left-hand sides (LHSs) of these equations shows that

$$\frac{V_3}{V_{3s}} = \sqrt{\frac{T_3}{T_{3s}}}$$

so that $V_3 > V_{3s}$, but without a great loss of accuracy the temperature ratio is often replaced by unity, and then V_{3s} is replaced by V_3.

If a stage is designed such that $V_1 = V_3$, then the kinetic energy terms in the numerator of Eq. (2.17) be canceled. If next the approximation $V_3 = V_{3s}$ is used, then Eq. (2.15) for total-to-total efficiency reduces to

$$\eta_t = \frac{h_1 - h_3}{h_1 - h_{3s}}$$

the more familiar definition of turbine efficiency from the study in the first course of thermodynamics. In a multistage turbine the exit state would need a different label. It will be denoted by label e when the distinction needs to be clarified.

■ EXAMPLE 2.9

Steam enters an adiabatic multistage turbine at static pressure of 80 bar, static temperature 520°C, and velocity 50 m/s. It leaves the turbine at pressure 0.35 bar, temperature 80°C, and velocity 200 m/s. Find the total temperature and pressure at the inlet, total temperature and pressure at the exit, total-to-total efficiency, total-to-static efficiency, and the specific work done.

Solution: Using steam tables static enthalpy and entropy of steam at the inlet and exit are

$$h_1 = 3447.8 \, \text{kJ/kg} \qquad s_1 = 6.7873 \, \text{kJ/(kg} \cdot \text{K)}$$
$$h_e = 2645.0 \, \text{kJ/kg} \qquad s_e = 7.7553 \, \text{kJ/(kg} \cdot \text{K)}$$

Stagnation enthalpies are

$$h_{01} = h_1 + \frac{1}{2}V_1^2 = 3447.8 + \frac{50^2}{2 \cdot 1000} = 3449.1 \, \text{kJ/kg}$$

$$h_{0e} = h_e + \frac{1}{2}V_2^2 = 2645.0 + \frac{200^2}{2 \cdot 1000} = 2665.0 \, \text{kJ/kg}$$

Had the flow been isentropic, the exit state would have corresponded to $p_e = 0.35$ bar and $s_{es} = s_1$. This is inside the vapor dome at quality

$$x_{es} = \frac{s_{es} - s_f}{s_g - s_f} = \frac{6.7873 - 0.9874}{7.7148 - 0.9874} = 0.8621$$

and the enthalpy at this state is

$$h_{es} = h_f + x_{es}(h_g - h_f) = 304.20 + 0.8621 \cdot (2630.7 - 304.20) = 2309.9 \, \text{kJ/kg}$$

Assuming that $V_{es} = V_e$ then gives

$$h_{0es} = h_{es} + \frac{1}{2}V_e^2 = 2309.9 + \frac{200^2}{2 \cdot 1000} = 2329.9 \, \text{kJ/kg}$$

and the total-to-total efficiency is

$$\eta_{tt} = \frac{h_{01} - h_{0e}}{h_{01} - h_{0es}} = \frac{3449.1 - 2665.0}{3449.1 - 2329.9} = 0.7006$$

The total-to-static efficiency is

$$\eta_{ts} = \frac{h_{01} - h_{0e}}{h_{01} - h_{es}} = \frac{3449.1 - 2665.0}{3449.1 - 2309.9} = 0.6883$$

and the definition of efficiency when kinetic energy changes are neglected is

$$\eta_t = \frac{h_1 - h_e}{h_1 - h_{es}} = \frac{3447.8 - 2645.0}{3447.8 - 2309.9} = 0.7055$$

The specific work delivered is

$$w = h_{01} - h_{0e} = 3449.1 - 2665.0 = 784.1 \, \text{kJ/kg}$$

∎

Consider next a single-stage centrifugal compressor. The flow leaving the impeller enters a *diffuser section*, and then a *volute*. These stationary parts of the machine are designed to decelerate the flow so that at the exit velocity is well matched with the desired flow velocity in the discharge pipe. Since kinetic energy from the impeller is utilized in this way, it is again appropriate to define the efficiency as the total-to-total efficiency. It is given by

$$\eta_{tt} = \frac{w_s}{w} = \frac{h_{03s} - h_{01}}{h_{03} - h_{01}} \qquad (2.18)$$

The process lines between the stagnation states and the corresponding static states are shown in Figure 2.8. Now, as for the turbine, the state 03s is assumed to be on the constant-pressure line p_{03}, and the sketch reflects this. Had the same amount of compression work been done reversibly the exit stagnation pressure would have been p_{03i}, which is also shown in the figure.

In ventilating blowers no use is made of the exit kinetic energy and in such applications the total-to-static efficiency is used. In these cases efficiency is defined as

$$\eta_{ts} = \frac{h_{3s} - h_{01}}{h_{03} - h_{01}}$$

Since h_{3s} is smaller than h_{03s}, the efficiency is likewise smaller, and the difference accounts for the wasted kinetic energy. To be sure, in ventilating a space, high velocity may be needed to blow off light particulate matter sitting on the floors or attached to walls. In this case a blower may be placed upstream of the ventilated space and *forced draft* used to remove the particles. In *induced draft* contaminated air is drawn from the ventilated space into a blower and the kinetic energy in the exhaust stream is lost to the surroundings.

■ **EXAMPLE 2.10**

Air is drawn into a fan of diameter $D = 95.4$ cm from atmosphere at pressure 101.325 kPa and temperature 288.0 K. The volumetric flow rate is $Q = 4.72 \, \text{m}^3/\text{s}$ of standard air, and the power to the fan is $\dot{W} = 2.52$ kW. The total-to-total efficiency

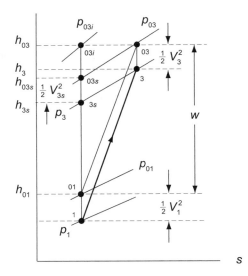

Figure 2.8 Thermodynamic states used to define a compressor efficiency.

of the fan is 0.8. (a) Find the total-to-static efficiency. (b) Find the stagnation pressure rise across the fan.

Solution: As air is drawn from the atmosphere at standard temperature and pressure into the blower, it undergoes a reversible adiabatic acceleration to the inlet of the blower. The inlet stagnation pressure is therefore $p_{01} = 101.325 \, \text{kPa}$, and the stagnation temperature is $T_{01} = 288.0 \, \text{K}$. The density of standard air is $\rho = 1.225 \, \text{kg/m}^3$, and this corresponds to the actual density in the atmosphere in this situation:

$$\rho_{01} = \frac{p_{01}}{RT_{01}} = \frac{101325}{287 \cdot 288.0} = 1.225 \, \text{kg/m}^3$$

The mass flow rate is therefore

$$\dot{m} = \rho_{01} Q = 1.225 \cdot 4.72 = 5.78 \, \text{kg/s}$$

The fan flow area and velocity are

$$A = \frac{1}{4} \pi D^2 = 0.715 \, \text{m}^2 \qquad V = \frac{Q}{A} = \frac{4.72}{0.715} = 6.6 \, \text{m/s}$$

The specific work is

$$w = \frac{\dot{W}}{\dot{m}} = \frac{2520}{5.78} = 435.8 \, \text{J/kg}$$

and the isentropic work is

$$w_s = \eta_{tt} w = 0.8 \cdot 435.8 = 348.7 \, \text{J/kg}$$

The total-to-static efficiency may be written as

$$\eta_{ts} = \frac{T_{3s} - T_{01}}{T_{03} - T_{01}} = \frac{T_{03s} - T_{01} - V^2/(2c_p)}{T_{03} - T_{01}} = \eta_{tt} - \frac{V^3}{2w} = 0.8 - \frac{6.6^2}{2 \cdot 436} = 0.75$$

EFFICIENCY

(b) The stagnation pressure rise can be calculated from

$$p_{03} - p_{01} = \rho w_s = 1.225 \cdot 348.7 = 427 \text{ Pa} = 43.6 \text{ mm H}_2\text{O}$$

The exit states were labeled with subscript 3 even though this machine does not have anything that functions as a stator.

■

In a multistage compressor with large pressure and temperature differences, the variability of specific heats with temperature needs to be factored in to obtain an accurate result. This is illustrated in the next example.

■ EXAMPLE 2.11

Air from the atmosphere flows into a multistage compressor at pressure 1 bar and temperature 300 K. The ratio of total pressures across the compressor is 30, and its total-to-total efficiency is 0.82. (a) Find the loss of stagnation pressure during this compression process, assuming specific heats to be constant. (b) Find also the loss in stagnation pressure assuming specific heats to vary with temperature.

Solution: (a) The inlet to the compressor is labeled as state 1, and its exit is denoted as e. The isentropic compression gives the stagnation temperature

$$T_{0es} = T_{01} \left(\frac{p_{0e}}{p_{01}}\right)^{(\gamma-1)/\gamma} = 300 \cdot 30^{1/3.5} = 792.8 \text{ K}$$

From the definition of efficiency

$$\eta_{tt} = \frac{T_{02s} - T_{01}}{T_{02} - T_{01}}$$

the exit temperature is

$$T_{0e} = T_{01} + \frac{1}{\eta_{tt}}(T_{0es} - T_{01}) = 300 + \frac{1}{0.82}(792.8 - 300) = 901.0 \text{ K}$$

If the same amount of work had been done isentropically, the pressure ratio would have been

$$\frac{p_{0ei}}{p_{01}} = \left(\frac{T_{0e}}{T_{01}}\right)^{\gamma/(\gamma-1)} = \left(\frac{901}{300}\right)^{3.5} = 46.94$$

Hence $p_{0ei} = 46.94$ bar and the loss of stagnation pressure is $\Delta p_{0L} = p_{0ei} - p_{0e} = 46.94 - 30 = 16.94$ bar.

(b) For variable specific heats, at $T_{01} = 300$ K, from air tables $p_{r1} = 1.386$ and $h_{01} = 300.19$ kJ/kg. Hence

$$p_{re} = p_{r1}\frac{p_{0e}}{p_{01}} = 1.386 \cdot 30 = 41.58$$

From air tables $T_{0es} = 771.32$ K and $h_{0es} = 790.56$ kJ/kg. Using the definition of total-to-total efficiency

$$\eta_{tt} = \frac{h_{0es} - h_{01}}{h_{0e} - h_{01}}$$

gives for the exit enthalpy the value

$$h_{0e} = h_{01} + \frac{h_{0es} - h_{01}}{\eta_{tt}} = 300.19 + \frac{790.56 - 300.19}{0.82} = 898.26 \,\text{kJ/kg}$$

From the air tables for this value $p_{rei} = 65.71$. It then follows that

$$\frac{p_{0ei}}{p_{01}} = \frac{p_{rei}}{p_{r1}} = \frac{65.71}{1.386} = 47.41$$

and $p_{0ei} = 47.41$ bar. Hence the loss of stagnation pressure is $\Delta p_{0L} = p_{0ei} - p_{0e} = 47.41 - 30.0 = 17.41$ bar. There is now some difference in the calculated results because the temperature range between inlet and exit states is large.

■

2.5.2 Thermodynamic losses

The effect of thermodynamic losses is illustrated by considering an increment on either a turbine or a compressor process line, as shown in Figure 2.9. From the Gibbs equation

$$T\,ds = dh - v\,dp$$

the slope of the constant-pressure line is

$$\left(\frac{\partial h}{\partial s}\right)_p = T$$

which shows that this slope is equal to the absolute temperature on an enthalpy–entropy (hs) diagram The enthalpy change between the end states may be considered to be made up of two parts. The irreversible change in enthalpy is $dh_f = T\,ds$, and the isentropic change is obtained by setting $ds = 0$, which gives $dh_s = v\,dp$. Substituting these back into the $T\,ds$ equation shows that in this notation

$$dh = dh_s + dh_f$$

For a compressor, all three terms are positive. For a turbine, dh_f is positive, but dh and dh_s have negative values. The irreversible process associated with dh_f is called *reheating*, or *internal heating*. Although the former term is in general use to describe this, the latter is better for it reflects what is happening physically. In other words, the irreversibilities cause an increase in temperature.

The nature of the irreversibilities may be further illustrated by considering a flow channel that extends from an inlet at location ℓ_1 to some general location ℓ. The first law of thermodynamics for this control volume is

$$\dot{Q} + \dot{m}\left(u_1 + p_1 v_1 + \frac{1}{2}V_1^2 + gz_1\right) = \dot{m}\left(u + pv + \frac{1}{2}V^2 + gz\right) + \dot{W}$$

Differentiating this with respect to ℓ and rearranging gives

$$\dot{m}\frac{du}{d\ell} = \dot{Q}' - \dot{m}\left[\frac{d(pv)}{d\ell} + \frac{1}{2}\frac{dV^2}{d\ell} + g\frac{dz}{d\ell}\right] - \dot{W}' \qquad (2.19)$$

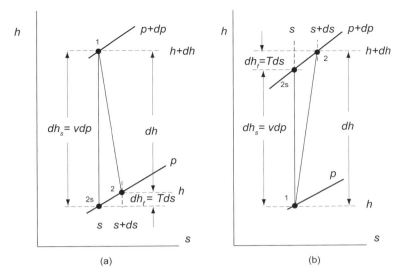

Figure 2.9 (a) An infinitesimal irreversible process in a turbine; (b) an infinitesimal irreversible process in a compressor.

in which the rate of heat transfer and work interaction per unit length along the element $d\ell$ have been defined as

$$\dot{Q}' = \frac{d\dot{Q}}{d\ell} \qquad \dot{W}' = \frac{d\dot{W}}{d\ell}$$

Clearly, in those parts of the flow in which there are no heat interactions $\dot{Q}' = 0$ and similarly $\dot{W}' = 0$ in those parts where there are no work interactions.

Differentiating the second law of thermodynamics

$$\dot{m}(s - s_1) = \int_{\ell_1}^{\ell} \frac{\dot{Q}'}{T} d\ell + \int_{\ell_1}^{\ell} \dot{s}_p' d\ell$$

with respect to ℓ gives

$$\dot{m}\frac{ds}{d\ell} = \frac{\dot{Q}'}{T} + \dot{s}_p' \qquad (2.20)$$

2.5.3 Incompressible fluid

For an incompressible fluid, density and its reciprocal specific volume are constant. For this kind of fluid the first Gibbs equation reduces to

$$T\frac{ds}{d\ell} = \frac{du}{d\ell}$$

Combining this with the second law in Eq. (2.20) gives

$$\dot{m}\frac{du}{d\ell} = \dot{Q}' + T\dot{s}_p' \qquad (2.21)$$

As was remarked earlier, the second term on the right shows that internal energy in incompressible flow *always increases* as a result of irreversibilities. The first term on the right side shows that internal energy increases by heat transfer into the fluid, but decreases when heat is lost to the surroundings. Integrating Eq. (2.21) gives

$$u_2 - u_1 = q + \frac{1}{\dot{m}} \int_{\ell_1}^{\ell_2} T\dot{s}'_p \, d\ell \qquad (2.22)$$

Substituting Eq. (2.21) into Eq. (2.19) puts the latter into the form

$$\dot{m}\left(\frac{d(pv)}{d\ell} + \frac{1}{2}\frac{dV^2}{d\ell} + g\frac{dz}{d\ell}\right) + \dot{W}' = -T\dot{s}'_p$$

Integrating this gives

$$p_1 v + \frac{1}{2}V_1^2 + gz_1 = p_2 v + \frac{1}{2}V_2^2 + gz_2 + w + \frac{1}{\dot{m}}\int_{\ell_1}^{\ell_2} T\dot{s}'_p \, d\ell \qquad (2.23)$$

In Eqs. (2.21) and (2.23) absolute temperature T multiplies the entropy production rate $\dot{s}'_p$ and it is the product $T\dot{s}'_p$, having the dimensions of energy flow rate per unit length of the channel, which represents a thermodynamic energy loss.

In the first law of thermodynamics, flow work, kinetic and potential energy, and external work, are all associated with *mechanical energy*. On the other hand, internal energy and heat interaction are *thermal energy* terms. Viewed from this perspective Eq. (2.22) may be said to be a *thermal energy balance* and Eq. (2.23) a *mechanical energy balance*. The term associated with entropy production represents an *irreversible conversion of mechanical energy into internal energy* and is the reason why it is also called a *thermodynamic energy loss*. In contrast to a conservation principle in which there are no terms that would represent conversion of one form of energy to another, in a balance equation such conversion terms are present. Further examination shows that heat transfer to, or from, an incompressible fluid changes only the internal energy and not the pressure, velocity, or elevation. These quantities change only as a result of work done or extracted, and they decrease as a result of irreversibilities in the flow. It is customary not to make a distinction between conservation and balance principles in practice, and often the principle of conservation of mass, for example, is called simply the *mass balance*. This practice will also be followed in this text.

For an *incompressible fluid* stagnation pressure has been shown to be given by

$$p_0 = p + \frac{1}{2}\rho V^2 + \rho g z$$

Making use of this relationship, Eq. (2.23) takes the form

$$p_{02} = p_{01} - \rho w - \frac{\rho}{\dot{m}}\int_{\ell_1}^{\ell_2} T\dot{s}'_p \, d\ell \qquad (2.24)$$

This shows that stagnation pressure changes because of a work interaction with the surroundings and it drops because of irreversibilities in the flow.

2.5.4 Compressible flows

For compressible flows the first law of thermodynamics is written for a flow extending from location ℓ_1 to an arbitrary location ℓ, as

$$\dot{Q} + \dot{m}\left(h_1 + \frac{1}{2}V_1^2 + gz_1\right) = \dot{W} + \dot{m}\left(h + \frac{1}{2}V^2 + gz\right)$$

or in terms of stagnation enthalpies as

$$\dot{Q} + \dot{m}h_{01} = \dot{W} + \dot{m}h_0$$

Differentiating gives

$$\dot{Q}' - \dot{W}' = \dot{m}\frac{dh_0}{d\ell} \qquad (2.25)$$

The second law of thermodynamics in the differential form has been shown to have the form

$$\dot{m}\frac{ds}{d\ell} = \frac{\dot{Q}'}{T} + \dot{s}_p'$$

Writing next the second Tds equation between stagnation states as

$$T_0\frac{ds}{d\ell} = \frac{dh_0}{d\ell} - v_0\frac{dp_0}{d\ell}$$

and substituting $ds/d\ell$ from this into the second law leads to

$$\dot{m}\frac{dh_0}{d\ell} - \dot{m}v_0\frac{dp_0}{d\ell} = \frac{T_0}{T}\dot{Q}' + T_0\dot{s}_p'$$

Using Eq. (2.25) to eliminate $dh_0/d\ell$ yields the equation

$$\dot{m}v_0\frac{dp_0}{d\ell} = \left(\frac{T_0}{T} - 1\right)\dot{Q}' - T_0\dot{s}_p' - \dot{W}' \qquad (2.26)$$

Turbomachinery flows are adiabatic so the heat transfer term may be dropped.

In an adiabatic flow integrating Eq. (2.26) gives

$$\int_{p_{01}}^{p_{02}} v_0\,dp_0 + w = -\frac{1}{\dot{m}}\int_{\ell_1}^{\ell_2} T_0\dot{s}_p'\,d\ell \qquad (2.27)$$

If the same amount of work had been done reversibly, then the exit stagnation pressure would have been different. Because the work done is the same, the pressure p_{02i} lies along the constant h_{02} line. A process line for this is shown in Figure 2.7, except that in that figure the exit enthalpy h_{03} corresponds to the enthalpy h_{02}.

Integrating next Eq. (2.26) for a reversible process gives

$$\int_{p_{01}}^{p_{02i}} v_0\,dp_0 + w_s = 0 \qquad (2.28)$$

Thus, since it has been stipulated that $w = w_s$, subtracting Eq. (2.27) from Eq. (2.28) gives

$$\int_{p_{02}}^{p_{02i}} v_0\,dp_0 = \frac{1}{\dot{m}}\int_{\ell_1}^{\ell_2} T_0\dot{s}_p'\,d\ell$$

Integration of this along the constant h_{02} line means that T_{02} remains constant and factors out after the substitution $v_0 = RT_0/p_0$. Then, carrying out the integration gives

$$R\ln\frac{p_{02i}}{p_{02}} = \frac{1}{\dot{m}}\int_{\ell_1}^{\ell_2} \dot{s}_p'\,d\ell$$

But in an adiabatic flow $\dot{m}\,ds = \dot{s}_p'\,d\ell$, and this equation reduces to

$$s_2 - s_1 = s_p = R\ln\frac{p_{02i}}{p_{02}}$$

This relates the entropy increase to a loss in stagnation pressure. A simpler development of this result is given in later chapters.

■ EXAMPLE 2.12

The inlet stagnation temperature to a multistage turbine is 1400 K, and the inlet stagnation pressure is 1000 kPa. The pressure ratio is 10, and the total-to-total efficiency of the turbine is 0.89. Assuming that gases flowing through the turbine have $\gamma = \frac{4}{3}$ and $R = 287 \text{ J/(kg} \cdot \text{K)}$, find the specific entropy production during the expansion assuming constant specific heats.

Solution: In this multistage turbine the inlet state is denoted by 1 and the exit by e. Assuming constant specific heats the definition of total-to-total efficiency reduces to

$$\eta_{tt} = \frac{T_{01} - T_{0e}}{T_{01} - T_{0es}} \quad \text{or} \quad \eta_{tt} = \frac{1 - T_{0e}/T_{01}}{1 - T_{0es}/T_{01}}$$

from which

$$\frac{T_{0e}}{T_{01}} = 1 - \eta_{tt}\left(1 - \frac{T_{0es}}{T_{01}}\right)$$

The isentropic temperature ratio is

$$\frac{T_{0es}}{T_{01}} = \left(\frac{p_{0e}}{p_{01}}\right)^{(\gamma-1)/\gamma} = \frac{1}{10^{1/4}} = 0.5623$$

so that

$$T_{0es} = 1400 \cdot 0.5623 = 787.3 \text{ K}$$

and

$$T_{0e} = 1400 \left[1 - 0.89(1 - 0.5623)\right] = 854.7 \text{ K}$$

Since the amount of work done is proportional to the difference between the stagnation temperatures, if this work had been done reversibly the exit pressure would reach the value p_{0ei} which is higher than before. It can be calculated from

$$\frac{p_{0ei}}{p_{01}} = \left(\frac{T_{0e}}{T_{01}}\right)^{\gamma/(\gamma-1)} = \left(\frac{854.7}{1400}\right)^4 = 0.1389$$

Hence $p_{0ei} = 1000 \cdot 0.1389 = 138.9 \text{ kPa}$. Since the pressure ratio is 10, the exit stagnation pressure is $p_{02} = 100 \text{ kPa}$, and the stagnation pressure loss is

$$p_{0ei} - p_{0e} = 138.6 - 100 = 38.6 \text{ kPa}$$

This represents 4.3% of the overall pressure difference $p_{01} - p_{02} = 900 \text{ kPa}$. The entropy production is calculated to be

$$s_p = R \ln \frac{p_{0ei}}{p_{0e}} = 287 \ln \frac{138.6}{100.0} = 9.4 \text{ J/(kg} \cdot \text{K)}$$

■

2.6 MOMENTUM BALANCE

In this section the use of momentum balance is illustrated in applications of interest in turbomachinery. In uniform steady flow in a channel the momentum balance reduces to

$$\dot{m}(\mathbf{V}_2 - \mathbf{V}_1) = \mathbf{F}_p + \mathbf{F}_v + \mathbf{F}_m \qquad (2.29)$$

in which $\mathbf{F}_p$ is a pressure force and $\mathbf{F}_v$ is a viscous force. The force $\mathbf{F}_m$ is present if the control volume cuts across the solid parts of the machine. If the control volume contains only fluid, this term is absent. Weights of the fluid and hardware have been omitted with the understanding that when stress analysis is carried out, they will be taken into account. The first illustration on the use of the momentum equation is to calculate the force a that deflection of a jet causes on a fixed vane.

■ EXAMPLE 2.13

Consider a jet of water that flows into a vane at an angle α_1. The vane is *equiangular* with $\alpha_2 = -\alpha_1$, and it turns the flow so that it leaves at a negative angle α_2, as shown in Figure 2.10. Positive angles are measured in the counterclockwise direction from the x axis. The jet velocity at the inlet is V, and pressure surrounding the jet and the vane is atmospheric. Find the y component of the force on the vane.

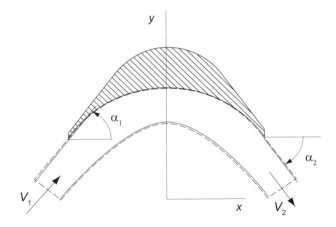

Figure 2.10 Turning of a flow by a vane.

Solution: The inlet to the vane is denoted as station 1 and the exit as station 2. Since the streamlines are straight, both at the inlet and at the exit, pressure at both of these locations is equal to the atmospheric pressure p_a across the jet. Then, with gravity neglected, Bernoulli equation shows that velocity at the exit is the same as at the inlet, so that $V_1 = V_2 = V$.

Momentum equation in the y direction gives

$$\dot{m}(V_{2y} - V_{1y}) = F_{py}$$

in which the force F_{py} is the y component of the pressure force exerted by the vane on the fluid. An equal and opposite force acts on the vane and it is denoted by R_y.

At the inlet $V_{1y} = V \sin \alpha_1$ and at the exit $V_{2y} = V \sin \alpha_2$, and for a negative value of α_2 this velocity component points in the minus y direction. The force on the blade is therefore

$$R_y = \dot{m} V (\sin \alpha_1 - \sin \alpha_2)$$

The mass flow rate is $\dot{m} = \rho V A$. Since the blade is equiangular, $\alpha_2 = -\alpha_1$. Substituting gives the force as

$$R_y = 2\rho A V^2 \sin \alpha_1$$

The force is largest when $\alpha_1 = 90°$ and the flow is turned $180°$. ∎

Consider again the flow shown in Figure 2.10, but now let the blade move in the y direction with velocity **U**, as shown in Figure 2.11. It is known that Newton's second law is valid in all coordinate systems that move at constant speed. Thus the momentum balance given by Eq. (2.29) is also valid for a control volume that moves with a uniform velocity **U**, if all velocities are to be replaced by relative velocities. In fact, noting that the relationship between the absolute velocity **V** and relative velocity **W** is given by

$$\mathbf{V} = \mathbf{W} + \mathbf{U}$$

and substituting this into Eq. (2.29) gives

$$\dot{m}(\mathbf{W}_2 + \mathbf{U} - \mathbf{W}_1 - \mathbf{U}) = \mathbf{F}_\mathrm{p} + \mathbf{F}_\mathrm{v} + \mathbf{F}_\mathrm{m}$$

or

$$\dot{m}(\mathbf{W}_2 - \mathbf{W}_1) = \mathbf{F}_\mathrm{p} + \mathbf{F}_\mathrm{v} + \mathbf{F}_\mathrm{m} \qquad (2.30)$$

In the next example the momentum balance is used to analyze the force on a moving blade.

■ **EXAMPLE 2.14**

Consider a waterjet directed at a blade that moves with speed U. The angle α_2 of the jet is such that the relative velocity meets the blade smoothly at the angle β_2. The blade is shaped such that it deflects the flow backward at an angle $\beta_3 = -\beta_2$, as is shown in Figure 2.11. (a) Find the work done on the blade per unit mass of the flow and the blade speed for maximum work by carrying out the analysis in a set of fixed coordinates. (b) Carry out the same analysis in a set of moving coordinates, and find the y component of the force on the blade.

Solution: (a) The analysis is carried out first in fixed coordinates. Station 1 is now the inlet to a nozzle (not shown) that issues the water at station 2 at velocity V_2 at angle α_2. Station 2 is also the inlet to the moving blade, and its exit is station 3. The force that the blade exerts on the fluid is

$$F_{py} = \dot{m}(V_3 \sin \alpha_3 - V_2 \sin \alpha_2)$$

and with $R_y = -F_{py}$ the force on the blade is

$$R_y = \dot{m}(V_2 \sin \alpha_2 - V_3 \sin \alpha_3)$$

Assuming that there are no losses, applying the first law to the control volume shown gives

$$\frac{p_\mathrm{a}}{\rho} + \frac{1}{2}V_2^2 = \frac{p_\mathrm{a}}{\rho} + \frac{1}{2}V_3^2 + w_s \qquad (2.31)$$

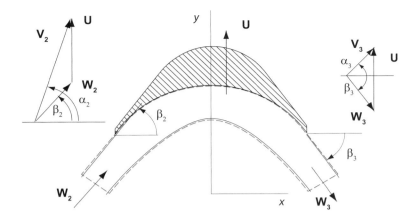

Figure 2.11 A water jet impinging on a moving blade.

and since the rate at which work is delivered to the blade is $\dot{W} = R_y U$, then

$$w_s = U(V_2 \sin\alpha_2 - V_3 \sin\alpha_3) \tag{2.32}$$

Substituting this into Eq. (2.31) gives

$$\frac{1}{2}V_2^2 = \frac{1}{2}V_3^2 + U(V_2 \sin\alpha_2 - V_3 \sin\alpha_3) \tag{2.33}$$

As is shown in the vector diagrams for the velocities in Figure 2.11, the x and y velocity components of absolute and relative velocity are related by

$$W_2 \cos\beta_2 = V_2 \cos\alpha_2 \qquad W_2 \sin\beta_2 = V_2 \sin\alpha_2 - U \tag{2.34}$$

Squaring each equation and adding them yields

$$W_2^2 = V_2^2 + U^2 - 2V_2 U \sin\alpha_2$$

which is a form of a law of cosines. A similar equation is obtained at the exit, namely

$$W_3 \cos\beta_3 = V_3 \cos\alpha_3 \qquad W_3 \sin\beta_3 = V_3 \sin\alpha_3 - U \tag{2.35}$$

from which another law of cosines is

$$W_3^2 = V_3^2 + U^2 - 2V_3 U \sin\alpha_3$$

Solving the laws of cosines for $V_2 \sin\alpha_2$ and $V_3 \sin\alpha_3$ and substituting them into Eq. (2.32) reduces it to

$$W_2^2 = W_3^2$$

so that $W_2 = W_3$. Thus the relative velocity at the exit is the same as at the inlet. This means that an observer in moving coordinates sees that the blade changes only the direction of the flow, but not its magnitude.

Substituting from Eqs. (2.34) and (2.35) into Eq. (2.32) gives

$$w_s = U[V_2 \sin\alpha_2 - (W_3 \sin\beta_3 + U)] = U(V_2 \sin\alpha_2 + W_2 \sin\beta_2 - U)$$

since $\beta_3 = -\beta_2$ and $W_3 = W_2$. Also $W_2 \sin \beta_2 = V_2 \sin \alpha_2 - U$, so that

$$w_s = 2U(V_2 \sin \alpha_2 - U)$$

From this equation it is seen that $w_s = 0$ when $U = 0$, or $U = V_2 \sin \alpha_2$. In the former case the load on the blade is too large and cannot be made to move. In the second case the load is too low, and the blade is *free-wheeling*. The condition for maximum power to the blade is obtained by differentiating w_s with respect to U and setting it to zero. This gives

$$\frac{dw_s}{dU} = V_2 \sin \alpha_2 - 2U = 0 \qquad \text{or} \qquad \frac{U}{V_2} = \frac{1}{2} \sin \alpha_2$$

For $\alpha_2 = 90°$ the maximum work is done by the jet on the blade when the blade moves at half the jet speed.

(b) If the moving coordinates are used, then

$$R_y = \dot{m}(W_2 \sin \beta_2 - W_3 \sin \beta_3)$$

Substituting $W_3 = W_2$ and $\beta_3 = -\beta_2$ gives

$$R_y = 2\dot{m}W_2 \sin \beta_2 = 2\dot{m}(V_2 \sin \alpha_2 - U)$$

and with the rate of work done $\dot{W} = R_y U$, the specific work becomes

$$w_s = 2U(V_2 \sin \alpha_2 - U)$$

as before.

■

Next, the momentum equation is applied to situations in which the results may be used to quantify thermodynamic losses.

■ **EXAMPLE 2.15**

Water with density $\rho = 1000 \text{ kg/m}^3$ and velocity $V_1 = 20 \text{ m/s}$ flows into a sudden expansion as shown in Figure 2.12. The supply pipe has a diameter $D_1 = 7 \text{ cm}$, and the pipe downstream has a diameter $D_2 = 14 \text{ cm}$. Find the increase in pressure $p_2 - p_1$.

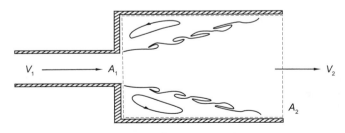

Figure 2.12 Flow in a channel with a sudden expansion.

Solution: Mass balance gives

$$A_1 V_1 = A_2 V_2$$

Thus
$$V_2 = V_1 \frac{D_1^2}{D_2^2} = 20 \left(\frac{7}{14}\right)^2 = 5.0 \, \text{m/s}$$

As the water enters a sudden expansion, it detaches from the boundaries and moves into the larger space as a *jet*. Regions of recirculating flow develop at the upstream corners. Flow speed in these corners is sufficiently low to make the pressure uniform. The pressure across the jet at the exit plane is therefore equal to that in the recirculating regions. The stream velocity rises from small backflow in the recirculating regions to a large forward velocity in the jet. This leads to appreciable viscous forces in the *free shear layers* forming the jet boundary. Such free shear layers are unstable to small disturbances and roll up into vortices. These cause mixing of the low velocity fluid in the recirculating zone with the fast flow in the jet. As a consequence, the jet spreads and fills the channel. In the mixing zone the shear forces along the walls influence pressure much less than the mixing, and therefore they may be neglected.

Applying the momentum balance in the x direction leads to

$$\rho A_2 V_2 (V_2 - V_1) = (p_1 - p_2) A_2$$

in which the mass balance, $\dot{m} = \rho V_1 A_2$ was used. Thus

$$p_2 - p_1 = \rho V_2 (V_1 - V_2)$$

and the numerical value for the pressure increase is

$$p_2 - p_1 = 1000 \cdot 5.0 (20 - 5.0) = 75 \, \text{kPa}$$

As the area A_2, is increased the exit velocity is reduced and finally becomes zero. In that case $p_2 = p_1$. This is the situation of a jet discharging to an atmosphere and then the exit pressure is equal to the atmospheric pressure.

The loss in stagnation pressure in a sudden expansion is

$$p_{01} - p_{02} = p_1 + \frac{1}{2}\rho V_1^2 - p_2 - \frac{1}{2}\rho V_2^2 = \frac{1}{2}\rho V_1^2 - \rho V_1 V_2 + \frac{1}{2}\rho V_2^2 = \frac{1}{2}\rho(V_1 - V_2)^2$$

or

$$p_{01} - p_{02} = \frac{1}{2} 1000 \, (20 - 5)^2 = 112.5 \, \text{kPa}$$

If the flow were to diffuse to the exit pipe without irreversibilities, there would be no loss of stagnation pressure and the exit pressure could be calculated from the Bernoulli equation. It would have the value

$$p_{2i} - p_1 = \frac{1}{2}\rho(V_1^2 - V_2^2) = \frac{1}{2} 1000 (20^2 - 5^2) = 187.5 \, \text{kPa}$$

above the inlet pressure. Contrasting this to the value 75 kPa calculated in the actual case shows that not all pressure is *recovered*, and the irreversibility can be regarded as a loss in pressure.

This underscores the importance of a well-designed diffuser to recover as much of the pressure as possible. The reduction of kinetic energy in diffusion goes into flow work on the fluid particles ahead, causing pressure to increase. If part of the kinetic energy is dissipated by viscous action in mixing, less is available for increasing the pressure. This is clearly true if the sudden expansion takes place into a vast reservoir, for then all the kinetic energy leaving the pipe will be dissipated in the reservoir and none is recovered. ∎

As a third example on the application of the momentum balance, consider how mixing of a stream in a constant-area duct changes the pressure. This requires the use of the balance equations in their integral forms.

■ **EXAMPLE 2.16**

An incompressible fluid flows in channel of cross-sectional area A, as shown in Figure 2.13. The flow is in the z direction with a nonuniform velocity profile $V_1 = V(1+f(x,y))$ at station 1 with the nonuniformity such that V is the average velocity. As a result of mixing, the flow enters station 2 with a uniform profile. Find the pressure change between stations 1 and 2. Work out the solution when the inlet velocity consists of two adjacent streams, one moving with velocity $V(1+f_a)$ and the other with velocity $V(1+f_b)$. Note that, since V is the average velocity, either f_a or f_b must have a negative value.

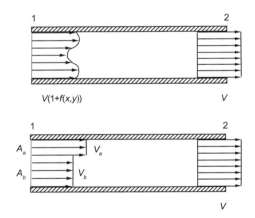

Figure 2.13 Mixing in a constant area channel.

Solution: Integral form of the mass balance

$$\dot{m} = \int_{A_1} \rho V_1 \, dA = \int_{A_2} \rho V_2 \, dA$$

applied to a control volume between stations 1 and 2 gives, with $A = A_1 = A_2$ and $V_2 = V$ this expression can be written as

$$\int_A V(1+f(x,y))dA = VA$$

Since V is the constant average velocity this yields the condition

$$\int_A f(x,y)dA = 0$$

The x component of the momentum equation applied to this control volume gives

$$\int_{A_2} \rho V_2^2 dA - \int_{A_1} \rho V_1^2 dA = (p_1 - p_2)A$$

Again, since at the exit velocity $V_2 = V$ is uniform, this can be written as

$$-\rho V^2 A + \rho \int_A V^2(1 + f(x,y))^2 dA = (p_2 - p_1)A$$

Wall shear has been neglected under the assumption that mixing influences the pressure change much more than does wall shear.

Expanding the integral leads to

$$\int_A (1 + f^2(x,y)) dA = \int_A (1 + 2f(x,y) + f^2(x,y)) dA = A + \int_A f^2(x,y) dA$$

for, as shown above, the middle term is zero. Denoting

$$\bar{f}^2 = \frac{1}{A} \int_A f^2(x,y) dA$$

the pressure increase is seen to be

$$p_2 - p_1 = \rho V^2 \bar{f}^2$$

With the velocities as shown on the bottom half of the figure, mass balance gives

$$A_a V_a + A_b V_b = AV$$

and since $A = A_a + A_b$, this reduces to

$$V = \frac{V_a A_a + V_b A_b}{A_a + A_b}$$

Writing
$$V_a = V(1 + f_a) \quad \text{and} \quad V_b = V(1 + f_b)$$

and solving for f_a and f_b gives

$$f_a = \frac{V_a}{V} - 1 = \frac{(V_a - V_b)A_b}{V_a A_a + V_b A_b}$$

and

$$f_b = \frac{V_b}{V} - 1 = \frac{(V_b - V_a)A_a}{V_a A_a + V_b A_b}$$

The value of $\bar{f}^2$ is the area-weighted average

$$\bar{f}^2 = \frac{f_a^2 A_a + f_b^2 A_b}{A_a + A_b} = \frac{(V_a - V_b)^2 A_a A_b}{(V_a A_a + V_b A_b)^2}$$

Hence the pressure increase is

$$p_2 - p_1 = \rho (V_a - V_b)^2 \frac{A_a A_b}{(A_a + A_b)^2}$$

As a special case, consider the situation in which $V_b = 0$. Then

$$V = \frac{V_a A_a}{A_a + A_b} \quad \text{and} \quad V_a - V = \frac{V_a A_b}{A_a + A_b}$$

and

$$p_2 - p_1 = \rho V(V_a - V) \tag{2.36}$$

This is the same result that was developed in Example 2.15. ∎

In these examples the momentum equation could be used to obtain information about the downstream pressure. This was possible because the viscous forces could be neglected. Such a situation also characterizes many turbomachinery flows in which the important force balance is between the inertia of the flow and pressure forces.

EXERCISES

2.1 Steam flows through a bank of nozzles shown in Figure 2.1 with wall thickness $t_2 = 2\,\text{mm}$, spacing $s = 4\,\text{cm}$, blade height $b = 2.5\,\text{cm}$, and exit angle $\alpha_2 = 68°$. The exit velocity $V_2 = 400\,\text{m/s}$, pressure is $p_2 = 1.5\,\text{bar}$, and temperature is $T_2 = 200°\text{C}$. Find the mass flow rate.

2.2 Air enters a compressor from atmosphere at pressure $102\,\text{kPa}$ and temperature $42°\text{C}$. Assuming that its density remains constant, determine the specific compression work required to raise its pressure to $140\,\text{kPa}$ in a reversible adiabatic process, given an exit velocity of $50\,\text{m/s}$.

2.3 Steam flows through a turbine at the rate of $\dot{m} = 9000\,\text{kg/h}$. The rate at which power is delivered by the turbine is $\dot{W} = 440\,\text{hp}$. The inlet total pressure is $p_{01} = 70\,\text{bar}$ and total temperature is $T_{01} = 420°\text{C}$. For a reversible and adiabatic process, find the total pressure and temperature leaving the turbine.

2.4 Water enters a pump as saturated liquid at total pressure of $p_{01} = 0.08\,\text{bar}$ and leaves it at $p_{02} = 30\,\text{bar}$. The mass flow rate is $\dot{m} = 10,000\,\text{kg/h}$ and assuming that the process takes place reversibly and adiabatically, determine the power required.

2.5 Liquid water at $700\,\text{kPa}$ and temperature $20°\text{C}$ flows at velocity $15\,\text{m/s}$. Find the stagnation temperature and stagnation pressure.

2.6 Water at temperature $T_1 = 20°\text{C}$ flows through a turbine with inlet velocity $V_1 = 3\,\text{m/s}$, static pressure $p_1 = 780\,\text{kPa}$, and elevation $z_1 = 2\,\text{m}$. At the exit the conditions are $V_2 = 6\,\text{m/s}$, $p_2 = 100\,\text{kPa}$, and $z_2 = 1.2\,\text{m}$. Find the specific work delivered by the turbine.

2.7 Air at static pressure $2\,\text{bar}$ and static temperature $300°\text{K}$ flows with velocity $60\,\text{m/s}$. Find total temperature and pressure.

2.8 Air at static temperature $300°\text{K}$ and static pressure $140\,\text{kPa}$ flows with velocity $60\,\text{m/s}$. Evaluate the total temperature and total pressure of air. Repeat the calculation assuming that the airspeed is $300\,\text{m/s}$.

2.9 Air undergoes an increase of $1.75\,\text{kPa}$ in total pressure through a blower. The inlet total pressure is one atmosphere, and the inlet total temperature is $21°\text{C}$. Evaluate the exit total temperature assuming that the process is reversible adiabatic. Evaluate the energy added to the air per unit mass flow.

2.10 Air enters a blower from the atmosphere where pressure is 101.3 kPa and temperature is 27°C. Its velocity at the inlet is 46 m/s. At the exit the total temperature is 28°C and the velocity is 123 m/s. Assuming that the flow is reversible and adiabatic, determine (a) the change in total pressure in millimeters of water and (b) the change in static pressure, also in millimeters of water.

2.11 The total pressure, static pressure, and the total temperature of air at a certain point in a flow are 700 kPa, 350 kPa, and 450 K, respectively. Find the velocity at that point.

2.12 Air has static pressure 2 bar and static temperature 300°K while flowing at speed 1000 m/s. (a) Assuming that air obeys the ideal gas law with constant specific heats, determine its stagnation temperature and stagnation pressure. (b) Repeat part (a) using the air tables.

2.13 At a certain location the velocity of air flowing in a duct is 321.5 m/s. At that location the stagnation pressure is 700 kPa and stagnation temperature is 450 K. What is the static density at this location?

2.14 Air flows in a circular duct of diameter 4 cm at the rate of 0.5 kg/s. The flow is adiabatic with stagnation temperature 288 K. At a certain location the static pressure is 110 kPa. Find the velocity at this location.

2.15 Saturated steam enters a nozzle at static pressure 14 bar at velocity 52 m/s. It expands isentropically to pressure 8.2 bar. Mass flow rate is $\dot{m} = 0.7$ kg/s. Find the exit area, assuming that (a) steam behaves as an ideal gas with $\gamma = 1.135$, and $c_p = 2731$ J/kg K; (b) the end state is calculated with properties obtained from the steam tables.

2.16 A fluid enters a turbine with total temperature of 330 K and total pressure of 700 kPa. The outlet total pressure is 100 kPa, and assume that the expansion process through the turbine is isentropic. Evaluate (a) the work per unit mass flow assuming that the fluid is incompressible with a density 1000 kg/m³, (b) and assuming that the fluid is air.

2.17 Air flows through a turbine that has a total pressure ratio 5 to 1. The total-to-total efficiency is 80%, and the flow rate is 1.5 kg/s. The desired output power is to be 250 hp (186.4 kW). Determine (a) the inlet total temperature, (b) the outlet total temperature, (c) the outlet static temperature given an exit velocity 90 m/s. (d) Then draw the process on a Ts diagram and determine the total-to-static efficiency of the turbine.

2.18 A blower has a change in total enthalpy of 6000 J/kg, an inlet total temperature 288 K, and an inlet total pressure 101.3 kPa. Find (a) the exit total temperature assuming that the working fluid is air, (b) the total pressure ratio across the machine, given a total-to-total efficiency of 75%.

2.19 A multistage turbine has a total pressure ratio of 2.5 across each of four stages. The inlet total temperature is $T_{01} = 1200$ K and the total-to-total efficiency of each stage is 0.87. Evaluate the overall total-to-total efficiency of the turbine by assuming that steam is flowing through it. Steam can be assumed to behave as a perfect gas with $\gamma = 1.3$. Why is the overall efficiency higher than the stage efficiency?

2.20 Gases from a combustion chamber enter a gas turbine at a total pressure of 700 kPa and a total temperature of 1100 K. The total pressure and total temperature at the exit of the turbine are 140 kPa and 780 K. Assuming that $\gamma = \frac{4}{3}$ is used for the mixture of combustion gases, which has a molecular mass of 28.97 kg/kmol, find the total-to-total efficiency and the total-to-static efficiency of the turbine, for an exit velocity of 210 m/s.

2.21 Air enters a compressor from atmosphere at 101.3 kPa, 288 K. It is compressed to a static pressure of 420 kPa, and at the exit its velocity is 300 m/s. The compressor total-to-total efficiency is 0.82. (a) Find the exit static temperature by assuming that $V_{2s} = V_2$. (b) Find the exit static temperature, without making the assumption that V_{2s} is equal to V_2.

2.22 Liquid water issues at velocity $V_1 = 20$ m/s from a bank of five oblique nozzles shown in Figure 2.14. The nozzles with wall thickness $t = 0.2$ cm are spaced $s = 4$ cm apart. The nozzle angle is $\alpha_1 = 70°$. Using the mass and momentum balance, (a) find the downstream velocity V_2, (b) find the pressure increase in the flow, (c) show how to deduce this result from Eq. 2.36. (d) Assuming that the thickness of the wall is vanishingly small, what is the change in pressure?

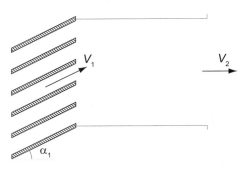

Figure 2.14 Nozzles with an oblique discharge.

2.23 Consider the flow shown in Figure 2.14. Prove that the kinetic energy lost in the flow as it moves to the downstream section is equal to that associated with the transverse component of the velocity. Neglect the wall thickness of the nozzles.

CHAPTER 3

COMPRESSIBLE FLOW THROUGH NOZZLES

In this chapter the dynamics and thermodynamics of compressible fluid flow through nozzles are discussed. First, the isentropic relations are developed and applied to a converging and a converging–diverging nozzle. After that normal shock relations are given. Then nozzle flows with friction are presented, various loss coefficients are introduced, and wet steam behavior is discussed. The last topic of the chapter is the Prandtl–Meyer expansion.

3.1 MACH NUMBER AND THE SPEED OF SOUND

Consider a stationary fluid in which a weak *pressure wave* travels to the right at velocity c, as is shown on the top part of Figure 3.1. Pressure and other thermodynamic properties change across the *wavefront*. Ahead of the front in the stagnant fluid velocity $V = 0$, and its pressure has the value p and its density is ρ. After the front has passed through a given location, let the velocity there be ΔV and pressure and density be $p + \Delta p$ and $\rho + \Delta \rho$, respectively. It is advantageous to shift to a frame of reference that moves to the right with speed c, for in that frame the front is stationary. Hence the balance principles in their steady form can be applied to a stationary control volume containing the front. In this frame fluid approaches the control volume with speed c from the right. Mass balance then gives

$$\rho c A = (\rho + \Delta \rho)(c - \Delta V) A \qquad (3.1)$$

Carrying out the multiplications on the right-hand side, (RHS) and assuming that the pressure wave is weak so that $\Delta V \Delta \rho$ can be neglected reduces this equation to

$$\Delta V = \frac{c \Delta \rho}{\rho} \qquad (3.2)$$

With the positive x direction pointing to the right, the x component of momentum equation

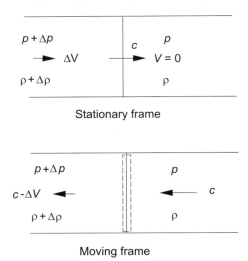

Figure 3.1 Sketch illustrating a weak pressure wave.

obtained from Eq. (2.29) and applied to this control volume, gives

$$\rho c A[c - (c - \Delta V)] = (p + \Delta p)A - pA$$

which reduces to

$$\rho c \Delta V = \Delta p$$

Substituting the expression for ΔV from Eq. (3.2) into this gives

$$c^2 = \frac{\Delta p}{\Delta \rho}$$

Since the pressure wave is assumed to be weak, the entire process may be assumed to be isentropic. In that case the speed of the wave is given by

$$c = \sqrt{\left(\frac{\partial p}{\partial \rho}\right)_s} \qquad (3.3)$$

This quantity is called the *speed of sound* because sound waves are weak pressure waves. For an ideal gas, for which $p\rho^{-\gamma}$ is constant in an isentropic process, taking logarithms and differentiating gives

$$\ln p - \gamma \ln \rho = \text{constant} \qquad \frac{dp}{p} - \gamma \frac{d\rho}{\rho} = 0$$

from which can be formed the partial derivative

$$\left(\frac{\partial p}{\partial \rho}\right)_s = \gamma \frac{p}{\rho} = \gamma RT$$

Speed of sound c for a perfect gas is therefore given by

$$c = \sqrt{\gamma RT} = \sqrt{\gamma \bar{R} T / \mathcal{M}}$$

For air $R = \bar{R}/\mathcal{M} = 8314/28.97 = 287\,\text{J}/(\text{kg}\cdot\text{K})$ and $\gamma = 1.4$ the speed of sound at $T = 300\,\text{K}$ is $c = 347\,\text{m/s}$. For combustion gases with $R = 287\,\text{J}/(\text{kg}\cdot\text{K})$ and $T = 1200\,\text{K}$ it is $c = 677.6\,\text{m/s}$. For gases of large molecular mass the speed of sound is small and the opposite is true for gases of low molecular mass. For example, the refrigerant R134a, or tetrafluoroethane, with a chemical formula CH_2FCF_3, has a molecular mass of $\mathcal{M} = 102.0\,\text{kg/kmol}$. Its ratio of specific heats is $\gamma = 1.14$. Hence at $T = 300\,\text{K}$ the speed of sound in R134a is only $c = 167\,\text{m/s}$. For helium at the same temperature, speed of sound is $c = 1019\,\text{m/s}$.

Mach number is defined to be the ratio of the local fluid velocity to the local sound speed

$$M = \frac{V}{c}$$

Subsonic flows have $M < 1$, *supersonic* ones have $M > 1$, and for *hypersonic* flows $M \gg 1$. Flows for which $M \sim 1$ are called *transonic*.

3.1.1 Mach number relations

In an ideal gas with constant specific heats the definition of stagnation enthalpy

$$h_0 = h + \frac{1}{2}V^2$$

can be recast as

$$T_0 = T + \frac{V^2}{2c_p} = T + \frac{(\gamma - 1)}{2\gamma R}V^2 = T\left(1 + \frac{\gamma - 1}{2}M^2\right)$$

from which

$$\frac{T_0}{T} = 1 + \frac{\gamma - 1}{2}M^2$$

From the definition of a stagnation state, it follows that

$$\frac{p_0}{p} = \left(\frac{T_0}{T}\right)^{\gamma/(\gamma-1)} \qquad \frac{\rho_0}{\rho} = \left(\frac{T_0}{T}\right)^{1/(\gamma-1)}$$

These can be written in terms of Mach number as

$$\frac{p_0}{p} = \left(1 + \frac{\gamma - 1}{2}M^2\right)^{\gamma/(\gamma-1)} \tag{3.4}$$

and

$$\frac{\rho_0}{\rho} = \left(1 + \frac{\gamma - 1}{2}M^2\right)^{1/(\gamma-1)} \tag{3.5}$$

These equations are in dimensionless form, and they represent the most economical way to show functional dependence of variables on the flow velocity.

Equation (3.4) for pressure can be expanded by the binomial theorem[1] for small values of Mach number. This leads to

$$\frac{p_0}{p} = 1 + \frac{\gamma}{2}M^2 + \frac{\gamma}{8}M^4 - \frac{\gamma(\gamma-2)}{48}M^6 + \cdots = 1 + \frac{\gamma}{2}M^2\left(1 + \frac{1}{4}M^2 + \cdots\right)$$

which, when only the first two terms are retained, can be rearranged as

$$\frac{p_0}{p} = 1 + \frac{\gamma V^2}{2\gamma RT} = 1 + \frac{1}{2}\frac{\rho}{p}V^2$$

so that

$$p_0 = p + \frac{1}{2}\rho V^2$$

For incompressible fluids this was taken to be the definition of stagnation pressure. In fact, it is seen to be approximately valid also for flows of compressible fluids when $M \ll 1$. In practice, this approximation is quite accurate if $M < 0.3$.

■ **EXAMPLE 3.1**

At a certain location in a flow of air static pressure has been measured to be $p = 2.4$ bar and stagnation pressure, $p_0 = 3$ bar. Measurement of the total temperature shows it to be $T_0 = 468$ K. Find the Mach number and flow rate per unit area.

Solution: Static temperature can be determined from

$$T = T_0\left(\frac{p}{p_0}\right)^{(\gamma-1)/\gamma} = 468\left(\frac{2.4}{3}\right)^{1/3.5} = 439.1\,\text{K}$$

Then, solving

$$\frac{T_0}{T} = 1 + \frac{\gamma-1}{2}M^2$$

for Mach number gives $M = 0.574$. With Mach number known, velocity can be determined from $V = Mc$. Speed of sound at this temperature is

$$c = \sqrt{\gamma RT} = \sqrt{1.4 \cdot 287 \cdot 439.1} = 420.0\,\text{m/s}$$

so that the velocity is

$$V = Mc = 241\,\text{m/s}$$

Static density is given by

$$\rho = \frac{p}{RT} = \frac{240000}{287 \cdot 439.1} = 1.904\,\text{kg/m}^3$$

Hence the mass flow rate per unit area is

$$\frac{\dot{m}}{A} = \rho V = 1.904 \cdot 241 = 458.9\,\text{kg}/(\text{s} \cdot \text{m}^2)$$

■

---

[1] Binomial theorem gives the expansion $(1+a)^n = 1 + na + \frac{n(n-1)}{2!}a^2 + \frac{n(n-1)(n-2)}{3!}a^3 + \cdots$.

3.2 ISENTROPIC FLOW WITH AREA CHANGE

Consider a one-dimensional isentropic gas flow in a converging–diverging nozzle as shown in Figure 3.2. Since the mass flow rate is

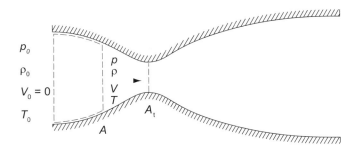

Figure 3.2 A converging–diverging nozzle.

$$\dot{m} = \rho V A$$

and $\dot{m}$ is constant, taking logarithms and then differentiating yields

$$\frac{d\rho}{\rho} + \frac{dA}{A} + \frac{dV}{V} = 0$$

Since in adiabatic flow h_0 is constant, differentiating

$$h_0 = h + \frac{1}{2}V^2$$

gives

$$dh = -V\,dV$$

Gibbs relation $T\,ds = dh - dp/\rho$ for isentropic flow leads to the relation

$$dh = \frac{1}{\rho}dp$$

Equating the last two expressions for enthalpy change gives

$$-V\,dV = \frac{1}{\rho}dp = \frac{1}{\rho}\left(\frac{\partial p}{\partial \rho}\right)_s d\rho = c^2 \frac{d\rho}{\rho}$$

Using this to eliminate density from the mass balance and simplifying it gives

$$(M^2 - 1)\frac{dV}{V} = \frac{dA}{A} \qquad (3.6)$$

From this it is seen that for subsonic flow, with $M < 1$, an increase in area decreases the flow velocity. Thus walls of a subsonic *diffuser diverge* in the downstream direction. For supersonic flow with $M > 1$ a *decrease in area leads to diffusion*. Since a *nozzle* increases the velocity of a flow, in a subsonic nozzle flow area decreases and in supersonic flow it increases in the flow direction.

In a continuously accelerating flow $dV > 0$, and Eq. (3.6) shows that at the *throat* where $dA = 0$, the flow is *sonic* with $M = 1$. If the flow continues its acceleration to a supersonic speed, the area must diverge after the throat. Such a converging–diverging nozzle, shown in Figure 3.2, is called *de Laval nozzle*. The assumptions made in arriving at these results are that the flow is steady and one-dimensional and that it is reversible and adiabatic. It has not been assumed that the fluid obeys the ideal gas law.

It was shown in the previous chapter that Mach number is a convenient parameter for expressing the relationship between the static and stagnation properties. By assuming ideal gas behavior and constant specific heats, the expressions

$$\frac{T_0}{T} = 1 + \frac{\gamma - 1}{2} M^2$$

$$\frac{p_0}{p} = \left(1 + \frac{\gamma - 1}{2} M^2\right)^{\gamma/(\gamma-1)}$$

$$\frac{\rho_0}{\rho} = \left(1 + \frac{\gamma - 1}{2} M^2\right)^{1/(\gamma-1)}$$

were obtained. Inverses of these ratios for a gas with $\gamma = 1.4$ are shown in Figure 3.3.

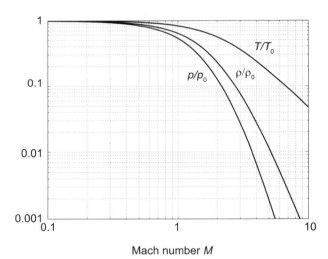

Figure 3.3 Pressure, density, and temperature ratios as functions of Mach number.

At *sonic condition*, denoted by the symbol (*), and for which $M = 1$, they reduce to

$$\frac{T^*}{T_0} = \frac{2}{\gamma + 1} = 0.8333$$

$$\frac{p^*}{p_0} = \left(\frac{2}{\gamma + 1}\right)^{\gamma/(\gamma-1)} = 0.5283$$

$$\frac{\rho^*}{\rho_0} = \left(\frac{2}{\gamma + 1}\right)^{1/(\gamma-1)} = 0.6339$$

The numerical values correspond to $\gamma = 1.4$.

Mass balance for a compressible flow, which obeys the ideal gas model, can be written as

$$\dot{m} = \rho V A = \frac{pAM}{RT}\sqrt{\gamma RT} = pAM\sqrt{\frac{\gamma}{RT}}$$

Multiplying and dividing the RHS by stagnation pressure and the square root of stagnation temperature, and expressing p/p_0 ratio in terms of temperature ratio T/T_0 gives

$$\dot{m} = \frac{p_0 AM\gamma}{\sqrt{c_p(\gamma-1)T_0}}\left(\frac{T_0}{T}\right)^{1/2-\gamma/(\gamma-1)}$$

which can be recast as

$$F = \frac{\dot{m}\sqrt{c_p T_0}}{Ap_0} = \frac{\gamma M}{\sqrt{\gamma-1}}\left(1+\frac{\gamma-1}{2}M^2\right)^{-(\gamma+1)/2(\gamma-1)} \quad (3.7)$$

This is called a *flow function*. Denoting the area at which the flow would reach $M = 1$ by A^*, the previous equation at this state gives

$$\frac{\dot{m}\sqrt{c_p T_0}}{A^* p_0} = \frac{\gamma}{\sqrt{\gamma-1}}\left(\frac{\gamma+1}{2}\right)^{-(\gamma+1)/2(\gamma-1)} \quad (3.8)$$

The ratio of the last two equations is

$$\frac{A}{A^*} = \frac{1}{M}\left(\frac{2}{\gamma+1}+\frac{\gamma-1}{\gamma+1}M^2\right)^{(\gamma+1)/2(\gamma-1)} \quad (3.9)$$

In the usual case area A^* is the throat area in a supersonic flow through a converging–diverging nozzle. But this equation is useful also when there is no location in the actual flow where $M = 1$ is reached. Then A^* can be regarded as a reference area. In the same manner in which stagnation properties are reached in a thought experiment in an isentropic deceleration of the flow to a rest state, so can the area A^* in a thought experiment be taken to be an area at which the sonic condition is reached in a hypothetical extension of a properly designed and operated variable area duct. If the velocity V^* denotes the velocity at the location where $M = 1$, it can be used as a reference velocity, and a velocity ratio can be written as

$$\frac{V}{V^*} = M\sqrt{\frac{T}{T^*}} = M\left(\frac{2}{\gamma+1}+\frac{\gamma-1}{\gamma+1}M^2\right)^{-1/2} \quad (3.10)$$

This and the area ratio are shown in Figure 3.4. Maximum flow rate per unit area takes place at the throat where $M = 1$. It is given by Eq. (3.8) as

$$\frac{\dot{m}}{A^*} = \frac{\gamma p_0}{\sqrt{c_p(\gamma-1)T_0}}\left(\frac{2}{\gamma+1}\right)^{(\gamma+1)/2(\gamma-1)} \quad (3.11)$$

■ **EXAMPLE 3.2**

Air flows through a circular duct of diameter $D = 10$ cm at the rate of $\dot{m} = 1.5$ kg/s. At a certain location, static pressure is $p = 120$ kPa and stagnation pressure is $T_0 = 318$ K. At this location, find the values for Mach number, velocity, and static density.

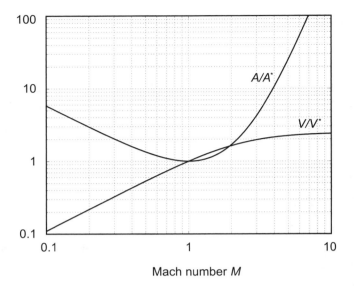

Figure 3.4 Area and velocity ratios as functions of Mach number.

Solution: Since the mass flow rate and diameter of the duct are known, mass balance

$$\dot{m} = \rho V A$$

can be recast into a form in which the known quantities of area, pressure, and stagnation temperature appear and Mach number is the only unknown. Thus

$$\dot{m} = \frac{p}{RT} M \sqrt{\gamma RT} A = pMA \sqrt{\frac{T_0}{T}} \sqrt{\frac{\gamma}{RT_0}}$$

or

$$\frac{\dot{m}}{pA} \sqrt{\frac{RT_0}{\gamma}} = M(1 + \frac{\gamma-1}{2} M^2)^{1/2}$$

Squaring both sides leads to a quadratic equation for M^2, which may be simplified and cast into the standard form:

$$M^4 + \frac{2}{\gamma-1} M^2 - \frac{2}{\gamma-1} \left(\frac{\dot{m}}{pA}\right)^2 \frac{RT_0}{\gamma} = 0$$

For the data given this reduces to

$$M^4 + 5M^2 - 0.828 = 0$$

and solving it gives $M = 0.40$. Static temperature is then

$$T = \frac{T_0}{1 + \frac{\gamma-1}{2} M^2} = \frac{318}{1.032} = 308.1 \, \text{K}$$

and the velocity and density are

$$V = M\sqrt{\gamma RT} = 140.8 \, \text{m/s} \qquad \rho = \frac{p}{RT} = \frac{120}{0.287 \cdot 308.1} = 1.357 \, \text{kg/m}^3$$

∎

3.2.1 Converging nozzle

A converging nozzle is shown in Figure 3.5. Consider a flow that develops from upstream stagnation state and in which backpressure p_b is controlled by a throttling valve located downstream of the nozzle. When the valve is closed there is no flow. With a slight opening of the valve pressure in the nozzle follows the line marked 1 and the flow leaves the nozzle at exit pressure $p_e = p_{b1}$. The mass flow rate corresponds to condition labeled 1 in the bottom right part of the figure. As the back pressure is reduced to p_{b2} pressure in the nozzle drops along the curve 2 and the mass flow rate has increased to a value indicated by the label 2. A further decrease in the back pressure increases the flow rate until the back pressure is reduced to the critical value $p_b = p^*$ at which point Mach number reaches unity at the exit plane. Further reduction of the exit pressure has no effect on the flow upstream, for the disturbances caused by further opening of the valve cannot propagate upstream of the throat when the velocity there has reached the sonic speed. The flow at this condition is said to be *choked* and its mass flow rate can no longer be increased. How the flow adjusts from this exit pressure to the value of back pressure cannot be analyzed by one-dimensional methods.

Flow rate through the nozzle can be determined at the choked condition if, in addition to the stagnation pressure and stagnation temperature, the throat area is known. This is illustrated in the next example.

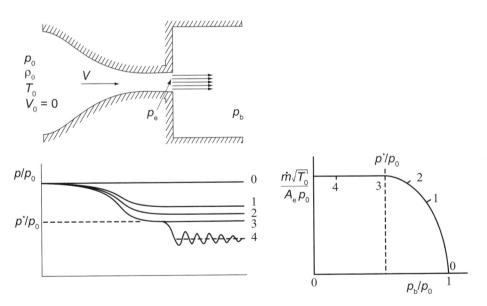

Figure 3.5 Flow through a converging nozzle.

■ **EXAMPLE 3.3**

Air at stagnation temperature $T_0 = 540\,\text{K}$ and stagnation pressure $p_0 = 200\,\text{kPa}$ flows isentropically in a converging nozzle, with exit area $A_t = 10\,\text{cm}^2$. (a) If the flow is choked, what are the exit pressure and the mass flow rate? (b) Assuming that the backpressure is $p_b = 160\,\text{kPa}$, find the flow rate.

Solution: (a). With the flow choked, the pressure ratio is

$$\frac{p_e}{p_0} = 0.5283$$

from which the exit pressure is determined to be $p_e = 106\,\text{kPa}$. The flow rate can be obtained by first calculating the flow function

$$F^* = \frac{\gamma}{\sqrt{\gamma-1}}\left(\frac{2}{\gamma+1}\right)^{(\gamma+1)/2(\gamma-1)}$$

which has the numerical value

$$F^* = \frac{1.4}{\sqrt{0.4}}\left(\frac{2}{2.4}\right)^3 = 1.281$$

Then the mass flow rate per unit area can be determined to be

$$\frac{\dot{m}}{A_t} = \frac{F^* p_0}{\sqrt{c_p T_0}} = \frac{1.281 \cdot 200{,}000}{\sqrt{1004.5 \cdot 540}} = 347.9\,\text{kg}/(\text{s}\cdot\text{m}^2)$$

so the flow rate is $\dot{m} = 0.348\,\text{kg/s}$.

(b). The second part of the example asks for the flow rate when back pressure is $p_b = 160\,\text{kPa}$. Since this pressure is larger than the critical value $106\,\text{kPa}$, the flow is no longer choked and $p_e = p_b$. The exit Mach number is obtained from

$$\frac{p_0}{p_e} = \left(1 + \frac{\gamma-1}{2}M_e^2\right)^{\gamma/(\gamma-1)}$$

Solving for M_e gives

$$M_e = \left[\frac{2}{\gamma-1}\left(\left(\frac{p_0}{p_e}\right)^{(\gamma-1)/\gamma} - 1\right)\right]^{1/2} = [5(1.25^{1/3.5} - 1)]^{1/2} = 0.574$$

The flow function at this Mach number is

$$F_e = \frac{\gamma M_e}{\sqrt{\gamma-1}}\left(1 + \frac{\gamma-1}{2}M_e^2\right)^{-(\gamma+1)/2(\gamma-1)}$$

and its numerical value is

$$F_e = \frac{1.4 \cdot 0.574}{\sqrt{0.4}}\left(1 + 0.2 \cdot 0.574^2\right)^{-3} = 1.049$$

The mass flow rate can then be determined from

$$\dot{m} = \frac{F p_0 A_t}{\sqrt{c_p T_0}} = \frac{1.049 \cdot 200{,}000 \cdot 10}{\sqrt{1004.5 \cdot 540} \cdot 100^2} = 0.285\,\text{kg/s}$$

■

3.2.2 Converging–diverging nozzle

Consider the operation of a converging–diverging nozzle in the same manner as was described for the converging one. The flow rate is adjusted by a regulating valve downstream of the nozzle. With the valve closed there is no flow and the pressure throughout equals the stagnation pressure. As the valve is opened slightly, flow is accelerated in the converging part of the nozzle and its pressure drops. It is then decelerated after the throat with rising pressure such that the exit plane pressure p_e reaches the backpressure p_b. This corresponds to case 1 shown in Figure 3.6. Further opening of the valve drops the backpressure and the flow rate increases until the valve is so far open that the Mach number has the value one at the throat and the pressure at the throat is equal to the critical pressure p^*. After the throat the flow diffuses and pressure rises until the exit plane is reached. The pressure variation is shown as condition 2 in the figure. If the valve is opened further, acoustic waves to signal what has happened downstream cannot propagate past the throat once the flow speed there is equal to the sound speed. The flow is now *choked* and no further adjustment in the mass flow rate is possible.

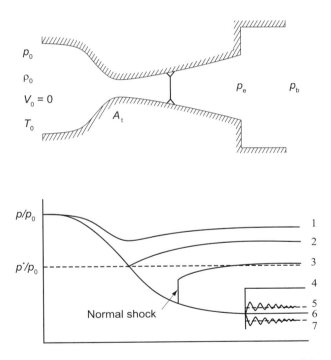

Figure 3.6 Supersonic nozzle with a shock in the diverging part of the nozzle.

The adjustment to the backpressure is now achieved through a *normal shock* and diffusion in the diverging part of the nozzle. This situation is shown as condition 3 in Figure 3.6. Flows with normal shocks are discussed in the next section. A weak normal shock appears just downstream of the throat for backpressures slightly lower than that at which the flow becomes choked, and as the backpressure is further reduced, the position of the shock moves further downstream until it reaches the exit plane, which is shown as condition 4 in the figure. After this any decrease in the backpressure cannot cause any change in the exit plane pressure. Condition 5 corresponds to an *overexpanded* flow, since the exit pressure

COMPRESSIBLE FLOW THROUGH NOZZLES

has dropped below the backpressure and the adjustment to the backpressure takes place after the nozzle through a series of *oblique shock waves* and *expansion fans*.

There is one value of backpressure for which the flow is isentropic and supersonic all the way to the exit plane, and at this condition the exit plane pressure reaches the value of the backpressure. This corresponds to one of the two solutions in the area ratio graph of Figure 3.4. It is also shown by line 6 in Figure 3.6. For backpressures below this value, the flow is said to be *underexpanded* as its pressure remains above the backpressure. The flow adjusts to the backpressure by expanding through a series of *oblique expansion waves* and *shock waves* as schematically shown by line 7 in the figure.

In flows through turbomachinery blade passages, the flow channel is not symmetric about its centerline and oblique shocks may appear in the flow channel itself. In aircraft propulsion the aim is to build lightweight machines with large mass flows. This requires small blade passages and large velocities, which leads to locally supersonic flow. The next example illustrates the conditions for isentropic supersonic flow.

■ EXAMPLE 3.4

Air flows isentropically in a converging–diverging nozzle, with a throat of area of $10 \, \text{cm}^2$, such that at the exit $M_e = 2$. The supply pressure and temperature at the inlet are 2 bar and 540 K, respectively, and the inlet velocity is negligibly small. (a) Find the fluid properties at the throat, (b) the exit area, pressure, and temperature, and (c) the flow rate.

Solution: (a) At the stagnation state density is

$$\rho_0 = \frac{p_0}{RT_0} = \frac{200}{0.287 \cdot 540} = 1.2905 \, \text{kg/m}^3$$

Since the flow is supersonic downstream of the throat, it is sonic at the throat. Hence

$$\begin{aligned}
p^* &= 0.5283 \, p_0 = 0.5283 \cdot 2 = 1.056 \, \text{bar} \\
T^* &= 0.8333 \, T_0 = 0.8333 \cdot 540 = 450.0 \, \text{K} \\
\rho^* &= 0.6339 \, \rho_0 = 0.6339 \cdot 1.32905 = 0.8180 \, \text{kg/m}^3 \\
V^* &= \sqrt{\gamma R T^*} = \sqrt{1.4 \cdot 287 \cdot 450.0} = 425.2 \, \text{m/s}
\end{aligned}$$

(b) At the exit plane, where $M_e = 2$, temperature is

$$T_e = \frac{T_0}{1 + \dfrac{\gamma - 1}{2} M_e^2} = \frac{540}{1 + 0.2 \cdot 4} = 300 \, \text{K}$$

and pressure and density are

$$p_e = p_0 \left(\frac{T_e}{T_0}\right)^{\gamma/(\gamma-1)} = 2 \left(\frac{300}{540}\right)^{3.5} = 25.56 \, \text{kPa}$$

$$\rho_e = \rho_0 \left(\frac{T_e}{T_0}\right)^{1/(\gamma-1)} = 1.2905 \left(\frac{300}{540}\right)^{2.5} = 0.2969 \, \text{kg/m}^3$$

Since the throat area is $A^* = 10 \, \text{cm}^2$, exit area is obtained by first calculating the area ratio

$$\frac{A_e}{A^*} = \frac{1}{M_e} \left(\frac{2}{\gamma+1} + \frac{\gamma-1}{\gamma+1} M_e^2\right)^{(\gamma+1)/2(\gamma-1)} = \frac{1.5^3}{2} = 1.6875$$

from which the exit area is $A_e = 16.875 \text{ cm}^2$.

(c) The mass flow rate is obtained from

$$\dot{m} = \rho^* A^* V^* = 0.8180 \cdot 0.001 \cdot 425.2 = 0.348 \text{ kg/s}$$

■

Examination of Figure 3.4 shows that for a given area ratio A/A^*, the Mach number can be supersonic or subsonic. The supersonic solution requires a low exit pressure, and this was examined in the previous example. To find the subsonic solution, Eq. (3.9) needs to be solved for Mach number when the area ratio is given. This can be carried out with Matlab's fzero function. Its syntax is

```
x=fzero(@(x) F(x),[x1,x2]);
```

This finds the value of x that satisfies $F(x) = 0$, with the zero in the range $[x_1, x_2]$.

To obtain the subsonic solution, in the following Matlab script Mach number is bracketed to the range $[0.05, 1.0]$. The variable k is used for γ, and a is the area ratio.

```
clear all;
a=1.6875; k=1.4;
M = fzero(@(M) farea(M,a,k),[0.05,1.0])
```

The function farea is defined as

```
function f = farea(M,a,k)
f = a-(1/M)*((2/(k+1))*(1+0.5*(k-1)*M^2))^(0.5*(k+1)/(k-1));
%The name of this M-file is farea.
```

The result is:

```
M=0.3722
```

A separate function file is not needed if this is written as:

```
a=1.6875; k=1.4;
M = fzero(@(M) a-(1/M)*((2/(k+1)) ...
    (1+0.5*(k-1)*M^2))^(0.5*(k+1)/(k-1)),[0.05,1.0])
```

3.3 NORMAL SHOCKS

In a converging–diverging duct two *isentropic* solutions can be found for a certain range of backpressures. If the backpressure is reduced slightly from that corresponding to the subsonic branch of the flow, a *normal shock* develops just downstream of the throat where the flow is now supersonic. It will be seen that the flow after the shock is subsonic and there is a jump in pressure across the shock. After the shock, the flow diffuses to the backpressure. A Schlieren photograph of a normal shock is shown in Figure 3.7. The shock is seen to interact with the boundary layers along the walls, and downstream of the shock this interaction influences the flow across the entire channel. Still, one-dimensional analysis gives good results even in this part of the flow.

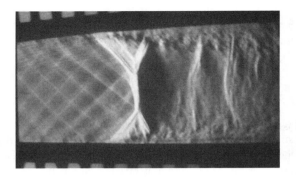

Figure 3.7 Interaction between a normal shock and wall boundary layers. (Photograph courtesy Professor D. Papamoschou.)

The flow through a shock can be analyzed by considering a control volume around the shock. Since the flow is adiabatic, the energy equation reduces to

$$h_x + \frac{1}{2}V_x^2 = h_y + \frac{1}{2}V_y^2 \tag{3.12}$$

where the subscript x denotes the upstream state and subscript y, the downstream state. Mass balance for this control volume yields

$$\frac{\dot{m}}{A} = \rho_x V_x = \rho_y V_y \tag{3.13}$$

and A is the area at the location of the shock. The momentum equation becomes

$$\dot{m}(V_y - V_x) = (p_x - p_y)A \tag{3.14}$$

since the wall friction can be neglected. Pressure increase across the shock is thus

$$p_y - p_x = \frac{\dot{m}}{A}(V_x - V_y) \tag{3.15}$$

Making use of the mass balance, this equation takes the form

$$p_x + \rho_x V_x^2 = p_y + \rho_y V_y^2 \tag{3.16}$$

Since the flow is adiabatic, the energy equation, if ideal gas behavior is assumed, may be written as

$$c_p T_x + \frac{1}{2}V_x^2 = c_p T_y + \frac{1}{2}V_y^2 \tag{3.17}$$

or as

$$T_{0x} = T_{0y} \tag{3.18}$$

From the definition of stagnation state the expression

$$\frac{T_{0x}}{T_x} = 1 + \frac{\gamma - 1}{2}M_x^2$$

is obtained, and a similar equation holds on the downstream side. Hence their ratio yields

$$\frac{T_y}{T_x} = \frac{1 + \frac{\gamma - 1}{2}M_x^2}{1 + \frac{\gamma - 1}{2}M_y^2} \tag{3.19}$$

Making use of the ideal gas relation $p = \rho RT$ and the mass balance $\rho_x V_x = \rho_y V_y$ in this equation gives

$$\frac{T_y}{T_x} = \frac{p_y \, \rho_x}{p_x \, \rho_y} = \frac{p_y \, V_y}{p_x \, V_x}$$

and, using $V = Mc$ to eliminate the velocities, leads to

$$\frac{T_y}{T_x} = \frac{p_y M_y c_y}{p_x M_x c_x} = \frac{p_y M_y}{p_x M_x}\sqrt{\frac{T_y}{T_x}}$$

from which

$$\frac{T_y}{T_x} = \left(\frac{p_y}{p_x}\right)^2 \left(\frac{M_y}{M_x}\right)^2 \tag{3.20}$$

Combining this with Eq. (3.19) gives

$$\frac{p_y}{p_x} = \frac{M_x}{M_y}\frac{\sqrt{1 + \frac{\gamma-1}{2}M_x^2}}{\sqrt{1 + \frac{\gamma-1}{2}M_y^2}} \tag{3.21}$$

For an ideal gas $\rho_x V_x^2 = \gamma p_x M_x^2$, and a similar equation holds on the downstream side. Substituting these into Eq. (3.16) gives

$$\frac{p_y}{p_x} = \frac{1 + \gamma M_x^2}{1 + \gamma M_y^2} \tag{3.22}$$

Equating Eqs. (3.21) and (3.22) gives

$$\frac{M_x\sqrt{1 + \frac{\gamma-1}{2}M_x^2}}{1 + \gamma M_x^2} = \frac{M_y\sqrt{1 + \frac{\gamma-1}{2}M_y^2}}{1 + \gamma M_y^2}$$

This is clearly satisfied if $M_x = M_y$, but in this case nothing interesting happens and the flow moves through the control volume undisturbed. Squaring both sides yields a quadratic equation in M_y^2. Solving it gives the result

$$M_y^2 = \frac{2 + (\gamma - 1)M_x^2}{2\gamma M_x^2 - (\gamma - 1)} \tag{3.23}$$

This equation relates the upstream and downstream Mach numbers across a shock, and the expression is plotted in Figure 3.8. It shows that upstream states have $M_x > 1$ and those downstream have $M_y < 1$. As will be shown below, only in this situation will entropy increase across the shock, as it must.

Pressure before and after the shock is obtained by substituting Eq. (3.23) into Eq. (3.22), giving the result

$$\frac{p_y}{p_x} = \frac{2\gamma}{\gamma+1}M_x^2 - \frac{\gamma-1}{\gamma+1} \tag{3.24}$$

This shows that if $M_x = 1$, there is no pressure jump. Defining the fractional increase in pressure as measure of the strength of the shock, the strength is defined as

$$\frac{p_y}{p_x} - 1 = \frac{2\gamma}{\gamma-1}(M_x^2 - 1) \tag{3.25}$$

The temperature ratio across a shock is obtained by substituting the value of M_y^2 from Eq. (3.23) into Eq. (3.20). The result is

$$\frac{T_y}{T_x} = \frac{[2\gamma M_x^2 - (\gamma - 1)][2 + (\gamma - 1)M_x^2]}{(\gamma + 1)^2 M_x^2} \quad (3.26)$$

The density ratio is

$$\frac{\rho_y}{\rho_x} = \frac{V_y}{V_x} = \frac{(\gamma + 1)M_x^2}{2 + (\gamma - 1)M_x^2} \quad (3.27)$$

Since $T_{0x} = T_{0y}$, Eq. (3.11) shows that

$$\frac{p_{0y}}{p_{0x}} = \frac{A_x^*}{A_y^*} \quad (3.28)$$

This equation is useful for finding the area at which a shock is located when the upstream Mach number is known.

The ratio of stagnation pressures across a shock is obtained from

$$\frac{p_{0y}}{p_{0x}} = \frac{p_{0y}}{p_y} \frac{p_y}{p_x} \frac{p_x}{p_{0x}} \quad (3.29)$$

which takes the form

$$\frac{p_{0y}}{p_{0x}} = \left(\frac{(\gamma + 1)M_x^2}{2 + (\gamma - 1)M_x^2}\right)^{\gamma/\gamma-1} \left(\frac{\gamma + 1}{2\gamma M_x^2 - (\gamma - 1)}\right)^{1/(\gamma-1)} \quad (3.30)$$

The changes in properties across the shock are shown in Figure 3.8.

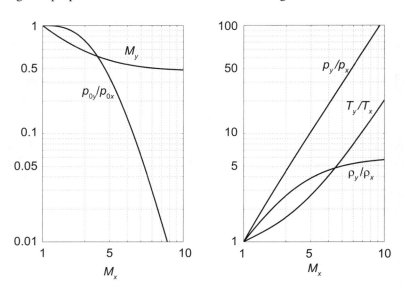

Figure 3.8 Normal shock relations for $\gamma = 1.4$.

3.3.1 Rankine–Hugoniot relations

A relationship between the pressure and density ratio across the shock can be obtained from the momentum equation

$$p_y - p_x = \rho_x V_x^2 - \rho_y V_y^2 = \rho_x V_x^2 \left(1 - \frac{\rho_x}{\rho_y}\right)$$

which, when solved for upstream velocity, gives

$$V^2 = \left[\frac{(p_y - p_x)\rho_y}{(\rho_y - \rho_x)\rho_x}\right]^{1/2} \qquad (3.31)$$

A similar expression is obtained downstream of the shock:

$$V_y = \left[\frac{(p_y - p_x)\rho_x}{(\rho_y - \rho_x)\rho_y}\right]^{1/2}$$

The energy equation across the shock is

$$h_x + \frac{1}{2}V_x^2 = h_y + \frac{1}{2}V_y^2$$

Since

$$h_x - h_y = c_p(T_x - T_y) = \frac{\gamma}{\gamma - 1}\left(\frac{p_x}{\rho_x} - \frac{p_y}{\rho_y}\right)$$

the energy equation can be written as

$$\frac{\gamma}{\gamma - 1}\frac{p_x}{\rho_x} + \frac{p_y - p_x}{2(\rho_y - \rho_x)}\frac{\rho_y}{\rho_x} = \frac{\gamma}{\gamma - 1}\frac{p_y}{\rho_y} + \frac{p_y - p_x}{2(\rho_y - \rho_x)}\frac{\rho_x}{\rho_y}$$

from which

$$\frac{p_x}{p_y} = \frac{\left(\frac{\gamma + 1}{\gamma - 1}\frac{\rho_y}{\rho_x}\right) - 1}{\frac{\gamma + 1}{\gamma - 1} - \frac{\rho_y}{\rho_x}} \qquad (3.32)$$

Solving this for the density ratio gives

$$\frac{\rho_y}{\rho_x} = \frac{\left(\frac{\gamma + 1}{\gamma - 1}\frac{p_y}{p_x}\right) + 1}{\frac{\gamma + 1}{\gamma - 1} + \frac{p_y}{p_x}} \qquad (3.33)$$

Equations (3.32) and (3.33) are known as Rankine–Hugoniot relations [62, 40].

The strength of the shock is obtained from the first of the Rankine–Hugoniot equations. It is

$$\frac{p_y}{p_x} - 1 = \frac{\frac{2\gamma}{\gamma - 1}\left(\frac{\rho_y}{\rho_x} - 1\right)}{\frac{2}{\gamma - 1} - \left(\frac{\rho_y}{\rho_x} - 1\right)} \qquad (3.34)$$

and similarly for the density ratio

$$\frac{\rho_y}{\rho_x} = \frac{1 + \frac{\gamma+1}{2\gamma}\left(\frac{p_y}{p_x} - 1\right)}{1 + \frac{\gamma-1}{2\gamma}\left(\frac{p_y}{p_x} - 1\right)} \tag{3.35}$$

Entropy change across a shock is given by integrating

$$T ds = du - p\, dv$$

across the shock. This gives

$$s_y - s_x = c_v \ln \frac{T_y}{T_x} + R \ln \frac{v_y}{v_x} = c_v \ln \left(\frac{T_y}{T_x}\right)\left(\frac{v_y}{v_x}\right)^{\gamma-1}$$

The temperature ratio is

$$\frac{T_y}{T_x} = \frac{p_y}{p_x}\frac{v_y}{v_x} = \frac{p_y}{p_x}\frac{\rho_x}{\rho_y}$$

so that the entropy change can also be written as

$$s_y - s_x = c_v \ln \left(\frac{p_y}{p_x}\right)\left(\frac{\rho_x}{\rho_y}\right)^{\gamma}$$

Substituting the expression for ρ_y/ρ_x from Eq. (3.34) to this and noting that $c_v = R/(\gamma-1)$ this equation can be recast as

$$\frac{s_y - s_x}{c_v} = \ln\left(\frac{p_y}{p_x}\right) - \gamma \ln\left[1 + \frac{\gamma+1}{2\gamma}\left(\frac{p_y}{p_x} - 1\right)\right] + \gamma \ln\left[1 + \frac{\gamma-1}{2\gamma}\left(\frac{p_y}{p_x} - 1\right)\right]$$

For weak shocks p_y/p_x is just slightly greater than one. For this reason, let $p_y/p_x = 1+\varepsilon$ and on substituting this in the previous equation and expanding in Taylor series for small value of ε, leads to [2]

$$\frac{s_y - s_x}{c_v} = \frac{\gamma^2-1}{12\gamma^2}\left(\frac{p_y}{p_x} - 1\right)^3 - \frac{\gamma^2-1}{8\gamma^2}\left(\frac{p_y}{p_x} - 1\right)^4 + \cdots$$

or

$$\frac{s_y - s_x}{R} = \frac{\gamma+1}{12\gamma^2}\left(\frac{p_y}{p_x} - 1\right)^3 - \frac{\gamma+1}{8\gamma^2}\left(\frac{p_y}{p_x} - 1\right)^4 + \cdots$$

Using Eq. (3.25) to express the shock strength in terms of Mach number gives

$$\frac{s_y - s_x}{R} = \frac{2}{3}\frac{\gamma}{(\gamma+1)^2}(M_x^2 - 1)^3 - \frac{2\gamma^2}{(\gamma+1)^3}(M_x^2 - 1)^4 + \cdots \tag{3.36}$$

This shows that were $M_x < 1$, entropy would decrease across the shock. Thus shocks are possible only for $M_x > 1$. Furthermore, entropy increases only slightly across weak shocks.

---

[2] For small values of ε series expansion yields $\ln(1+\epsilon) = \epsilon - \epsilon^2/2 + \epsilon^3/3 - \epsilon^4/4 + \cdots$.

The stagnation pressure change across weak shocks can also be developed by writing $M_x^2 = 1 + \varepsilon$ in

$$\frac{p_{0y}}{p_{0x}} = \left(\frac{(\gamma+1)M_x^2}{2+(\gamma-1)M_x^2}\right)^{\gamma/\gamma-1}\left(\frac{\gamma+1}{2\gamma M_x^2-(\gamma-1)}\right)^{1/(\gamma-1)}$$

and expanding the resulting expression in for small values of ε. The result, when written in terms of $M_x^2 - 1$, is

$$\frac{p_{0y}}{p_{0x}} = 1 - \frac{2}{3}\frac{\gamma}{(\gamma+1)^2}(M_x^2-1)^3 + \frac{2\gamma^2}{(\gamma+1)^3}(M_x^2-1)^4 + \cdots \quad (3.37)$$

This is an important result as it shows that flows through weak shocks experience only a small loss in stagnation pressure. In fact, Eq. (3.37) may be shown to be accurate to 1% for $M_x < 1.2$, and for $M_x = 1.2$ it gives $p_{0y}/p_{0x} = 0.986$, or a stagnation pressure drop of less than 2%. The significance of this result for turbomachinery design is that in transonic flows with shocks stagnation pressure losses are relatively small.

3.4 INFLUENCE OF FRICTION IN FLOW THROUGH STRAIGHT NOZZLES

There are various ways in which the irreversibilities caused by friction have been taken into account in studies of nozzle flow. These are discussed in this section. First, a polytropic efficiency is introduced, and it is then related to a static enthalpy loss coefficient, which, in turn, is related to a loss of stagnation pressure. Next, nozzle efficiency and the velocity coefficient are discussed. After this the equations for compressible flow in a variable-area duct with wall friction are given. In the discussion that follows, the flow is adiabatic and no work is done. Therefore the stagnation temperature remains constant, and assuming constant specific heats and ideal gas behavior, the relation

$$\frac{T_0}{T} = 1 + \frac{\gamma-1}{2}M^2 \quad (3.38)$$

remains valid for adiabatic flow even when friction is present.

3.4.1 Polytropic efficiency

The concept of polytropic efficiency follows from examining the Tds equation

$$Tds = dh - v\,dp$$

for an isentropic process

$$dh_s = v\,dp \quad (3.39)$$

and a nonisentropic one. A *polytropic efficiency* of an incremental expansion process is defined as

$$\eta_p = \frac{dh}{dh_s}$$

so that $dh = \eta_p dh_s$. The process is shown in Figure 2.9a in Chapter 2. Substituting this into Eq. (3.39) and making use of the ideal gas relation gives

$$dh = c_p\,dT = \eta_p v\,dp = \eta_p \frac{RT}{p}dp$$

From this follows the relation

$$\frac{dT}{T} = \frac{\eta_p(\gamma-1)}{\gamma}\frac{dp}{p} \tag{3.40}$$

A polytropic index n is now introduced via the equation

$$\frac{n-1}{n} = \frac{\eta_p(\gamma-1)}{\gamma} \quad \text{so that} \quad \eta_p = \left(\frac{n-1}{n}\right)\left(\frac{\gamma}{\gamma-1}\right) \tag{3.41}$$

and Eq. (3.41) can also be written as

$$n = \frac{\gamma}{\eta_p + \gamma(1-\eta_p)} \tag{3.42}$$

Assuming that η_p and hence also n remain constant along the entire expansion path integrating Eq. (3.40) yields

$$\frac{T_2}{T_1} = \left(\frac{p_2}{p_1}\right)^{(n-1)/n} \tag{3.43}$$

Rewriting this as

$$\frac{n-1}{n} = \frac{\ln T_2/T_1}{\ln p_2/p_1}$$

gives an equation from which the polytropic exponent may be calculated, if the inlet and exit pressures and temperatures have been determined experimentally. Real gas effects have been incorporated into the theory by Shultz [69], Mallen and Saville [55], and Huntington [42].

If the inlet state is a stagnation state, then, writing Eq. (3.43) as

$$\frac{T_{01}}{T_2} = \left(\frac{p_{01}}{p_2}\right)^{(n-1)/n} \tag{3.44}$$

and making use of the ideal gas relation in Eq. (3.44) it follows that

$$\frac{\rho_{01}}{\rho_2} = \left(\frac{T_{01}}{T_2}\right)^{1/(n-1)} \qquad \frac{p_{01}}{p_2} = \left(\frac{\rho_{01}}{\rho}\right)^n$$

Finally, using Eq. (3.38) the pressure and density ratios may be written as

$$\frac{p_{01}}{p_2} = \left(1 + \frac{\gamma-1}{2}M_2^2\right)^{n/(n-1)} \qquad \frac{\rho_{01}}{\rho_2} = \left(1 + \frac{\gamma-1}{2}M_2^2\right)^{1/(n-1)}$$

The flow velocity is

$$V = M\sqrt{\gamma RT} = M\sqrt{\gamma RT_0}\left(1 + \frac{\gamma-1}{2}M^2\right)^{-(1/2)}$$

which can be used to express the mass flow rate per unit area at the throat in the form

$$\frac{\dot{m}\sqrt{c_p T_{01}}}{p_{01}A_2} = \frac{\gamma}{\sqrt{\gamma-1}}M_2\left(1 + \frac{\gamma-1}{2}M_2^2\right)^{-(n+1)/2(n-1)} \tag{3.45}$$

INFLUENCE OF FRICTION IN FLOW THROUGH STRAIGHT NOZZLES

With the conditions at the inlet fixed, M_2 is the only variable in this equation. Differentiating with respect to M_2 gives that value of Mach number at the throat for which the maximum flow rate is achieved. This operation leads to

$$1 + \frac{\gamma - 1}{2} M_2^2 - \frac{(n+1)(\gamma - 1)}{2(n-1)} M_2^2 = 0$$

which, when simplified and solved for M_2, gives

$$M_t = \sqrt{\frac{n-1}{\gamma - 1}} \qquad (3.46)$$

in which the subscript indicates that this is the Mach number at the throat at a choked condition. Two alternative forms are

$$M_t = \sqrt{\frac{\eta_p}{\eta_p + \gamma(1 - \eta_p)}} \qquad M_t = \sqrt{1 - (1 - \eta_p)n} \qquad (3.47)$$

It is seen that Mach number at the throat is slightly less than one. Making use of this value of Mach number the critical pressure ratio becomes

$$\frac{p_t}{p_{01}} = \left(\frac{2}{n+1}\right)^{n/(n-1)} \qquad (3.48)$$

This has the same form as the expression for isentropic flow when γ replaced by n. Substituting M_t from Eq. (3.47) into Eq. (3.45) gives

$$\left(\frac{\dot{m}}{A_t}\right)_{max} = \sqrt{2\eta_p n p_{01} \rho_{01} \left(\frac{2}{n+1}\right)^{(n+1)/(n-1)}} \qquad (3.49)$$

Velocity at the throat at this condition is

$$V_t = \sqrt{2c_p(T_{01} - T_t)} = \sqrt{\frac{2\gamma}{\gamma - 1} RT_t \left(\frac{T_{01}}{T_t} - 1\right)} = \sqrt{\eta_p n RT_t}$$

in which the relation $T_{01}/T_t = (n+1)/2$ was used. Alternatively, velocity at the throat may be determined from $V_t = M_t \sqrt{\gamma R T_t}$.

■ EXAMPLE 3.5

Air in a reservoir, with temperature 540 K and pressure 200 kPa, flows into a converging nozzle with a polytropic efficiency $\eta_p = 0.98$. The throat area is $A_t = 10 \, \text{cm}^2$. (a) If the flow is choked, what are the exit pressure and the mass flow rate? (b) Given that the backpressure is $p_b = 160$ kPa, find the mass flow rate.

Solution: (a) The polytropic exponent is

$$n = \frac{\gamma}{\eta_p + \gamma(1 - \eta_p)} = \frac{1.4}{0.98 + 1.4 \cdot 0.02} = 1.389$$

and the Mach number can then be determined to be

$$M_t = \sqrt{\frac{n-1}{\gamma - 1}} = \sqrt{\frac{1.389 - 1}{1.4 - 1}} = 0.986$$

With the flow choked, pressure and temperature at the throat are

$$p_t = p_{01}\left(\frac{2}{n+1}\right)^{n/(n-1)} = 200\left(\frac{2}{1.389+1}\right)^{1.389/0.389} = 106\,\text{kPa}$$

$$T_t = \frac{2T_{01}}{n+1} = \frac{2\cdot 540}{1.389+1} = 452.1\,\text{K}$$

Mass flux at the throat at choked condition is

$$\frac{\dot{m}}{A_t} = p_{01}\sqrt{\frac{\gamma}{RT_{01}}}M_2\left(\frac{n+1}{2}\right)^{-(n+1)/2(n-1)} = 343.42\,\text{kg}/(\text{s}\cdot\text{m}^2)$$

and, with $A_t = 10\,\text{cm}^2$, flow rate is

$$\dot{m} = 0.343\,\text{kg/s}$$

(b) For $p_b = 160\,\text{kPa}$ the flow is not choked. Hence M_t is calculated from

$$M_t = \sqrt{\frac{2}{\gamma-1}\left[\left(\frac{p_{01}}{p_b}\right)^{(n-1)/n} - 1\right]} = 0.5678$$

and the mass flux at the throat is obtained from

$$\frac{\dot{m}}{A_t} = \rho_t V_t = p_{01}\sqrt{\frac{\gamma}{RT_{01}}}M_t\left(1 + \frac{\gamma-1}{2}M_t^2\right)^{-(n+1)/2(n-1)} = 281.7\,\text{kg}/(\text{s}\cdot\text{m}^2)$$

Hence the mass flow rate is

$$\dot{m} = 0.282\,\text{kg/s}$$

By comparing this to the calculation in Example 3.3 for an isentropic flow, the mass flow rate is seen to be slightly smaller for the polytropic process. ∎

Since the sonic state does not appear anywhere in the actual flow, it serves as a reference state. At the sonic state the static properties may be calculated from the stagnation state upstream by using the following equations:

$$\frac{T^*}{T_{01}} = \frac{2}{\gamma+1} \qquad \frac{p^*}{p_{01}} = \left(\frac{2}{\gamma+1}\right)^{n/(n-1)} \qquad \frac{\rho^*}{\rho_{01}} = \left(\frac{2}{\gamma+1}\right)^{1/(n-1)}$$

Since T_{01} remains constant, the relationship between the static temperature at the sonic state and its value at the inlet is

$$\frac{T^*}{T_1} = \frac{T^*}{T_{01}}\frac{T_{01}}{T_1} = \left(\frac{2}{\gamma+1}\right)\left(1 + \frac{\gamma-1}{2}M_1^2\right)$$

or

$$\frac{T^*}{T_1} = \frac{2}{\gamma+1} + \frac{\gamma-1}{\gamma+1}M_1^2$$

The pressure ratio is clearly

$$\frac{p^*}{p_1} = \left(\frac{2}{\gamma+1} + \frac{\gamma-1}{\gamma+1}M_1^2\right)^{n/(n-1)}$$

and the density ratio is

$$\frac{\rho^*}{\rho_1} = \left(\frac{2}{\gamma+1} + \frac{\gamma-1}{\gamma+1}M_1^2\right)^{1/(n-1)}$$

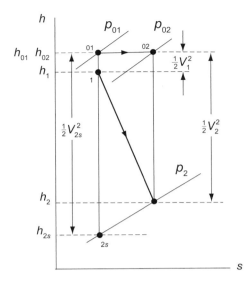

Figure 3.9 The thermodynamic states for a flow through a nozzle with friction.

3.4.2 Loss coefficients

In addition to the polytropic efficiency, there are other measures of irreversibility in nozzle flow. The first of these is the loss coefficient for static enthalpy, defined as

$$\zeta = \frac{h_2 - h_{2s}}{h_{02} - h_2} = \frac{h_2 - h_{2s}}{\frac{1}{2}V_2^2} \qquad (3.50)$$

where the end states are as shown in Figure 3.9. The numerator is the change in enthalpy owing to internal heating.

This may also be written as

$$\zeta = \frac{h_{02} - h_{2s} - (h_{02} - h_2)}{h_{02} - h_2} = \frac{V_{2s}^2 - V_2^2}{V_2^2} = \frac{1}{c_v^2} - 1$$

in which $c_v = V_2/V_{2s}$ is called a *velocity coefficient*. For given inlet conditions and exit pressure, the static enthalpy loss coefficient may be related to the polytropic efficiency, as the next example illustrates.

■ **EXAMPLE 3.6**

Air in a reservoir, with temperature 540 K and pressure 200 kPa, flows in a converging nozzle with a polytropic efficiency $\eta_p = 0.98$. (a) Find the static enthalpy loss coefficient, given an exit pressure of $p_2 = 160$ kPa. (b) Find the velocity coefficient. (c) Find the loss in stagnation pressure.

Solution: (a) As in the previous example, the polytropic exponent is

$$n = \frac{\gamma}{\eta_p + \gamma(1 - \eta_p)} = \frac{1.4}{0.98 + 1.4 \cdot 0.02} = 1.389$$

The exit temperature is therefore

$$T_2 = T_{01}\left(\frac{p_2}{p_{01}}\right)^{(n-1)/n} = 540\left(\frac{160}{200}\right)^{1.389/0.389} = 507.29\,\text{K}$$

and the Mach number at the exit is

$$M_2 = \sqrt{\frac{2}{\gamma-1}\left(\frac{T_{01}}{T_2}-1\right)} = \sqrt{\frac{2}{0.4}\left(\frac{540}{506.29}-1\right)} = 0.5678$$

The exit velocity is therefore

$$V_2 = M_2\sqrt{\gamma R T_2} = 0.5678\sqrt{1.4 \cdot 287 \cdot 507.3} = 256.34\,\text{m/s}$$

The temperature T_{2s} is obtained from the expression for isentropic expansion

$$T_{2s} = T_{01}\left(\frac{p_2}{p_{01}}\right)^{(k-1)/k} = 540\left(\frac{160}{200}\right)^{1/3.5} = 506.65\,\text{K}$$

Hence the static enthalpy loss coefficient is

$$\zeta = \frac{2c_p(T_2 - T_{2s})}{V_2^2} = \frac{2 \cdot 1004.5 \cdot (507.29 - 506.65)}{256.34^2} = 0.01976$$

(b) The velocity coefficient is then

$$c_V = \frac{1}{\sqrt{1+\zeta}} = 0.9903$$

(c) The loss of stagnation pressure is obtained by noting that

$$\frac{p_{02}}{p_{01}} = \frac{p_{02}}{p_2}\frac{p_2}{p_{01}} = \left(\frac{T_{02}}{T_2}\right)^{\gamma/(\gamma-1)}\left(\frac{T_{2s}}{T_{02}}\right)^{\gamma/(\gamma-1)} = \left(\frac{T_{2s}}{T_2}\right)^{\gamma/(\gamma-1)}$$

Hence

$$p_{02} = p_{01}\left(\frac{T_{2s}}{T_2}\right)^{\gamma/(\gamma-1)} = 200\left(\frac{506.65}{507.29}\right)^{3.5} = 199.11\,\text{kPa}$$

The loss in stagnation pressure is $\Delta p_0 = p_{01} - p_{02} = 890\,\text{Pa}$. Because the values of p_{01} and p_{02} are nearly equal, on subtracting one from the other, a number of significant figures are lost. If the value of the static enthalpy loss coefficient is known, the polytropic efficiency may be calculated by reversing the steps in this example. ∎

The ratio of stagnation pressures p_{02}/p_{01} may also be developed by integrating the Gibbs equation

$$T\,ds = dh - v\,dp$$

along the constant p_2 line. This gives

$$s_2 - s_1 = c_p \ln\frac{T_2}{T_{2s}}$$

INFLUENCE OF FRICTION IN FLOW THROUGH STRAIGHT NOZZLES 81

Similarly, integrating it between states 01 and 02 along the constant h_0 line leads to

$$s_2 - s_1 = R \ln \frac{p_{01}}{p_{02}}$$

Hence

$$\frac{p_{02}}{p_{01}} = \left(\frac{T_{2s}}{T_2}\right)^{\gamma/(\gamma-1)} \qquad (3.51)$$

This may also be written as

$$\frac{T_{2s}}{T_2} = \left(\frac{p_{02}}{p_{01}}\right)^{(\gamma-1)/\gamma} = \left(\frac{p_{01} - \Delta p_0}{p_{01}}\right)^{(\gamma-1)/\gamma} = \left(1 - \frac{\Delta p_0}{p_{01}}\right)^{(\gamma-1)/\gamma}$$

in which $\Delta p_0 = p_{01} - p_{02}$. Assuming that $\Delta p_0/p_{01} \ll 1$ and making use of the binomial theorem, the following approximation can be used

$$\frac{T_{2s}}{T_2} = 1 - \frac{\gamma-1}{\gamma}\frac{\Delta p_0}{p_{01}}$$

which may be also recast as

$$\frac{T_2 - T_{2s}}{T_2} = \frac{\gamma-1}{\gamma}\frac{\Delta p_0}{p_{01}}$$

and from which

$$h_2 - h_{2s} = \frac{RT_2 \Delta p_0}{p_{01}} = \frac{\Delta p_0/\rho_{01}}{1 + \frac{\gamma-1}{2}M_2^2}$$

so that the static enthalpy loss coefficient is

$$\zeta = \frac{\Delta p_0/\rho_{01}}{\frac{1}{2}V_2^2\left(1 + \frac{\gamma-1}{2}M_2^2\right)}$$

For small values of exit Mach number this reduces to

$$\zeta = \frac{\Delta p_0/\rho_{01}}{\frac{1}{2}V_2^2}$$

and this is often called a *stagnation pressure loss coefficient*.
Another measure of the loss of stagnation pressure is given by

$$Y_p = \frac{p_{01} - p_{02}}{p_{02} - p_2}$$

which may be written as

$$Y_p = \frac{p_{01}/p_{02} - 1}{1 - p_2/p_{02}}$$

The pressure ratios in this expression are

$$\frac{p_{02}}{p_2} = \left(\frac{T_{02}}{T_2}\right)^{\gamma/(\gamma-1)}$$

and
$$\frac{p_{01}}{p_{02}} = \left(\frac{T_{2s}}{T_{01}}\right)^{\gamma/(\gamma-1)}$$

and the definition of static enthalpy loss coefficient, Eq. (3.50), can be rewritten as
$$\frac{T_{2s}}{T_2} = 1 - \zeta\left(\frac{T_{02}}{T_2} - 1\right)$$

Substituting these into the expression for the stagnation pressure loss coefficient gives it the form
$$Y_p = \frac{\left[1 - \zeta\left(\frac{T_{02}}{T_2} - 1\right)\right]^{-\gamma/(\gamma-1)} - 1}{1 - \left(\frac{T_{02}}{T_2}\right)^{-\gamma/(\gamma-1)}}$$

or
$$Y_p = \frac{\left(1 - \zeta\frac{\gamma-1}{2}M_2^2\right)^{-\gamma/(\gamma-1)} - 1}{1 - \left(1 + \frac{\gamma-1}{2}M_2^2\right)^{-\gamma/(\gamma-1)}}$$

For small M_2 this reduces to $Y_p = \zeta$. The development indicates that the value of ζ is not dependent on Mach number, but the loss of stagnation pressure is, and therefore also the value of Y_p. For this reason ζ ought to be favored over other measures of irreversiblity [18]. However, it is worthwhile to be familiar with the various loss coefficients, as they have been and still are in use in the analysis of turbine nozzles.

3.4.3 Nozzle efficiency

The nozzle efficiency is defined as
$$\eta_N = \frac{h_1 - h_2}{h_1 - h_{2s}}$$

or, since the stagnation enthalpy remains constant so that $h_1 + V_1^2/2 = h_2 + V_2^2/2$, this relation can be rewritten as
$$\eta_N = \frac{V_2^2 - V_1^2}{2(h_1 - h_{2s})}$$

Similarly, when $h_1 + V_1^2/2 = h_{2s} + V_{2s}^2/2$ is used to rewrite the isentropic enthalpy change in the denominator in terms of kinetic energy differences, nozzle efficiency takes the alternative form
$$\eta_N = \frac{V_2^2 - V_1^2}{V_{2s}^2 - V_1^2}$$

If the fluid enters the nozzle from a large reservoir where $V_1 = 0$, nozzle efficiency becomes
$$\eta_N = \frac{V_2^2}{2(h_{01} - h_{2s})} = \frac{V_2^2}{V_{2s}^2} = c_v^2$$

Thus nozzle efficiency can be interpreted as a ratio of the actual increase in kinetic energy of the flow to that in reversible adiabatic flow. Nozzle efficiency takes into account the losses in the entire nozzle, from its inlet to its exit. If the nozzle is a converging–diverging type and the flow is subsonic, then most of the losses take place in the diverging part in which the flow diffuses to a low velocity. Nozzle efficiency of a converging nozzle has a value very close to unity.

For a flow that starts from the stagnation state nozzle efficiency

$$\eta_N = \frac{V_2^2}{2c_p(T_0 - T_{2s})}$$

may be further manipulated into the form

$$\eta_N = \frac{\gamma - 1}{2} \frac{M_2^2}{\left(1 - \dfrac{T_{2s}}{T_{01}}\right)} \frac{T_2}{T_{01}}$$

Expressing the isentropic temperature ratio T_{2s}/T_{01} and T_{01}/T_2 as

$$\frac{T_{2s}}{T_{01}} = \left(\frac{p_2}{p_{01}}\right)^{(\gamma-1)/\gamma} \qquad \frac{T_{01}}{T_2} = 1 + \frac{\gamma - 1}{2} M_2^2$$

and solving the resulting equation for the pressure ratio gives

$$\frac{p_2}{p_{01}} = \left[1 - \frac{(\gamma - 1)M_2^2}{\eta_N(2 + (\gamma - 1)M_2^2)}\right]^{\gamma/(\gamma-1)} \tag{3.52}$$

The following development of an expression for mass flow rate through a converging nozzle makes use of this equation. In such a nozzle the exit area A_2 is equal to the throat area A_t, and the mass flux, which is the mass flow rate per unit area, at the throat is given by

$$\frac{\dot{m}}{A_t} = \rho_t V_t = p_t M_t \sqrt{\frac{\gamma}{RT_{01}}} \sqrt{\frac{T_{01}}{T_t}}$$

or

$$\frac{\dot{m}}{A_t} = p_{01} \sqrt{\frac{\gamma}{RT_{01}}} \frac{p_t}{p_{01}} M_t \sqrt{1 + \frac{\gamma - 1}{2} M_t^2}$$

Substituting the pressure ratio from Eq. (3.52) into this gives

$$\frac{\dot{m}}{A_t} = p_{01} \sqrt{\frac{\gamma}{RT_{01}}} \left[1 - \frac{(\gamma - 1)M_t^2}{\eta_N(2 + (\gamma - 1)M_t^2)}\right]^{\gamma/(\gamma-1)} \left[M_t^2 \left(1 + \frac{\gamma - 1}{2} M_t^4\right)\right]^{1/2} \tag{3.53}$$

In this equation square of Mach number appears and therefore differentiating this equation with respect to respect to M_t^2 and setting the result to zero gives the value of M_t for maximum mass flux at the throat. Carrying out the differentiation gives the equation

$$(\gamma - 1)^2(1 - \eta_N)M_t^4 - [(\gamma - 1)3\eta_N - 2\gamma]M_t^2 - 2\eta_N = 0$$

From this the throat Mach number at the condition of maximum mass flux is obtained as

$$M_t = \left[\frac{(\gamma-1)(3\eta_N - 1) - 2\gamma + \sqrt{[(\gamma-1)(3\eta_N - 1) - 2\gamma]^2 + 8\eta_N(1 - \eta_N)(\gamma - 1)^2}}{2(\gamma - 1)^2(1 - \eta_N)}\right]^{1/2} \tag{3.54}$$

If the nozzle efficiency is known, then polytropic efficiency can be calculated by equating the Mach number obtained from Eq. (3.54) to $M_t = \sqrt{(n-1)/(\gamma-1)}$. For $\gamma = 1.4$ and $\eta_N = 0.98$ the polytropic exponent becomes $n = 1.3868$.

Neither the polytropic efficiency nor the overall nozzle efficiency reveals how the length and shape of the nozzle influence the magnitude of irreversibilities in the flow. These issues are discussed in the next section.

3.4.4 Combined Fanno flow and area change

Compressible flow with friction in constant-diameter ducts goes by the name *Fanno flow* and for this flow the momentum equation is used to relate pressure and velocity change to wall friction. There are two closely related friction factors in use. The *Fanning friction factor* is defined as

$$\bar{f} = \frac{\tau_w}{\frac{1}{2}\rho V^2}$$

The *Darcy friction factor* is 4 times the Fanning value, $f = 4\bar{f}$. Care must be exercised that the right one is used. The Darcy friction factor can be calculated for turbulent flow from the *Colebrook* formula

$$\frac{1}{\sqrt{f}} = -2\log_{10}\left(\frac{\varepsilon/D_h}{3.7} + \frac{2.51}{\text{Re}\sqrt{f}}\right) \tag{3.55}$$

in which ε is a root-mean-square (RMS) roughness of the walls and D_h is a hydraulic diameter, equal to $D_h = 4A/C$. Here A is a cross-sectional area and C is a wetted perimeter.

Since this equation is nonlinear, the value of f for given Re has to be determined iteratively; that is, an initial guess can be obtained by assuming Reynolds number Re to be so large that the second term in the parentheses may be neglected. The value of f obtained from this calculation is then substituted on the RHS. In this way a new value of f is then found, and it is sufficiency accurate that the iterations can be stopped. In the equations that follow, the Fanning friction factor is used, but the equations are left in a form in which 4 appears explicitly.

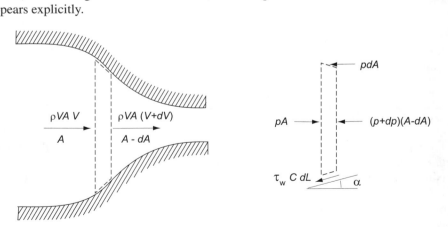

Figure 3.10 A converging nozzle with friction.

To analyze the combined effects of friction and area change, equation of continuity and energy are used together with the momentum equation. The x-component of the momentum balance for the control volume, shown in Figure 3.10, gives

$$\rho A V (V + dV - V) = pA - p\, dA - (p + dp)(A - dA) - \tau_w C \cos \alpha \, dL$$

Since in the sketch the flow area decreases, the downstream area is written as $A - dA$. In order to draw the vector $p\,dA$ in the correct direction for a converging channel, the area change must be assumed to have the sign that is consistent with the sketch.

Making use of the relations $C = 4A/D_h$ and $\cos \alpha \, dL = dx$ in this equation and neglecting the small term $dp\,dA$, reduces it to the form

$$\rho V\,dV = -dp - \frac{4\tau_w}{D_h} dx$$

Introduction of the Fanning friction factor puts it into the form

$$\rho V\,dV = -dp - \frac{4\bar{f}}{D_h} \frac{1}{2} \rho V^2 dx$$

Using the ideal gas relation and definition of Mach number to establish the equality $\rho V^2 = \gamma p M^2$ gives, after each term has been divided by p, the equation

$$\frac{dp}{p} + \frac{\gamma M^2}{2} \frac{dV^2}{V^2} + \frac{\gamma M^2}{2} 4\bar{f} \frac{dx}{D_h} = 0 \qquad (3.56)$$

To see the effect of area ratio and friction on the flow, this equation is next recast into a form in which the first two terms are expressed in terms of area ratio and Mach number. The second term is considered first. Since $V^2 = M^2 \gamma RT$, taking logarithms and differentiating this gives

$$\frac{dV^2}{V^2} = \frac{dM^2}{M^2} + \frac{dT}{T}$$

Next, since in adiabatic flow T_0 is constant, taking logarithms and differentiating the adiabatic relation $T_0 = T(1 + \frac{\gamma-1}{2} M^2)$ gives

$$\frac{dT}{T} = -\frac{\frac{\gamma-1}{2} M^2}{1 + \frac{\gamma-1}{2} M^2} \frac{dM^2}{M^2} \qquad (3.57)$$

and eliminating dT/T between the last two equations gives

$$\frac{dV^2}{V^2} = \frac{1}{1 + \frac{\gamma-1}{2} M^2} \frac{dM^2}{M^2} \qquad (3.58)$$

Next, an equation between Mach number, area ratio, and friction factor is obtained by taking logarithms and differentiating the mass balance $\dot{m} = \rho A V$. This gives

$$\frac{d\rho}{\rho} + \frac{dA}{A} + \frac{dV}{V} = 0$$

A similar operation on the ideal gas relation $p = \rho RT$ leads to

$$\frac{dp}{p} = \frac{d\rho}{\rho} + \frac{dT}{T}$$

Eliminating the term involving density between these two equations gives

$$\frac{dp}{p} = -\frac{dA}{A} - \frac{1}{2}\frac{dV^2}{V^2} + \frac{dT}{T}$$

Substituting this expression for dp/p into Eq. (3.56) gives

$$-\frac{dA}{A} - \frac{1}{2}(1 - \gamma M^2)\frac{dV^2}{V^2} + \frac{dT}{T} + \frac{\gamma M^2}{2}4\bar{f}\frac{dx}{D} = 0$$

Substituting Eqs. (3.57) and (3.58) into this yields the form

$$\frac{1 - M^2}{2 + (\gamma - 1)M^2}\frac{dM^2}{M^2} = -\frac{dA}{A} + \frac{\gamma M^2}{2}4\bar{f}\frac{dx}{D} \qquad (3.59)$$

from which the qualitative behavior of the flow can be seen. For sufficiently low backpressure this equation shows that in a converging duct with $dA/A < 0$ both terms on the right are positive. Hence in a flow that begins from a stagnation state, the Mach number increases in the flow direction, as was also true in isentropic flow. If the nozzle has a throat with $dA = 0$, the right side is still positive and the Mach number must still increase as it passes through the throat. This means that the Mach number is less than unity at the throat. In a supersonic nozzle the flow may reach $M = 1$ in the diverging part, with $dA/A > 0$, when the terms on the RHS exactly cancel. Stagnation pressure may be calculated by solving

$$\frac{dp_0}{p_0} = -\frac{\gamma}{2}M^2\,4\bar{f}\frac{dx}{D_{\mathrm{h}}} \qquad (3.60)$$

which shows that it drops only because of friction.

In steam turbines high-pressure steam is admitted into the turbine from a *steam chest*, to which it has entered via a regulated valve system. From the steam chest it flows first through a nozzle row arranged as shown in Figure 3.11. After leaving the nozzles it enters an interblade gap and then a set of rotor blades. Steam enters the nozzles in the axial direction, and the nozzles turn it into the general direction of the wheel velocity.

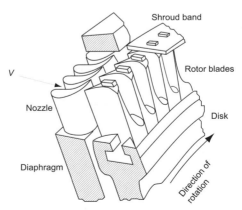

Figure 3.11 Steam turbine nozzles and blades. (Adapted from Keenan [47].)

Curvature of the nozzle passage does not introduce new complications into the analysis of frictional flow except, of course, at the initial stage when the geometry is laid out

To illustrate turning of steam into the direction of the turbine wheel rotation, in the next example, and as shown in Figure 3.12, the nozzle shape is a combination of a circular arc followed by a straight-line segment. The length of the arc is chosen such that the following straight line continues tangentially from the arc, and its direction is such that the flow leaves the nozzle at the correct angle.

Wet steam may be assumed to follow an ideal gas model with adiabatic index from Zeuner's equation $\gamma = 1.035 + 0.1x$. But usually as steam expands, it remains in a supersaturated state, provided the state does not drop too far into the two phase-region. In such a case isentropic flow is better modeled with an adiabatic index $\gamma = 1.3$. This calculation is illustrated in the next example.

■ **EXAMPLE 3.7**

Consider steam flow through the nozzle shown in Figure 3.12. The nozzle is rectangular, 3 cm in height, and its width at the inlet is 5.12 cm. The nozzle walls are made up of a circular arc of radius $R = 2.85$ cm and a straight section at the nozzle angle $\alpha = 75°$. At the inlet steam is dry and saturated with pressure $p_1 = 275$ kPa and $M_1 = 0.1$. The friction factor is assumed to be $4\bar{f} = 0.032$. Find the steam conditions through the nozzle, assuming that it remains supersaturated as an ideal gas with $\gamma = 1.3$.

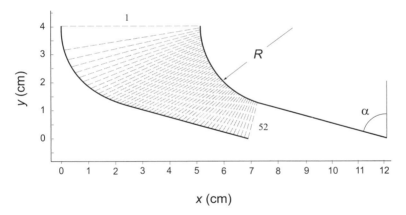

Figure 3.12 Offset nozzle and its grid.

Solution: Since at the inlet steam is saturated vapor at $p_i = 275$ kPa, its temperature is $T_i = 403.7$ K and specific volume is $v_i = 0.6578 \, \text{m}^3/\text{kg}$. The speed of sound at the inlet is
$$c_i = \sqrt{\gamma R T_i} = \sqrt{1.3 \cdot 461.5 \cdot 403.7} = 492.1 \, \text{m/s}$$
Hence the speed at the inlet is $V_i = c_i M_i = 49.2 \, \text{m/s}$ and the mass flow rate is
$$\dot{m} = \frac{V_i A_i}{v_i} = \frac{49.3 \cdot 3 \cdot 5.12}{0.6578 \cdot 100^2} = 0.116 \, \text{kg/s}$$

In order to check the value of friction factor, the size of the hydraulic diameter is needed. For the inlet section it is
$$D_h = \frac{4A}{C} = \frac{2Lb}{L+b} = \frac{2 \cdot 0.0512 \cdot 0.03}{0.0512 + 0.03} = 0.0378 \, \text{m}$$

With steam viscosity $1.322 \cdot 10^{-5}\,\text{kg}/(\text{m}\cdot\text{s})$, the Reynolds number comes out to be

$$\text{Re} = \frac{V_i D_{\text{hi}}}{v_i \mu} = \frac{49.3 \cdot 0.0378}{0.6578 \cdot 1.322 \cdot 10^{-5}} = 214{,}300$$

The value of the friction factor from the Colebrook formula is seen to be about $4\bar{f} = 0.032$, for a pipe with relative roughness $\epsilon/D_{\text{hi}} = 0.006$. If the roughness of the pipe is known, a more accurate value can be determined and the value clearly varies along the flow path. This variation is ignored, and the value $4\bar{f} = 0.032$ is used in the calculations.

To establish cross sections for the flow channel, the circular arc of the left wall was divided into angular increments of $3°$. The increment dy between the last two points was chosen as an approximate vertical separation for the points along the straight-line segment. The actual number of points was chosen such that the realized vertical separation was closest to this value of dy. This procedure resulted in 52 grid points. Next, the circular arc along the right wall was divided into 51 arcs of equal length, and the corresponding points on the left boundary were recalculated by choosing 52 points equally spaced in the value of their y coordinate. The locations of the corresponding x coordinates were then obtained by interpolation, based on the previously calculated base points along the left boundary. A sample grid is shown in Figure 3.12.

It is assumed that the flow properties are uniform on each cross section. This construction makes the flow path for the first element somewhat longer than the others, but the change in the flow properties is rather small in this region in which the cross-sectional area is large. Clearly, there are other ways in which to divide the flow nozzle into suitable sections.

The governing equations can now be solved, for example, by the Euler method, in which derivatives are replaced by forward differences. Equation (3.59) in finite-difference form is

$$m_{i+1} = m_i - \frac{[2+(\gamma-1)m_i]m_i}{1-m_i}\frac{(A_{i+1}-A_i)}{A_i} + \frac{\gamma m_i^2[2+(\gamma-1)m_i]}{2(1-m_i)}4\bar{f}\frac{x_{i+1}-x_i}{D_i}$$

in which $m_i = M_i^2$.

To check the convergence, in addition to calculating the base with 51 elements, the number of elements was increased also to 166 and then again successively roughly doubled to 4642 elements. Accuracy to two significant figures can be obtained with about 180 grid points. Three significant figures takes over 1000 elements. The results are shown in Table 3.1, in which M_e and p_e/p_1 are the exit values for isentropic flow and $M_e\text{f}$ and $p_e\text{f}/p1$ are for a flow with friction.

The areas at the inlet and outlet are

$$A_1 = 15.360\,\text{cm} \qquad A_e = 3.976\,\text{cm}$$

Hence, with $M_1 = 0.1$ at the inlet, the flow function is

$$F_1 = \frac{\dot{m}\sqrt{c_p T_0}}{A p_0} = \frac{\gamma M_1}{\sqrt{\gamma-1}}\left(1+\frac{\gamma-1}{2}M_1^2\right)^{-(\gamma+1)/2(\gamma-1)} = 0.2360$$

Therefore the flow function at the exit is

$$F_e = F_1 \frac{A_1}{A_e} = 0.9118 = \frac{\gamma M_e}{\sqrt{\gamma-1}}\left(1+\frac{\gamma-1}{2}M_e^2\right)^{-(\gamma+1)/2(\gamma-1)}$$

Table 3.1 Convergence of the solution for steam flow through a nozzle

N	M_e	M_{ef}	p_e/p_1	p_{ef}/p_1
51	0.3914	0.4018	0.9104	0.8942
166	0.4156	0.4238	0.9009	0.8828
340	0.4207	0.4293	0.8985	0.8798
689	0.4232	0.4321	0.8973	0.8783
1386	0.4245	0.4335	0.8967	0.8776
2782	0.4251	0.4341	0.8964	0.8772
4642	0.4254	0.4344	0.8962	0.8771

An exit Mach number for isentropic flow is obtained from the expression for the flow function. This gives $M_e = 0.4258$ and a normalized pressure $p_e/p_1 = 0.8960$. The numerical solution is seen to agree with this.

Plots of Mach number and normalized pressure for isentropic flow and for frictional flow are shown in Figure 3.13. Friction is seen to increase the Mach number, as was also seen in the previous example. Similarly, pressure drops more rapidly in frictional flow.

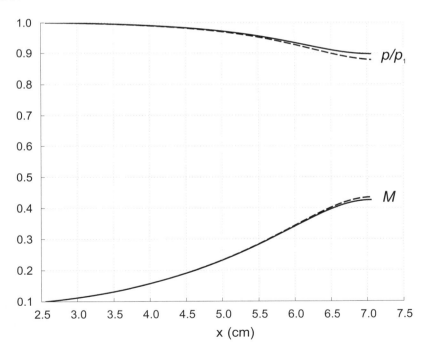

Figure 3.13 Mach numbers and normalized pressure for a steam flow through a nozzle. At the inlet $M_1 = 0.1$ and $p_1 = 275\,\text{kPa}$. Dashed lines correspond to frictional flow and solid lines, to isentropic flow.

The plots in this figure are shown with axial distance on the abscissa, and therefore they do not show clearly how the variables change along the streamline through the centerline of the nozzle. In particular, in the entrance region the flow moves approximately in the negative y direction and a small increase in x coordinate corresponds to a large increase in the path length. Hence the Mach number appears to increase more rapidly than it would have if the path length had been used on the abscissa.

■

In the foregoing example, length of the nozzle was taken into account explicitly. Since the irreversibilities are clearly a function of both the surface roughness and the length of the flow passage, this is an improvement over assigning a polytropic exponent to the process or by estimating the nozzle efficiency by past experience. However, an objection may be raised in the use of friction factors, obtained experimentally from flow of incompressible fluids in pipes, to *compressible flow with large area change*. In addition, experiments have shown that flows through curved nozzles develop secondary flows and these increase the losses. To account for them it has been suggested that the friction factor might be increased by some factor, but this procedure is not satisfactory, since it does not take into account the amount of turning. But lack of better alternative has forced such choices on the designer in the past. Today, it is possible to carry out *computational fluid dynamics* CFD simulations to take account of frictional effects better than the one-dimensional analysis discussed here yields. Nevertheless, it is still worthwhile to carry out a one-dimensional analysis by hand and by use of effective software, such as Matlab and EES, for such methods increase intuition, which is difficult to gain by CFD alone.

3.5 SUPERSATURATION

Consider again a steam flow through a nozzle, with steam dry and saturated at the inlet. As the steam accelerates through the nozzle, its pressure drops, and, if the process were to follow a path of thermodynamic equilibrium states, some of the steam would condense into water droplets. The incipient condensation may begin from crevices along the walls, in which case it is said to be by *heterogeneous nucleation*. The word *nucleation* suggests that the droplets start by molecular processes at nucleation sites and that incipient nucleation processes are distinct from those that cause the droplet to grow after it has reached a finite size. *Homogeneous nucleation* may begin at dust particles, or ions, carried in the vapor. At the nucleation sites the intermolecular forces that bind a vapor molecule to a site are stronger than those between two isolated vapor molecules. Once a molecular cluster is formed, the surface molecules form a distinct layer on which intermolecular forces on the liquid side are sufficiently strong to keep the molecules in this layer from evaporating into the vapor phase. The macroscopic manifestation of the distinct structure of a surface layer is the *surface tension* of liquids.

Kinetic theory of gases and liquids shows that there is a distribution of energy among the molecules, some having higher, some lower energy, than others. The more energetic molecules in the liquid are more prone to leave the liquid surface, and the molecules in the vapor phase with lower than average kinetic energy are more likely to condense. In a droplet of small size the phase boundary is curved, and then net force on a surface molecule originates from fewer neighbors than when the phase boundary is flat. As a consequence the smaller the liquid droplet, the weaker is the binding of the surface molecules. Hence liquid in small droplets is more volatile, and its vaporization takes place at lower temperature than it would if the phase boundaries were flat. In other words, at any given temperature vapor is formed more readily from smaller droplets than from large ones, and they can evaporate into a saturated vapor. This leads to *supersaturation*.

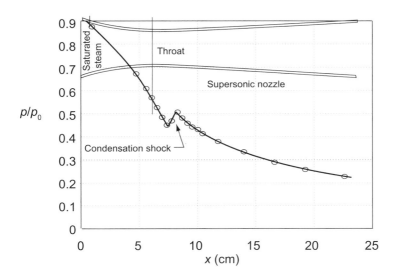

Figure 3.14 Illustration of a condensation shock from Binnie and Woods [8].

Both the evaporation and droplet formation by homogeneous nucleation take place at conditions not allowed by thermodynamic equilibrium. The practical effect is that steam flowing through a nozzle will not readily condense by homogeneous nucleation, and its temperature may drop quite far below the saturation temperature before nucleation takes place. Under these conditions the vapor is also said to be *undercooled* or *subcooled*.

When clean, dry, and saturated steam enters a nozzle, it remains supersaturated to a lower value of quality than if it were contaminated with foreign particles or ions. For clean steam the limit of supersaturation is marked by the Wilson line with a quality of 0.96 when the inlet steam is dry and saturated and has a pressure of 0.1 bar, and the quality drops to 0.95 along the Wilson line as the inlet pressure is increased to 14 bar.

When steam conditions pass the Wilson line, a *condensation shock* is formed. Binnie and Woods [8] have measured the pressure change across such a condensation shock, and their results are shown in Figure 3.14. They also carried out calculations to predict the pressure rise across the shock. More extensive analysis of condensation shocks has been carried out by Guha [31].

For purposes of calculation, the Wilson line will be assumed to correspond to constant quality of $x = 0.955$. By this measure the Wilson line is reached by isentropic expansion when enthalpy is 143.5 kJ/kg below the saturation line at pressure of 0.1 bar and 115 kJ/kg at 14 bar. Supersaturated steam above $x = 0.955$ can be assumed to behave as an ideal gas with $\gamma = 1.3$. Thus dry saturated steam at inlet temperature T_1 and pressure p_1, when it expands isentropically to pressure p_2, reaches a temperature

$$T_2 = T_1 \left(\frac{p_2}{p_2}\right)^{(\gamma-1)/\gamma}$$

The saturation pressure corresponding to temperature T_2 is denoted by p_{ss} and the ratio of the pressure p_2 to which the expansion takes place and p_{ss} is called the *degree of supersaturation*. It is given by

$$S = \frac{p_2}{p_{ss}} \tag{3.61}$$

EXAMPLE 3.8

Steam expands from condition $p_1 = 10$ bar and $T_1 = 473.2$ K isentropically through a nozzle to pressure $p_2 = 3.60$ bar. Find the degree of supersaturation and the amount of undercooling.

Solution: At the inlet condition the entropy of steam is $s_1 = 6.693 \text{ kJ}/(\text{kg} \cdot \text{K})$. At pressure $p_2 = 3.60$ bar and entropy $s_2 = 6.693 \text{ kJ}/(\text{kg} \cdot \text{K})$ the steam quality is $x_2 = 0.9541$. Thus the quality is close to the Wilson line. Assuming that the steam is supersaturated, its temperature is

$$T_2 = T_1 \left(\frac{p_2}{p_1}\right)^{(\gamma-1)/\gamma} = 473.2 \left(\frac{3.60}{10}\right)^{0.3/1.3} = 373.6 \text{ K}$$

Saturation pressure corresponding to this temperature is $p_{ss} = 1.028$ bar. Hence the degree of supersaturation is

$$S = \frac{3.600}{1.028} = 3.502$$

The saturation temperature corresponding to $p_2 = 3.6$ bar is $T_u = 413.0$ K, and the amount of undercooling is 39.4 K.

∎

3.6 PRANDTL–MEYER EXPANSION

3.6.1 Mach waves

Consider a source of small disturbances that moves with supersonic speed to the left, as shown in Figure 3.15. The source produces spherical acoustic waves that propagate

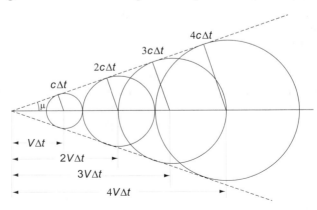

Figure 3.15 Illustration of the formation of Mach cone

outward with speed c. Next, consider five instances of time, the present time and four preceding instances of time, separated by equal time increments. At time $-4\Delta t$ the source was at location $4V\Delta t$ to the right of the present location and the wavefront which formed at time $-4\Delta t$ has moved a distance $4c\Delta t$ from the source. Similar reasoning applies to disturbances formed at $-3\Delta t$, $-2\Delta t$, and $-\Delta t$. The spherical wavefronts generated at these times are shown in the figure. Examination of the figure reveals that a *region of influence* of the disturbances is inside a cone with cone angle μ given by

$$\sin\mu = \frac{c}{V} = \frac{1}{M} \tag{3.62}$$

If a fluid moves to the right at speed V and meets a body at rest, the acoustic signal from the body is again a spherical wave. It travels upstream with the absolute velocity $c - V$ and downstream with velocity $c + V$. In supersonic flow $c - V$ is negative and the disturbance cannot influence the flow outside a cone with cone angle μ given by Eq. (3.62). This is called a *Mach cone*.

In a two-dimensional flow in which the source of small disturbances is a line perpendicular to the plane of the paper, the cone becomes a wedge. The region inside the cone, or the wedge, is called a *zone of action*, that outside is a *zone of silence*. The dividing surface between these zones is called a *Mach wave*.

3.6.2 Prandtl–Meyer theory

In a supersonic flow over an exterior corner, shown in Figure 3.16a, as the flow turns, Mach waves emanating from the sharp corner form an *expansion fan*. Since the flow is supersonic and moves to a larger area, its Mach number increases and pressure drops.

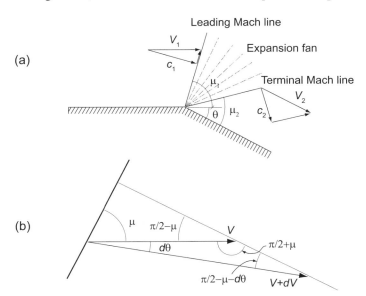

Figure 3.16 Supersonic expansion of flow over a convex corner.

The expansion fan is located in the region between the Mach waves oriented at angles $\sin\mu_1 = c_1/V_1 = 1/M_1$ and $\sin\mu_2 = c_2/V_2 = 1/M_2$, with μ_2 defining the terminal Mach wave at which V is parallel to the downstream wall. Such an expansion fan is said to be *centered* about the corner. Upstream of the leading Mach wave, pressure is uniform and the incremental pressure drop across a given Mach wave is the same regardless of where the flow crosses it. If the expansion fan is considered to consist of a discrete number of Mach waves, then the wedges between successive Mach waves are regions of constant thermodynamic properties.

Turning of the flow across one Mach wave is shown in Figure 3.16b. From the law of sines

$$\frac{V + dV}{V} = \frac{\sin(\pi/2 + \mu)}{\sin(\pi/2 - \mu - d\theta)}$$

or

$$1 + \frac{dV}{V} = \frac{\cos \mu}{\cos \mu \cos d\theta - \sin \mu \sin d\theta}$$

The angle $d\theta$ is small and is assumed to *increase in the clockwise* direction so that the previous equation can be written as

$$1 + \frac{dV}{V} = \frac{\cos \mu}{\cos \mu - d\theta \sin \mu} = \frac{1}{1 - d\theta \tan \mu}$$

Again making use of the smallness of $d\theta$, this can be expanded by binomial theorem, and the following equation is obtained:

$$\frac{dV}{V} = d\theta \tan \mu$$

Since $\tan \mu = c/\sqrt{V^2 + c^2} = 1/\sqrt{M^2 - 1}$, this can be written as

$$d\theta = \sqrt{M^2 - 1}\frac{dV}{V} \qquad (3.63)$$

Taking logarithms and differentiating $V = M\sqrt{\gamma RT}$ gives

$$\frac{dV}{V} = \frac{dM}{M} + \frac{1}{2}\frac{dT}{T} = \frac{1}{2}\frac{dM^2}{M^2} + \frac{1}{2}\frac{dT}{T}$$

The same operations on $T_0 = T[1 + (\gamma - 1)M^2/2)]$ give

$$-\frac{dT}{T} = \frac{\frac{\gamma - 1}{2}M^2}{1 + \frac{\gamma - 1}{2}M^2}$$

Using this to eliminate dT/T from the previous equation leads to

$$\frac{dV}{V} = \frac{1}{2\left(1 + \frac{\gamma - 1}{2}M^2\right)}\frac{dM^2}{M^2}$$

This can now be substituted into Eq. (3.63), which is then written as

$$d\theta = \frac{\sqrt{M^2 - 1}}{2\left(1 + \frac{\gamma - 1}{2}M^2\right)}\frac{dM^2}{M^2}$$

Next defining,
$$\nu(M) = \int_1^M \frac{\sqrt{M^2-1}}{2\left(1+\frac{\gamma-1}{2}M^2\right)} \frac{dM^2}{M^2}$$

and integrating gives
$$\nu(M) = \sqrt{\frac{\gamma+1}{\gamma-1}} \tan^{-1} \sqrt{\frac{\gamma-1}{\gamma+1}(M^2-1)} - \tan^{-1}\sqrt{M^2-1} \qquad (3.64)$$

so that a flow that expands to a state at which the Mach number is M_2 turns by an amount
$$\theta_2 - \theta_1 = \nu_2 - \nu_1$$

If the coordinates are aligned such that $\theta_1 = 0$, then
$$\nu_2 = \nu_1 + \theta_2$$

Two common situations are encountered. First, the wall along which the flow moves has a convex corner of known magnitude. Hence the angle θ_2 is known and the angle ν_2 can be determined and then the M_2 calculated from Eq. (3.64). The second situation is one in which the flow leaves as a jet from a nozzle to a space in which the backpressure is known. The next example illustrates the flow over a known convex corner.

■ **EXAMPLE 3.9**

Consider a supersonic air flow over a convex corner with angle $\theta_2 = 10°$, when the inflow moves in the direction of $\theta_1 = 0°$. The upstream Mach number is $M_1 = 1.46$, pressure is $p_1 = 575.0$ kPa, and temperature is $T_1 = 360.0$ K. Find the Mach number, temperature, and pressure after the expansion is complete.

Solution: The solution is obtained by the following Matlab script.

```
M1=1.46; k=1.4; thetadeg=10; theta=thetadeg*pi/180;
mu1=atan(1/(sqrt(M1^2-1)))*180/pi;
nu1=sqrt((k+1)/(k-1))*atan(sqrt((k-1)*(M1^2-1)/(k+1))) ...
    -atan(sqrt(M1^2-1));
nu2=nu1+theta;
M2 = fzero(@(M2) nu2+sqrt((k+1)/(k-1))* ...
    atan(sqrt((k-1)*(M2^2-1)/(k+1))) ...
    -atan(sqrt(M2^2-1)),[1.4,4]);
Result:
mu1=43.23 deg
M2=1.800
```

The angle of the leading Mach wave is $\mu_1 = 43.23°$. After that, the Prandtl–Meyer function at the inlet is calculated and when converted to degrees, it is $\nu_1 = 10.73°$. The Prandtl–Meyer function for complete turning is obtained as

$$\nu_2 = \nu_1 + \theta_2 = 10.73° + 10° = 20.73° \qquad \nu_2 = 0.3618 \text{ radian}$$

Since the Prandtl–Meyer function is implicit in the downstream Mach number, its value is obtained by invoking Matlab's `fzero` function. An assumed range of downstream Mach numbers is given as $[1.4, 4]$, or something similar. The value of Mach number after the expansion is $M_2 = 1.8$.

This problem can also be solved by EES software, and for that the syntax is simpler. Only the following statements are needed:

```
M1=1.46;  k=1.4;   theta2=10 [deg]
nu1=sqrt((k+1)/(k-1))*arctan(sqrt((k-1)*(M1^2-1)/(k+1)))
     -arctan(sqrt(M1^2-1));
nu2=nu1+theta2;
nu2=sqrt((k+1)/(k-1))*arctan(sqrt((k-1)*(M2^2-1)/(k+1)))
     -arctan(sqrt(M2^2-1));
```

In this script the two equations that are split into two lines must be placed on a single line in EES. The second statement is a nonlinear equation for the unknown M_2. Its root is found by EES's solution engine. It is possible to give the program an initial guess if the default value is not satisfactory.

To find the temperature after expansion, stagnation temperature is first determined. It is obtained from

$$\frac{T_0}{T_1} = 1 + \frac{\gamma - 1}{2} M_1^2 = 1 + 0.2 \cdot 1.46^2 = 1.426 \qquad T_0 = 513.48\,K$$

Downstream temperature is calculated from

$$\frac{T_0}{T_2} = 1 + \frac{\gamma - 1}{2} M_2^2 \qquad \text{so that} \qquad T_2 = 311.55\,K$$

Downstream pressure is therefore

$$p_2 = p_1 \left(\frac{T_2}{T_1}\right)^{\gamma/(\gamma-1)} = 575.0 \left(\frac{311.55}{360.00}\right)^{3.5} = 346.7\,\text{kPa}$$

■

The second application of Prandtl–Meyer theory is from Kearton [46], who considers a steam nozzle such as that shown in Figure 3.17. Assuming isentropic and choked flow, the Mach number at the throat is unity. Hence the speed of sound and the velocity are equal and the Mach waves are perpendicular to the flow and therefore aligned with the exit cross section of the nozzle. In addition, $\nu(M_1) = 0$. Next, an angle ϕ is defined to be that between the leading Mach wave and a Mach wave at any location in the expansion fan. Hence

$$\phi = \frac{\pi}{2} - \mu + \theta$$

in which θ is the angle by which the flow has turned at this location. Since $\nu(M) = \nu(M_1) + \theta$, then

$$\phi = \nu(M) + \frac{\pi}{2} - \tan^{-1}\frac{1}{\sqrt{M^2 - 1}}$$

or

$$\phi = \sqrt{\frac{\gamma+1}{\gamma-1}} \tan^{-1}\sqrt{\frac{\gamma-1}{\gamma+1}(M^2-1)} + \frac{\pi}{2} - \tan^{-1}\sqrt{M^2-1} - \tan^{-1}\frac{1}{\sqrt{M^2-1}}$$

But from a right triangle it is seen that

$$\tan^{-1}\sqrt{M^2-1} + \tan^{-1}\frac{1}{\sqrt{M^2-1}} = \frac{\pi}{2}$$

so the expression for ϕ reduces to

$$\phi\sqrt{\frac{\gamma-1}{\gamma+1}} = \tan^{-1}\sqrt{\frac{\gamma-1}{\gamma+1}(M^2-1)} \qquad (3.65)$$

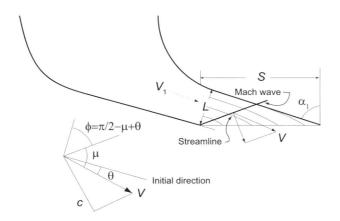

Figure 3.17 Expansion of steam from a choked nozzle.

The term inside the square root on the RHS of Eq. (3.65) can be replaced by a pressure ratio, for

$$\frac{p_0}{p} = \left(1 + \frac{\gamma-1}{2}M^2\right)^{\gamma/(\gamma-1)}$$

and therefore

$$\frac{\gamma-1}{\gamma+1}(M^2-1) = \frac{2}{\gamma+1}\left[\left(\frac{p_0}{p}\right)^{(\gamma-1)/\gamma} - 1\right]$$

Substitution yields

$$\phi\sqrt{\frac{\gamma-1}{\gamma+1}} = \tan^{-1}\sqrt{\frac{2}{\gamma+1}\left[\left(\frac{p_0}{p}\right)^{(\gamma-1)/\gamma} - 1\right]} \qquad (3.66)$$

Denoting

$$\delta = \phi\sqrt{\frac{\gamma-1}{\gamma+1}}$$

if follows that

$$\tan\delta = \sqrt{\frac{2}{\gamma+1}\left[\left(\frac{p_0}{p}\right)^{(\gamma-1)/\gamma} - 1\right]}$$

and therefore

$$\cos^2\delta = \frac{\gamma+1}{2}\left(\frac{p}{p_0}\right)^{(\gamma-1)/\gamma} \qquad (3.67)$$

This can be reduced further using the identity $2\cos^2 x = 1 + \cos 2x$. Hence

$$1 + \cos(2\delta) = (\gamma+1)\left(\frac{p}{p_0}\right)^{(\gamma-1)/\gamma}$$

Rewriting this in terms of ϕ gives

$$1 + \cos\left(2\sqrt{\frac{\gamma-1}{\gamma+1}}\phi\right) = (\gamma+1)\left(\frac{p}{p_0}\right)^{(\gamma-1)/\gamma} \tag{3.68}$$

which can be solved for the pressure ratio. The final form is

$$\frac{p}{p_0} = \left[\frac{1 + \cos\left(2\sqrt{\frac{\gamma-1}{\gamma+1}}\phi\right)}{\gamma+1}\right]^{\gamma/(\gamma-1)}$$

Streamlines can be calculated by considering a control volume of a wedge-shaped region with an inlet at the throat and exit coinciding with an arbitrary Mach wave in the expansion fan. The lateral boundary is taken to be a streamline across which there is no flow. Since the component of velocity perpendicular to a Mach wave is the local sonic velocity, it is this component that carries the flow trough the inflow and outflow boundaries of the chosen control volume. For this reason, the continuity equation reduces to

$$\rho_1 A_1 c_1 = \rho A c$$

Since the nozzle is rectangular, $A = rb$, in which b is the nozzle height and r is the distance from the corner to the chosen streamline. At the throat the area is $A_1 = r_1 b$, and r_1 is the distance from the corner to the same streamline. Hence

$$\frac{r}{r_1} = \frac{\rho_1 c_1}{\rho c} = \frac{p_1}{p}\sqrt{\frac{T}{T_1}}$$

In isentropic flow $T/T_1 = (p/p_1)^{(\gamma-1)/\gamma}$, and this equation reduces to

$$\frac{r}{r_1} = \frac{p_1}{p}\left(\frac{p}{p_1}\right)^{(\gamma-1)/2\gamma} = \left(\frac{p}{p_1}\right)^{-(\gamma+1)/2\gamma}$$

From Eq. (3.67) the expression

$$\cos\delta = \left(\frac{\gamma+1}{2}\right)^{1/2}\left(\frac{p}{p_0}\right)^{(\gamma-1)/2\gamma} \tag{3.69}$$

is obtained. With sonic conditions at station 1, the pressure ratio there is

$$\frac{p_0}{p_1} = \left(\frac{\gamma+1}{2}\right)^{\gamma/(\gamma-1)} \qquad \text{so that} \qquad \left(\frac{\gamma+1}{2}\right)^{1/2} = \left(\frac{p_0}{p_1}\right)^{(\gamma-1)/2\gamma}$$

and Eq. (3.69) can be recast as

$$\frac{p}{p_1} = \left[\cos\left(\sqrt{\frac{\gamma-1}{\gamma+1}}\phi\right)\right]^{2\gamma/(\gamma-1)}$$

Hence the distance from the corner to the streamline bounding the control volume is

$$\frac{r}{r_1} = \frac{1}{\left(\cos\sqrt{\frac{\gamma-1}{\gamma+1}}\phi\right)^{(\gamma+1)/(\gamma-1)}} \tag{3.70}$$

Some streamlines have been plotted to Figure 3.17. Now, letting r_1 denote the width of the nozzle at the throat, the corresponding streamline is seen to leave the nozzle before its end. If the nozzle shape were to coincide with the contour of this streamline, the Prandtl–Meyer flow would be an exact solution of the equations for inviscid compressible flow through a nozzle of this shape. Since the nozzle wall is straight, the solution presented is an *approximation* to the actual situation.

Equation (3.65) may be recast into the form

$$\tan\left(\sqrt{\frac{\gamma-1}{\gamma+1}}\phi\right) = \sqrt{\frac{\gamma-1}{\gamma+1}}\sqrt{M^2-1}$$

and since $1/\sqrt{M^2-1} = \tan\mu$, the following relation is obtained

$$\tan\mu = \sqrt{\frac{\gamma-1}{\gamma+1}}\cot\sqrt{\frac{\gamma-1}{\gamma+1}}\phi \qquad (3.71)$$

For a given exit pressure lower than the critical one, the angle ϕ can be found from either Eq. (3.66) or from Eq. (3.68), and Eq. (3.71) is then used to determine the angle of the terminal Mach wave. After that, the amount of turning of the flow is obtained from

$$\theta_2 = \phi - \frac{\pi}{2} + \mu_2$$

The flow direction is given by $\alpha_2 = \alpha_1 - \theta_2$.

■ **EXAMPLE 3.10**

Consider steam flow from a low-pressure nozzle such as shown in Figure 3.17, with nozzle angle $\alpha = 65°$. At the inlet of the nozzle steam is saturated vapor at pressure $p_0 = 20\,\text{kPa}$. Steam exhausts into the interblade space, where pressure is $8\,\text{kPa}$. Find the angle θ by which the flow turns on leaving the nozzle, the far downstream velocity, and its direction.

Solution: Since at the inlet steam is saturated, its adiabatic index, according to Zeuner's equation, is $\gamma = 1.135$. Denoting station 1 to be the throat and station 2 the exit where the backpressure is $p_2 = 8\,\text{kPa}$, the ratio of backpressure to stagnation pressure is $p_2/p_0 = 8/20 = 0.4$, and therefore the flow is choked. The Mach number after the expansion is

$$M_2 = \sqrt{\frac{2}{\gamma-1}\left[\left(\frac{p_0}{p_2}\right)^{(\gamma-1)/\gamma} - 1\right]} = 1.306$$

With this value of Mach number the temperature at the exit is

$$T_2 = T_0 \left(\frac{p_2}{p_0}\right)^{(\gamma-1)/\gamma} = 333.2 \left(\frac{8}{20}\right)^{0.135/1.135} = 298.8\,\text{K}$$

The sonic speed at the exit is

$$c_2 = \sqrt{nRT_2} = \sqrt{1.135 \cdot 8314 \cdot 298.8/18} = 395.8\,\text{m/s}$$

Hence $V_2 = M_2 c_2 = 516.9 \, \text{m/s}$. Angle ϕ_2 can be calculated using Eq. (3.65), with the result that $\phi_2 = 47.44°$. The Mach angle at the exit is

$$\mu_2 = \sin^{-1}\left(\frac{1}{M_2}\right) = \tan^{-1}\frac{1}{\sqrt{(M_2^2 - 1)}} = 49.97°$$

Hence the amount of turning is

$$\theta_2 = \phi_2 + \mu_2 - \frac{\pi}{2} = 47.44° + 49.97° - 90° = 7.41°$$

The flow angle after turning is complete is

$$\alpha_2 = \alpha_1 - \theta_2 = 65° - 7.41° = 57.59°$$

The extent of the jet after it has reached the backpressure is

$$\frac{r}{r_1} = \frac{1}{\left(\cos\sqrt{\frac{\gamma-1}{\gamma+1}}\phi\right)^{(\gamma+1)/(\gamma-1)}} = 1.412$$

∎

3.7 FLOW LEAVING A TURBINE NOZZLE

There is a second way to calculate the exit flow that does not rely on the Prandtl–Meyer theory. It is illustrated next for a flow from the steam nozzle shown in Figure 3.18. The mass balance equation gives the following relationship

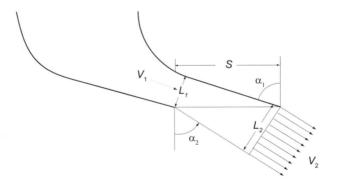

Figure 3.18 Steam nozzle analysis by mass balance.

$$\dot{m} = \frac{A_1 V_1}{v_1} = \frac{A_2 V_2}{v_2}$$

in which the areas can be related to angles by the geometric relations

$$\cos \alpha_1 = \frac{L_1}{s} \qquad \cos \alpha_2 = \frac{L_2}{s}$$

Since $A_1/A_2 = L_1/L_2$, the jet angle α_2 at the exit can be calculated from

$$\cos \alpha_2 = \frac{V_1 v_2}{V_2 v_1} \cos \alpha_1 \tag{3.72}$$

EXAMPLE 3.11

Consider the steam flow from a low-pressure nozzle as shown in Figure 3.17, with nozzle angle $\alpha = 65°$. At the inlet of the nozzle steam is saturated vapor with $\gamma = 1.135$, at pressure $p_0 = 20\,\text{kPa}$. Steam exhausts to the interblade space, where pressure is $8\,\text{kPa}$. Using the continuity equation, find the angle θ by which the flow turns on leaving the nozzle, the far downstream velocity, and its direction α_2.

Solution: The data given are the same as in the previous example, in which the values $T_2 = 298.8\,\text{K}$ and $V_2 = 516.9\,\text{m/s}$ were determined. Stagnation-specific volume, obtained from the steam tables, is $v_0 = 7.66\,\text{m}^3/\text{kg}$. At the throat $M_1 = 1$, so that

$$v_1 = v_0 \left(\frac{\gamma+1}{2}\right)^{1/(\gamma-1)} = 7.66 \left(\frac{2.135}{2}\right)^{1/0.135} = 12.43\,\text{m}^3/\text{kg}$$

$$T_1 = \frac{2T_0}{\gamma+1} = \frac{2 \cdot 332.2}{1.135+1} = 312.1\,\text{K}$$

Hence

$$V_1 = \sqrt{\gamma R T_1} = \sqrt{\frac{1.135 \cdot 8314 \cdot 312.1}{18}} = 404.5\,\text{m/s}$$

Far downstream the specific volume is

$$v_2 = \frac{RT_2}{p_2} = \frac{8.314 \cdot 298.8}{18 \cdot 8} = 17.25\,\text{m}^3/\text{kg}$$

and the velocity was determined earlier to be $V_2 = 516.9\,\text{m/s}$. Hence

$$\cos\alpha_2 = \frac{V_1 v_2}{V_2 v_1} \cos\alpha_1 = \frac{404.5 \cdot 17.25}{516.9 \cdot 12.43} \cos(65°) = 0.460$$

and the flow angle is $\alpha_2 = 62.7°$, and the flow turns by only $\theta = 2.3°$. This is $5°$ less than what is obtained by the Prandtl–Meyer theory.

∎

There is a limit on the extent to which the flow can turn. This limit is reached when the axial component of velocity of the jet reaches sonic speed. The condition may be analyzed by writing the mass flow rate in terms of the flow function $F(M)$. This results in the equations

$$\dot{m}\sqrt{c_p T_0} = F(M_1) A_1 p_{01} = F(M_2) A_2 p_{02}$$

so that

$$F(M_2) = F(M_1) \frac{A_1 p_{01}}{A_2 p_{02}}$$

For a nozzle of constant height

$$\frac{A_1}{A_2} = \frac{L_1}{L_2} = \frac{s \cos\alpha_1}{s \cos\alpha_2} = \frac{\cos\alpha_1}{\cos\alpha_2}$$

The axial component of velocity at sonic speed can be written as

$$V_{x2} = V_2 \cos\alpha_2 = M_2 \sqrt{\gamma R T_2} \cos\alpha_2 = \sqrt{\gamma R T_2}$$

so that $\cos\alpha_2 = 1/M_2$.

Substituting and simplifying gives

$$\frac{\gamma}{\sqrt{\gamma-1}}\left(1+\frac{\gamma-1}{2}M_2^2\right)^{-(\gamma+1)/2(\gamma-1)} = F(M_1)\cos\alpha_1\frac{p_{01}}{p_{02}}$$

from which M_2 may be determined.

■ EXAMPLE 3.12

Steam with $\gamma = 1.3$ flows from a low-pressure nozzle shown in Figure 3.17, with nozzle angle $\alpha = 65°$. The throat at the exit plane is choked. Find the limiting Mach number and the value of α_2 for the flow.

Solution: With flow choked at the throat, $M_1 = 1$, and

$$F^* = \frac{\gamma}{\sqrt{\gamma-1}}\left(\frac{\gamma+1}{2}\right)^{-(\gamma+1)/2(\gamma-1)} = \frac{1.3}{\sqrt{0.3}}(1+0.5\cdot 0.3)^{-23/6} = 1.389$$

Since $\cos\alpha_2 = 1/M_2$, assuming no losses so that $p_{02} = p_{01}$, the relation

$$F(M_2)\cos\alpha_2 p_{02} = F^*\cos\alpha_1 p_{01}$$

can be written as

$$\frac{\gamma}{\sqrt{\gamma-1}}\left(1+\frac{\gamma-1}{2}M_2^2\right)^{-(\gamma+1)/2(\gamma-1)} = F^*\cos\alpha_1$$

With $\alpha_1 = 65°$, solving this for M_2^2 gives

$$M_2^2 = \frac{2}{\gamma-1}\left[\left(\frac{\gamma}{\sqrt{\gamma-1}}\frac{1}{F^*\cos\alpha_1}\frac{p_{02}}{p_{01}}\right)^{2(\gamma-1)/(\gamma+1)} - 1\right] = 2.9314$$

so that $M_2 = 1.712$. Therefore

$$\alpha_2 = \cos^{-1}\left(\frac{1}{M_2}\right) = \cos^{-1}\left(\frac{1}{1.712}\right) = 54.26°$$

This is an $8.4°$ greater amount of turning than was calculated in the previous example.

■

It has been mentioned that a flow for which Prandtl–Meyer analysis is valid requires a nozzle in the shape of a streamline, which together with that along the opposite wall, define the flow stream. If this is not the case, the flow is more complicated, consisting of expansion waves that reflect from the adjacent nozzle wall and the jet boundary of the flow that leaves the nozzle. An alternating set of oblique shocks and expansions fans form in the jet as it moves downstream.

The turbine nozzle may also be designed such that the throat is upstream of the exit plane, in which case the flow may become supersonic at the exit plane. Then, in an *underexpanded* expansion, the flow adjusts to the backpressure through a set of shocks emanating from the trailing edge. The stagnation pressure losses are small if the Mach number is just slightly supersonic, as Eq. (3.37) shows. Such complex flows are beyond the scope of the present text.

EXERCISES

3.1 Conditions in an air reservoir are $680\,\text{kPa}$ and $560\,\text{K}$. From there the air flows isentropically though a convergent nozzle to a backpressure of $101.3\,\text{kPa}$. Find the velocity at the exit plane of the nozzle.

3.2 Air flows in a converging duct. At a certain location, where the area is $A_1 = 6.5\,\text{cm}^2$, pressure is $p_1 = 140\,\text{kPa}$ and Mach number is $M_1 = 0.6$. The mass flow rate is $\dot{m} = 0.25\,\text{kg/s}$. (a) Find the stagnation temperature. (b) If the flow is choked, what is the size of the throat area? (c) Give the percent reduction in area from station 1 to the throat. (d) Find the pressure at the throat.

3.3 Air flows in a convergent nozzle. At a certain location, where the area is $A_1 = 5\,\text{cm}^2$, pressure is $p_1 = 240\,\text{kPa}$ and temperature is $T_1 = 360\,\text{K}$. Mach number at this location is $M_1 = 0.4$. Find the mass flow rate.

3.4 The area of a throat in a circular nozzle is $A_t = 1\,\text{cm}^2$. For a choked flow find the diameter where $M_1 = 0.5$. Determine the Mach number value at a location where the diameter is $D_2 = 1.941\,\text{cm}$. Assume the flow to be isentropic and $\gamma = 1.4$.

3.5 In a location in a circular nozzle where the area is $A_1 = 4$, Mach number has the value $M_1 = 0.2$. Find the diameter at a location where $M = 0.6$.

3.6 Air flows through a circular duct $15\,\text{cm}$ in diameter with a flow rate $2.25\,\text{kg/s}$. The total temperature and static pressure at a certain location in the duct are $30°\text{C}$ and $106\,\text{kPa}$, respectively. Evaluate (a) the flow velocity, (b) the static temperature, (c) the total pressure, and (d) the density at this location.

3.7 Conditions in an air reservoir are $380\,\text{kPa}$ and $460\,\text{K}$. From there the air flows through a convergent nozzle to a backpressure of $101.3\,\text{kPa}$. The polytropic efficiency of the nozzle is $\eta_p = 0.98$. Find, (a) exit plane pressure, (b) exit plane temperature, and (c) the velocity at the exit plane of the nozzle.

3.8 Air issues from a reservoir at conditions $260\,\text{kPa}$ and $540\,\text{K}$ into a converging nozzle. The nozzle efficiency is estimated to be $\eta_N = 0.986$. The backpressure is $p_b = 101.3\,\text{kPa}$. Find, (a) exit Mach number, (b) exit plane temperature, (c) exit plane pressure, and (d) exit velocity.

3.9 At the inlet to a nozzle the conditions are $M_1 = 0.3$, $p_{01} = 320\,\text{kPa}$, and $T_{01} = 430\,\text{K}$. The flow is irreversible with polytropic exponent $n = 1.396$. Show that

$$\frac{T_{02}}{T_2} = \left(\frac{p_{01}}{p_1}\right)^{\gamma/(\gamma-1)} \left(\frac{p_1}{p_2}\right)^{(n-1)/n}$$

Find the Mach number at a location where $p_2 = 210\,\text{kPa}$.

3.10 Flow from a reservoir with $p_{01} = 260\,\text{kPa}$ and $T_{01} = 530\,\text{K}$ flows through a nozzle. It is estimated that the static enthalpy loss coefficient is $\zeta = 0.020$. The exit pressure is $p_2 = 180\,\text{kPa}$. (a) Find the exit Mach number. (b) Find the polytropic efficiency of the nozzle.

3.11 A two-dimensional nozzle has a shape

$$y = \frac{4}{5} + \frac{x(x-1)}{2(x+2)}$$

The nozzle stretches from $0 < x/a < 2(\sqrt{2}-1)$. The throat is at $x_t/a = 2(\sqrt{2}-1)$. The scale factor a is chosen such that the half-width of the nozzle at $x = 0$ is $4a/5$. Assume that $4\bar{f} = 0.02$ and the inlet Mach number is $M_1 = 0.5$. Calculate and plot p/p_1 as a function of x. Calculate the Mach number along the nozzle and graph it on the same plot.

3.12 Steam enters a nozzle from a steam chest at saturated vapor state at pressure $p_0 = 0.8\,\text{bar}$. It expands isentropically through a steam nozzle. Find the degree of supersaturation when it crosses the Wilson line at $x = 0.96$.

3.13 Consider a supersonic flow over a convex corner with angle $\theta_2 = 5°$, when the inflow moves in the direction of $\theta_1 = 0°$. The upstream Mach number is $M_1 = 1.1$, pressure is $p_1 = 130\,\text{kPa}$, and $T_1 = 310\,\text{K}$. Find, (a) Mach number, (b) temperature, and (c) pressure after the expansion is complete.

3.14 Consider the steam flow from a low-pressure nozzle at an angle $\alpha = 65°$. At the inlet of the nozzle steam is saturated vapor at pressure $p_0 = 18\,\text{kPa}$. Steam exhausts into the interblade space, where pressure is $7\,\text{kPa}$. Find the angle θ by which the flow turns on leaving the nozzle, the far downstream velocity, and its direction.

3.15 Consider the steam flow from a low-pressure nozzle at angle $\alpha = 65°$. At the inlet of the nozzle steam is saturated vapor at pressure $p_0 = 18\,\text{kPa}$. Steam exhausts to the interblade space, where pressure is $7\,\text{kPa}$. Using the continuity equation, find the angle θ by which the flow turns on leaving the nozzle, the far downstream velocity, and its direction α_2.

CHAPTER 4

PRINCIPLES OF TURBOMACHINE ANALYSIS

In this chapter the fundamental equation for turbomachinery analysis is developed from the moment of momentum balance. It gives an expression for the shaft torque in terms of the difference in the rate at which angular momentum of the working fluid leaves and enters a properly chosen control volume. Since power delivered (or absorbed) by a turbomachine is a product of torque and angular speed, a relationship between the flow rates of angular momentum at the inlet and exit, the rotational speed of the shaft, and power is obtained. The equation derived this way is called the *Euler equation of turbomachinery*. It is the most important equation in the study of this subject.

In earlier chapters power transferred to, or from, a turbomachine was expressed as the product of mass flow rate and a change in stagnation enthalpy. By equating the expression for work from the Euler equation of turbomachinery to the change in stagnation enthalpy, concepts from fluid mechanics become linked to thermodynamics. This link is central to understanding the performance of turbomachines.

In applying the momentum of momentum balance to a *stationary control volume* angular momentum is usually expressed in terms of absolute velocity of the fluid. In the analysis of the rotating blades velocity relative to the rotor is also needed. From it, together with the absolute velocity and the blade velocity, one can construct a velocity triangle. These velocity triangles are discussed first in this chapter. They are followed by the development of the Euler equation for turbomachinery. After that the work delivered, or absorbed, is recast in an alternative form and a concept of degree of reaction is developed. One measure of the effectiveness at which work transfer takes place in a turbomachine is called *utilization*. Although this concept is not extensively used today, it is introduced and its relationship to energy transfer and reaction is developed. The final section is on the theory of scaling and similitude, both of which are useful in determining the performance of one turbomachine from the known performance of a similar one.

4.1 VELOCITY TRIANGLES

The velocity vector of a fluid particle that flows through a turbomachine is most conveniently expressed by its components in cylindrical coordinates. The vector sum of radial and axial components

$$\mathbf{V}_m = V_r \mathbf{e}_r + V_z \mathbf{e}_z \tag{4.1}$$

is called the *meridional velocity*, for it lies on the meridional plane, which is a radial plane containing the axis of rotation. The various velocity components are shown in Figure 4.1.

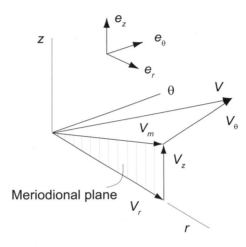

Figure 4.1 Meridional and tangential components of absolute velocity.

For axial machines the radial component of velocity is small and can be ignored, making the meridional velocity equal to the axial velocity. Similarly, at the outlet of a centrifugal compressor, or a radial pump, the axial velocity vanishes and the meridional velocity then equals the radial velocity.

The *absolute velocity* $\mathbf{V}$ is the sum of the *relative velocity* $\mathbf{W}$ and the velocity of the frame, or *blade velocity* $\mathbf{U}$. They are related by the vector equation

$$\mathbf{V} = \mathbf{W} + \mathbf{U} \tag{4.2}$$

By the usual construction this gives a *velocity triangle*, shown in Figure 4.2.

The angle that the absolute velocity makes with the meridional direction is denoted by α, and the angle that the relative velocity makes with this direction is β. These are called the *absolute* and *relative flow angles*.

From Eq. (4.2) and Figure 4.2 it is seen that the meridional components yield

$$V_m = W_m \tag{4.3}$$

and the tangential components are given by

$$V_u = W_u + U \tag{4.4}$$

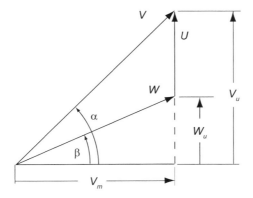

Figure 4.2 A typical velocity triangle.

These velocities are related to the meridional velocity by

$$V_u = V_m \tan \alpha \qquad W_u = W_m \tan \beta \qquad (4.5)$$

It is convenient to denote the tangential component by the subscript u and to take it positive when it is in the direction of the blade motion. Tangential components are associated with the blade forces; meridional components, with the rate at which fluid flows through the machine.

■ **EXAMPLE 4.1**

Consider the velocity diagram shown in Figure 4.3. The magnitude of the absolute velocity is $V_1 = 240 \, \text{m/s}$, and the flow angle is $\alpha_1 = -20°$. The blade speed is $U = 300 \, \text{m/s}$. Find the magnitude of the relative velocity and its flow angle.

Solution: The axial velocity is given by

$$V_{x1} = V_1 \cos \alpha_1 = 240 \cos(-20°) = 225.5 \, \text{m/s}$$

and $W_{x1} = V_{x1}$. The tangential components of the absolute and relative velocities are calculated as

$$V_{u1} = V_1 \sin \alpha_1 = 240 \sin(-20°) = -82.1 \, \text{m/s}$$

$$W_{u1} = V_{u1} - U = -82.1 - 300 = -382.1 \, \text{m/s}$$

Hence the magnitude of the relative velocity is

$$W_1 = \sqrt{W_{x1}^2 + W_{u1}^2} = \sqrt{225.5^2 + 382.2^2} = 443.7 \, \text{m/s}$$

and the flow angle of the relative velocity becomes

$$\beta_1 = \tan^{-1}\left(\frac{W_{u1}}{W_{x1}}\right) = \tan^{-1}\left(\frac{-382.1}{225.5}\right) = -59.4°$$

■

The foregoing example illustrates the sign convention for angles. Positive angles are measured from the meridional direction, and they increase in counterclockwise direction. Negative angles become more negative in the clockwise direction. Strict adherence to this convention will be followed, and this makes computer calculations easy to implement.

108 PRINCIPLES OF TURBOMACHINE ANALYSIS

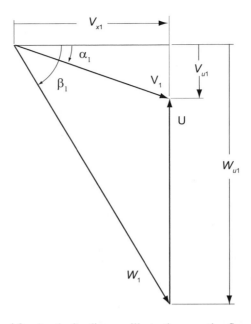

Figure 4.3 A velocity diagram illustrating negative flow angles.

4.2 MOMENT OF MOMENTUM BALANCE

Consider a flow through a pump, shown in the schematic diagram in Figure 4.4. To apply the moment of momentum equation, a control volume is chosen to include both the pump impeller and the fluid. The velocity vector is written in cylindrical coordinates as

$$\mathbf{V} = V_r \mathbf{e}_r + V_\theta \mathbf{e}_\theta + V_z \mathbf{e}_z \tag{4.6}$$

in terms of the unit vectors $\mathbf{e}_r, \mathbf{e}_\theta$, and $\mathbf{e}_z$. A working equation for the angular momentum balance for a uniform steady flow is

$$\dot{m}(\mathbf{r}_2 \times \mathbf{V}_2 - \mathbf{r}_1 \times \mathbf{V}_1) = \mathbf{T}_m + \mathbf{T}_f \tag{4.7}$$

On the right side, $\mathbf{T}_m$ is the torque the shaft exerts on the impeller and $\mathbf{T}_f$ is a contribution from fluid pressure and viscous stresses. The z component of this equation is obtained by taking its scalar product with $\mathbf{e}_z$. Thus

$$\dot{m}\, \mathbf{e}_z \cdot (\mathbf{r}_2 \times \mathbf{V}_2 - \mathbf{r}_1 \times \mathbf{V}_1) = \mathbf{e}_z \cdot \mathbf{T}_m = T$$

Owing to symmetry about the axis of rotation, pressure forces do not contribute to the axial torque, as they have radial and axial components only. Viscous forces act in the direction opposite to rotation and increase the required torque in a shaft of a compressor and decrease it in a turbine. These are neglected, or T is taken to be the net torque after they have been subtracted, or added. Rotation is taken to be clockwise when the pump is viewed in the flow direction. Hence the rotation vector is $\mathbf{\Omega} = \Omega \mathbf{e}_z$. In order for the shaft to rotate the pump impeller in this direction, torque must be given by $\mathbf{T}_m = \mathbf{e}_z T$, and thus $\mathbf{e}_z \cdot \mathbf{T}_m = T$.

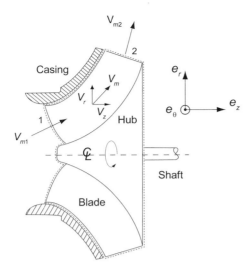

Figure 4.4 A schematic of a pump and a flow through it.

In cylindrical coordinates the radius vector is $\mathbf{r} = r\mathbf{e}_r + z\mathbf{e}_z$, so that

$$\mathbf{r} \times \mathbf{V} = \begin{vmatrix} \mathbf{e}_r & \mathbf{e}_\theta & \mathbf{e}_z \\ r & 0 & z \\ V_r & V_\theta & V_z \end{vmatrix} = -\mathbf{e}_r z V_\theta - \mathbf{e}_\theta (rV_z - zV_r) + \mathbf{e}_z r V_\theta$$

and

$$\mathbf{e}_z \cdot (\mathbf{r} \times \mathbf{V}) = rV_\theta$$

Hence the angular momentum equation becomes

$$T = \dot{m}(r_2 V_{\theta 2} - r_1 V_{\theta 1})$$

4.3 ENERGY TRANSFER IN TURBOMACHINES

The power delivered *to a turbomachine* is given by

$$\dot{W} = \mathbf{T} \cdot \boldsymbol{\Omega} = T\Omega = \dot{m}\Omega(r_2 V_{\theta 2} - r_1 V_{\theta 1})$$

The blade speeds are $U_1 = r_1 \Omega$, and $U_2 = r_2 \Omega$, and r_1 and r_2 are the mean radii at the inlet and outlet. Dividing this equation by the mass flow rate gives an expression for the work done per unit mass,

$$w = U_2 V_{\theta 2} - U_1 V_{\theta 1} \tag{4.8}$$

This is the *Euler equation for turbomachinery*.

As was already done in Figure 4.2, it is common to relabel the various terms and call the axial component of velocity V_x and denote the component of the velocity in the direction of the blade motion as V_u. In this notation there is no need to keep track of whether the rotor moves in clockwise or counterclockwise direction. The sense of rotation, of course, depends also on whether the rotor is viewed from the upstream or downstream direction. With these changes in notation, the Euler equation for turbomachinery may be written as

$$w = U_2 V_{u2} - U_1 V_{u1} \tag{4.9}$$

This gives the work done by the shaft on the rotor, and it is thus applicable to a compressor and a pump. For turbines, however, power is delivered by the machine, and the sign of the work would need to be changed. Since it is generally known whether the machine is power-absorbing or power-producing, work transfer will be taken as positive, and the Euler turbine equation is written as

$$w = U_2 V_{u2} - U_3 V_{u3} \tag{4.10}$$

For turbines, since a stage consists of a stator followed by a rotor, the inlet to the stator is designated as location 1, the inlet to the rotor is location 2, and the exit from the rotor is location 3.

For an axial turbomachine $U_1 = U_2 = U$. Work delivered by a stage is then given by

$$w = U(V_{u2} - V_{u3})$$

Its calculation is illustrated in the next example.

■ **EXAMPLE 4.2**

The shaft of small turbine turns at $20000\,\text{rpm}$, and the blade speed is $U = 250\,\text{m/s}$. The axial velocity leaving the stator is $V_{x2} = 175\,\text{m/s}$. The angle at which the absolute velocity leaves the stator blades is $\alpha_2 = 67°$, the flow angle of the relative velocity leaving the rotor is $\beta_3 = -60°$, and the absolute velocity leaves the rotor at the angle $\alpha_3 = -20°$. These are shown in Figure 4.5. Find (a) the mean radius of the blades, (b) the angle of the relative velocity entering the rotor, (c) the magnitude of the axial velocity leaving the rotor, (d) the magnitude of the absolute velocity leaving the stator, and (e) the specific work delivered by the stage.

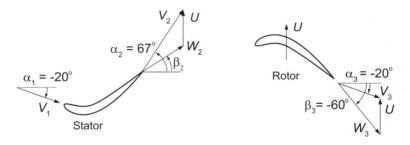

Figure 4.5 An axial turbine stage.

Solution: (a) The mean radius of the rotor is

$$r = \frac{U}{\Omega} = \frac{250 \cdot 60}{20,000 \cdot 2\pi} = 11.94\,\text{cm}$$

With the axial velocity and flow angle known the tangential component of the velocity is

$$V_{u2} = V_{x2} \tan \alpha_2 = 175 \tan(67°) = 412.3\,\text{m/s}$$

and therefore

$$W_{u2} = V_{u2} - U = 412.3 - 250 = 162.3\,\text{m/s}$$

Since $W_{x2} = V_{x2}$

$$\beta_2 = \tan^{-1}\left(\frac{W_{u2}}{W_{x2}}\right) = \tan^{-1}\left(\frac{162.3}{175}\right) = 42.8°$$

At the exit of the rotor

$$V_{u3} = W_{u3} + U \qquad V_{x3}\tan\alpha_3 = V_{x3}\tan\beta_3 + U$$

so that

$$V_{x3} = \frac{U}{\tan\alpha_3 - \tan\beta_3} = \frac{250}{\tan(-20°) - \tan(-60°)} = 182.7\,\text{m/s}$$

(b) The absolute velocity is obtained by first calculating the tangential component

$$V_{u3} = V_{x3}\tan\alpha_3 = 182.7\tan(-20°) = -66.5\,\text{m/s}$$

and then

$$V_3 = \sqrt{V_{x3}^2 + V_{u3}^2} = \sqrt{182.7^2 + 66.5^2} = 194.4\,\text{m/s}$$

(c) Specific work done is

$$w = U(V_{u2} - V_{u3}) = 250(412.3 + 66.5) = 119.7\,\text{kJ/kg}$$

As the flow expands through the turbine, its density decreases, and to accommodate this, the product of axial velocity and cross-sectional area needs to increase. Often axial machines are designed to keep the axial velocity constant. In this example the axial velocity increases by a small amount, and the cross-sectional area must be adjusted to account for the increase in velocity and a decrease in density. ∎

Work delivered by an axial turbine with constant axial velocity across a stage can be written as

$$w = U(V_{u2} - V_{u3}) = UV_x(\tan\alpha_2 - \tan\alpha_3)$$

Dividing by U^2 gives

$$\frac{w}{U^2} = \frac{V_x}{U}(\tan\alpha_2 - \tan\alpha_3)$$

Defining the blade-loading coefficient and flow coefficient as

$$\psi = \frac{w}{U^2} \qquad\qquad \phi = \frac{V_x}{U} \qquad\qquad (4.11)$$

gives a nondimensional version of this equation:

$$\psi = \phi(\tan\alpha_2 - \tan\alpha_3)$$

The blade loading coefficient is an appropriate term for ψ, since it is the blade force times the blade velocity that gives the work. Also, the flow coefficient ϕ is a ratio of the axial velocity to blade velocity and is thus a measure of the flow rate through the machine. Much use will be made of these nondimensional parameters, for they are independent of the size of machine, and their values for best designs have been established over many years of practice.

As another example of the general use of the Euler equation for turbomachinery, analysis of a centrifugal pump is considered next.

EXAMPLE 4.3

Water at $20°C$ leaves a pump impeller with an absolute velocity of $13.94\,\text{m/s}$ at the angle $72.1°$. The blade speed at the exit is $25.17\,\text{m/s}$, and the shaft speed is $3450\,\text{rpm}$. The absolute velocity is axial at the inlet. The flow rate is $18.0\,\text{L/s}$. Find (a) the magnitude of the relative velocity and its flow angle β_2, (b) the power required, and (c) the outlet blade radius and the blade height assuming that the open area at the periphery is 93% of the total area. The pump is shown in Figure 4.6.

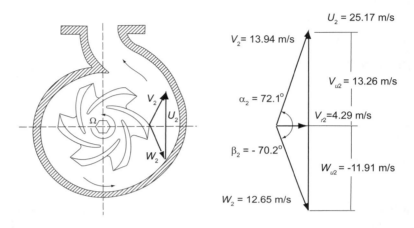

Figure 4.6 Pump exit and its velocity diagram.

Solution: (a) The tangential component of the absolute velocity at the exit is given by

$$V_{u2} = V_2 \sin \alpha_2 = 13.94 \sin(72.1°) = 13.26\,\text{m/s}$$

and its meridional component, which is radial here, is

$$V_{r2} = V_2 \cos \alpha_2 = 13.94 \cos(72.1°) = 4.29\,\text{m/s}$$

The tangential component of the relative velocity is determined as

$$W_{u2} = V_{u2} - U = 13.26 - 25.17 = -11.91\,\text{m/s}$$

Since the radial component of the relative velocity is $W_{r2} = V_{r2} = 4.29\,\text{m/s}$, the angle of the relative flow can be calculated as

$$\beta_2 = \tan^{-1}\left(\frac{W_{u2}}{W_{r2}}\right) = \tan^{-1}\left(\frac{-11.91}{4.29}\right) = -70.2°$$

The magnitude of the relative velocity is then

$$W_2 = \sqrt{W_{r2}^2 + W_{u2}^2} = \sqrt{4.29^2 + 11.91^2} = 12.65\,\text{m/s}$$

A velocity triangle can now be completed. The flow angle of the relative velocity is approximately equal to the blade angle χ, and in this pump the impeller blades curve backward; that is, they are curved in the direction opposite to blade rotation.

(b) Since the flow is axial at the inlet, $V_{u1} = 0$, and the work done is

$$w = U_2 V_{u2} = 25.17 \cdot 13.26 = 333.88 \text{ J/kg}$$

With density of water at $20°C$ equal to $\rho = 998 \text{ kg/m}^3$, the mass flow rate is

$$\dot{m} = \rho Q = 998 \cdot 0.018 = 17.964 \text{ kg/s}$$

and the power required is $\dot{W} = \dot{m} w = 17.964 \cdot 333.88 = 6.0 \text{ kW}$.

(c) The outlet radius is

$$r_2 = \frac{U_2}{\Omega} = \frac{25.17 \cdot 30}{3450 \cdot \pi} = 6.97 \text{ cm}$$

With the flow rate $Q = 0.018 \text{ m}^3/\text{s}$, the outlet area is

$$A_2 = \frac{Q}{V_{r2}} = \frac{0.018}{4.29} = 42.0 \text{ cm}^2$$

The blade height is then

$$b_2 = \frac{A_2}{0.93 \cdot 2\pi \cdot r_2} = \frac{42.0}{0.93 \cdot 2\pi \cdot 6.97} = 1.03 \text{ cm}$$

The blade-loading coefficient and flow coefficient are defined in terms of the tip speed of the blade at the exit:

$$\psi = \frac{w}{U_2^2} = \frac{333.88}{25.17^2} = 0.527 \qquad \phi = \frac{V_{r2}}{U_2} = \frac{4.29}{25.17} = 0.170$$

∎

4.3.1 Trothalpy and specific work in terms of velocities

Since no work is done in the stator, total enthalpy remains constant across it. In this section an analogous quantity to the total enthalpy is developed for the rotor. Specifically, consider a mixed-flow compressor in which the meridional velocity at the inlet is not completely axial and at the exit from the blades not completely radial. The work done by the rotor blades is

$$w = h_{02} - h_{01} = U_2 V_{u2} - U_1 V_{u1} \tag{4.12}$$

When this equation is written as

$$h_{01} - U_1 V_{u1} = h_{02} - U_2 V_{u2}$$

the quantity

$$I = h_0 - U V_u$$

is seen to be constant across the impeller. It can also be written as

$$I = h + \frac{1}{2}V^2 - U V_u = h + \frac{1}{2}V_m^2 + \frac{1}{2}V_u^2 - U V_u$$

Adding and subtracting $U^2/2$ to complete the square gives

$$I = h + \frac{1}{2}V_m^2 + \frac{1}{2}(V_u - U)^2 - \frac{1}{2}U^2 = h + \frac{1}{2}V_m^2 + \frac{1}{2}W_u^2 - \frac{1}{2}U^2$$

or since $V_m = W_m$, and $W^2 = W_m^2 + W_u^2$, it follows that

$$I = h_1 + \frac{1}{2}W_1^2 - \frac{1}{2}U_1^2 = h_2 + \frac{1}{2}W_2^2 - \frac{1}{2}U_2^2 \qquad (4.13)$$

is constant across the impeller. The quantity I is called *trothalpy*.[3]

Solving Eq. (4.13) for h_1 and h_2 and substituting them into the equation for work

$$w = h_{02} - h_{01} = h_2 + \frac{1}{2}V_2^2 - h_1 + \frac{1}{2}V_1^2 \qquad (4.14)$$

gives, after I has been canceled, the following equation:

$$w = \frac{1}{2}V_2^2 - \frac{1}{2}W_2^2 + \frac{1}{2}U_2^2 - \left(\frac{1}{2}V_1^2 - \frac{1}{2}W_1^2 + \frac{1}{2}U_1^2\right)$$

Rearranging gives the form

$$w = \frac{1}{2}(V_2^2 - V_1^2) + \frac{1}{2}(U_2^2 - U_1^2) + \frac{1}{2}(W_1^2 - W_2^2) \qquad (4.15)$$

Equating this and Eq. (4.14) leads to

$$h_2 - h_1 = \frac{1}{2}(U_2^2 - U_1^2) + \frac{1}{2}(W_1^2 - W_2^2) \qquad (4.16)$$

From Eq. (4.14) it is seen that the work done in a centrifugal pump increases the kinetic energy and the static enthalpy. Equation (4.16) shows first that the static enthalpy increase involves moving the fluid into a larger radius, resulting in increased pressure. The second term causes an increase in pressure as the relative velocity is reduced; that is, diffusion with $W_2 < W_1$ leads to pressure recovery. The pressure is increased further in the volute of a centrifugal pump where diffusion of the absolute velocity takes place. Since this diffusion is against an adverse pressure gradient, the kinetic energy at the exit of the impeller cannot be so large that its deceleration through the volute causes separation of boundary layers and a great increase in irreversibility. The use of these concepts is illustrated in the next example.

■ **EXAMPLE 4.4**

A small centrifugal pump with an impeller radius $r_2 = 4.5$ cm operates at 3450 rpm. Blades at the exit are curved back at an angle $\beta_2 = -65°$. Radial velocity at the exit is $V_{r2} = W_{r2} = 3.0$ m/s. Flow at the inlet is axial with velocity $V_1 = 4.13$ m/s. The mean radius of the impeller at the inlet is $r_1 = 2.8$ cm. (a) Find the work done using Eq. (4.12). (b) Calculate the kinetic energy change of the relative velocity, absolute velocity, and that associated with the change in the blade speed and calculate work done using Eq. (4.15). Confirm that the two methods give the same answer.

Solution: (a) Blade speed at the exit is

$$U_2 = r_2 \Omega = \frac{0.045 \cdot 3450 \cdot 2\pi}{60} = 16.26 \text{ m/s}$$

---

[3] This quantity is commonly called *rothalpy*, a compound word combining the terms *rotation* and *enthalpy*. Its construction does not conform to the established rules for formation of new words in the English language, namely that the roots of the new word originate from the same language. The word *trothalpy* satisfies this requirement as *trohos* is the Greek root for *wheel* and *enthalpy* is to put heat in, whereas *rotation* is derived from Latin *rotare*.

Since $W_{r2} = V_{r2}$, the tangential component of the relative velocity is

$$W_{u2} = V_{r2} \tan \beta_2 = 3.0 \tan(-65°) = -6.43 \,\mathrm{m/s}$$

Tangential component of the absolute velocity is then

$$V_{u2} = U_2 + W_{u2} = 16.26 - 6.43 = 9.83 \,\mathrm{m/s}$$

Since the flow at the inlet is axial $V_{u1} = 0$, the inlet does not contribute to the work done as calculated by the Euler equation for turbomachinery, which reduces to

$$w = U_2 V_{u2} = 16.26 \cdot 9.83 = 159.7 \,\mathrm{J/kg}$$

(b) Magnitudes of the relative and absolute velocities at the exit are given by

$$W_2 = \sqrt{W_{u2}^2 + W_{r2}^2} = \sqrt{6.43^2 + 3.0^2} = 7.10 \,\mathrm{m/s}$$

$$V_2 = \sqrt{V_{u2}^2 + V_{r2}^2} = \sqrt{9.83^2 + 3.0^2} = 10.27 \,\mathrm{m/s}$$

Since the flow is axial at the inlet. $V_{x1} = V_1 = 4.13 \,\mathrm{m/s}$. The blade speed is

$$U_1 = r_1 \Omega = \frac{0.028 \cdot 3450 \pi}{30} = 10.11 \,\mathrm{m/s}$$

Tangential components of the expression relating absolute and relative velocity give

$$W_{u1} = V_{u1} - U_1 = 0 - 10.11 = -10.11 \,\mathrm{m/s}$$

and therefore the magnitude of the relative velocity is

$$W_1 = \sqrt{W_{u1}^2 + W_{r1}^2} = \sqrt{10.11^2 + 4.13^2} = 10.92 \,\mathrm{m/s}$$

The kinetic energy changes are

$$\frac{1}{2}(V_2^2 - V_1^2) = \frac{1}{2}(10.27^2 - 4.13^2) = 44.21 \,\mathrm{m^2/s^2} = 44.21 \,\mathrm{J/kg}$$

$$\frac{1}{2}(U_2^2 - U_1^2) = \frac{1}{2}(16.26^2 - 10.11^2) = 80.99 \,\mathrm{m^2/s^2} = 80.99 \,\mathrm{J/kg}$$

$$\frac{1}{2}(W_1^2 - W_2^2) = \frac{1}{2}(10.93^2 - 7.10^2) = 34.51 \,\mathrm{m^2/s^2} = 34.51 \,\mathrm{J/kg}$$

Their sum checks with the direct calculation of the work done. ∎

Since there is no swirl at the inlet, $V_{u1} = 0$, the work done is independent of the inlet conditions. This means that when work is represented in terms of kinetic energy changes, terms involving inlet velocities must cancel. Velocity triangle at the inlet is a right triangle with W_1 as its hypothenuse, so that $V_1^2 + U_1^2 = W_1^2$. Hence in this case

$$w = \frac{1}{2}(V_2^2 + U_2^2 - W_2^2)$$

and using the law of cosines gives $w = U_2 V_{u2}$, as it should.

4.3.2 Degree of reaction

Degree of reaction, or *reaction* for short, is defined as the change in static enthalpy across the rotor divided by the static enthalpy change across the entire stage. For the turbine this is given as

$$R = \frac{h_2 - h_3}{h_1 - h_3}$$

Since an enthalpy change is proportional to a pressure change, the degree of reaction can be regarded in terms of pressure changes. In compressors pressure increases downstream and in order to keep the adverse pressure gradient small, the value of reaction provides a means to assess the strength of this gradient.

Work delivered by a rotor in a turbine is

$$w = h_{02} - h_{03} = h_2 + \frac{1}{2}V_2^2 - h_3 - \frac{1}{2}V_3^2$$

Since for nozzles (or stator) $h_{01} = h_{02}$, work can also be written as

$$w = h_1 - h_3 + \frac{1}{2}\left(V_1^2 - V_3^2\right)$$

Solving the last two equations for static enthalpy differences and substituting them into the definition of reaction gives

$$R = \frac{\frac{1}{2}\left(V_3^2 - V_2^2\right) + w}{\frac{1}{2}\left(V_3^2 - V_1^2\right) + w} \qquad (4.17)$$

Substituting Eq. (4.15) for work into this and simplifying leads to

$$R = \frac{U_2^2 - U_3^2 + W_3^2 - W_2^2}{V_2^2 - V_1^2 + U_2^2 - U_3^2 + W_3^2 - W_2^2} \qquad (4.18)$$

In a flow in which $V_1 = V_2$, the reaction $R = 1$. Such a machine is a pure reaction machine. A lawn sprinkler, rotating about an axis is such a machine, for all the pressure drops take place in the sprinkler arms. They turn as a *reaction* to the momentum leaving them.

The steam turbine shown in Figure 3.11 is an axial machine in which $U_2 = U_3$ and its reaction is zero when $W_2 = W_3$. For the rotor buckets shown, the blade angles are equal but opposite in sign and by adjustment of the flow area to account for the increase in specific volume, the magnitude of the relative velocity can be made constant across the rotor. Hence, since the trothalpy is also constant across the rotor, enthalpy change across it vanishes and the reaction becomes $R = 0$.

■ **EXAMPLE 4.5**

Consider an axial turbine stage with blade speed $U = 350\,\text{m/s}$ and axial velocity $V_x = 280\,\text{m/s}$. Flow enters the rotor at angle $\alpha_2 = 60°$. It leaves the rotor at angle $\alpha_3 = -30°$. Assume a stage for which $\alpha_1 = \alpha_3$ and a constant axial velocity. Find the velocities and the degree of reaction.

Solution: Since axial velocity is constant and the flow angles are equal at both the entrance and exit of the stage, the velocity diagrams at the inlet of the stator and the exit of the rotor are identical. From a velocity triangle, such as shown in Figure 4.2, the tangential velocities are:

$$V_{u2} = V_x \tan \alpha_2 = 280.0 \tan(60°) = 484.97 \, \text{m/s}$$

$$V_{u3} = V_x \tan \alpha_1 = 280.0 \tan(-30°) = -161.66 \, \text{m/s}$$

and work done is

$$w = U(V_{u2} - V_{u3}) = 350(484.97 + 161.66) = 226.32 \, \text{kJ/kg}$$

Tangential components of the relative velocities are

$$W_{u2} = V_{u2} - U = 484.97 - 350.00 = 134.97 \, \text{m/s}$$

$$W_{u3} = V_{u3} - U = -161.66 - 350.00 = -511.66 \, \text{m/s}$$

Hence

$$V_2 = \sqrt{V_{u2}^2 + V_x^2} = \sqrt{484.97^2 + 280.0^2} = 560.00 \, \text{m/s}$$

$$V_3 = \sqrt{V_{u3}^2 + V_x^2} = \sqrt{161.66^2 + 280.0^2} = 323.32 \, \text{m/s}$$

$$W_2 = \sqrt{W_{u2}^2 + W_x^2} = \sqrt{134.97^2 + 280.0^2} = 310.83 \, \text{m/s}$$

$$W_3 = \sqrt{W_{u3}^2 + W_x^2} = \sqrt{511.66^2 + 280.0^2} = 583.26 \, \text{m/s}$$

Since $U_2 = U_3$, the expression for reaction is

$$R = \frac{W_3^2 - W_2^2}{2w} = \frac{583.26^2 - 310.83^2}{2 \cdot 226{,}320} = 0.538$$

A reaction ratio close to one-half is often used to make the enthalpy drop, and thus also the pressure drop, in the stator and the rotor nearly equal. ∎

4.4 UTILIZATION

A measure of how effectively a turbine rotor converts the available kinetic energy at its inlet to work is called *utilization*, and a *utilization factor* is defined as the ratio

$$\varepsilon = \frac{w}{w + \frac{1}{2}V_3^2} \qquad (4.19)$$

The denominator is the available energy consisting of what is converted to work and the kinetic energy that leaves the turbine. This expression for utilization equals unity if the exit kinetic energy is negligible. But the exit kinetic energy cannot vanish completely because the flow has to leave the turbine. Hence utilization factor is always less than one. Maximum utilization is reached by turning the flow so much that the swirl component vanishes; that is, for the best utilization the exit velocity vector should lie on the meridional plane.

Making the appropriate changes in Eq. (4.15) to make it applicable to a turbine and substituting it into Eq. (4.19), gives an expression for utilization

$$\varepsilon = \frac{V_2^2 - V_3^2 + U_2^2 - U_3^2 + W_3^2 - W_2^2}{V_2^2 + U_2^2 - U_3^2 + W_3^2 - W_2^2} \qquad (4.20)$$

in terms of velocities alone.

Next, from Eq. (4.17) it is easy to see that the work delivered is also

$$w = -\frac{V_3^2}{2} + \frac{1}{2}\left(\frac{V_2^2 - RV_1^2}{1 - R}\right) \quad (4.21)$$

Substituting this into Eq. (4.19) gives

$$\varepsilon = \frac{V_2^2 - RV_1^2 - (1-R)V_3^2}{V_2^2 - RV_1^2} \quad (4.22)$$

In the situation in which $R = 1$ and therefore also $V_2 = V_1$, this expression become indeterminate. It is valid for other values of R.

In a usual design of a multistage axial turbine the exit velocity triangle is identical to the velocity triangle at the inlet of a stage. Under this condition $V_1 = V_3$ and $\alpha_1 = \alpha_3$, and the utilization factor simplifies to

$$\varepsilon = \frac{V_2^2 - V_3^2}{V_2^2 - RV_3^2} \quad (4.23)$$

The expression for work reduces to

$$w = \frac{V_2^2 - V_3^2}{2(1-R)} \quad (4.24)$$

With velocities expressed in terms of their tangential and axial components, this becomes

$$w = \frac{V_{x2}^2 + V_{u2}^2 - (V_{x3}^2 + V_{u3}^2)}{2(1-R)}$$

or

$$w = \frac{V_2^2 \sin^2\alpha_2 + V_{x2}^2 - (V_3^2 \sin^2\alpha_3 + V_{x3}^2)}{2(1-R)} \quad (4.25)$$

At maximum utilization $\alpha_3 = 0$ and the work is $w = UV_{u2}$. Equating this to the work given in Eq. (4.25) leads to the equality

$$UV_2 \sin\alpha_1 = \frac{V_2^2 \sin^2\alpha_2 + V_{x2}^2 - V_{x3}^2}{2(1-R)}$$

from which follows the relation

$$\frac{U}{V_2} = \frac{(V_{x2}^2 - V_{x3}^2)/V_2^2 + \sin^2\alpha_2}{2(1-R)\sin\alpha_2} \quad (4.26)$$

The left-hand side (LHS) is a speed ratio. It is denoted by $\lambda = U/V_2$. Since $V_{x2}/V_2 = \cos\alpha_2$, this reduces to

$$\lambda = \frac{1 - V_{x3}^2/V_2^2}{2(1-R)\sin\alpha_2} \quad (4.27)$$

from which the ratio

$$\frac{V_{x3}^2}{V_2^2} = 1 - 2(1-R)\lambda\sin\alpha \quad (4.28)$$

is obtained. For maximum utilization, $V_3 = V_{x3}$, and solving Eq. (4.23) for this ratio gives

$$\frac{V_{x3}^2}{V_2^2} = \frac{1 - \varepsilon_m}{1 - \varepsilon_m R}$$

Here the subscript m designates the condition of maximum utilization. Equating the last two expressions and solving for ε_m gives

$$\varepsilon_m = \frac{2\lambda \sin \alpha_2}{1 - 2R\lambda \sin \alpha_2} \tag{4.29}$$

If a stage is designed such that $V_{x3} = V_{x2}$, then the speed ratio in Eq. (4.27) may be written as follows:

$$\lambda = \frac{1 - V_{x2}^2/V_2^2}{2(1 - R)\sin \alpha_2} = \frac{\sin \alpha_2}{2(1 - R)} \tag{4.30}$$

Substituting this into Eq. (4.29) and simplifying gives

$$\varepsilon_m = \frac{\sin^2 \alpha_2}{1 - R \cos^2 \alpha_2} \tag{4.31}$$

This is shown in Figure 4.7.

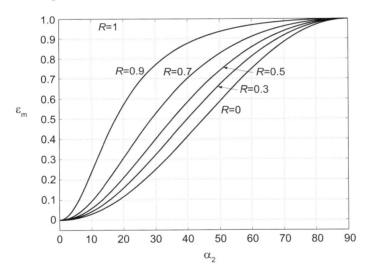

Figure 4.7 Maximum utilization factor for various degrees of reaction as a function of the nozzle angle.

It was mentioned earlier that a rotary lawn sprinkler is a pure reaction machine with $R = 1$. Its utilization is therefore unity for all nozzle angles. Inspection of Figure 4.7, as well as Eq. (4.31), shows that maximum utilization factor increases from zero to unity, when the nozzle angle α_2 increases from zero to $\alpha_2 = 90°$. Hence large nozzle angles give high utilization factors. Typically the first stage of a steam turbine has $R = 0$, with a nozzle angle in the range from $65°$ to $78°$.

Many turbines are designed with a 50% stage reaction. For such a stage $\beta_3 = -\alpha_2$ and $\alpha_3 = -\beta_2$. Also $V_3^2 = W_2^2$. Work delivered by a 50% reaction stage is

$$w = U(V_{u2} - V_{u3}) = U(V_{u2} - V_x \tan \beta_3) = U(V_{u2} + V_x \tan \alpha_2)$$

or
$$w = U(V_{u2} + W_{u2}) = U(V_{u2} + V_{u2} - U) = U(2V_2 \sin\alpha_2 - U)$$

The quantity $w + V_3^2/2$ becomes
$$w + \frac{1}{2}V_3^2 = w + \frac{1}{2}W_2^2 = w + \frac{1}{2}(V_x^2 + (V_{u2} - U)^2)$$

hence
$$w + \frac{1}{2}V_3^2 = w + \frac{1}{2}(V_2^2 - 2UV_2\sin\alpha_2 + U^2)$$

and the utilization factor from Eq. (4.19) is given as
$$\varepsilon = \frac{2U(2V_2\sin\alpha_2 - U)}{2U(V_2\sin\alpha_2 - U) + V_2^2 + U^2}$$

In nondimensional form this is
$$\varepsilon = \frac{2\lambda(2\sin\alpha_2 - \lambda)}{2\lambda(\sin\alpha_2 - \lambda) + 1 + \lambda^2} \tag{4.32}$$

Figure 4.8 gives a graphical representation of this relation. By differentiating this with

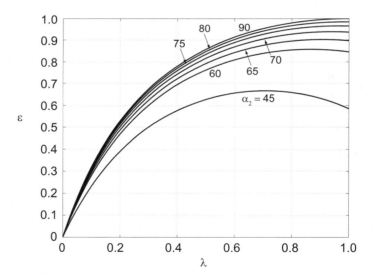

Figure 4.8 Utilization factor for an axial turbine with a 50% reaction stage.

respect to λ, shows that the maximum utilization factor is at $\lambda = \sin\alpha_2$, and the maximum utilization is given by
$$\varepsilon_m = \frac{2\sin^2\alpha_2}{1 + \sin^2\alpha_2}$$

which is consistent with Eq. (4.31).

EXAMPLE 4.6

Combustion gases flow from a stator of an axial turbine with absolute speed $V_2 = 500\,\text{m/s}$ at angle $\alpha_2 = 67°$. The relative velocity is at an angle $\beta_2 = 30°$ as it enters the rotor and at $\beta_3 = -65°$ as it leaves the rotor. (a) Find the utilization factor, and (b) the reaction. Assume the axial velocity to be constant.

Solution: (a) The axial and tangential velocity components at the exit of the nozzle are

$$V_x = V_2 \cos\alpha_2 = 500 \cos(67°) = 195.37\,\text{m/s}$$

$$V_{u2} = V_2 \sin\alpha_2 = 500 \sin(67°) = 460.25\,\text{m/s}$$

Since $W_x = V_x$, the tangential component of the relative velocity is

$$W_{u2} = W_x \tan\beta_2 = 195.37 \tan(30°) = 112.80\,\text{m/s}$$

so that

$$W_2 = \sqrt{W_x^2 + W_{u2}^2} = \sqrt{195.37^2 + 112.80^2} = 225.59\,\text{m/s}$$

Next, the blade speed is obtained as

$$U = V_{u2} - W_{u2} = 460.25 - 112.80 = 347.46\,\text{m/s}$$

Since the axial velocity remains constant, at the rotor exit the tangential component of the relative velocity is obtained as

$$W_{u3} = W_x \tan\beta_3 = 195.37 \tan(-65°) = -418.96\,\text{m/s}$$

so that

$$W_3 = \sqrt{W_x^2 + W_{u3}^2} = \sqrt{195.37^2 + 418.96^2} = 462.27\,\text{m/s}$$

Tangential component of the exit velocity is then obtained as

$$V_{u3} = W_{u3} + U = -418.96 + 347.46 = -71.50\,\text{m/s}$$

At the exit

$$\tan\alpha_3 = \frac{V_{u3}}{V_x} = \frac{-71.50}{195.37} = -0.366 \quad \text{so that} \quad \alpha_3 = -20.1°$$

and the absolute velocity at the exit is

$$V_3 = \sqrt{V_x^2 + V_{u3}^2} = \sqrt{195.37^2 + 71.50^2} = 208.04\,\text{m/s}$$

To calculate the utilization factor using its definition Eq. (4.19), work is first determined to be

$$w = U(V_{u2} - V_{u3}) = 347.46 \cdot (460.25 + 71.50) = 184{,}763\,\text{J/kg}$$

the utilization factor then becomes

$$\varepsilon = \frac{w}{w + \tfrac{1}{2}V_3^2} = \frac{184763}{184763 + 0.5 \cdot 208.04} = 0.895$$

(b) Reaction is obtained from

$$R = \frac{W_3^2 - W_2^2}{2w} = \frac{462.27^2 - 225.59^2}{2 \cdot 184763} = 0.44$$

■

As a second example, consider an axial turbine stage in which both utilization factor and reaction are given, together with the nozzle angle and efflux velocity from the nozzle.

■ **EXAMPLE 4.7**

An axial turbine operates at reaction $R = 0.48$ with utilization factor $\varepsilon = 0.82$. Superheated steam leaves the nozzles at speed $V_2 = 430\,\text{m/s}$ in the direction $\alpha_2 = 60.6°$. Find the (a) work delivered by the stage and (b) relative flow angles at the inlet and exit of the rotor. Assume the axial velocity to be constant through the stage.

Solution: Since the axial velocity is constant, the expression for the utilization factor may be written as

$$\varepsilon = \frac{V_2^2 - V_3^2}{V_2^2 - RV_3^2} = \frac{1 - \cos^2 \alpha_2 / \cos^2 \alpha_3}{1 - R \cos^2 \alpha_2 / \cos^2 \alpha_3}$$

Solving for the ratio of cosines gives

$$\frac{\cos \alpha_2}{\cos \alpha_3} = \sqrt{\frac{1 - \varepsilon}{1 - R\varepsilon}} = \sqrt{\frac{1 - 0.82}{1 - 0.48 \cdot 0.82}} = 0.5448$$

Thus

$$\cos \alpha_3 = \frac{\cos(60.6°)}{0.5448} = 0.901 \quad \text{and} \quad \alpha_3 = \pm 25.7°$$

There are two solutions. Which to choose? Inspection of Figure 4.8 shows that the curves of constant nozzle angle are concave downward so that for given utilization factor there are two speed ratios that satisfy the flow conditions. It is clear, however, that the speed ratio must be less than one, for blades cannot move faster than the flow. This is not yet a sufficient guideline for the correct choice for the sign, but after calculations have been carried out for both angles, the proper angle becomes clear after the fact. In addition, turbine blades typically turn the flow over 80°, and on this basis the negative angle may be tentatively chosen as being the correct one.

Next, the velocity components at the inlet to the blades are calculated:

$$V_x = V_2 \cos \alpha_2 = 430 \cos(60.6°) = 112.1\,\text{m/s}$$

$$V_{u2} = V_2 \sin \alpha_2 = 430 \sin(60.6°) = 374.6\,\text{m/s}$$

With the axial velocity constant, the tangential component at the exit is:

$$V_{u3} = V_x \tan \alpha_3 = 311.1 \tan(-25.7°) = -101.6\,\text{m/s}$$

The magnitude of the absolute velocity at the exit is thus

$$V_3 = \sqrt{V_x^2 + V_{u3}^2} = 234.3\,\text{m/s}$$

(a) Work delivered by the turbine may now calculated from

$$\varepsilon = \frac{w}{w + \frac{1}{2}V_3^2} \quad \text{or} \quad w = \frac{\varepsilon V_3^2}{2(1-\varepsilon)} = \frac{0.82 \cdot 234.3^2}{2(1-0.82)} = 125{,}015\,\text{kJ/kg}$$

(b) The blade speed is obtained from the expression

$$w = U(V_{u2} - V_{u3}) \qquad U = \frac{w}{V_{u2} - V_{u3}} = \frac{125{,}015}{374.6 + 101.6} = 262.5\,\text{m/s}$$

Thus $\lambda = U/V_2 = 262.5/430 = 0.61$, and since this number is less than one, the negative angle gives the correct solution.

It is worthwhile also to calculate the extent by which the blades turn the relative velocity. Tangential components of the relative velocity at the inlet and exit of the blade row are

$$W_{u2} = V_{u2} - U = 374.6 - 262.5 = 112.1\,\text{m/s}$$

$$W_{u3} = V_{u3} - U = -101.6 - 262.5 = -364.1\,\text{m/s}$$

The flow angles of the relative velocity are finally

$$\beta_2 = \tan^{-1}\left(\frac{W_{u2}}{W_x}\right) = \tan^{-1}\left(\frac{112.1}{211.1}\right) = 28.0°$$

$$\beta_3 = \tan^{-1}\left(\frac{W_{u3}}{W_x}\right) = \tan^{-1}\left(\frac{-364.1}{211.1}\right) = -59.9°$$

and the amount of turning is $28° + 59.9° = 87.9°$.

For the positive exit flow angle, $\alpha_3 = 25.7°$, and the tangential velocity becomes

$$V_{u3} = V_x \tan\alpha_3 = 211.1 \tan(25.7°) = 101.6\,\text{m/s}$$

and therefore the magnitude of V_3 is the same as before; so is the work delivered since the utilization factor is the same. The blade speed, however, is changed, as it is now calculated to be

$$U = \frac{w}{V_{u2} - V_{u3}} = \frac{125015}{374.6 - 101.6} = 457.9\,\text{m/s}$$

Consequently the blade speed ratio is $\lambda = U/V_2 = 457.9/430 = 1.06$, a value greater than one. Therefore this angle is the incorrect one. Proceeding with the calculation, the tangential velocities of the relative motion are:

$$W_{u2} = V_{u2} - U = 374.6 - 457.9 = -83.3\,\text{m/s}$$

$$W_{u3} = V_{u3} - U = -101.6 - 457.9 = -356.3\,\text{m/s}$$

Calculating the flow angles of the relative velocity gives

$$\beta_2 = \tan^{-1}\left(\frac{W_{u2}}{W_x}\right) = \tan^{-1}\left(\frac{-83.3}{211.1}\right) = -21.6°$$

$$\beta_3 = \tan^{-1}\left(\frac{W_{u3}}{W_x}\right) = \tan^{-1}\left(\frac{-356.3}{211.1}\right) = -59.3°$$

Now the flow turns only $-21.6° + 59.3° = 37.7°$. This small amount of turning is typical of compressors, but not of turbines. Another way is to check the value of blade-loading coefficient. It is $\psi = w/U^2 = 125,015/262.5^2 = 1.82$ for the negative angle. Experience shows that blade-loading coefficients in the range $1 < \psi < 2.5$ give good designs.

■

The examples in this chapter have illustrated the principles of turbomachinery analysis. Some of the thermodynamic properties that appear in the examples are extensive, and some are intensive. Few are nondimensional, and of these the ones encountered so far include the blade-loading coefficient, flow coefficient or speed ratio, reaction, utilization factor, and Mach number. In addition, flow angles of the absolute and relative velocities are nondimensional quantities. Once the nondimensional parameters, including the flow angles, have been chosen, a choice is made for the magnitude of the exit velocity from the nozzles or the value of the axial velocity. The blade speed can then be calculated. In order to complete the aerothermodynamic analysis, thermodynamic losses need to be estimated. After this, all the intensive parameters will be known.

For this much of the analysis there was no need to introduce any extensive variables. But for example, a rate at which a liquid needs to be pumped, or a power delivered by a turbine are typical design specifications. The size of the machine depends on these extensive variables. Thus the cross-sectional flow areas are calculated with these specifications in mind together with the size of rotor or impeller. Their diameter and the blade speed determine the rotational speed of the shaft. In large machines rotational speeds are low; in small machines they are high. Undoubtedly a design iteration needs to be carried out so that the machine conforms to a class of successful past designs. This includes also a stress analysis and vibrational characteristics of the blades, disks, and shafts. The next section introduces other aspects of the use of nondimensional variables.

4.5 SCALING AND SIMILITUDE

The aim of scaling analysis is to compare the performance of two turbomachines of similar design. Thus it is also used to relate the performance of a model turbomachine to its prototype. Both tasks are carried out in terms of proper nondimensional variables. In this section the conventional nondimensional groups for turbomachinery are introduced, scaling analysis of a model and a prototype is reviewed, and performance characteristics of a compressor and a turbine are presented.

4.5.1 Similitude

Similitude broadly refers to similarity in geometry and flow in two turbomachines. More precisely, *dynamic similarity* is obtained if the ratios of force components at corresponding points in the flow through these machines are equal. A necessary condition for dynamic similarity is *kinematic similarity*, which means that streamline patterns in two machines are the same. To achieve this, the two machines must be *geometrically similar*. This means that they differ only in scale. Proportionality of viscous force components implies that the Reynolds number is the same for the two machines. To obtain full dynamic similarity, the two flows must have similar density distributions, for then inertial forces are proportional at two corresponding points in kinematically similar flows. This is trivially satisfied for an incompressible fluid of uniform density, but for compressible fluids Mach numbers must be the same at two corresponding points in the flow. The definition of Mach number involves temperature, which together with pressure determine the value of density. Thus, in flows in which the Mach numbers match, forces at corresponding points in kinematically similar flows are proportional to each others and the flows are said to be dynamically similar.

In courses on fluid dynamics, systematic methods are presented for finding dimensionless groups. They consist of deciding first what the important variables are, and grouping them in categories of *geometric parameters*, *fluid properties*, *operational variables*, and *performance variables*.

The most obvious geometric variable is the diameter D of the rotor. Density ρ and viscosity μ are the two most common fluid properties encountered in turbomachinery flows, and since the fluid particles move along *curved paths* through the machine, the flow is dominated by *inertial effects*. This means that pressure force is in balance with inertial force and viscous forces are small when compared to these. Since the inertial term is proportional to density, in turbomachinery flows density is a more important fluid property than viscosity.

The rotational speed Ω of the rotor is the most important operational variable. It is conventionally given in revolutions per minute, and in many performance plots it is *not converted* to the standard form of radians per second. The performance variables include the volumetric flow rate Q and the reversible work done per unit mass w_s, and quantities such as the power $\dot{W}$, related to them.

4.5.2 Incompressible flow

The meridional velocity in a turbomachine accounts for the rate at which fluid flows through a machine. Thus the ratio V_m/U is a measure of the flow rate. As has been seen already, this ratio is used in theoretical analysis, but in testing it is converted and expressed in terms of more readily measurable quantities. The meridional velocity times the flow area equals the volumetric flow rate, and the blade speed is equal to radius times the rotational speed. Then, with V_m proportional to Q/D^2 and blade speed proportional to ΩD, the combination

$$\phi_d = \frac{Q}{\Omega D^3} \qquad (4.33)$$

is dimensionless. It is called a *flow coefficient*.

A nondimensional variable that includes the fluid viscosity μ will lead to some form of Reynolds number, such as $\text{Re} = \rho V_m D/\mu$, for example. The usual form, however, is

$$\text{Re} = \frac{\rho \Omega D^2}{\mu} = \frac{\Omega D^2}{\nu}$$

in which ν is the kinematic viscosity and the blade speed, proportional to ΩD, is used in place of the meridional velocity. If the boundary layers on the flow passage do not separate, viscous forces remain important only near the walls. Thus, as the machine size increases, boundary-layer regions become a smaller part of the flow. One contributing factor to losses in turbulent boundary layers is the size of wall roughness, which depends on manufacturing methods used. Hence, for example the casing and the impeller of two pumps of different size, but manufactured the same way may have about equal roughness. Since more of the flow through the larger pump does not contact the walls of the flow channel, this part of the flow does not experience as great a loss as does the flow through the boundary layer. This concept goes by the name *scale effect*, and it makes the performance variables quite independent of Reynolds number, provided the machine is sufficiently large. It is for this reason that in large machines Reynolds number effects are ignored in preliminary design and also that efficiency increases with machine size. This discussion of the influence of Reynolds number applies at a design condition. However, when turbomachines are operated at off-design conditions, boundary layers may separate. In worst cases such operation can be catastrophic. Thus the Reynolds number enters the theory of turbomachines indirectly through its influence on the behavior of boundary layers.

The most important performance variable is the work done on the fluid, or delivered by the machine. Its nondimensional form is the *work coefficient*

$$\psi_d = \frac{w_s}{\Omega^2 D^2}$$

in which w_s is the isentropic work and the product ΩD in the denominator is proportional to the blade speed, so that the denominator has the units of energy per unit mass, as does isentropic work.

In the Bernoulli equation

$$p_1 + \frac{1}{2}\rho V_1^2 + \rho g z_1 = p_2 + \frac{1}{2}\rho V_2^2 + \rho g z_2$$

the kinetic energy term shows that in this form the units of each term are energy per unit volume, since density has replaced mass. Furthermore, weight equals mass times gravitational acceleration, and specific weight of a fluid is defined as ρg. Dividing each term by ρg gives

$$\frac{p_1}{\rho g} + \frac{1}{2g}V_1^2 + z_1 + \frac{p_2}{\rho g} + \frac{1}{2g}V_2^2 + z_2$$

and each term has the dimensions of energy per unit weight of the fluid, which can be reduced to a length, as the potential energy term shows. From hydraulic practice it is common to call the first term in this equation a *pressure head*. The second term has the name *kinetic energy* or *dynamic head*, and the third term is an *elevation head*. The sum is called the *total head*.

The first law of thermodynamics applied to a flow through a pump gives

$$\frac{p_1}{\rho} + \frac{1}{2}V_1^2 + g z_1 + w_s = \frac{p_2}{\rho} + \frac{1}{2}V_2^2 + g z_2$$

which can also be written as

$$w_s = gH$$

and here H is the change in the total head across the pump. Pump manufacturers report the performance of pumps in terms of their head, making the values independent of the fluid being pumped. They also report a value for efficiency, so that actual work may be determined as $w = w_s/\eta$.

From the definition of total pressure, isentropic work for an incompressible fluid can also be calculated from

$$w_s = \frac{p_{02} - p_{01}}{\rho}$$

which shows that if the total pressure change across the pump is reported, then, to obtain the work done, density of the fluid being pumped needs to be known.

Also from the hydraulic practice comes the custom of expressing pressure as a height of a mercury, or a water column, particularly in fans and blowers in which the pressure rise is small. The conversion is carried out by the manometer formula

$$\Delta p = \rho_m g d \tag{4.34}$$

in which ρ_m is the density of the manometer fluid and d is the manometer deflection. As a consequence of the discussion above, for a pump, the work coefficient may be written as

$$\psi_d = \frac{gH}{\Omega^2 D^2}$$

Power coefficient may be introduced as

$$\hat{P}_d = \frac{\dot{W}}{\rho \Omega^3 D^5}$$

Here $\dot{W}$ is the actual rate at which work is done. Hence, multiplying the denominator of ψ_d by the mass flow rate and dividing it by a quantity with dimensions of mass flow rate, namely, by $\rho \Omega D D^2$, gives this form, except that efficiency of the machine must also be taken into account. It is easy to see that the power coefficients for a turbine with efficiency η_t and compressor with efficiency η_c are

$$\hat{P}_d = \eta_t \psi_d \phi_d \qquad\qquad \hat{P}_d = \frac{\psi_d \phi_d}{\eta_c}$$

respectively.

For dynamically similar situations two machines have the same values of the nondimensional parameters ψ_d and ϕ_d. They are also expected to have the same efficiency if the scale effect is neglected. This means that for machines 1 and 2 the following relationships are true:

$$\left(\frac{w_s}{\Omega^2 D^2}\right)_1 = \left(\frac{w_s}{\Omega^2 D^2}\right)_2 \qquad\qquad \left(\frac{Q}{\Omega D^3}\right)_1 = \left(\frac{Q}{\Omega D^3}\right)_2$$

■ **EXAMPLE 4.8**

Liquid water with density $\rho = 998 \text{ kg/m}^3$ flows through an axial flow pump, with a rotor diameter of 30 cm at a rate of 200 m³/h. The pump operates at 1600 rpm, and its efficiency is $\eta_p = 0.78$. The pump work is 180 J/kg. If a second pump in the same series has a diameter of 20 cm and operates at 3200 rpm, at the condition of same efficiency, find (a) flow rate, (b) the total pressure increase across it, and (c) the input power.

Solution: (a) Since the pumps are geometrically similar and their efficiencies are the same, dynamic similarity may be assumed. Thus

$$\frac{Q_1}{\Omega_1 D_1^3} = \frac{Q_2}{\Omega_2 D_2^3} \qquad Q_2 = Q_1 \frac{\Omega_2 D_2^3}{\Omega_1 D_1^3} = 200 \frac{3200 \cdot 20^3}{1600 \cdot 30^3} = 118.5 \text{ m}^3/\text{h}$$

and the mass flow rate is

$$\dot{m}_2 = \rho Q_2 = \frac{998 \cdot 118.5}{3600} = 32.86 \text{ kg/s}$$

(b) With equal work coefficient and efficiency, it follows that

$$\frac{w_{s1}}{\Omega_1^2 D_1^2} = \frac{w_{s2}}{\Omega_2^2 D_2^2} \qquad w_2 = w_1 \frac{\Omega_2^2 D_2^2}{\Omega_1^2 D_1^2} = 180 \, \frac{3200^2 \cdot 20^2}{1600^2 \cdot 30^2} = 320 \, \text{J/kg}$$

The total pressure rise is given by

$$\Delta p_{02} = \rho w_2 \eta \quad \text{so that} \quad \Delta p_{02} = 998 \cdot 320 \cdot 0.78 = 249.1 \, \text{kPa}$$

(c) Pumping power is

$$\dot{W}_2 = \dot{m}_2 \, w_2 = 32.86 \cdot 320 = 10.5 \, \text{kW}$$

Since the ratio of rotational speeds and flow rates are used in the calculations, there was no need to covert them to standard units. ■

4.5.3 Shape parameter or specific speed

Flow coefficient and work coefficient can be combined in such a way that the diameter is eliminated. Raising the result to a power such that it becomes *directly proportional to the rotational speed*, gives a parameter that is called the *specific speed*, which is given by

$$\Omega_s = \left(\frac{Q}{\Omega D^3}\right)^{1/2} \left(\frac{\Omega^2 D^2}{gH}\right)^{3/4} = \frac{\Omega Q^{1/2}}{(gH)^{3/4}} \qquad (4.35)$$

This equation is of relevance to pumps and hydraulic turbines, since the working fluid for them is a liquid. It shows that machines with low flow rates and high pressure rise have low Ω_s. In centrifugal machines the inlet area, which is close to the shaft, is relatively small and to keep the inlet velocity within a desirable range, the flow rate is relatively low. The centrifugal action causes a large pressure rise in such machines. Both make the specific speed low. In axial pumps and turbines a large flow rate is possible, as the annulus area is far from the axis and is therefore large. This leads to a high specific speed. The shape of the machines thus changes from a radial to an axial type as the shape parameter increases. Thus a better name for Ω_s would be a *shape parameter*. Diameter does not appear in the definition of the specific speed, but since the velocity must be kept within reasonable limits it appears implicity through the flow rate, with a larger flow rate requiring a larger machine. Shapes for pumps are shown in Figure 4.9. The figure shows that large machines have higher efficiencies owing to the scale effect discussed above.

4.5.4 Compressible flow analysis

For compressible flows temperature must be listed among the fundamental dimensions. It appears in the definition of Mach number, $M = V/c$, and with c the speed of sound, equal to $c = \sqrt{\gamma RT}$ for an ideal gas, the ratio of specific heats appears as an additional parameter.

Flows with Mach numbers less than 0.3 can be approximated as incompressible. Hence in low-Mach-number flows, the influence of Mach number will be slight. Testing of centrifugal compressors with refrigerants as working fluids also shows that their performance depends only weakly on the ratio of specific heats [17]. This is useful to notice, for their performance maps generated for air are not expected to lead to large errors when they are used for other gases.

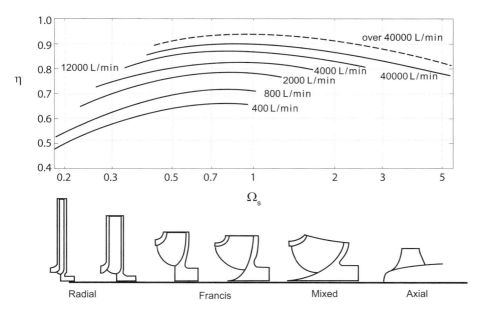

Figure 4.9 Shapes of pumps with increasing specific speed. (Modified from a graph in Stepanoff [74].)

In compressible flows the volumetric flow rate is replaced by the mass flow rate, as only the latter is constant through a machine. Conventional practice is to replace the volumetric flow rate with $\dot{m}/\rho_{01}$, where ρ_{01} is the inlet stagnation density. The other parameters include a modified blade Mach number $\Omega D/c_{01}$, Reynolds number $\rho_{01}\Omega D^2/\mu$, and the ratio of specific heats. In the definition of c_{01} the inlet stagnation temperature is used.

The functional relationship between the reversible work and these parameters can be expressed as

$$\frac{\Delta h_{0s}}{\Omega^2 D^2} = f\left(\frac{\dot{m}}{\rho_{01}\Omega D^3}, \frac{\Omega D}{c_{01}}, \frac{\rho_{01}\Omega D^2}{\mu}, \gamma\right) \tag{4.36}$$

For an ideal gas it was shown in Chapter 2 that

$$\frac{\Delta h_{0s}}{c_{01}^2} = \frac{1}{\gamma - 1}\left[\left(\frac{p_{02}}{p_{01}}\right)^{(\gamma-1)/\gamma} - 1\right]$$

which may be recast as

$$\frac{\Delta h_{0s}}{\Omega^2 D^2} = \frac{c_{01}^2}{\Omega^2 D^2}\frac{1}{\gamma - 1}\left[\left(\frac{p_{02}}{p_{01}}\right)^{(\gamma-1)/\gamma} - 1\right]$$

Since the blade Mach number and the ratio of specific heats are already taken as independent parameters, the ideal work coefficient can be replaced by a *stagnation pressure ratio*. This is done, regardless of whether the gas is ideal.

The flow coefficient may be written as

$$\frac{\dot{m}}{\rho_{01}\Omega D^3} = \frac{\dot{m}}{\rho_{01}c_{01}D^2}\frac{c_{01}}{\Omega D} = \frac{c_{01}}{\Omega D}\frac{\dot{m}\sqrt{RT_{01}}}{p_{01}\sqrt{\gamma}D^2}$$

Again, since the blade Mach number and γ are already counted as independent parameters, inspection of the right side shows that the flow coefficient may be modified to

$$\phi_d = \frac{\dot{m}\sqrt{RT_{01}}}{p_{01}D^2}$$

The power coefficient is manipulated into the form

$$\frac{\dot{W}}{\rho_{01}\Omega^3 D^5} = \frac{\dot{m}c_p \Delta T_0}{\rho_{01}\Omega^3 D^5} = \frac{\dot{m}}{\rho_{01}\Omega D^3}\frac{1}{\gamma-1}\frac{c_{01}^2}{\Omega^2 D^2}\frac{\Delta T_0}{T_{01}}$$

The first factor on the right is the original flow coefficient. It is multiplied by a factor dependent only on γ and the reciprocal of the blade Mach number squared. Since all these factors have been taken into account separately, the power coefficient may be replaced by $\Delta T_0/T_{01}$. Hence

$$\frac{p_{02}}{p_{01}} = f_1\left(\frac{\dot{m}\sqrt{RT_{01}}}{p_{01}D^2}, \frac{\Omega D}{\sqrt{\gamma RT_{01}}}, \frac{\rho_{01}\Omega D^2}{\mu}, \gamma\right) \quad (4.37)$$

and

$$\frac{\Delta T_0}{T_{01}} = f_2\left(\frac{\dot{m}\sqrt{RT_{01}}}{p_{01}D^2}, \frac{\Omega D}{\sqrt{\gamma RT_{01}}}, \frac{\rho_{01}\Omega D^2}{\mu}, \gamma\right) \quad (4.38)$$

Efficiency is another performance variable and is functionally related to the parameters listed on the right in the equations above. For an ideal gas undergoing compression, it can be calculated from

$$\eta = \frac{T_{02s} - T_{01}}{T_{02} - T_{01}} = \frac{T_{01}}{\Delta T_0}\left[\left(\frac{p_{02}}{p_{01}}\right)^{(\gamma-1)/\gamma} - 1\right]$$

For a particular design and fluid, the geometric parameters and γ are fixed. This allows the flow coefficient and the blade Mach number to be replaced by

$$\phi_m = \frac{\dot{m}\sqrt{T_{01}}}{p_{01}} \qquad \Omega_m = \frac{\Omega}{\sqrt{T_{01}}} \quad (4.39)$$

These are *not dimensionless*. Alternatively, the *corrected* mass flow rate and the rotational speed

$$\dot{m}_c = \frac{\dot{m}\sqrt{T_{01}/T_{0r}}}{p_{01}/p_{0r}} \qquad \Omega_c = \frac{\Omega}{\sqrt{T_{01}/T_{0r}}} \quad (4.40)$$

are used, in which subscript r refers to a reference condition.

4.6 PERFORMANCE CHARACTERISTICS

Use of the performance map is illustrated in this section by representative compressor and turbine maps for an automotive turbocharger. It is used to precompress air before it is inducted to an internal combustion engine, thereby allowing a larger mass flow rate than is possible in a naturally aspirated engine. A turbocharger is shown in Figure 4.10. It consists of a centrifugal compressor and a radial inflow turbine. The exhaust gases from the engine drive the turbine. Shaft speeds vary from 60,000 to 200,000 rpm in automotive applications.

Figure 4.10 A turbocharger. (Photo courtesy NASA.)

4.6.1 Compressor performance map

To characterize the performance of a compressor, the pressure ratio is typically plotted as a function of the flow coefficient, as is shown in Figure 4.11. Here the flow rate and the rotational speed are modified to a corrected flow rate and a corrected shaft speed

$$Q_c = Q\sqrt{\frac{T_{0r}}{T_{01}}} \qquad \Omega_c = \Omega\sqrt{\frac{T_{0r}}{T_{01}}}$$

This particular *compressor map* is for a centrifugal compressor of an automotive turbocharger manufactured by BorgWarner Turbo Systems, similar to that shown in Figure 4.10. Air is drawn in from stagnant atmosphere with reference pressure $p_{0r} = 0.981$ bar and reference temperature $T_{0r} = 293$ K. These are the nominal inlet stagnation properties.

Efficiency curves are superimposed on the plot on a family of curves at constant corrected speed, given in rpm. The constant speed curves terminate at a line called a *surge line*. If the flow rate decreases beyond this, the blades will stall. Severe stall leads to a condition known as *surge*. Under surge conditions the flow may actually *reverse* direction, leading to a possible *flameout* in a jet engine.

In an automotive application the operating speed of the turbomachine follows the engine speed of the internal combustion engine. When the shaft speed is increased, the operating condition moves *across* the constant speed curves in the general direction parallel to the surge line to lower efficiency. At large flow rates the flow in the blade passages will *choke* and this is indicated by the sharp drop in the constant-speed curves.

4.6.2 Turbine performance map

A sample plot of turbine characteristics is shown in Figure 4.12 for the radial inflow turbine of the same BorgWarner turbocharger. Inlet to the turbine is identified by the label 3, and its exit is a state 4. Pressure ratio is given as the ratio of stagnation pressures at states 3 and 4. Inlet *reference* temperature has a value $T_{03r} = 873$ K. The rotational speed of the shaft is corrected by the square root of the ratio of the reference temperature to its actual value at the inlet. This arises from the square root dependence of speed of sound on temperature.

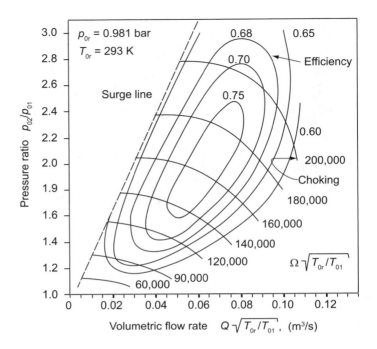

Figure 4.11 Characteristics of a centrifugal compressor. (Courtesy BorgWarner Turbo Systems.)

Since the exit pressure from the turbine is close to the atmospheric value, the pressure ratio is determined by the inlet pressure to the turbine, which in turn is related to engine pressure. A high pressure ratio leads to choking of the turbine. Thus an increase in the pressure ratio no longer increases the flow rate and the lines of constant speed remain flat as seen in Figure 4.12.

Efficiency curves are also flat near choking conditions, for the aerodynamic design is optimized for these conditions. This is in contrast to the small *envelope* of high efficiency at low pressure ratios when the turbine can still accommodate a large change in the mass flow rate as the pressure ratio is increased.

Although it would be desirable to have a consistent representation of the dimensionless parameters, this is not yet a common practice. Hence, the dimensions and units in each of the parameters need to be examined for each performance map encountered.

EXERCISES

4.1 Steam enters a rotor of an axial turbine with an absolute velocity $V_2 = 320 \text{ m/s}$ at an angle $\alpha_2 = 73°$. The axial velocity remains constant. The blade speed is $U = 165 \text{ m/s}$. The rotor blades are equiangular so that $\beta_3 = -\beta_2$, and the magnitude of the relative velocity remains constant across the rotor. Draw the velocity triangles. Find (a) the relative flow angle β_2, (b) the magnitude of the velocity V_3 after the flow leaves the rotor, and (c) the flow angle α_3 that V_3 makes with the axial direction.

4.2 Water with density 998 kg/m^3 flows in a centrifugal pump at the rate of 22 L/s. The impeller radius is $r_2 = 7.7 \text{ cm}$, and the blade width at the impeller exit is $b_2 = 0.8 \text{ cm}$. If the flow angles at the impeller exit are $\alpha_2 = 67°$ and $\beta_2 = -40°$, what is the rotational speed of the shaft in rpm?

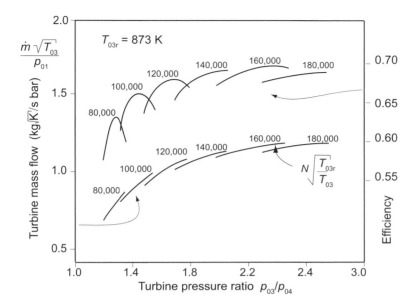

Figure 4.12 Characteristics of a radial inflow turbine. (Courtesy BorgWarner Turbo Systems.)

4.3 In a velocity diagram at the inlet of a turbine the angle of the absolute velocity is $60°$ and the flow angle of the relative velocity is $-51.7°$. Draw the velocity diagram and find the value of U/V and V_x/U.

4.4 A small axial-flow turbine has an output power of 37 kW when handling 1 kg of air per second with an inlet total temperature of 335 K. The total-to-total efficiency of the turbine is 80%. The rotor operates at 50,000 rpm and the mean blade diameter is 10 cm. Evaluate (a) the average driving force on the turbine blades, (b) the change in the tangential component of the absolute velocity across the rotor, and (c) the required total pressure ratio across the turbine.

4.5 The exit flow angle of stator in an axial steam turbine is $68°$. The flow angle of the relative velocity leaving the rotor is $-67°$. Steam leaves the stator at $V_2 = 120\,\text{m/s}$, and the axial velocity is $V_{x2} = 0.41U$. At the exit of the rotor blades the axial steam velocity is $V_{x3} = 0.42U$. The mass flow rate is $\dot{m} = 2.2\,\text{kg/s}$. Find (a) the flow angle entering the stator, assuming it to be the same as the absolute flow angle leaving the rotor; (b) the flow angle of the relative velocity entering the rotor; (c) the reaction; and (d) the power delivered by the stage.

4.6 The axial component of airflow leaving a stator in an axial-flow turbine is $V_{x2} = 175\,\text{m/s}$ and its flow angle is $64°$. The axial velocity is constant, the reaction of the stage is $R = 0.5$, and the blade speed is $U = 140\,\text{m/s}$. Since the reaction is 50%, the relationships between the flow angles are $\beta_2 = -\alpha_3$ and $\alpha_2 = -\beta_3$. Find the flow angle of the velocity entering the stator.

4.7 The airflow leaving the rotor of an axial-flow turbine is $V_{x3} = 140\,\text{m/s}$ and its flow angle is $0°$. The axial velocity is constant and equal to the blade speed. The inlet flow angle to the rotor is $\alpha_2 = 60°$. Find the reaction.

4.8 A large centrifugal pump operates at 6000 rpm and produces a head of 800 m while the flow rate is 30,000 L/min. (a) Find the value of the specific speed. (b) Estimate the efficiency of the pump.

4.9 A fan handles air at the rate of 500 L/s second when operating at 1800 rpm. (a) What is the flow rate if the same fan is operated at 3600 rpm? (b) What is the percentage increase in total pressure rise of the air assuming incompressible flow? (c) What is the power input required at 3600 rpm relative to that at 1800 rpm? Assume that the operating point of the fan in terms of the dimensionless parameters is the same in both cases.

4.10 An axial-flow pump having a rotor diameter of 20 cm handles water at the rate of 60 L/s when operating at 3550 rpm. The corresponding increase in total enthalpy of the water is 120 J/kg and the total-to-total efficiency is 75%. Suppose that a second pump in the same series is to be designed to handle water having a rotor diameter of 30 cm and operating at 1750 rpm. For this second pump what will be the predicted values for (a) the flow rate, (b) the change in the total pressure of water, and (c) the input power?

4.11 A small centrifugal pump handles water at the rate of 6 L/s with input power of 5 hp and total-to-total efficiency of 70%. Suppose that the fluid being handled is changed to gasoline having specific gravity 0.70. What are the predicted values for (a) flow rate, (b) input power, and (c) total pressure rise of the gasoline?

4.12 A blower handling air at the rate of 240 L/s at the inlet conditions of 103.1 kPa for total pressure and 288 K for total temperature. It produces a pressure rise of air equal to 250 mm of water. If the blower is operated at the same rotational speed, but with an inlet total pressure and total temperature of 20 kPa and 253 K. What are (a) the predicted value for the mass flow rate and (b) the total pressure rise?

4.13 Consider a fan with a flow rate of 1500 cfm, [cubic feet per minute (ft^3/min)] and a shaft speed of 3600 rpm. If a similar fan one half its size is to have the same tip speed, what will the flow rate be at a dynamically similar operating condition? What is the ratio of power consumption of the second fan compared to the first one?

4.14 A fan operating at 1750 rpm at a volumetric flow of 4.25 m^3/s develops a head of 153 mm of water. It is required to build a larger, geometrically similar fan that will deliver the same head at the same efficiency as the existing fan, but at the rotational speed of 1440 rpm. (a) Determine the volume flow rate of the larger fan. (b) If the diameter of the original fan is 40 cm, what is the diameter of the larger fan? (c) What are the specific speeds of these fans?

4.15 The impeller of a centrifugal pump, with an outlet radius $r_2 = 8.75$ cm and a blade width $b_2 = 0.7$ cm, operates at 3550 rpm and produces a pressure rise of 522 kPa at a flow rate of 1.5 L/min. Assume that the inlet flow is axial and that the pump efficiency is 0.63. (a) Find the specific speed. (b) Show that Eq. (4.15) for work reduces to $w = (V_{u2}^2 + U_2^2 - W_{u2}^2)/2$, and calculate the work two ways and confirm that they are equal.

CHAPTER 5

STEAM TURBINES

5.1 INTRODUCTION

The prime mover in a steam power plant is a steam turbine that converts part of the thermal energy of steam at high pressure and temperature to shaft power. Other components of the plant are a steam generator, a condenser, and feedwater pumps and heaters. The plants operate on various modifications of the Rankine cycle. The basic Rankine cycle, operating between $40°$C and $565°$C, has a Carnot efficiency of 37%. Modifications, including superheating, reheating, and feedwater heating, increase the efficiency by approximately an additional 10%.

Most large power plants have two reheats and three or more turbines. The turbines are said to be *compounded* when steam passes through each of them in series. The high-pressure (HP) turbine receives steam from the steam generator. After leaving this turbine the steam is reheated and then enters an intermediate-pressure (IP) turbine, also called a *reheat turbine*, through which it expands to an intermediate pressure. After the second reheat the rest of the expansion takes place through a low-pressure (LP) turbine, from which it enters a condenser at a pressure below the atmospheric value.

A turbine from which the steam leaves at quality near 90% is called a *condensing turbine*. An *extraction* turbine has ports from which steam is extracted for feedwater heating. An *induction* turbine receives steam at intermediate pressures for additional power generation.

In a *noncondensing or backpressure* turbine, steam leaves at superheated conditions and the thermal energy in the exhaust steam is used in various industrial processes. A well-designed *combined heat and power* plant generates appropriate amount of power to drop the steam temperature and pressure to values that meet the process heating needs. District heating is an application in which steam is used at even lower temperatures than in many industrial processes. An important consideration in providing the heating needs of an economy is to match the source to the application. Combined heat and power plants are designed with this in mind.

In modern coal-burning power plants axial steam turbines are typically housed in three, four, or five casings. Many LP turbines are *double-flow* type, and their single casing accommodates a pair of turbines in which steam flows in opposite directions to balance axial forces on the turbine shaft. When two or more turbines are connected to a common shaft, they are said to operate in *tandem*, and the plant is said to be a tandem compound type. If the steam is directed to a set of turbines on different shafts, then the system is said to be *cross-compounded*. For example, a 1000-MW power plant could have an HP turbine and an IP turbine on a single shaft and three LP turbines on a different shaft. Rotational speeds are typically either 3600 or 1800 rpm; the lower speed for the LP turbines, that have larger rotors in order to accommodate the larger volumetric flow rate of steam at low pressure. The large volumetric flow rate at the low pressure end requires very long blades. As an example, a General Electric/Toshiba LP steam turbine running at 3600 rpm has blades 1016 mm in length and a flow area $8.2\,\text{m}^2$ in the last stage. When this is designed to run at 3000 rpm for generating electricity at 50 Hz, the blades of the last stage are designed to be 1220 mm long, with a corresponding flow area $11.9\,\text{m}^2$. For a 26-stage steam turbine the hub-to-casing radius ratio for the last stage may have a value 0.42, whereas for the first stage a typical value is 0.96.

Outlet pressures from the boiler vary from a subcritical 10 MPa to a supercritical 30 MPa, or more. The condenser pressure is below the atmospheric pressure, typically about 8 kPa, which corresponds to a saturation temperature of $41°C$. This makes the overall pressure ratio equal to 1250 for a conventional plant and 3 times this for supercritical plants. In a 400-MW power plant the HP turbine provides about 100-MW, and power at about an equal rate is delivered by the IP turbine. A double-flow LP turbine delivers the remaining 200-MW. The pressure ratios are 4.5 doe the HP turbine and 3 for the IP turbine. The LP turbine has ports for extracting steam to feedwater heaters at pressure ratios ranging from 1.5 to 4.5.

Owing to the large inlet pressure, design of the HP turbine differs from that of the others. The first stage is designed for low reaction, so that most of the pressure drop takes place at the nozzles feeding this stage. This brings the steam to a very high velocity as it enters the first rotor. However, the pressure is now sufficiently low that leakage flow through seals is tolerable. Later stages are designed for higher reaction, and in IP and LP turbines the reaction is close to 50%.

Table 5.1 lists some typical designs and rated power outputs for coal-burning steam power plants. Designations such as 1SF and 3DF refer to one single-flow and three double-flow turbines [23]. At the preliminary design stage steam inlet pressure to the HP turbine is specified. Intermediate pressures at which reheating takes place are then calculated according to how much moisture is allowed at the exit of the HP and IP turbines. A similar decision is made for the reheat (RH) turbine to determine the appropriate pressure at the inlet of the LP turbines.

The steam turbine industry is very large. The annual worldwide electricity generation from steam plants is 38.5 EJ (exajoules) equal to 1.2 TW (terawatts) of generated power. This means that on the order of 10,000 steam turbines are in use. The major manufactures include GE power systems in the United States, Siemens, Alstom, and Ansaldo Energia in western Europe, Mitsubishi Heavy Industries, Hitachi and Toshiba in Japan. Many of the

IMPULSE TURBINES   137

Table 5.1 Fossil steam power turbine arrangements

Output (MW)	Reheats	Steam pressure (MPa)	HP	IP	RH	LP
50–150	0 or 1	10.1	1SF	-	-	1SF
150–200	1	12.5	1SF	1SF	-	1DF
250–450	1	16.6	1SF	1SF	-	1DF
450–600	1	16.6	1SF	1SF	-	2DF
450–600	1	24.2	1SF	1SF	-	2DF
600–850	1	16.6	1DF	1DF	-	2DF
600–850	1 or 2	24.2	1SF	1SF	1DF	2DF
850–1100	1	16.6	1DF	1DF	-	3DF
850–1100	1 or 2	24.2	1SF	1SF	1DF	3DF

steam turbines in eastern and central Europe are supplied by LMZ in Russia and Skoda in the Czech Republic.

World's wide production of electricity during the year 2008 was 18,800 billion kWh. In standard units this is 67.7 EJ. In *thermal plants* electricity generation is fueled by coal, natural gas, oil, and uranium. Coal provides 27 EJ; natural gas, 13 EJ; and oil, 3 EJ of the generated electricity. Nuclear fuels provide 10 EJ. The remaining 12 EJ comes from hydropower and a small amount from wind. In coal, fuel oil, and nuclear power plants the working fluid is water and these *steam plants* provide 57% of the generated electricity. The other 43% comes in about equal amounts from hydropower and from gas turbine power plants fueled by natural gas [43]. Since these figures relate to production of electricity, they do not take into account the energy value of coal, gas, and uranium that needs to be mined. A rough value for thermal efficiency of fossil fuel plants is 38% and therefore the energy content of the fuels delivered to the plants is a factor of 2.6 larger. Ordinarily the energy requirement for mining and transporting the fuel needed by power stations is not factored into the energy evaluation, as it should be.

5.2 IMPULSE TURBINES

This section begins with a discussion of impulse turbines and how a single-stage turbine is compounded to multiple stages by two methods. These are called *pressure compounding*, or Rateau staging, and *velocity compounding*, or Curtis staging. Both are used to reduce the shaft speed, which in a single-stage impulse turbine may be intolerably high.

5.2.1 Single-stage impulse turbine

Carl Gustaf Patrik de Laval (1845–1913) of Sweden in 1883 developed an impulse turbine consisting of a set of nozzles and a row of blades, as shown in Figure 5.1. This turbine is designed to undergo the entire pressure drop in the nozzles and none across the rotor. For sufficiently low exit pressure, *converging-diverging* nozzles accelerate the steam to a supersonic speed. The angle of the relative velocity approaching the rotor blades is β_2, and the exit angle has a negative value β_3. For *equiangular* blades $\beta_3 = -\beta_2$. This gives the blades a *bucket* shape because at the design condition the actual metal angles of the blades are close to the flow angles of the relative velocity.

The blades change the direction of the momentum of the flow, and this gives an *impulsive force* to the blades. This is the origin of the name for this turbine, for it appears as if the fluid particles were executing trajectories similar to a ball striking a wall and caused to bounce back by an impulsive force. But if the surface forces on the blades are examined, it is clear that the difference between the high pressure on the concave side of the blade and the low pressure on the convex side is the actual cause of the blade force. A combination of a nozzle row and a rotor row make up a *stage*. For this reason the de Laval turbine is also called a *single-stage impulse turbine*.

Work done on the blades is given by

$$w = h_{02} - h_{03} = U(V_{u2} - V_{u3})$$

and since the relative stagnation enthalpy is constant across the rotor

$$h_2 + \frac{1}{2}W_2^2 = h_3 + \frac{1}{2}W_3^2 \tag{5.1}$$

As seen from the hs diagram in Figure 5.1 irreversibilities cause the static enthalpy to increase from h_2 to h_3 and then Eq. (5.1) shows that W_3 is less than W_2. For equiangular blades, this means that the tangential and axial components of the relative velocity must decrease in the same proportion.

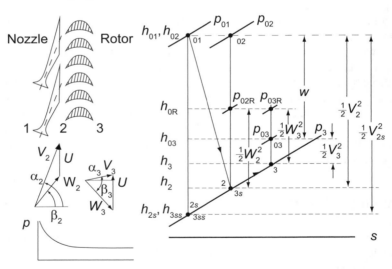

Figure 5.1 Single-stage impulse turbine and its Mollier chart.

It is assumed that steam flows into the nozzles from a steam chest in which velocity is negligibly small. The nozzle efficiency, as shown in the previous chapter, is given by

$$\eta_N = \frac{h_{01} - h_2}{h_{01} - h_{2s}} = \frac{h_{02} - h_2}{h_{02} - h_{2s}}$$

and the second expression follows since $h_{01} = h_{02}$. This can also be written in the form

$$\eta_N = \frac{V_2^2}{V_{2s}^2} = c_N^2$$

in which the velocity coefficient for the nozzle has been defined as $c_N = V_2/V_{2s}$. The loss of stagnation pressure across the nozzles is $\Delta p_{0LN} = p_{01} - p_{02}$. Examination of Figure 5.1 shows that the velocity coefficient is defined such that it represents the loss of kinetic energy in the nozzles.

For the rotor $c_R = W_3/W_{3s}$, but since the state $3s$ is the same as state 2, the velocity coefficient for the rotor is also $c_R = W_3/W_2$. The loss of stagnation pressure is given by $\Delta p_{0LR} = p_{02R} - p_{03R}$. The constant pressure lines p_{02R} and p_{03R} are shown in Figure 5.1. The velocity coefficient c_R has been defined such that the decrease in kinetic energy of the relative velocities represents a thermodynamic loss. The sum of the separate stagnation pressure losses across the nozzles and the rotor gives a different and correct value for the total loss than what was calculated in Chapter 2, where this loss in stagnation pressure was determined for the entire stage.

The efficiency of the rotor is defined as

$$\eta_R = \frac{h_{02} - h_{03}}{h_{02} - h_{3s}}$$

because the kinetic energy leaving the rotor is assumed to be wasted. Since $h_2 = h_{3s}$ this can also be written as

$$\eta_R = \frac{w}{\frac{1}{2} V_2^2}$$

The product

$$\eta_{ts} = \eta_N \eta_R$$

can be written as

$$\eta_{ts} = \frac{h_{02} - h_2}{h_{02} - h_{2s}} \frac{h_{02} - h_{03}}{h_{02} - h_{3s}} = \frac{h_{01} - h_{03}}{h_{01} - h_{3ss}}$$

since $h_2 = h_{3s}$ and $h_{2s} = h_{3ss}$ and the standard definition of the total-to-static efficiency is recovered. With this introduction, the principles learned in Chapter 3 are next applied to a single-stage impulse steam turbine.

■ **EXAMPLE 5.1**

Steam flows from nozzles at the rate 0.2 kg/s and speed 900 m/s. It then enters the rotor of single-stage impulse turbine with equiangular blades. The flow leaves the nozzles at an angle of $70°$, the mean radius of the blades is 120 mm, and the rotor speed is $18,000 \text{ rpm}$. The frictional loss in the rotor blades is 15% of the kinetic energy of the relative motion entering the rotor. (a) Draw the velocity diagrams at the inlet and outlet of the rotor with properly calculated values of the inlet and outlet flow angles for the relative and absolute velocities. (b) Find the power delivered by the turbine. (c) Find the rotor efficiency.

Solution: (a) The blade speed is

$$U = r\Omega = \frac{0.12 \cdot 18{,}000 \cdot 2\pi}{60} = 226.2 \text{ m/s}$$

and the axial velocity is

$$V_{x2} = V_2 \cos \alpha_2 = 900 \cos(70°) = 307.8 \text{ m/s}$$

and the tangential component of the absolute velocity at the inlet to the rotor is

$$V_{u2} = V_2 \sin \alpha_2 = 900 \sin(70°) = 845.7 \text{ m/s}$$

The tangential component of the relative velocity entering the rotor is therefore

$$W_{u2} = V_{u2} - U = 845.7 - 226.2 = 619.5 \,\mathrm{m/s}$$

so that the relative flow speed, since $W_{x2} = V_{x2}$, comes out to be

$$W_2 = \sqrt{W_{2x2}^2 + W_{u2}^2} = \sqrt{307.8^2 + 619.5^2} = 691.8 \,\mathrm{m/s}$$

The inlet angle of the relative velocity is then

$$\beta_2 = \tan^{-1}\left(\frac{W_{u2}}{W_{x2}}\right) = \tan^{-1}\left(\frac{619.5}{307.8}\right) = 63.6°$$

Since 15% of the kinetic energy of the relative motion is lost, at the exit the kinetic energy of the relative flow is

$$\frac{1}{2}W_3^2 = \frac{1}{2}(1 - 0.15)W_2^2 \quad \text{and} \quad W_3 = \sqrt{0.85 \cdot 691.8^2} = 637.8 \,\mathrm{m/s}$$

For equiangular blades, $\beta_3 = -\beta_2$, and the axial velocity leaving the rotor is

$$W_{x3} = W_3 \cos \beta_3 = 637.8 \cos(-63.6°) = 283.8 \,\mathrm{m/s}$$

and the tangential component of the relative velocity is

$$W_{u3} = W_3 \sin \beta_3 = 637.8 \sin(-63.6°) = -571.2 \,\mathrm{m/s}$$

The tangential component of the absolute velocity becomes

$$V_{u3} = U + W_{u3} = 226.2 - 571.2 = -345.0 \,\mathrm{m/s}$$

Hence, with $V_{x3} = W_{x3}$, the flow angle at the exit is

$$\alpha_3 = \tan^{-1}\left(\frac{V_{u3}}{V_{x3}}\right) = \tan^{-1}\left(\frac{-345.0}{283.8}\right) = -50.6°$$

and the velocity leaving the rotor is

$$V_3 = \sqrt{V_{x3}^2 + V_{u3}^2} = \sqrt{283.8^2 + 345^2} = 446.7 \,\mathrm{m/s}$$

The velocity triangles at the inlet and outlet are shown in Figure 5.2.

(b) The specific work done on the blades is obtained as

$$w = U(V_{u2} - V_{u3}) = 226.2(845.7 + 345.0) = 269.3 \,\mathrm{kJ/kg}$$

and the power delivered is

$$\dot{W} = \dot{m}w = 0.2 \cdot 269.3 = 53.87 \,\mathrm{kW}$$

(c) The kinetic energy leaving the nozzle is

$$\frac{1}{2}V_2^2 = \frac{1}{2}900^2 = 405.0 \,\mathrm{kJ/kg}$$

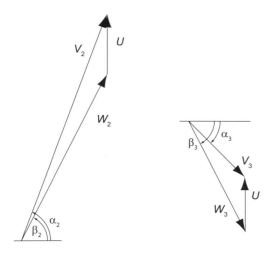

Figure 5.2 Velocity diagrams for a single stage impulse turbine.

and the rotor efficiency therefore becomes

$$\eta_R = \frac{w}{V_2^2/2} = \frac{269.3}{405.0} = 0.665 \quad \text{or} \quad 66.5\%$$

Of the difference $1 - \eta_R$ a fraction is lost as irreversibilities and the rest as kinetic energy leaving the rotor. The latter is obtained by calculating the ratio $V_3^2/2$ to $V_2^2/2$, which is $(446.7/900)^2 = 0.246$. Hence loses from irreversibilities are $1 - 0.665 - 0.246 = 0.089$, or about 9%. It turns out that the blade speed in this example is too low for optimum performance, as will be shown next. ∎

As the theory of turbomachinery advanced, measures more general than the velocity coefficients to account for irreversibility replaced them. A useful measure is the increase in the static enthalpy by internal heating. The loss coefficients ζ_N and ζ_R are defined by the equations

$$h_2 - h_{2s} = \frac{1}{2}\zeta_N V_2^2 \qquad h_3 - h_{3s} = \frac{1}{2}\zeta_R W_3^2 \qquad (5.2)$$

with the thermodynamic states as shown in Figure 5.1. Since the stagnation enthalpy is constant for the flow through the nozzle, it follows that

$$h_{2s} + \frac{1}{2}V_{2s}^2 = h_2 + \frac{1}{2}V_2^2$$

which is rearranged to the form

$$h_2 - h_{2s} = \frac{1}{2}\zeta_N V_2^2 = \frac{1}{2}V_{2s}^2 - \frac{1}{2}V_2^2$$

With $V_{s2} = V_2/c_N$, this equation can be expressed as

$$c_N = \frac{V_2}{V_{2s}} = \frac{1}{\sqrt{1+\zeta_N}} \qquad (5.3)$$

Since the relative stagnation enthalpy is constant for the rotor,

$$h_{3s} + \frac{1}{2}W_2^2 = h_3 + \frac{1}{2}W_3^2$$

and this equation can be rewritten in the form

$$h_3 - h_{3s} = \frac{1}{2}\zeta_R W_3^2 = \frac{1}{2}W_2^2 - \frac{1}{2}W_3^2$$

from which

$$c_R = \frac{W_3}{W_2} = \frac{1}{\sqrt{1+\zeta_R}} \tag{5.4}$$

The stagnation pressure loss across the nozzles is $\Delta p_{0LN} = p_{01} - p_{02}$, and the stagnation pressure loss across the rotor is $\Delta p_{0LR} = p_{02R} - p_{03R}$. The relative stagnation pressures are determined by first calculating the relative stagnation temperatures, which are obtained from

$$T_{02R} = T_2 + \frac{W_2^2}{2c_p} \qquad T_{03R} = T_3 + \frac{W_3^2}{2c_p}$$

and then relative stagnation pressures are determined from

$$\frac{p_{02R}}{p_2} = \left(\frac{T_{02R}}{T_2}\right)^{\gamma/(\gamma-1)} \qquad \frac{p_{03R}}{p_3} = \left(\frac{T_{03R}}{T_3}\right)^{\gamma/(\gamma-1)}$$

To calculate the total-to-static efficiency, from $\eta_{ts} = \eta_N \eta_R$ the nozzle efficiency is first determined from

$$\eta_N = c_N^2 = \frac{1}{1+\zeta_N} \tag{5.5}$$

and then the rotor efficiency needs to be found. This is done by manipulating its definition into the form

$$\frac{1}{\eta_R} - 1 = \frac{h_{03} - h_{3s}}{h_{02} - h_{03}} = \frac{\frac{1}{2}V_3^2 + (h_3 - h_{3s})}{w}$$

which can be recast as

$$\frac{1}{\eta_R} - 1 = \frac{V_3^2 + \zeta_R W_3^2}{2w} \tag{5.6}$$

Then, using

$$V_{u2} = U + W_{u2} \qquad V_{u3} = U + W_{u3}$$

the work done on the blades may be written as

$$w = U(V_{u2} - V_{u3}) = U(W_{u2} - W_{u3}) = U(W_2 \sin\beta_2 - W_3 \sin\beta_3)$$

For equiangular blades $\beta_3 = -\beta_2$ and making use of the relationship $W_3 = W_2/\sqrt{1+\zeta_R}$ this equation takes the form

$$w = UW_2\left(1 + \frac{1}{\sqrt{1+\zeta_R}}\right)\sin\beta_2$$

or, with further substitution of $W_2 \sin\beta_2 = V_2 \sin\alpha_2 - U$, it is

$$w = U\frac{(1+\sqrt{1+\zeta_R})}{\sqrt{1+\zeta_R}}(V_2 \sin\alpha_2 - U) \tag{5.7}$$

The numerator of Eq. (5.6) still needs to be expressed in terms of V_2 and α_2, as was done for work. After the component equations for the velocities

$$W_3 \cos\beta_3 = V_3 \cos\alpha_3 \qquad W_3 \sin\beta_3 + U = V_3 \sin\alpha_3$$

are squared and added, the relationship

$$W_3^2 + 2UW_3 \sin \beta_3 + U^2 = V_3^2$$

is obtained. Making use of Eq. (5.4), the term $V_3^2 + \zeta_R W_3^2$ can be written as

$$V_3^2 + \zeta_R W_3^2 = \frac{W_2^2}{1 + \zeta_R} - \frac{2UW_2 \sin \beta_2}{\sqrt{1 + \zeta_R}} + U^2 + \frac{\zeta_R W_2^2}{1 + \zeta_R}$$

or as

$$V_3^2 + \zeta_R W_3^2 = W_2^2 - \frac{2UW_2 \sin \beta_2}{\sqrt{1 + \zeta_R}} + U^2$$

The relative velocity W_2 is next expressed in terms of V_2 and α_2. Again, the component equations for the definition of relative velocity give

$$W_2 \cos \beta_2 = V_2 \cos \alpha_2 \qquad W_2 \sin \beta_2 = V_2 \sin \alpha_2 - U$$

which, when squared and added, lead to

$$W_2^2 = V_2^2 - 2UV_2 \sin \alpha_2 + U^2$$

Hence the expression for $V_3^2 + \zeta_R W_3^2$ takes the final form

$$V_3^2 + \zeta_R W_3^2 = \frac{(V_2^2 - 2UV_2 \sin \alpha_2 + 2U^2)(1 + \zeta_R) - 2U(V_2 \sin \alpha_2 - U)\sqrt{1 + \zeta_R}}{1 + \zeta_R}$$

The equation for the efficiency can now be written as

$$\frac{1}{\eta_R} - 1 = \frac{(V_2^2 - 2UV_2 \sin \alpha_2 + 2U^2)(1 + \zeta_R) - 2U(V_2 \sin \alpha_2 - U)\sqrt{1 + \zeta_R}}{2U(1 + \zeta_R + \sqrt{1 + \zeta_R})(V_2 \sin \alpha_2 - U)}$$

Introducing the speed ratio $\lambda = U/V_2$ into this gives

$$\frac{1}{\eta_R} - 1 = \frac{(1 - 2\lambda \sin \alpha_2 + 2\lambda^2)(1 + \zeta_R) - 2\lambda(\sin \alpha_2 - \lambda)\sqrt{1 + \zeta_R}}{2\lambda(1 + \zeta_R + \sqrt{1 + \zeta_R})(\sin \alpha_2 - \lambda)}$$

The rotor efficiency now can be expressed as

$$\eta_R = \frac{2\lambda(1 + \zeta_R + \sqrt{1 + \zeta_R})(\sin \alpha_2 - \lambda)}{1 + \zeta_R} \qquad (5.8)$$

and the stage efficiency as

$$\eta_{ts} = \eta_N \eta_R = \frac{2\lambda(1 + \zeta_R + \sqrt{1 + \zeta_R})(\sin \alpha_2 - \lambda)}{(1 + \zeta_N)(1 + \zeta_R)} \qquad (5.9)$$

By making use of Eqs. (5.3) and (5.4) this can also be written as

$$\eta_{ts} = 2\lambda c_N^2 (1 + c_R)(\sin \alpha_2 - \lambda) \qquad (5.10)$$

The blade speed at which the stage efficiency reaches its maximum value is obtained by differentiating this with respect to λ and setting the result to zero. This gives

$$\frac{d\eta_{ts}}{d\lambda} = 2c_N^2 (1 + c_R)(\sin \alpha_2 - 2\lambda) = 0$$

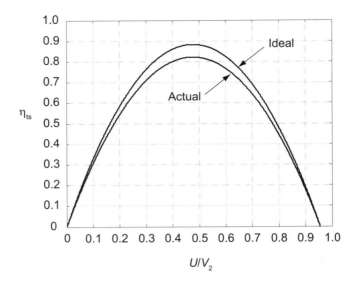

Figure 5.3 Efficiency of a single-stage impulse turbine: ideal and actual with $c_N = 0.979$, $c_R = 0.940$, and $\alpha_2 = 70°$.

and the maximum efficiency is obtained when the speed ratio is

$$\lambda = \frac{U}{V_2} = \frac{1}{2}\sin\alpha_2$$

This equation is independent of the velocity coefficients. For typical nozzle angles, in the range from 65° to 75°, the speed ratio $\lambda = U/V_2$ is about 0.47, so that the blade speed at this optimum condition is about one-half of the exit velocity from the nozzles. The turbine efficiency at this value of U/V_2 is

$$\eta_{ts_{opt}} = \frac{1}{2}c_N^2(1 + c_R)\sin^2\alpha_2$$

so that for $c_N = 0.979$, $c_R = 0.940$, and $\alpha_2 = 70°$, the stage efficiency at the optimal condition is $\eta_{ts_{opt}} = 0.821$. Figure 5.3 shows the stage efficiency for these parameters and for an ideal case, with $c_N = 1$ and $c_R = 1$.

■ **EXAMPLE 5.2**

Steam leaves the nozzles of a single-stage impulse turbine at the speed 900 m/s. Even though the blades are not equiangular, the blade speed is set at the optimum for equiangular blades when the nozzles are at the angle 68°. The velocity coefficient of the nozzles is $c_N = 0.97$, and for the rotor blades it is $c_R = 0.95$. The absolute value of the relative flow angle at the exit of the rotor is 3° greater than the corresponding inlet flow angle. Find (a) the total-to-static efficiency, and (b) find again the total-to-static efficiency of the turbine, assuming that it operates at the same conditions, but has equiangular blades. If one efficiency is higher than the other, explain the reason; if they are the same, give an explanation for this as well.

IMPULSE TURBINES

Solution: (a) With the optimum operating blade speed determined from

$$U = \frac{1}{2}V_2 \sin \alpha_2 = \frac{1}{2} 900 \sin(68°) = 417.2 \, \text{m/s}$$

the velocity components at the exit from the nozzles are

$$V_{x2} = V_2 \cos \alpha_2 = 900 \cos(68°) = 337.2 \, \text{m/s}$$

$$V_{u2} = V_2 \sin \alpha_2 = 900 \sin(68°) = 834.5 \, \text{m/s}$$

the velocity components of the relative velocity are

$$W_{x2} = V_{x2} = 337.2 \, \text{m/s} \qquad W_{u2} = V_{u2} - U = 834.5 - 417.2 = 417.3 \, \text{m/s}$$

The magnitude of the relative velocity is

$$W_2 = \sqrt{W_{x2}^2 + W_{u2}^2} = \sqrt{337.2^2 + 417.3^2} = 536.4 \, \text{m/s}$$

The angle at which the relative velocity enters the rotor row is

$$\beta_2 = \tan^{-1}\left(\frac{W_{u2}}{W_{x2}}\right) = \tan^{-1}\left(\frac{417.3}{337.2}\right) = 51.06°$$

and the exit angle is $\beta_3 = -51.06° - 3° = -54.06°$, and the magnitude of the relative velocity is $W_3 = c_R W_2 = 0.95 \cdot 536.4 = 509.6 \, \text{m/s}$. Work delivered by the rotor is

$$w = U(W_{u2} - W_{u3}) = U(1 + c_R C) W_2 \sin \beta_2$$

in which $C = \sin|\beta_3|/\sin \beta_2 = 1.041$ and $c_R = 0.95$. Therefore the work delivered is $w = 346.23 \, \text{kJ/kg}$. Since the exit kinetic energy is wasted, its value is needed. The exit velocity components are

$$V_{x3} = W_{x3} = W_3 \cos \beta_3 = 509.6 \cos(-54.06°) = 299.1 \, \text{m/s}$$

$$V_{u3} = W_3 \sin \beta_3 + U = 509.6 \sin(-54.06°) + 417.2 = 4.6 \, \text{m/s}$$

Hence

$$V_3 = \sqrt{V_{x3}^2 + V_{u3}^2} = \sqrt{299.1^2 + 4.6^2} = 299.1 \, \text{m/s}$$

In the calculation of rotor efficiency the rotor loss coefficient is $\zeta_R = 1/c_R^2 - 1 = 0.208$, so that

$$\eta_R = \frac{2w}{2w + V_3^2 + \zeta_R W_3^2} = \frac{2 \cdot 346,230}{2 \cdot 346,230 + 299.1^2 \cdot 0.1026 \cdot 509.6^2} = 0.855$$

and since $\eta_N = c_N^2 = 0.97^2 = 0.941$ the total-to-static efficiency becomes

$$\eta_{ts} = \eta_N \eta_R = 0.941 \cdot 0.855 = 0.805$$

(b) If the blades were equiangular and the turbine were to operate at its optimal condition, the total-to-static efficiency would be

$$\eta_{ts} = \frac{1}{2}c_N^2(1 + c_R)\sin^2 \alpha_2 = 0.789$$

The loss of efficiency for equiangular blades is caused by the exit kinetic energy now being larger than before. When the blades are not equiangular, even when the turbine is not operated at its optimum blade speed, it has a high efficiency because the exit velocity is nearly axial and the turbine has a high utilization. For equiangular blades the exit flow angle is slightly larger and therefore the flow is faster, with more of the kinetic energy leaving the stage.

■

5.2.2 Pressure compounding

The optimum blade speed for a single stage impulse turbine is about one half the exit velocity from the nozzles. Such a high blade speed requires a high shaft speed, which may lead to large blade stresses. To reduce the shaft speed, two or more single-stage impulse turbines are arranged in series and the steam is then expanded partially in each of the set of nozzles. This decreases the velocity from the nozzles and thus the blade speed for optimal performance. This arrangement, shown in Figure 5.4, is called *pressure compounding*, or *Rateau staging*, after Auguste Camille Edmond Rateau (1863–1930) of France. Between any two rotors there is a nozzle row. The pressure drop takes place in the nozzles and none across the rotor. As the steam expands, its specific volume increases and a larger flow area is needed in order to keep the increase in velocity moderate. One approach is to keep the mean radius of the wheel constant and to increase the blade height. When this is done the blade speed at the mean radius remains the same for all stages and the velocities leaving and entering a stage can be made equal. Such a stage is called a *repeating stage*, or a *normal stage*.

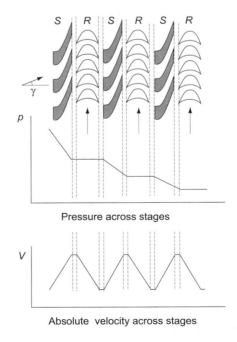

Figure 5.4 Sketch of a multistage pressure-compounded impulse turbine and the pressure drop and velocity variation across each stage.

Consider a multistage pressure-compounded impulse turbine with repeating stages. Unlike in the single-stage turbine, the flow now enters the second set of nozzles at a velocity close to the exit velocity from the preceding stage, and the function of the nozzle is to increase it further. The process lines are shown in Figure 5.5. The stagnation states 03,

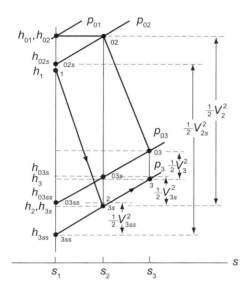

Figure 5.5 The process lines for a pressure-compounded impulse stage.

03s, and 03ss are reached from the static states 3, 3s, and 3ss, which are on the constant pressure line p_3. As was discussed earlier, it does not necessarily follow that the states 03s and 03ss are on the constant-pressure line p_{03} since the magnitudes of the velocities V_{3s} and V_{3ss} in

$$h_{03s} = h_{3s} + \frac{1}{2}V_{3s}^2 \qquad h_{03ss} = h_{3ss} + \frac{1}{2}V_{3ss}^2$$

are not known. However, a consistent theory can be developed if it is *assumed* that the stagnation states 03s and 03ss are on the constant pressure line p_{03} and their thermodynamic states are then fixed by the known value of pressure and entropy. The previous equations then fix the magnitudes of V_{3s} and V_{3ss} to definite values.

With this assumption the Gibbs equation

$$T\,ds = dh - v\,dp \qquad ds = c_p \frac{dT}{T} - R\frac{dp}{p}$$

when integrated between states 03s and 03 and then between states 03ss and 03s along the constant pressure line p_{03} and similarly along the corresponding states along the constant pressure line p_3 give

$$s_2 - s_1 = c_p \ln \frac{T_{03s}}{T_{03ss}} \qquad s_3 - s_2 = c_p \ln \frac{T_{03}}{T_{03s}}$$

and

$$s_2 - s_1 = c_p \ln \frac{T_{3s}}{T_{3ss}} \qquad s_3 - s_2 = c_p \ln \frac{T_3}{T_{3s}}$$

from which

$$\frac{T_{03}}{T_{03s}} = \frac{T_3}{T_{3s}} \quad \text{and} \quad \frac{T_{03s}}{T_{03ss}} = \frac{T_{3s}}{T_{3ss}} \quad (5.11)$$

and from these it follows that

$$\frac{T_{03}}{T_3} = \frac{T_{03s}}{T_{3s}} = \frac{T_{03ss}}{T_{3ss}}$$

Expressing these temperature ratios in terms of Mach numbers yields

$$\frac{T_{03}}{T_3} = 1 + \frac{\gamma - 1}{2} M_3^2 \quad \frac{T_{03s}}{T_{3s}} = 1 + \frac{\gamma - 1}{2} M_{3s}^2 \quad \frac{T_{03ss}}{T_{3ss}} = 1 + \frac{\gamma - 1}{2} M_{3ss}^2 \quad (5.12)$$

so that $M_3 = M_{3s} = M_{3ss}$.

The stage efficiency for a pressure-compounded stage is the total-to-total efficiency

$$\eta_{tt} = \frac{h_{01} - h_{03}}{h_{01} - h_{03ss}}$$

which can be recast into the form

$$\frac{1}{\eta_{tt}} - 1 = \frac{h_{03} - h_{03ss}}{h_{01} - h_{03}} = \frac{h_{03} - h_{03s} + h_{03s} - h_{03ss}}{h_{01} - h_{03}} \quad (5.13)$$

Subtracting one from each side of both Eqs. (5.11), multiplying by c_p, and rearranging gives

$$h_{03} - h_{03s} = \frac{T_{03s}}{T_{3s}}(h_3 - h_{3s}) \quad h_{03s} - h_{03ss} = \frac{T_{03ss}}{T_{3ss}}(h_{3s} - h_{3ss})$$

Substituting these into Eq. (5.13) leads to

$$\frac{1}{\eta_{tt}} - 1 = \frac{\frac{T_{03s}}{T_{3s}}(h_3 - h_{3s}) + \frac{T_{03ss}}{T_{3ss}}(h_{3s} - h_{3ss})}{w} \quad (5.14)$$

or, since the temperature ratios are the same, according to Eqs. (5.12), this expression for efficiency becomes

$$\frac{1}{\eta_{tt}} - 1 = \frac{\left(1 + \frac{\gamma - 1}{2} M_3^2\right)(h_3 - h_{3s} + h_2 - h_{2s})}{w} \quad (5.15)$$

The Mach number at the exit of the rotor is quite low and is often set to zero. It is, of course, easily determined once the work and exit velocity have been calculated. Then

$$T_{03} = T_{01} - \frac{w}{c_p} \quad T_3 = T_{03} - \frac{V_3^2}{2c_p} \quad \text{and} \quad M_3 = \frac{V_3}{\sqrt{\gamma R T_3}}$$

Ignoring the Mach number and introducing the static enthalpy loss coefficients gives

$$\frac{1}{\eta_{tt}} - 1 = \frac{\zeta_R W_3^2 + \zeta_N V_2^2}{2w} \quad (5.16)$$

The relationship $W_3^2 = c_R^2 W_2^2$ may be substituted into this, giving

$$\frac{1}{\eta_{tt}} - 1 = \frac{\zeta_R c_R^2 W_2^2 + \zeta_N V_2^2}{2w}$$

or

$$\eta_{tt} = \frac{2w}{\zeta_R c_R^2 W_2 + \zeta_N V_2^2 + 2w} \quad (5.17)$$

The relative velocity W_2 can be written as

$$W_2^2 = V_2^2 - 2V_2 U \sin \alpha_2 + U^2$$

so that

$$\frac{1}{\eta_{tt}} - 1 = \frac{\zeta_R c_R^2 (V_2^2 - 2V_2 U \sin \alpha_2 + U^2) + \zeta_N V_2^2}{2w}$$

The work delivered by the stage is

$$w = U(V_{u2} - V_{03}) = U(W_{u2} - W_{u3}) = U(1 + c_R)W_{u2} = U(1 + c_R)(V_2 \sin \alpha_2 - U)$$

Substituting $\lambda = U/V_2$ into the previous expression for the stage efficiency, it takes the form

$$\frac{1}{\eta_{tt}} - 1 = \frac{\zeta_R c_R^2 (1 - 2\lambda \sin \alpha_2 + \lambda^2) + \zeta_N}{2\lambda(1 + c_R)(\sin \alpha_2 - \lambda)}$$

Defining the quantity f_L as

$$f_L = \frac{\lambda^2 - 2\lambda \sin \alpha_2 + 1 + \zeta_N (1 + \zeta_R)/\zeta_R}{2(1 + \zeta_R + \sqrt{1 + \zeta_R}/\zeta_R)(\lambda \sin \alpha_2 - \lambda^2)}$$

in which the relation $c_R = 1/\sqrt{1 + \zeta_R}$ has been used, the stage efficiency can now be written as

$$\eta_{tt} = \frac{1}{1 + f_L}$$

from which it is clear that f_L is a measure of the losses. The maximum value for the efficiency of a pressure-compounded stage is obtained by minimizing f_L. Thus differentiating it with respect to λ and setting the result to zero gives

$$\lambda^2 - \frac{2(\zeta_R + \zeta_N(1 + \zeta_R))}{\zeta_R \sin \alpha_2} \lambda + \frac{\zeta_R + \zeta_N(1 + \zeta_R)}{\zeta_R} = 0$$

Of the two roots

$$\lambda_{opt} = \frac{U}{V_2} = \frac{\zeta_R + \zeta_N(1 + \zeta_R) + \sqrt{(\zeta_R + \zeta_N(1 + \zeta_R))(\zeta_R + \zeta_N(1 + \zeta_R) + \zeta_R \sin^2 \alpha_2)}}{\zeta_R \sin \alpha_2}$$

is the correct one, as the second root leads to U/V_2 ratio greater than unity. This would mean that the wheel moves faster than the approaching steam. The stage efficiency now can be written as

$$\eta_{tt} = \frac{2\lambda(1 + c_R)(\sin \alpha_2 - \lambda)}{2\lambda(1 + c_R)(\sin \alpha_2 - \lambda) + \zeta_R c_R^2 (\lambda^2 - 2\lambda \sin \alpha_2 + 1) + \zeta_N} \quad (5.18)$$

The stage efficiencies for various nozzle angles are shown in Figure 5.6 for both a pressure-compounded stage and for a single-stage impulse turbine with exit kinetic energy wasted. The efficiency curves for the pressure-compounded stage are quite flat at the top and naturally higher as the exit kinetic energy is used at the inlet of the next stage.

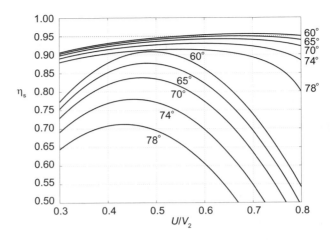

Figure 5.6 Stage efficiencies of single stage impulse turbines with nozzle angles in the range from $60°$ to $78°$ with $\zeta_N = 0.02$ and $\zeta_R = 0.14$; the exit kinetic energy is wasted for the set of graphs with lower efficiency, and the family of graphs of higher efficiency are total-to-total efficiencies applicable to a pressure-compounded turbine stage.

5.2.3 Blade shapes

Some details of the construction of impulse blades are considered next. The equiangular blades are shown in the sketch Figure 5.7. The concave side of the blade is circular, drawn with its center a distance $c \cot \beta_2 / 2$ below the midchord point. Here c is the length of the chord and the radius of the circular arc is given by

$$R = \frac{c}{2 \sin \beta_2}$$

To establish the geometric dimensions of the blade, a line segment of length equal to the interblade spacing is marked off from the center along the line of symmetry. This point becomes the center of the circular arc of the concave side of blade j. The convex side of the blade i consists of a circular arc that is drawn from the same origin and its extent is such that a straight-line segment in the direction of the blade angle meets the exit at a location that gives the correct spacing to the blades; that is, the radius of the arc is chosen such that this line segment is tangent to the arc at point a. This point is chosen at the location of the intersection of a perpendicular from the trailing edge of blade j to this line segment. The blade at the inlet is made quite sharp, and at the outlet the blade may also have a straight segment extending past the conventional exit plane. In a multistage turbine the extent of the straight segment controls the spacing between the exit of the rotor and the inlet to the next set of nozzles. These nozzles are usually designed to have an axial entry. If the turbine operates at design the conditions and the absolute velocity at the exit is axial, then the steam flows smoothly into these nozzles. At off-design conditions, the flow angle at the entry will not match the metal angle of the nozzles, leading to increased *losses in the nozzles*, particularly for blades with sharp edges. In order to improve steam turbine's operation at a fractional flow rate, absolute values of the flow angles, at both inlet and exit, are made larger by $2°$ or $3°$ and in a multistage turbine for the blades next to the last stage this may be $4°$ or $5°$. For the last stage the range from $5°$ to $10°$ is used [46].

Impulse blading is designed to ensure equal pressures at the inlet and exit of the rotor. However, owing to irreversibilities, temperature increases across the rotor, and this causes the specific volume to increase. Since mass flow rate is constant, mass balance gives

$$\dot{m} = \frac{A_2 W_2 \cos \beta_2}{v_2} = \frac{A_3 W_3 \cos \beta_3}{v_3}$$

in which $V_{x2} = W_2 \cos \beta_2$ and $V_{x3} = W_3 \cos \beta_3$ have been used. Since $\beta_3 = -\beta_2$ and $W_3 < W_2$, then with $v_3 > v_2$, the flow area has to be increased. This is done by increasing the height of the blade. However, it is also possible to alter the exit angle as was mentioned above.

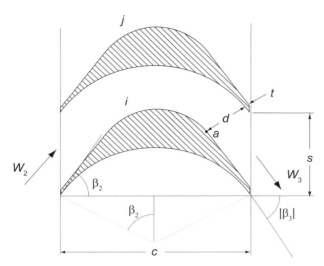

Figure 5.7 Equiangular bucket blade shape. (Modified from Kearton [46].)

For blades that are not equiangular the absolute value of the outlet blade angle in most impulse turbines is larger than the inlet angle. For them the radius of the concave surface of the blade is given by

$$R = \frac{c}{\sin \beta_2 + \sin |\beta_3|}$$

The offset between the leading and trailing edges in this case is

$$x = R(\cos \beta_2 - \cos |\beta_3|)$$

With $|\beta_3| = \beta_2 + 3°$ the bisector of the blade profile will lean to the right, as shown in Figure 5.8. The channel width at the exit is given by $d = b(s \cos |\beta_3| - t)$, in which t is the trailing edge blade thickness. For a flow with mass flow rate $\dot{m}$ and specific volume v_3 mass balance gives

$$\dot{m} v_3 = W_3 Z b (s \cos |\beta_3| - t) \tag{5.19}$$

in which Z is the number of blades in the rotor. For a given spacing of the blades, their thickness and number, and for a specified mass flow rate and exit specific volume, this equations shows that only two of the three parameters: blade height b, relative velocity W_3, and flow angle $|\beta_3|$, may be chosen independently. If it is possible to accommodate the increase in specific volume in the downstream direction by an increase in the blade height, then this equation shows that increasing $|\beta_3|$ decreases the channel width d, and this leads to an increase in the relative velocity W_3. Equation (5.1) then shows that since the trothalpy is constant, the static enthalpy decreases. A drop in static enthalpy along the flow is associated with a drop of pressure, as process lines on a Mollier diagram show. The

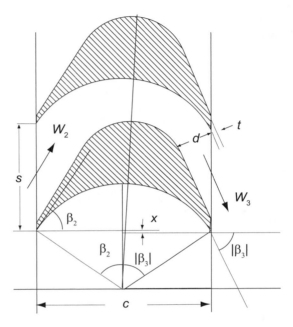

Figure 5.8 Bucket construction details for unequal blade angles. (Modified from Kearton [46].)

acceleration of the flow increases the force on the blade, for by the momentum principle

$$F_u = \dot{m}(V_{u2} - V_{u3}) = \dot{m}(W_{u2} - W_{u3})$$

and since W_{u3} is negative, an increase in its magnitude increases the force component F_u. It has become conventional to call this additional force a *reaction force* in analogy to the thrust force given to a rocket being a reaction to the exit momentum leaving the rocket nozzle. Reaction turbines are discussed more fully in the next chapter.

5.2.4 Velocity compounding

A second way of *compounding* a turbine was developed by the American Charles Gordon Curtis (1860–1953). In his design steam first enters an impulse stage, and as it leaves this stage, it enters a stator row of equiangular vanes. They redirect it to the second rotor row of equiangular blades, but of course with a different magnitude for their angles than in the first row. All the pressure drop takes place in the upstream nozzles, and thus no further reduction of pressure takes place as the steam moves through the downstream stages. There are practical reasons for not fitting the turbine by more than four stages. Namely, work done by later stages drops rapidly. In this kind of Curtis staging, velocity is said to be compounded from one stage to the next.

Consider an n-stage Curtis turbine with equal velocity coefficients c_v for each blade row. Analysis of the first stage is the same as for a single-stage impulse turbine. Work delivered is
$$w_{1n} = U(V_{u2} - V_{u3}) = U(W_{u2} - W_{u3})$$
As was shown above for equiangular blades, $W_{u3} = -c_v W_{u2}$, and work delivered by the first stage may be written as
$$w_{1n} = U(1 + c_v)W_{u2}$$
In the same way work delivered by the second and third stages are
$$w_{2n} = U(1 + c_v)W_{u4}$$
and
$$w_{3n} = U(1 + c_v)W_{u6}$$
If the relative velocity W_{u4} is related to W_{u2} and W_{u6} is related to W_{u4}, work from each stage can be expressed in terms of W_{u2}. With $V_{u4} = -c_v V_{u3}$ for equiangular stator blades, W_{u4} can be written as
$$W_{u4} = V_{u4} - U = -c_v V_{u3} - U = -c_v(W_{u3} + U) - U = -c_v(-c_v W_{u2} + U) - U$$
in which $W_{u3} = -c_v W_{u2}$ has been used. Hence the final result is
$$W_{u4} = c_v^2 W_{u2} - (1 + c_v)U$$
Similarly
$$W_{u6} = c_v^2 W_{u4} - (1 + c_v)U$$
Substituting W_{u4} from the previous expression into this gives
$$W_{u6} = c_v^4 W_{u2} - (1 + c_v)(1 + c_v^2)U$$
Work delivered by each of the three rotors is then
$$w_{1n} = U(1 + c_v)(V_2 \sin \alpha_2 - U)$$
$$w_{2n} = U(1 + c_v)c_v^2(V_2 \sin \alpha_2 - U) - (1 + c_v)(1 + c_v)U^2$$
$$w_{3n} = U(1 + c_v)c_v^4(V_2 \sin \alpha_2 - U) - (1 + c_v)(1 + c_v + c_v^2 + c_v^3)U^2$$
Work delivered by the next stage is easily shown to be
$$w_{4n} = U(1 + c_v)c_v^6(V_2 \sin \alpha_2 - U) - (1 + c_v)(1 + c_v + c_v^2 + c_v^3 + c_v^4 + c_v^5)U^2$$
Inspection of these shows that work delivered by the nth stage is[4]
$$w_{nn} = U(1 + c_v)c_v^{2n-2}(V_2 \sin \alpha_2 - U) - (1 + c_v)\sum_{i=1}^{2n-2} c_v^{i-1}U^2$$
which can also be written as
$$w_{nn} = U(1 + c_v)c_v^{2n-2}(V_2 \sin \alpha_2 - U) - \left(\frac{1 + c_v}{1 - c_v}\right)(1 - c_v^{2n-2})U^2 \qquad (5.20)$$

[4]That this conjecture is true can be shown by first proving by mathematical induction the three term recurrence relationship $w_{i+2,n} - (1 + c_v^2)w_{i+1,n} + c_v^2 w_{i,n} = 0$ between the stages and solving this difference equation.

Optimum operating conditions are now developed for turbines with different numbers of wheels. Work delivered by the single wheel of single-stage turbine is given by

$$w_1 = U(1 + c_v)(V_2 \sin \alpha_2 - U)$$

and the optimum blade speed was shown in the beginning of this chapter to be

$$\frac{U}{V_2} = \frac{1}{2} \sin \alpha_2$$

At the optimum speed work from a single-stage turbine is

$$w_1 = \frac{1 + c_v}{4} V_2^2 \sin^2 \alpha_2$$

and if $c_v = 1$, this is

$$w_1 = \frac{1}{2} V_2^2 \sin^2 \alpha_2$$

Work delivered by a two-wheel Curtis turbine is, $w_2 = w_{12} + w_{22}$, or

$$w_2 = U(1 + c_v)(1 + c_v)(V_2 \sin \alpha_2 - U) - (1 + c_v)(1 + c_v)U^2$$

and when this is differentiated with respect to U and the result is set to zero, the optimum blade speed is found to be

$$\frac{U}{V_2} = \frac{(1 + c_v^2) \sin \alpha_2}{2(2 + c_v + c_v^2)}$$

For $c_v = 1$ this reduces to

$$\frac{U}{V_2} = \frac{1}{4} \sin \alpha_2$$

Hence a two-wheel Curtis turbine can be operated at *about one-half* the shaft speed of a single-stage impulse turbine. At the optimum speed, work delivered by a two-wheel turbine is as follows:

$$w_2 = \frac{(1 + c_v)(1 + c_v^2)^2}{4(2 + c_v + c_v^2)} V_2^2 \sin^2 \alpha_2$$

For $c_v = 1$ this reduces to

$$w_2 = \frac{1}{2} V_2^2 \sin^2 \alpha_2$$

which is *exactly the same* as in a single-stage turbine. For a three-wheel Curtis turbine work delivered is $w_3 = w_{13} + w_{23} + w_{33}$, and the expression for the work, when written in full, is

$$w_3 = U(1+c_v)(1+c_v^2+c_v^4)(V_2 \sin \alpha_2 - U) - (1+c_v)(1+c_v+c_v^2+c_v^3)U^2 - (1+c_v)(1+c_v)U^2$$

Differentiating this to determine the value of blade speed for which work is maximum gives

$$\frac{U}{V_2} = \frac{(1 + c_v^2 + c_v^4) \sin \alpha_2}{2[(1 + c_v^2 + c_v^4) + (1 + c_v + c_v^2 + c_v^3) + (1 + c_v)]}$$

and for $c_v = 1$ this reduces to

$$\frac{U}{V_2} = \frac{1}{6} \sin \alpha_2$$

and the work delivered at this speed is

$$w_3 = \frac{(1+c_v)(1+c_v^2+c_v^4)^2}{4(3+2c_v+2c_v^2+c_v^3+c_v^4)}V_2^2\sin^2\alpha_2$$

Finally, the optimum blade speed for a four-wheel turbine is

$$\frac{U}{V_2} = \frac{(1+c_v^2+c_v^4+c_v^6)\sin\alpha_2}{2[(1+c_v^2+c_v^4+c_v^6)+(1+c_v+c_v^2+c_v^3+c_v^4+c_v^5)+(1+c_v+c_v^2+c_v^3)+(1+c_v)]}$$

and work delivered at this speed is

$$w_4 = \frac{(1+c_v)(1+c_v^2+c_v^4+c_v^6)^2}{4(4+3c_v+3c_v^2+2c_v^3+2c_v^4+c_v^5+c_v^6)}V_2^2\sin^2\alpha_2$$

Although velocity compounding with four wheels have been built in the past, they are no longer in use.

If $c_v = 1$ for a two wheel turbine the ratio of the work done is $3:1$ between the first and second stage. If further stages are included, the work ratios become $5:3:1$ and $7:5:3:1$ for three and four stage turbines respectively, and the optimum blade speeds drop to $U = \sin\alpha_2 V_2/6$ and $U = \sin\alpha_2 V_2/8$. As has been shown, addition of successive stages does not increase the amount of work delivered by the turbine in the ideal case, and its advantage lies entirely in the reduction of the shaft speed. When irreversibilites are taken into account turbines with multiple stages deliver less work than does a single-stage impulse turbine.

■ **EXAMPLE 5.3**

Consider a velocity-compounded two-stage steam turbine. The velocity at the inlet to the nozzle is axial and it leaves the nozzle with speed $V_2 = 850\,\text{m/s}$ at angle $\alpha_2 = 67°$. The blade speed is $U = 195.6\,\text{m/s}$. The velocity coefficient for the nozzle is $c_N = 0.967$ and for the rotors they are $c_{R1} = 0.939$ and $c_{R2} = 0.971$. For the stator between the rotors it is $c_S = 0.954$. The rotors and the stator are equiangular. Find the efficiency of the turbine.

Solution: The axial and tangential velocity components are

$$V_{x2} = V_2\cos\alpha_2 = 850\cos(67°) = 332.1\,\text{m/s}$$

$$V_{u2} = V_2\sin\alpha_2 = 850\sin(67°) = 782.4\,\text{m/s}$$

The relative velocity components are $W_{x2} = V_{x2} = 332.1\,\text{m/s}$ and

$$W_{u2} = V_{u2} - U = 782.4 - 195.6 = 586.8\,\text{m/s}$$

so that

$$W_2 = \sqrt{W_{x2}^2 + W_{u2}^2} = \sqrt{332.1^2 + 586.8^2} = 674.3\,\text{m/s}$$

and the flow angle becomes

$$\beta_2 = \tan^{-1}\left(\frac{W_{u2}}{W_{x2}}\right) = \tan^{-1}\left(\frac{586.8}{332.1}\right) = 60.49°$$

The flow angle of the relative velocity leaving the first rotor is $\beta_3 = -60.49°$, and its relative velocity is

$$W_3 = c_{R1} W_2 = 0.939 \cdot 674.3 = 633.2 \, \text{m/s}$$

so that its components are

$$W_{x3} = W_3 \cos \beta_3 = 633.2 \cos(-60.49°) = 311.9 \, \text{m/s}$$

$$W_{u3} = W_3 \sin \beta_3 = 633.2 \sin(-60.49°) = -551.0 \, \text{m/s}$$

The axial component of the absolute velocity entering the stator is $V_{x3} = W_{x3} = 311.9 \, \text{m/s}$, and its tangential component is

$$V_{u3} = W_{u3} + U = -551.0 + 195.6 = -355.4 \, \text{m/s}$$

Therefore

$$V_3 = \sqrt{V_{x3}^2 + V_{u3}^2} = \sqrt{311.9^2 + 355.4^2} = 472.9 \, \text{m/s}$$

and the flow angle is

$$\alpha_3 = \tan^{-1}\left(\frac{V_{u3}}{V_{x3}}\right) = \tan^{-1}\left(\frac{-355.4}{311.9}\right) = -48.74°$$

The flow angle leaving the stator is $\alpha_4 = -\alpha_3 = 48.74°$ and the magnitude of the velocity is

$$V_4 = c_S V_3 = 0.954 \cdot 472.9 = 451.1 \, \text{m/s}$$

The components are

$$V_{x4} = V_4 \cos \alpha_4 = 451.1 \cos(48.74°) = 297.5 \, \text{m/s}$$

$$V_{u4} = V_4 \sin \alpha_4 = 451.1 \sin(48.74°) = 339.1 \, \text{m/s}$$

The axial component of the relative velocity is $W_{x4} = V_{x4} = 297.5 \, \text{m/s}$, and its tangential component is

$$W_{u4} = V_{u4} - U = 339.1 - 195.6 = 143.5 \, \text{m/s}$$

so that

$$W_4 = \sqrt{W_{x4}^2 + W_{u4}^2} = \sqrt{297.5^2 + 143.5^2} = 330.3 \, \text{m/s}$$

and the flow angle is

$$\beta_4 = \tan^{-1}\left(\frac{W_{u4}}{W_{x4}}\right) = \tan^{-1}\left(\frac{143.5}{297.5}\right) = 25.75°$$

At the inlet of the second rotor relative velocity is at the angle $\beta_5 = -\beta_4 = -25.75°$, and its relative velocity is

$$W_5 = c_{R2} W_4 = 0.971 \cdot 330.3 = 320.7 \, \text{m/s}$$

so that its components are

$$W_{x5} = W_5 \cos \beta_5 = 320.7 \cos(-25.75°) = 288.9 \, \text{m/s}$$

$$W_{u5} = W_5 \sin \beta_5 = 320.7 \sin(-25.75°) = -139.3 \, \text{m/s}$$

The axial component of the absolute velocity leaving the second rotor is $V_{x5} = W_{x5} = 288.89 \, \text{m/s}$, and its tangential component is

$$V_{u5} = W_{u5} + U = -139.3 + 195.6 = 56.3 \, \text{m/s}$$

For the exit velocity, this gives the value

$$V_5 = \sqrt{V_{x3}^2 + V_{u3}^2} = \sqrt{288.9^2 + 56.3^2} = 294.3 \, \text{m/s}$$

and the flow angle is

$$\alpha_5 = \tan^{-1}\left(\frac{V_{u5}}{V_{x5}}\right) = \tan^{-1}\left(\frac{56.3}{288.9}\right) = 11.03°$$

Work delivered by the two stages are

$$w_{12} = U(V_{u2} - V_{u3}) = 195.6(782.4 + 355.4) = 222.6 \, \text{kJ/kg}$$

$$w_{22} = U(V_{u4} - V_{u5}) = 195.6(339.1 - 56.3) = 55.3 \, \text{kJ/kg}$$

so the total work is

$$w_2 = w_{12} + w_{22} = 222.6 + 55.3 = 277.9 \, \text{kJ/kg}$$

If all the velocity coefficients had been equal to $c_v = 0.96$, the work would have been

$$w_2 = \frac{(1+c_c)(1+c_v)^2}{4(2+c_v+c_v^2)} V_s \sin^2\alpha_2 = 285.4 \, \text{kJ/kg}$$

The total-to-total efficiency is

$$\frac{1}{\eta_{tt}} - 1 = \frac{V_5^2 + \zeta_{R2}W_5^2 + \zeta_S V_4^2 + \zeta_{R1}W_3^2 + \zeta_N V_2^2}{2w}$$

With

$$\zeta_N = \frac{1}{c_N^2} - 1 = \frac{1}{0.967^2} - 1 = 0.0694 \qquad \zeta_S = \frac{1}{c_S^2} - 1 = \frac{1}{0.954^2} - 1 = 0.0988$$

and

$$\zeta_{R1} = \frac{1}{c_{R1}^2} - 1 = \frac{1}{0.939^2} - 1 = 0.1341 \qquad \zeta_{R2} = \frac{1}{c_{R2}^2} - 1 = \frac{1}{0.971^2} - 1 = 0.0606$$

the reciprocal of efficiency is

$$\frac{1}{\eta_{tt}} = 1 + \frac{294.3^2 + 0.0606 \cdot 320.7^2 + 0.0988 \cdot 451.1^2 + 0.1341 \cdot 633.2^2 + 0.0694 \cdot 850^2}{2 \cdot 277,890}$$

$$= 1.390$$

so that $\eta_{tt} = 0.719$.

5.3 STAGE WITH ZERO REACTION

A stage design that is closely related to the impulse stage is one with zero reaction. As Eq. (4.18) shows, for such a stage $W_3 = W_2$, and since trothalpy does not change across the rotor, neither does the static enthalpy. If the axial velocity is constant, then the blades need to be equiangular with $\beta_3 = -\beta_2$. The processes lines are shown in Figure 5.9. If the

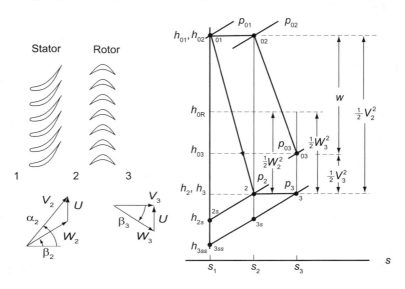

Figure 5.9 Process lines for a turbine with 0% reaction.

exit kinetic energy is wasted, the stage efficiency is the total-to-static efficiency:

$$\eta_{ts} = \frac{h_{01} - h_{03}}{h_{01} - h_{3ss}}$$

This is now rewritten in the form

$$\frac{1}{\eta_{ts}} - 1 = \frac{h_{03} - h_{3ss}}{h_{01} - h_{03}} = \frac{\frac{1}{2}V_3^2 + h_3 - h_{3ss}}{h_{01} - h_{03}}$$

which can be recast further as

$$\frac{1}{\eta_{ts}} - 1 = \frac{\frac{1}{2}V_3^2 + h_3 - h_{3s} + h_{3s} - h_{3ss}}{h_{01} - h_{03}} = \frac{V_3^2 + \zeta_R W_3^2 + \zeta_N V_2^2}{2w}$$

The work delivered by the stage is

$$w = U(V_{u2} - V_{u3}) = U(W_{u2} - W_{u3}) = 2UW_{u2} = 2U(V_2 \sin\alpha_2 - U)$$

As before, the component equations for velocities are

$$W_3 \cos\beta_3 = V_3 \cos\alpha_3 \qquad W_3 \sin\beta_3 + U = V_3 \sin\alpha_2$$

and squaring and adding them gives

$$V_3^2 = W_3^2 + 2UW_3 \sin\beta_3 + U^2$$

Since $W_3 = W_2$ and $\beta_3 = -\beta_2$, this can be written as
$$V_3^2 = W_2^2 - 2UW_2 \sin \beta_2 + U^2$$

Similarly
$$W_2 \cos \beta_2 = V_2 \cos \alpha_2 \qquad W_2 \sin \beta_2 = V_2 \sin \alpha_2 - U$$
which when squared and added give
$$W_2^2 = V_2^2 - 2UV_2 \sin \alpha_2 + U^2$$

When these are included in the expression for efficiency, it takes the form
$$\frac{1}{\eta_{ts}} - 1 = \frac{V_2^2 - 4UV_2 \sin \alpha_2 + 4U^2 + \zeta_R(V_2^2 - 2UV_2 \sin \alpha_2 + U^2) + \zeta_N V_2}{4U(V_2 \sin \alpha - U)}$$

This can be written as
$$\eta_{ts} = \frac{1}{1 + f_L}$$
in which, after the substitution $\lambda = U/V_2$ has been made, f_L is given by
$$f_L = \frac{1 - 4\lambda \sin \alpha_2 + 4\lambda^2 + \zeta_R(1 - 2\lambda \sin \alpha_2 + \lambda^2) + \zeta_N}{4\lambda(\sin \alpha - \lambda)}$$

The maximum efficiency is obtained by minimizing the loss f_L. Thus differentiating f_L with respect to λ and setting it to zero yields
$$\lambda^2 - \frac{2(1 + \zeta_R + \zeta_N)}{\zeta_R \sin \alpha_2}\lambda + \frac{1 + \zeta_R + \zeta_N}{\zeta_R} = 0$$
and the maximum efficiency is at the speed ratio
$$\lambda = \frac{1 + \zeta_R + \zeta_N - \sqrt{(1 + \zeta_R + \zeta_N)(1 + \zeta_R + \zeta_N - \zeta_R \sin^2 \alpha_2)}}{\zeta_R \sin \alpha_2}$$

The efficiency may be written as
$$\eta_{ts} = \frac{4\lambda(\sin \alpha_2 - \lambda)}{1 + \zeta_R(\lambda^2 - 2\lambda \sin \alpha_2 + 1) + \zeta_N}$$

These results are shown as the lower set of curves in Figure 5.10. The efficiencies of a zero reaction stage for various nozzle angles are slightly lower than those for the pressure-compounded impulse stage shown in Figure 5.6. The graphs for a 0% *repeating stage* are also shown. Both are denoted by η_s, which is to interpreted appropriately, either as repeating stage, or as single stage with kinetic energy wasted.

The efficiency of a repeating stage is obtained from
$$\eta_{tt} = \frac{4\lambda(\sin \alpha_2 - \lambda)}{(\zeta_R - 4)\lambda^2 + (4 - 2\zeta_R)\lambda \sin \alpha_2 + \zeta_R + \zeta_N}$$
with maxima at speed ratios
$$\lambda = \frac{\zeta_R + \zeta_S - \sqrt{(\zeta_R + \zeta_N)(\zeta_R + \zeta_S - \zeta_R \sin^2 \alpha_2)}}{\zeta_R \sin \alpha_2}$$

These are left to be worked out as an exercise.

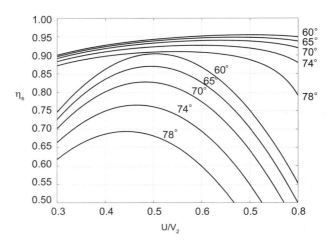

Figure 5.10 Stage efficiencies of single-stage and pressure-compounded zero-reaction turbines with nozzle angles in the range from 60° to 78° with $\zeta_N = 0.02$ and $\zeta_R = 0.14$; the exit kinetic energy is wasted for the set of graphs with lower efficiency, and the family of graphs of higher efficiency are total-to-total efficiencies applicable to a pressure-compounded turbine stage.

5.4 LOSS COEFFICIENTS

A simple correlation for the loss coefficients was developed by Soderberg [72]. In the definition

$$h - h_s = \frac{1}{2}\zeta V^2$$

V replaced by V_2 for the stator and by W_3 for the rotor. The loss coefficients are calculated from

$$\zeta = 0.04 + 0.06 \left(\frac{\varepsilon}{100}\right)^2$$

in which ε is the amount of turning of the flow. For the nozzles the amount of turning is $\varepsilon_N = \alpha_2 - \alpha_3$, and for the rotor it is $\varepsilon_R = \beta_2 - \beta_3$. In both expressions the angles are in degrees. Soderberg's correlation is based on steam turbine designs, which commonly have axial entry into the nozzles, but it gives good results for the flow through the rotor as well, for the loss appears to depend mainly on the deflection of the flow.

■ **EXAMPLE 5.4**

Steam enters the nozzles of single-stage impulse turbine axially and leaves from the nozzles with speed $V_2 = 555\,\text{m/s}$ at angle $\alpha_2 = 74°$. The blade speed is $U = 260\,\text{m/s}$. The exit flow angle of the relative velocity from the rotor is $\beta_3 = -65°$. What is the efficiency of the stage if the exit kinetic energy is wasted?

Solution: The tangential and axial velocities at the exit of the nozzles are

$$V_{u2} = V_2 \sin \alpha_2 = 555 \cdot \sin(74°) = 533.5\,\text{m/s}$$
$$V_{x2} = V_2 \cos \alpha_2 = 555 \cdot \cos(74°) = 153.0\,\text{m/s}$$

The components of the relative velocity at this location are

$$W_{u2} = V_{u2} - U = 533.5 - 260 = 273.5\,\text{m/s} \qquad W_{x2} = V_{x2} = 153.0\,\text{m/s}$$

Hence
$$W_2 = \sqrt{W_{x2}^2 + W_{u2}^2} = \sqrt{153.0^2 + 273.5^2} = 313.4 \, \text{m/s}$$
and the flow angle of the relative velocity is
$$\beta_2 = \tan^{-1}\left(\frac{W_{u2}}{W_{x2}}\right) = \tan^{-1}\left(\frac{273.5}{153.0}\right) = 60.78°$$
The amount of turning by the nozzles and by the rotor blades are
$$\varepsilon_N = \alpha_2 - \alpha_1 = 74° \qquad \varepsilon_R = \beta_2 - \beta_3 = 60.78° + 65° = 125.78°$$
The static enthalpy loss coefficients are
$$\zeta_N = 0.04 + 0.06\left(\frac{74}{100}\right)^2 = 0.07286 \qquad \zeta_R = 0.04 + 0.06\left(\frac{125.78}{100}\right)^2 = 0.1349$$
and the velocity coefficients are therefore
$$c_N = \frac{1}{\sqrt{1+\zeta_N}} = 0.9654 \qquad c_R = \frac{1}{\sqrt{1+\zeta_R}} = 0.9387$$
At the exit of the rotor the relative velocity has the magnitude
$$W_3 = c_R W_2 = 0.9387 \cdot 313.4 = 294.2 \, \text{m/s}$$
and its components are
$$W_{u3} = W_3 \sin\beta_3 = 294.2 \sin(-65°) = -266.6 \, \text{m/s}$$
$$W_{x3} = W_3 \cos\beta_3 = 294.2 \cos(-65°) = 124.3 \, \text{m/s}$$
The components of the absolute velocity at the exit are
$$V_{u3} = U + W_{u3} = 260 - 266.6 = -6.6 \, \text{m/s} \qquad V_{x3} = W_{x3} = 124.5 \, \text{m/s}$$
Hence
$$V_3 = \sqrt{V_{x3}^2 + V_{u3}^2} = \sqrt{124.3^2 + 6.6^2} = 124.5 \, \text{m/s}$$
and the flow angle is
$$\alpha_3 = \tan^{-1}\left(\frac{V_{u3}}{V_{x3}}\right) = \tan^{-1}\left(\frac{-6.6}{134.3}\right) = -3.04°$$
The work delivered by the turbine is
$$w = U(V_{u2} - V_{u3}) = 260(533.5 + 6.60) = 140,430 \, \text{J/kg}$$
The nozzle efficiency is
$$\eta_N = c_N^2 = 0.9654^2 = 0.9321$$
and the rotor efficiency is
$$\eta_R = \frac{2w}{2w + V_3^2 + \zeta_R W_3^2} = \frac{2 \cdot 140,426}{2 \cdot 140,426 + 124.5^2 + 0.1349 \cdot 294.2^2} = 0.9118$$
The turbine efficiency is therefore
$$\eta_{ts} = \eta_N \eta_R = 0.9321 \cdot 0.9118 = 0.850$$

■

EXERCISES

5.1 Steam leaves the nozzles of a de Laval turbine with the velocity $V_2 = 1000\,\text{m/s}$. The flow angle from the nozzle is $\alpha_2 = 70°$. The blade velocity is $U = 360\,\text{m/s}$, and the mass flow rate is $800\,\text{kg/h}$. Take the rotor velocity coefficient to be $c_R = 0.8$. The rotor blade is equiangular. Draw the velocity diagrams and determine (a) the flow angle of the relative velocity at the rotor, (b) the relative velocity of the steam entering the blade row, (c) the tangential force on the blades, (d) the axial thrust on the blades, (e) the power developed, and (e) the rotor efficiency.

5.2 The diameter of a wheel of a single-stage impulse turbine is $1060\,\text{mm}$ and shaft speed, $3000\,\text{rpm}$. The nozzle angle is $72°$, and the ratio of the blade speed to the speed at which steam issues from the nozzles is 0.42. The ratio of the relative velocity leaving the blades is 0.84 of that entering the blades. The outlet flow angle of the relative velocity is $3°$ more than the inlet flow angle. The mass flow rate of steam is $7.23\,\text{kg/s}$. Draw the velocity diagram for the blades and determine (a) the axial thrust on the blades, (b) the tangential force on the blades, (c) power developed by the blade row, and (d) rotor efficiency.

5.3 The wheel diameter of a single-stage impulse steam turbine is $400\,\text{mm}$, and the shaft speed is $3000\,\text{rpm}$. The steam issues from nozzles at velocity $275\,\text{m/s}$ at the nozzle angle of $70°$. The rotor blades are equiangular, and friction reduces the relative velocity as the steam flows through the blade row to 0.86 times the entering velocity. Find the power developed by the wheel when the axial thrust is $F_x = 120\,\text{N}$.

5.4 Steam issues from the nozzles of a single-stage impulse turbine with the velocity $400\,\text{m/s}$. The nozzle angle is at $74°$. The absolute velocity at the exit is $94\,\text{m/s}$, and its direction is $-8.2°$. Assuming that the blades are equiangular, find (a) the power developed by the blade row when the steam flow rate is $7.3\,\text{kg/s}$ and (b) the rate of irreversible energy conversion per kilogram of steam flowing through the rotor.

5.5 Carry out the steps in the development of the expression for ratio of the optimum blade speed to the steam velocity for a single-stage impulse turbine with equiangular blades. Note that this expression is independent of the velocity coefficient. Carry out the algebra to obtain the expression for the rotor efficiency at this condition. (a) Find the numerical value for the velocity ratio when the nozzle angle is $76°$. (b) Find the rotor efficiency at this condition, assuming that $c_R = 0.9$. (c) Find the flow angle of the relative velocity entering the blades at the optimum condition.

5.6 Steam flows from a set of nozzles of a single-stage impulse turbine at $\alpha_2 = 78°$ with the velocity $V_2 = 305\,\text{m/s}$. The blade speed is $U = 146\,\text{m/s}$. The outlet flow angle of the relative velocity is $3°$ greater than its inlet angle, and the velocity coefficient is $c_R = 0.84$. The nozzle velocity coefficient is $c_N = 1$. The power delivered by the wheel is $1000\,\text{kW}$. Draw the velocity diagrams at the inlet and outlet of the blades. Calculate the mass flow rate of steam.

5.7 Steam flows from a set of nozzles of a single-stage impulse turbine at an angle $\alpha_2 = 70°$. (a). Find the maximum total-to-static efficiency given velocity coefficients $c_R = 0.83$ and $c_N = 0.98$. (b) If the rotor efficiency is 90% of its maximum value, what are the possible outlet flow angles for the relative velocity.

5.8 The nozzles of a single-stage impulse turbine have a wall thickness $t = 0.3\,\text{cm}$ and height $b = 15\,\text{cm}$. The mean diameter of the wheel is $1160\,\text{mm}$ and the nozzle angle is $\alpha_2 = 72°$. The number of nozzles in a ring is 72. The specific volume of steam at the exit of the nozzles is $15.3\,\mathrm{m^3/kg}$ and the velocity there is $V_2 = 366\,\mathrm{m/s}$. (a) Find the mass flow rate of steam through the steam nozzle ring. (b) Find the power developed by the blades for an impulse wheel of equiangular blades, given that the velocity coefficient is $c_R = 0.86$ and $c_N = 1.0$. The shaft turns at 3000 rpm.

5.9 The isentropic static enthalpy change across a stage of a single-stage impulse turbine is $\Delta h_s = 22\,\mathrm{kJ/kg}$. The nozzle exit angle is $\alpha_2 = 74°$. The mean diameter of the wheel is 148 cm and the shaft turns at 1500 rpm. The blades are equiangular with a velocity coefficient of $c_R = 0.87$. The nozzle velocity coefficient is $c_N = 0.98$. (a) Find the steam velocity at the exit from the nozzles. (b) Find the flow angles of the relative velocity at the inlet and exit of the wheel. (c) Find the overall efficiency of the stage.

5.10 An impulse turbine has a nozzle angle $\alpha_2 = 72°$ and steam velocity $V_2 = 244\,\mathrm{m/s}$. The velocity coefficient for the rotor blades is $c_R = 0.85$, and the nozzle efficiency is $\eta_N = 0.92$. The output power generated by the wheel is $\dot{W} = 562\,\mathrm{kW}$ when the mass flow rate is $\dot{m} = 23\,\mathrm{kg/s}$. Find the total-to-static efficiency of the turbine.

5.11 A two-row velocity-compounded impulse wheel is part of a steam turbine with many other stages. The steam velocity from the nozzles is $V_2 = 580\,\mathrm{m/s}$, and the mean speed of the blades is $U = 116\,\mathrm{m/s}$. The flow angle leaving the nozzle is $\alpha_2 = 74°$, and the flow angle of the relative velocity leaving the first set of rotor blades is $\beta_3 = -72°$. The absolute velocity of the flow as it leaves the stator vanes between the two rotors is $\alpha_4 = 68°$, and the outlet angle of the relative velocity leaving the second rotor is $\beta_5 = -54°$. The steam flow rate is $\dot{m} = 2.4\,\mathrm{kg/s}$. The velocity coefficient is $c_v = 0.84$ for both the stator and the rotor row. (a) Find the axial thrust from each wheel. (b) Find the tangential thrust from each wheel. (c) Find the total-to-static efficiency of the rotors defined as the work out divided by the kinetic energy available from the nozzles.

5.12 A velocity-compounded impulse wheel has two rows of moving blades with a mean diameter of $D = 72\,\mathrm{cm}$. The shaft rotates at 3000 rpm. Steam issues from the nozzles at angle $\alpha_2 = 74°$ with velocity $V_2 = 555\,\mathrm{m/s}$. The mass flow rate is $\dot{m} = 5.1\,\mathrm{kg/s}$. The energy loss through each of the moving blades is 24% of the kinetic energy entering the blades, based on the relative velocity. Steam leaves the first set of moving blades at $\beta_3 = -72°$ the guide vanes between the rows at $\alpha_4 = 68°$ and the second set of moving blades at $\beta_5 = -52°$. (a) Draw the velocity diagrams and find the flow angles at the blade inlets both for absolute and relative velocities. (b) Find the power developed by each row of blades. (c) Find the rotor efficiency as a whole.

5.13 Steam flows from the nozzles of a 0% repeating stage at an angle $\alpha_2 = 69°$ and speed $V_2 = 450\,\mathrm{m/s}$ and enters the rotor with blade speed moving at $U = 200\,\mathrm{m/s}$. Find (a) its efficiency when the loss coefficients are calculated from Soderberg's correlation and (b) the work delivered by the stage.

5.14 For a repeating stage the efficiency of a 0% reaction, by neglecting the temperature factors show that the approximate form of the total-to-total efficiency is

$$\eta_{tt} = \frac{4\lambda(\sin\alpha_2 - \lambda)}{(\zeta_R - 4)\lambda^2 + (4 - 2\zeta_R)\lambda\sin\alpha_2 + \zeta_R + \zeta_N}$$

and its maximum values is at the condition $\lambda = U/V_2$ given by

$$\lambda = \frac{\zeta_R + \zeta_S - \sqrt{(\zeta_R + \zeta_N)(\zeta_R + \zeta_S - \zeta_R \sin^2\alpha_2)}}{\zeta_R \sin\alpha_2}$$

CHAPTER 6

AXIAL TURBINES

In the previous chapter the impulse stages of steam turbines were analyzed. This chapter extends the development of axial turbine theory to reaction turbines. These include gas turbines and all except the leading stages of steam turbines. The extent of the global steam turbine industry was mentioned in the last chapter. Gas turbine industry is even larger, owing to the use of gas turbine in a jet engine. Gas turbines are also used for electric power generation in central station power plants. In addition, they drive the large pipeline compressors that transmit natural gas across continents and provide power on oil-drilling platforms.

The chapter begins with the development of the working equations for the reaction stages. These relate the flow angles of the absolute and relative velocities to the degree of reaction, flow coefficient, and the blade-loading coefficient. Three-dimensional aspects of the flow are considered next. Then semiempirical theories are introduced to calculate the static enthalpy rise caused by internal heating, which is then used to develop an expression for the stage efficiency. After this the equations used to calculate the stagnation pressure losses across the stator and the rotor are developed.

6.1 INTRODUCTION

Two adjacent blades of an axial reaction turbine are shown in Figure 6.1. Their spacing along the periphery of the disk is called the *pitch*. The pitch increases in the radial direction from the *hub* of the rotor to its *casing*. The nominal value of the pitch is at the mean radius.

The lateral boundaries of the flow channel are along the *pressure* and *suction* sides of the blades and the *endwalls* along the hub and the casing.

The flow is from front to back in Figure 6.1. The blade *chord* is the straight distance from the leading edge of the blade to its trailing edge. Its projection in the axial direction is the *axial chord*. The path of a fluid particle, as it passes through the *blade passage*, is curved and thus longer than the chord.

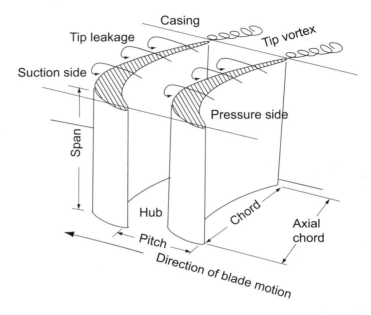

Figure 6.1 Flow channel between two adjacent turbine blades.

The annular region formed from the blade passage areas is called the *flow annulus*. The annulus area is calculated as

$$A = 2\pi r_\mathrm{m}(r_\mathrm{c} - r_\mathrm{h}) = \pi(r_\mathrm{c}^2 - r_\mathrm{h}^2)$$

if the mean radius r_m is taken as the arithmetic average

$$r_\mathrm{m} = \frac{1}{2}(r_\mathrm{c} + r_\mathrm{h})$$

of the casing radius r_c and the hub radius r_h. The blade height, or span, is $b = r_\mathrm{c} - r_\mathrm{h}$ and $2\pi r_\mathrm{m} = Zs$, in which Z is the number of blades and s is the mean pitch or spacing of the blades. Therefore the annulus area is also $A = Zsb$.

An alternative is to define a mean radius such that the flow area from it to the hub and to the casing are equal. This definition leads to the equality

$$\pi(\bar{r}_\mathrm{m}^2 - r_\mathrm{h}^2) = \pi(r_\mathrm{c}^2 - \bar{r}_\mathrm{m}^2)$$

which, when solved for $\bar{r}_\mathrm{m}$, gives

$$\bar{r}_\mathrm{m} = \sqrt{\frac{r_\mathrm{c}^2 + r_\mathrm{h}^2}{2}}$$

In using this RMS value of the radius, the annulus area is clearly

$$A = \pi(r_c^2 - r_h^2) = 2\pi(\bar{r}_m^2 - r_h^2) = 2\pi(r_c^2 - \bar{r}_m^2)$$

The distance between the tip of the rotor blade and the casing is called a *tip clearance*. This is kept small in order to prevent *tip leakage flow* in the rotor. Because the tip clearance is small, in the discussion that follows the distinction between the casing radius and tip radius is usually ignored. The stator blades, as shown in Figure 6.2, are fixed to the casing, and their tips are near the hub of the rotor blades. In many designs, their tips are fastened to a diaphragm that extends inward. At the end of the diaphragm a labyrinth seal separates it from the rotating shaft. The seal prevents the leakage flow that is caused by the pressure difference across the stator. Since the seal is located close to the shaft, the flow area for the possible leakage flow is small.

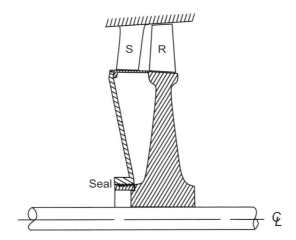

Figure 6.2 A stage of an axial turbine.

Axial turbines are commonly designed such that the axial velocity remains constant, or nearly so. Therefore as the gas expands through the turbine the annulus area must increase from stage to stage. This flaring of the annulus is accomplished by changing the hub radius, the casing radius, or both. If both are changed the mean radius can be kept constant.

6.2 TURBINE STAGE ANALYSIS

Consider a turbine stage as shown schematically in Figure 6.3. It consists of a *stator followed by a rotor*. As in the previous chapter on steam turbines, the inlet to the stage is station 1 and the outlet from the stator is station 2, which is is also the inlet to the rotor. The outlet from the rotor, and hence the stage, is station 3. For a *normal stage* in a multistage machine the magnitude and direction of the velocity at the outlet of the rotor are the same as those at the inlet to the stator. Pressure, temperature, and density naturally change from stage to stage.

Work delivered by a turbine stage is given by the Euler equation for turbomachinery

$$w = U(V_{u2} - V_{u3}) = U(W_{u2} - W_{u3}) \tag{6.1}$$

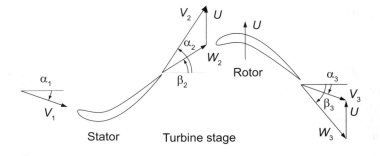

Figure 6.3 Velocity triangles for a turbine stage.

For the situation shown in Figure 6.3, the inlet flow angle of the absolute velocity is negative as the flow enters the stator. For the rotor a *deflection* is the difference in the swirl velocities $V_{u2} - V_{u3} = W_{u2} - W_{u3}$. It is also measured by the amount of turning, $\beta_2 - \beta_3$. The amount of turning across the stator is given by $\alpha_2 - \alpha_1$. Clearly, if a stage is to deliver a large amount of the work, for a given blade velocity, turning across the rotor must be large. A typical value is 70°, and it rarely exceeds 90°. A large deflection also means that the average pressure difference between the pressure and suction sides of a blade must be large. Such blades are said to be *heavily loaded*.

In order to achieve a large amount of turning in the rotor, the stator must also turn the flow, but in the opposite direction. The velocity diagrams in Figure 6.3 show that, as the stator deflects the flow toward the direction of rotation, the stream velocity increases. Since the stagnation enthalpy remains constant across the stator, it follows that

$$h_1 + \frac{1}{2}V_1^2 = h_2 + \frac{1}{2}V_2^2 \qquad h_1 - h_2 = \frac{1}{2}\left(V_2^2 - V_1^2\right)$$

and the increase in kinetic energy leads to a drop in the static enthalpy. This expression may be written as

$$(u_1 - u_2) + p_1 v_1 - p_2 v_2 = \frac{1}{2}\left(V_2^2 - V_1^2\right)$$

It shows that the increase in kinetic energy comes from conversion of internal energy and from the difference in the flow work done in pushing the fluid into and out of the flow passage. This may also be written in a differential form. By considering station 2 to be an arbitrary location, and differentiating, yields

$$-\frac{du}{d\ell} - \frac{d(pv)}{d\ell} = \frac{1}{2}\frac{dV^2}{d\ell} = V\frac{dV}{d\ell}$$

in which $d\ell$ is an element of length along the flow path. This shows that a drop in the internal energy increases the kinetic energy of the flow, as does the net pv work term in this small section of the channel. That both terms have the same sign is clear for an ideal gas, for then $du = c_v \, dT$ and $d(pv) = R \, dT$ and since internal energy drops in the flow direction, so does temperature and pv. The ratio of these contributions is

$$\frac{du}{d(pv)} = \frac{1}{\gamma - 1} = 3$$

with the numerical value corresponding to $\gamma = \frac{4}{3}$. Thus the conversion of internal energy contributes more to the increase in kinetic energy than the flow work.

As the gas passes through the rotor, it is directed back toward the axis, reducing its kinetic energy. The work delivered by the stage is given by

$$w = h_{02} - h_{03} = h_2 - h_3 + \frac{1}{2}\left(V_2^2 - V_3^2\right) \qquad (6.2)$$

With the reaction defined as the static enthalpy drop across the rotor divided by the static enthalpy drop across the stage, for positive stage reaction $h_2 > h_3$. An exception to this is an expansion at constant pressure in impulse blades. Equation (6.2) may be written as

$$w = u_2 - u_3 + p_2 v_2 - p_3 v_3 + \frac{1}{2}\left(V_2^2 - V_3^2\right) \qquad (6.3)$$

and, since each term is expected to be positive, each contributes to the work delivered by the turbine. This is illustrated in the following example.

■ EXAMPLE 6.1

Consider the flow of combustion gases, with $\gamma = \frac{4}{3}$ and $R = 287\,\text{J}/(\text{kg}\cdot\text{K})$, through a normal turbine stage such that the flow angle at the exit of the rotor is the same as that entering the stator, and $\alpha_1 = \alpha_3 = -14.4°$. The inlet total temperature is $T_{01} = 1200\,\text{K}$. The axial velocity is constant $V_x = 280\,\text{m/s}$. The flow leaves the stator at angle $\alpha_2 = 57.7°$. The mean radius of the rotor is $r = 17\,\text{cm}$, and the rotor turns at $20{,}000$ rpm. (a) Find the work done and the drop in stagnation temperature across the stage. (b) Determine the flow angles of the relative velocity at the inlet and exit of the rotor. (c) Calculate the contribution of internal energy and flow work in increasing the kinetic energy through the stator. (d) Calculate the contributions of internal energy, flow work, and kinetic energy to work delivered by the stage.

Solution: (a,b) The specific heats at constant pressure and volume for the gas are

$$c_p = \frac{\gamma R}{\gamma - 1} = 4 \cdot 287 = 1148\,\text{J}/(\text{kg}\cdot\text{K}) \qquad c_v = c_p - R = 1148 - 287 = 861\,\text{J}/(\text{kg}\cdot\text{K})$$

The blade velocity is

$$U = r\Omega = \frac{0.17 \cdot 20{,}000 \cdot \pi}{30} = 356.0\,\text{m/s}$$

The tangential component and the magnitude of the absolute velocity leaving the stator are

$$V_{u2} = V_x \tan \alpha_2 = 280 \tan(57.7°) = 442.9\,\text{m/s}$$

$$V_2 = \sqrt{V_x^2 + V_{u2}^2} = \sqrt{280^2 + 442.9^2} = 524.0\,\text{m/s}$$

The tangential component and the flow angle of the relative velocity at the rotor exit are

$$W_{u2} = V_{u2} - U = 442.9 - 356.0 = 86.9\,\text{m/s}$$

$$\beta_2 = \tan^{-1}\left(\frac{W_{u2}}{W_x}\right) = \tan^{-1}\left(\frac{86.9}{280}\right) = 17.2°$$

At the exit of the rotor for a normal stage $\alpha_3 = \alpha_1$ and the velocities are

$$V_{u3} = V_x \tan \alpha_3 = 280 \tan(-14.4°) = -71.9\,\text{m/s}$$

$$V_3 = \sqrt{V_x^2 + V_{u3}^2} = \sqrt{280^2 + 71.9^2} = 289.1 \, \text{m/s}$$

The tangential component and the flow angle of the relative velocity there are

$$W_{u3} = V_{u3} - U = -71.9 - 356.0 = -427.9 \, \text{m/s}$$

$$\beta_3 = \tan^{-1}\left(\frac{W_{u3}}{W_x}\right) = \tan^{-1}\left(\frac{-427.9}{280}\right) = -56.8°$$

Work delivered by the turbine is

$$w = U(V_{u2} - V_{u3}) = 356.0\,(442.9 + 71.9) = 183.3 \, \text{kJ/kg}$$

and the stagnation temperature drop across the rotor is

$$\Delta T_0 = \frac{w}{c_p} = \frac{183.3}{1.148} = 159.7 \, \text{K}$$

(c) At the inlet to the stator the static temperature is given by

$$T_1 = T_{01} - \frac{V_1^2}{2\,c_p} = 1200 - \frac{289.1^2}{2 \cdot 1148} = 1200 - 36.4 = 1163.6 \, \text{K}$$

At the exit of the stator the static temperature is

$$T_2 = T_{02} - \frac{V_2^2}{2\,c_p} = 1200 - \frac{524.0^2}{2 \cdot 1148} = 1200 - 119.6 = 1080.4 \, \text{K}$$

so that

$$u_1 - u_2 = c_v(T_1 - T_2) = 0.861\,(1163.6 - 1080.4) = 71.6 \, \text{kJ/kg}$$

and

$$p_1 v_1 - p_2 v_2 = R(T_1 - T_2) = 0.287\,(1163.6 - 1080.4) = 23.9 \, \text{kJ/kg}$$

Increase in the kinetic energy across the stator is

$$\frac{1}{2}\left(V_2^2 - V_1^2\right) = \frac{1}{2}(524.0^2 - 289.1^2) = 95.5 \, \text{kJ/kg}$$

which also equals the sum of the previous two terms.

(d) Since the stagnation temperature drop across the rotor is $\Delta T_0 = 159.7 \, \text{K}$, the stagnation temperature after the rotor is

$$T_{03} = T_{02} - \Delta T_0 = 1200 - 159.7 = 1040.3 \, \text{K}$$

and

$$T_3 = T_{03} - \frac{V_3^2}{2\,c_p} = 1040.3 - \frac{289.1^2}{2 \cdot 1148} = 1040.3 - 36.4 = 1003.9 \, \text{K}$$

The contributions to work are

$$u_2 - u_3 = c_v(T_2 - T_3) = 0.861\,(1080.4 - 1003.9) = 65.8 \, \text{kJ/kg}$$

and

$$p_2 v_2 - p_3 v_3 = R(T_2 - T_3) = 0.287\,(1080.4 - 1003.9) = 21.9\,\text{kJ/kg}$$

In a normal stage the increase in kinetic energy across the stator is equal to its decrease across the rotor. Hence the decrease in kinetic energy across the rotor is

$$\frac{1}{2}\left(V_2^2 - V_3^2\right) = \frac{1}{2}(524.0^2 - 289.1^2) = 95.5\,\text{kJ/kg}$$

The sums of internal energy changes and pv work, and the change in kinetic energy across the rotor, add up to the work delivered by the stage. ∎

6.3 FLOW AND LOADING COEFFICIENTS AND REACTION RATIO

The work delivered by a stage is given by

$$w = U(V_{u2} - V_{u3}) = U(W_{u2} - W_{u3})$$

which, if $V_x = W_x$ is constant across the stage, may be written as

$$w = UV_x(\tan \alpha_2 - \tan \alpha_3) = UV_x(\tan \beta_2 - \tan \beta_3) \qquad (6.4)$$

Let $\phi = V_x/U$ denote *a flow coefficient* and $\psi = w/U^2$ a *blade-loading coefficient*. Then, dividing both sides of this equation by U^2 gives the Euler turbine equation in a nondimensional form as

$$\psi = \phi(\tan \alpha_2 - \tan \alpha_3) \qquad (6.5)$$

Other names for the blade-loading coefficient are *work coefficient* and *loading factor*. In addition to ψ and ϕ, a third nondimensional quantity of importance in the theory is the *reaction ratio R*, introduced previously. It was defined as the *ratio of the static enthalpy change across the rotor to that across the entire stage*. Hence

$$R = \frac{h_2 - h_3}{h_1 - h_3} = \frac{h_1 - h_3 - (h_1 - h_2)}{h_1 - h_3} = 1 - \frac{h_1 - h_2}{h_1 - h_3} \qquad (6.6)$$

Reaction naturally falls into the range $0 \le R \le 1$, but it was seen to be slightly negative for a pure impulse stage. This equation shows that the reaction is zero, if the entire static enthalpy drop takes place in the stator.

Recalling that the total enthalpy of the relative motion, given by Eq. (4.13), remains constant across a rotor in an axial stage, it follows that

$$h_2 + \frac{1}{2}W_2^2 = h_3 + \frac{1}{2}W_3^2$$

or

$$h_2 - h_3 = \frac{1}{2}\left(W_3^2 - W_2^2\right)$$

Hence, if $W_2 = W_3$ the reaction is zero. Across the stator the stagnation enthalpy is constant so that

$$h_1 - h_2 = \frac{1}{2}\left(V_2^2 - V_1^2\right)$$

The value of V_1 is smallest for an axial entry, and this equation shows that the static enthalpy drop across the stator increases and reaction decreases for increasing V_2. In addition, a large deflection of the flow across the stator leads to a large V_2.

It is useful also to think of R in the incompressible limit, for then in an isentropic flow across the stator internal energy remains constant and in the pv work only the pressure changes. Thus pressure changes are directly proportional to changes in static enthalpy. Hence in this limit a large reaction means a small pressure drop across the stator and a large decrease in pressure across the rotor.

The reaction may be related to the flow angles by noting first that $V_1^2 = V_{x1}^2 + V_{u1}^2$ and $V_2^2 = V_{x2}^2 + V_{u2}^2$. Then, for constant axial velocity $V_{x1} = V_{x2}$, and the change in enthalpy across the stator may be written as

$$h_1 - h_2 = \frac{1}{2}(V_{u2}^2 - V_{u1}^2) = \frac{1}{2}V_x^2(\tan^2 \alpha_2 - \tan^2 \alpha_1)$$

For a *normal stage* $V_1 = V_3$, and therefore $h_1 - h_3 = h_{01} - h_{03}$. With $w = h_{01} - h_{03}$ Eq. (6.6) for the reaction may be written as

$$R = 1 - \frac{V_x^2}{2}\frac{(\tan^2 \alpha_2 - \tan^2 \alpha_3)}{\psi U^2}$$

or as

$$R = 1 - \frac{\phi^2}{2\psi}(\tan^2 \alpha_2 - \tan^2 \alpha_3) \tag{6.7}$$

Substituting ψ from Eq. (6.5) into this gives

$$R = 1 - \frac{1}{2}\phi(\tan \alpha_2 + \tan \alpha_3) \tag{6.8}$$

Next α_2 is eliminated, again using Eq.(6.5), and the important result

$$\psi = 2(1 - R - \phi \tan \alpha_3) \tag{6.9}$$

is obtained. It shows that a decreasing R increases the loading. A small R means that the pressure drop across the rotor is small, but the large loading is the result of a large deflection. In the stator the flow leaves at high speed at large angle α_2. The high kinetic energy obtained this way becomes available for doing work on the rotor blades. The flow is then deflected back toward the axis and beyond to a negative value of α_3, so that the last term in this equation is positive. Hence, for R fixed, an increase in the absolute value of α_3, obtained by increasing it in the direction opposite to U, leads to a large deflection and a large value for the blade-loading factor ψ. Thus a fairly low value of R and high turning gives heavily loaded blades and a compact design.

Equations (6.5) and (6.8), written as

$$\tan \alpha_2 - \tan \alpha_3 = \frac{\psi}{\phi} \tag{6.10}$$

$$\tan \alpha_2 + \tan \alpha_3 = \frac{2 - 2R}{\phi} \tag{6.11}$$

when solved for the unknown angles, give

$$\tan \alpha_3 = \frac{1 - R - \psi/2}{\phi} \qquad \tan \alpha_2 = \frac{1 - R + \psi/2}{\phi} \tag{6.12}$$

Experienced turbomachinery designers choose the flow and loading coefficients and the degree of reaction at the outset and then determine the flow angles from these equations. These are true only for a normal stage. If the axial velocity does not remain constant, the proper equations need to be redeveloped from the fundamental concepts.

Similar expressions are next developed for the flow angles of the relative velocity. The Euler turbine equation may be written as

$$w = U(W_{u2} - W_{u3}) = UV_x(\tan \beta_2 - \tan \beta_3)$$

which after dividing through by U^2 gives

$$\psi = \phi(\tan \beta_2 - \tan \beta_3) \qquad (6.13)$$

To arrive at the second equation relating the relative flow angles to the nondimensional parameters, the reaction ratio

$$R = \frac{h_2 - h_3}{h_1 - h_3}$$

is converted into an appropriate form. Since the stagnation enthalpy of the relative motion is constant across the rotor, the relation

$$h_2 - h_3 = \frac{1}{2}W_3^2 - \frac{1}{2}W_2^2$$

follows. Then, since $W_{x2} = W_{x3}$, the term on the right may be rewritten in the form

$$h_2 - h_3 = \frac{1}{2}(W_{x3}^2 + W_{u3}^2 - W_{x2}^2 - W_{u2}^2) = \frac{1}{2}(W_{u3}^2 - W_{u2}^2) = \frac{1}{2}V_x^2(\tan^2 \beta_3 - \tan^2 \beta_2)$$

In addition, for a normal stage, $h_1 - h_3 = h_{01} - h_{03} = w = U^2\psi$. When this and the previous equation are substituted into the definition of reaction ratio, it becomes

$$R = \frac{V_x^2}{2}\frac{(\tan^2 \beta_3 - \tan^2 \beta_2)}{U^2\psi} \qquad \text{or} \qquad R = \frac{\phi^2}{2\psi}(\tan^2 \beta_3 - \tan^2 \beta_2)$$

Substituting ψ from Eq. (6.13) into this gives

$$R = -\frac{\phi}{2}(\tan \beta_3 + \tan \beta_2)$$

This and Eq. (6.13) written as

$$\tan \beta_2 - \tan \beta_3 = \frac{\psi}{\phi} \qquad (6.14)$$

$$\tan \beta_2 + \tan \beta_3 = -\frac{2R}{\phi} \qquad (6.15)$$

when solved for the flow angles of the relative velocity, give

$$\tan \beta_3 = -\frac{R + \psi/2}{\phi} \qquad \tan \beta_2 = -\frac{R - \psi/2}{\phi} \qquad (6.16)$$

The flow angles may now be determined if the value of the parameters ϕ, ψ, and R are specified. With four equations and seven variables, any three may be specified and the other four calculated from them. One such calculation is illustrated in the next example.

EXAMPLE 6.2

Combustion gases with $\gamma = \frac{4}{3}$ and $c_p = 1148\,\text{J}/(\text{kg}\cdot\text{K})$ flow through an axial turbine stage with $\phi = 0.80$ as the design value for the flow coefficient and $\psi = 1.7$ for the blade-loading coefficient. The stage is normal with flow into the stator at angle $\alpha_1 = -21.2°$. The absolute velocity of the gases leaving the stator is $V_2 = 463\,\text{m/s}$. The inlet stagnation temperature is $T_{01} = 1200\,\text{K}$, and the total-to-total efficiency is 0.89. (a) Find the flow angles for a normal stage, and the amount of turning by the stator and the rotor. (b) Calculate the work delivered by the stage, and the drop in the stagnation temperature. (c) Determine the static pressure ratio across the stage.

Solution: (a) With $\alpha_3 = \alpha_1$, solving

$$\psi = 2(1 - R - \phi\tan\alpha_3)$$

for R gives

$$R = 1 - \frac{\psi}{2} - \phi\tan\alpha_3 = 1 - 0.85 - 0.8\tan(-21.2°) = 0.46$$

The remaining flow angles are

$$\tan\alpha_2 = \frac{1 - R + \psi/2}{\phi} = 1.737 \qquad \alpha_2 = 60.08°$$

$$\tan\beta_3 = \frac{-(R + \psi/2)}{\phi} = -1.638 \qquad \beta_3 = -58.59°$$

$$\tan\beta_2 = \frac{-(R - \psi/2)}{\phi} = 0.487 \qquad \beta_2 = 25.97°$$

The amounts of turning by the stator and the rotor are

$$\alpha_2 - \alpha_3 = 81.27° \qquad \beta_2 - \beta_3 = 84.57°$$

(b) The axial velocity is

$$V_x = V_2 \cos\alpha_2 = 463\cos(60.08°) = 231.0\,\text{m/s}$$

and the blade speed is

$$U = \frac{V_x}{\phi} = \frac{231.0}{0.8} = 288.7\,\text{m/s}$$

Hence the work done may be calculated from

$$w = \psi U^2 = 1.7 \cdot 288.7^2 = 141.7\,\text{kJ/kg}$$

and the drop in stagnation temperature is

$$\Delta T_0 = \frac{w}{c_p} = \frac{141.7}{1.148} = 123.4\,\text{K}$$

The isentropic work is obtained by dividing the work w by the stage efficiency, which is the total-to-total efficiency,

$$w_s = \frac{w}{\eta_{tt}} = \frac{141.7}{0.89} = 159.2\,\text{kJ/kg}$$

(c) The static temperature at the inlet is

$$T_1 = T_{01} - \frac{V_1^2}{2c_p} = 1200 - \frac{247.8^2}{2 \cdot 1148} = 1173.3\,\text{K}$$

Neither the static nor the stagnation pressure is known at the inlet, but their ratio can be calculated from

$$\frac{p_1}{p_{01}} = \left(\frac{T_1}{T_{01}}\right)^{\gamma/(\gamma-1)} = \left(\frac{1173.3}{1200}\right)^4 = 0.9139$$

The stagnation temperature at the exit of the stage comes out to be

$$T_{03} = T_{01} - \frac{w}{c_p} = 1200 - \frac{141.7}{1.148} = 1076.6\,\text{K}$$

The exit velocity from the stage is

$$V_3 = \frac{V_x}{\cos\alpha_3} = \frac{231.0}{\cos(-21.2°)} = 247.8\,\text{m/s}$$

and therefore the static temperature at the exit is

$$T_3 = T_{03} - \frac{V_3^2}{2c_p} = 1094.7 - \frac{247.8^2}{1148} = 1049.8\,\text{K}$$

This can be used to calculate the ratio of static to stagnation pressure at the exit:

$$\frac{p_3}{p_{03}} = \left(\frac{T_3}{T_{03}}\right)^{\gamma/(\gamma-1)} = \left(\frac{1048.9}{1076.5}\right)^4 = 0.9043$$

The stagnation temperature at the isentropic end state is determined to be

$$T_{03s} = T_{01} - \frac{w_s}{c_p} = 1200 - \frac{159.2}{1.1148} = 1061.3\,\text{K}$$

and the stagnation pressure ratio across the stage is therefore

$$\frac{p_{01}}{p_{03}} = \left(\frac{T_{01}}{T_{03s}}\right)^{\gamma/(\gamma-1)} = \left(\frac{1200}{1061.3}\right)^4 = 1.635$$

The ratio of static pressures across the stage can now be determined:

$$\frac{p_1}{p_3} = \frac{p_1}{p_{01}}\frac{p_{01}}{p_{03}}\frac{p_{03}}{p_3} = \frac{0.9138 \cdot 1.635}{0.9043} = 1.652$$

The representative values for an axial turbine stage $\psi = 1.7$, $\phi = 0.8$ and $R = 0.46$ give these values for the stagnation and static pressure ratios and the amount of turning comes out to be about $80°$–$85°$. In addition, the stagnation pressure drop of $\Delta T_0 = 123.4\,\text{K}$ is representative for a stage.

∎

6.3.1 Fifty percent (50%) stage

A 50% reaction stage has equal static enthalpy drops across the stator and the rotor. For such a stage, Eq. (6.9) reduces to

$$\psi = 1 - 2\phi \tan \alpha_3 \qquad (6.17)$$

The blades and velocity triangles are shown in Figure 6.4. To achieve a high efficiency the flow angle at the inlet is kept only slightly negative, but if some of the efficiency is sacrificed to achieve higher performance, the inlet flow angle may reach $\alpha_1 = -45°$. For such a stage, a flow coefficient may have a value of $\phi = 0.75$, which then gives $\psi = 2.5$. Gas turbines for aircraft are designed for high performance and low weight. Hence the number of stages is kept as low as possible and materials that withstand high stresses are used for the turbine blades.

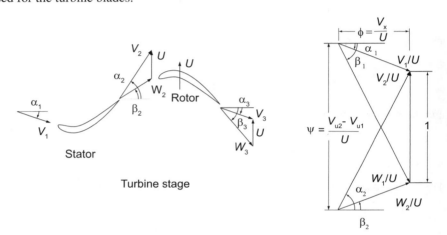

Figure 6.4 Blading for a 50% reaction turbine.

For a 50% reaction stage Eqs. (6.12) and (6.16) for absolute and relative flow angle reduce to

$$\tan \alpha_3 = \frac{1-\psi}{2\phi} \qquad \tan \alpha_2 = \frac{1+\psi}{2\phi} \qquad (6.18)$$

and

$$\tan \beta_3 = -\frac{1+\psi}{2\phi} \qquad \tan \beta_2 = -\frac{1-\psi}{2\phi} \qquad (6.19)$$

From these it is seen that

$$\tan \alpha_3 = -\tan \beta_2 \qquad \tan \alpha_2 = -\tan \beta_1 \qquad (6.20)$$

With their absolute values less than 90°, these equations are satisfied if $\alpha_3 = -\beta_2$ and $\alpha_2 = -\beta_3$. Since $W_x = V_x$, it follows that

$$W_3^2 = W_x^2 + W_{u3}^2 = V_x^2 + W_x^2 \tan^2 \beta_3$$
$$= V_x^2 + V_x^2 \tan^2 \alpha_2 = V_x^2 + V_{u2}^2 = V_2^2$$

By a similar argument it can be shown that $W_2^2 = V_3^2$. Hence

$$W_3 = V_2 \qquad\qquad W_2 = V_3$$

FLOW AND LOADING COEFFICIENTS AND REACTION RATIO

A *combined velocity diagram* is shown also in Figure 6.4. It is constructed by drawing the inlet and exit velocity diagrams with a common side of distance U. If all velocities are then divided by U, the diagram is *normalized* with a right side of length unity. The blade-loading coefficient ψ can be identified as the vertical distance between the left vertices of the two triangles. For a constant axial velocity, widths of the triangles are the same and in a normalized diagram equal to the flow coefficient ϕ. The results for the angles and velocities show that the velocity triangles for 50% reaction stage are symmetric.

■ EXAMPLE 6.3

Combustion gases, with $\gamma = \frac{4}{3}$ and $R = 287\,\text{kJ}/(\text{kg·K})$, flow through a 50% reaction stage of an axial turbine, that has a total-to-total efficiency $\eta_{tt} = 0.91$ and design flow coefficient $\phi = 0.80$. The flow into the stator is at an angle $\alpha_1 = -14.0°$, and the axial velocity is $V_x = 240\,\text{m/s}$ and constant across the stage. The inlet stagnation temperature is $T_{01} = 1200\,\text{K}$. (a) Find the flow angles for a normal stage, and the amount of turning by the stator and the rotor. (b) Find the work delivered by the stage and the drop in the stagnation temperature. (c) Show that $1/\eta_{ts} = 1/\eta_{tt} + \phi^2/2\psi \cos^2 \alpha_3$ and calculate the total-to-static efficiency.

Solution: (a) The blade-loading coefficient is

$$\psi = 1 - 2\phi \tan \alpha_3 = 1 - 2 \cdot 0.8 \tan(-14.0°)) = 1.40$$

The remaining flow angles are

$$\tan \alpha_2 = \frac{1-\psi}{2\phi} = 1.50 \quad \text{so that} \quad \alpha_2 = 56.3°$$

and

$$\beta_3 = -\alpha_2 = -56.3° \qquad \beta_2 = -\alpha_3 = 14.0°$$

The amount turning by the stator and rotor are

$$\alpha_2 - \alpha_3 = 70.3° \qquad \beta_2 - \beta_1 = 70.3°$$

(b) The blade speed is

$$U = \frac{V_x}{\phi} = \frac{240}{0.8} = 300.0\,\text{m/s}$$

Hence the work delivered may be calculated from

$$w = \psi U^2 = 1.40 \cdot 300.0^2 = 125.9\,\text{kJ/kg}$$

and the drop in stagnation temperature becomes $\Delta T_0 = w/c_p = 109.7\,\text{K}$, since $c_p = 1148\,\text{J}/(\text{kg} \cdot \text{K})$. The isentropic work is

$$w_s = \frac{w}{\eta_{tt}} = \frac{125.9}{0.91} = 138.4\,\text{kJ/kg}$$

(c) The total-to-static efficiency is obtained from

$$\eta_{ts} = \frac{h_{01} - h_{03}}{h_{01} - h_{3s}} = \frac{h_{01} - h_{03}}{h_{01} - h_{03s} + V_{3s}^2/2} = \frac{w}{w_s + V_{3s}^2/2}$$

Assuming that $V_{3s} = V_3$ so that

$$\frac{1}{\eta_{ts}} = \frac{w_s + V_x^2/2\cos^2\alpha_3}{w} = \frac{1}{\eta_{tt}} + \frac{\phi^2}{2\psi\cos^2\alpha_3}$$

$$= \frac{1}{0.91} + \frac{0.8^2}{2 \cdot 1.40 \cos^2(-14°)} = 1.342$$

The total-to-static efficiency is therefore $\eta_{ts} = 0.745$.

■

6.3.2 Zero percent (0%) reaction stage

Consider a stage for which $R = 0$. Equation (6.16) then shows that

$$\tan\beta_3 = -\tan\beta_2 \quad \text{or} \quad \beta_3 = -\beta_2$$

Assuming that the axial velocity constant, it follows from these that $W_2 = W_3$. If the flow angles were to be equal to the blade angles, then the blade would have a symmetric bucket shape, as shown in Figure 6.5. With low reaction the blades are heavily loaded. This stage

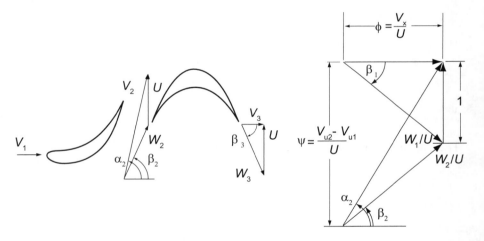

Figure 6.5 Blades for a 0% reaction stage.

reaction and the impulse stage were discussed in Chapter 5 for steam turbines, and the differences between them were shown to be slight. Since the flow enters the nozzles from a steam chest, the inlet to the nozzles is naturally axial, and if most of the pressure drop takes place across the nozzles, the stage reaction is close to zero. Here the 0% reaction stage is considered as a special case of a stage with an arbitrary reaction.

The normalized velocity diagram for a stage with $\alpha_3 = 0$ is shown on the right hand side (RHS) of Figure 6.5. For a normal stage with axial entry and with $R = 0$, the relation

$$\psi = 2(1 - R + \phi\tan\alpha_3)$$

reduces to $\psi = 2$. Thus the line indicating the blade loading is twice as long as blade speed line.

EXAMPLE 6.4

A normal stage with $R = 0$ operates with axial entry. The nozzle turns the flow by $64°$. (a) Find the flow coefficient. (b) What is the discharge velocity from the nozzles, if the axial velocity is $V_x = 240 \text{ m/s}$? (c) Calculate the work delivered by the stage and the drop in the stagnation temperature, given gases with $c_p = 1148 \text{ J}/(\text{kg} \cdot \text{K})$.

Solution: (a) From
$$\psi = 2(1 - R - \phi \tan \alpha_1)$$
a stage with $R = 0$ and axial entry has $\psi = 2$. Next, solving
$$\tan \alpha_2 = \frac{1 + \psi/2}{\phi}$$
for ϕ, gives
$$\phi = 2 \cot \alpha_2 = 2 \cot(64°) = 0.9755$$

(b) The discharge velocity is
$$V_2 = \frac{V_x}{\cos \alpha_2} = \frac{240}{\cos(64°)} = 547.5 \text{ m/s}$$

(c) The blade speed is
$$U = \frac{V_x}{\phi} = \frac{240}{0.9755} = 246.0 \text{ m/s}$$
and the work delivered and the drop in the stagnation temperature are
$$w = \psi U^2 = 121.1 \text{ kJ/kg} \qquad \Delta T_0 = \frac{w}{c_p} = \frac{121.1}{1.148} = 105.5 \text{ K}$$

∎

6.3.3 Off-design operation

When a turbine is operated away from its design conditions, the incidence of the flow entering the blades changes, and this will increase the thermodynamic losses in the flow. The angle at which the flow leaves the stator tends not to change, however; nor does the angle of the relative velocity leaving the rotor. By recasting the Euler turbine equation
$$w = U(V_{u2} - V_{u3})$$
in terms of the exit angles gives an equation that shows how the turbine performs under off-design conditions. Replacing the exit velocity by
$$V_{u3} = U + W_{u3}$$
gives
$$w = U(-U - W_{u3} + V_{u2}) = -U^2 + UV_x(\tan \alpha_2 - \tan \beta_3)$$
Dividing though by U^2 yields
$$\psi = -1 + \phi(\tan \alpha_2 - \tan \beta_3) \qquad (6.21)$$

The departure angle from the stator α_2 is positive, and the exit angle of the relative velocity β_3 from the rotor is usually negative. Therefore the term in parentheses is positive. As these angles are close to the metal angles that are set by the design, they are fixed and the trigonometric terms in Eq. (6.21) tend to remain constant when the machine is operated away from its design condition. Thus Eq. (6.21) represents an operating line through the design point.

Figure 6.6b shows this operating line. As the flow rate is increased beyond its design value, the flow will eventually choke, and the flow coefficient no longer increases. In Figure 6.6a the velocity triangles illustrate how the incidence changes as the flow rate decreases. The design condition is denoted by subscript d and off-design, by o. With the exit flow angles held constant, the value of $\psi + 1$ drops in proportion to a drop in ϕ, and as the sketch shows, both α_3 and β_2 decrease in absolute value as the flow coefficient is reduced. This means that incidence of the flow to the vanes and blades decreases. Although the improper incidence leads to larger losses, these can be reduced by making the leading edges of the blades and vanes well rounded.

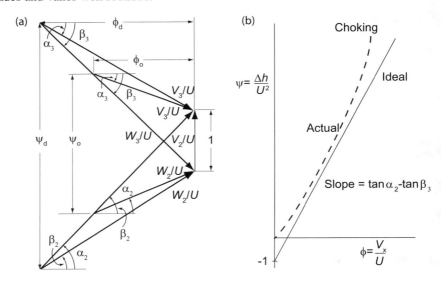

Figure 6.6 Off-design performance comparison.

■ EXAMPLE 6.5

In an axial turbine stage, flow leaves the stator at $V_2 = 350$ m/s in the direction $\alpha_2 = 60°$. Its blade-loading coefficient is $\psi = 1.8$ and flow coefficient is $\phi = 0.7$. Assuming that the axial velocity is reduced by 25 m/s from its design condition, find the percent reduction in the reaction.

Solution: The axial velocity at the design condition is

$$V_x = V_2 \cos \alpha_2 = 350 \cos(60°) = 175 \text{ m/s}$$

and the reaction is

$$R = 1 + \frac{\psi}{2} - \phi \tan \alpha_2 = 1 + 0.9 - 0.7 \tan(60°) = 0.688$$

Hence the angle of the relative flow leaving the rotor is

$$\tan \beta_3 = -\frac{R + \psi/2}{\phi} = -\frac{0.688 + 0.9}{0.7} = -2.2678 \qquad \beta_3 = -66.2°$$

Reduction of the axial velocity by 25 m/s gives $V_{xn} = 150 \text{ m/s}$, and with the blade speed constant the *new* value of flow coefficient is

$$\phi_n = \frac{V_{xn}}{V_x}\phi = \frac{150}{175}0.7 = 0.6$$

Assuming that the flow angles α_2 and β_3 remain constant, the new value for blade-loading coefficient is

$$\psi_n = -1 + \frac{\phi_n}{\phi}(\psi + 1) = -1 + \frac{0.6}{0.7}(1.8 + 1) = 1.4$$

A new value for reaction is then obtained from the equality

$$\tan \beta_3 = -\frac{R + \psi/2}{\phi} = -\frac{R_n + \psi_n/2}{\phi_n}$$

and it gives

$$R_n = -\frac{\psi_n}{2} + \frac{\phi_n}{\phi}\left(R + \frac{\psi}{2}\right) = -\frac{1.4}{2} + \frac{0.6}{0.7}\left(0.688 + \frac{1.8}{2}\right) = 0.661$$

Hence the percent reduction for the reaction is $(0.688 - 0.661)/0.688 = 0.039$ or about 4% when the axial velocity is reduced by 14.3%.

■

The values of the flow coefficients in the foregoing examples were chosen to be in their typical range $0.5 < \phi < 1.0$ and the blade-loading coefficients were chosen to be in their range of $1.4 < \psi < 2.2$. The reaction turned out to be generally close to 0.5, except, of course, for the zero-reaction stage. The typical stagnation temperature drop across the rotor is $120 \text{ K} - 150 \text{ K}$.

6.4 THREE-DIMENSIONAL FLOW

In the last stages of gas turbines, and certainly for steam turbines that exhaust to below atmospheric pressure, the blades are long in order to accommodate the large volumetric flow rate. Since the blade speed increases with radius, a simple approach is to construct velocity triangles at each element of the blade. As a consequence, the blade loading and reaction may vary considerably along the span of the blade. But this approach does not take into account the pressure variation properly. The aim of this section is to take into account the influence of the pressure increase from the hub to the casing in a flow with a swirl velocity and as a result also of the variation of the reaction and the blade loading along the span of long blades.

6.5 RADIAL EQUILIBRIUM

Consider a flow in which fluid particles move on cylindrical surfaces. Applying the momentum balance in the radial direction to the control volume shown in Figure 6.7 gives

$$-2V_u \sin\left(\frac{d\theta}{2}\right)\rho V_u\, dr = pr\, d\theta - (p + dp)(r + dr)d\theta + 2\left(p + \frac{1}{2}dp\right)dr \sin\left(\frac{d\theta}{2}\right)$$

The left side represents the rate at which the radial component of momentum leaves the control volume and the right side is the net pressure force. Viscous terms have been neglected. Noting that $\sin(d\theta/2) \approx \theta/2$ for small $d\theta$, simplifying, and dropping higher-order terms reduces this equation to

$$-\frac{1}{\rho}\frac{dp}{dr} + \frac{V_u^2}{r} = 0 \qquad (6.22)$$

The first term represents the net pressure force on a fluid particle of unit mass. The second term is the centrifugal force. Since the second term is positive, the pressure gradient must increase in the radial direction so that the sum of the two terms vanishes. Therefore, if the flow has a swirl component, its pressure must increase from the hub to the casing.

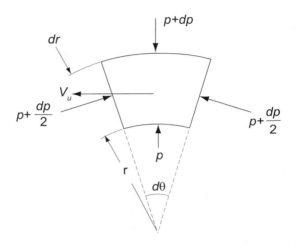

Figure 6.7 Radial equilibrium condition on a fluid element.

If the radial velocity is small, the definition of stagnation enthalpy can be written as

$$h_0 = h + \frac{1}{2}(V_x^2 + V_u^2) \qquad (6.23)$$

Differentiating gives

$$\frac{dh_0}{dr} = \frac{dh}{dr} + V_x\frac{dV_x}{dr} + V_u\frac{dV_u}{dr} \qquad (6.24)$$

The Gibbs equation $Tds = dh - dp/\rho$ can also be written as

$$T\frac{ds}{dr} = \frac{dh}{dr} - \frac{1}{\rho}\frac{dp}{dr} \qquad (6.25)$$

and substituting into this dh/dr from Eq. (6.24) and dp/dr from Eq. (6.22) gives

$$\frac{dh_0}{dr} - T\frac{ds}{dr} = V_x\frac{dV_x}{dr} + V_u\frac{dV_u}{dr} + \frac{V_u^2}{r}$$

which can be rewritten as

$$\frac{dh_0}{dr} - T\frac{ds}{dr} = V_x\frac{dV_x}{dr} + \frac{V_u}{r}\frac{d}{dr}(rV_u)$$

If neither h_0 nor s varies with r, then the left side is zero and this equation reduces to

$$V_x \frac{dV_x}{dr} + \frac{V_u}{r} \frac{d}{dr}(rV_u) = 0 \tag{6.26}$$

For a given the radial variation of V_u, the variation of the *axial velocity* with r can be determined by solving this equation.

6.5.1 Free vortex flow

Let the radial variation of the tangential velocity be given by $rV_u = K$. Then the second term in Eq. (6.26) vanishes, and V_x is seen to be constant. The work done on the blades by the fluid passing through a streamtube at radial location r of the rotor is given by

$$w = U(V_{u2} - V_{u3}) = \Omega r(\frac{K_2}{r} - \frac{K_3}{r}) = \Omega(K_2 - K_3)$$

Since this is independent of r, the work done is the same at each radial location. The meanline analysis that has been used in the previous chapters is thus justified if the tangential velocity distribution follows $V_u = K/r$.

The degree of reaction has been shown to be

$$R = 1 - \frac{1}{2}\phi(\tan\alpha_2 + \tan\alpha_3) = 1 - \frac{1}{2U}(V_x \tan\alpha_2 + V_x \tan\alpha_3) = 1 - \frac{1}{2U}(V_{u2} + V_{u3})$$

or

$$R = 1 - \frac{K_2 + K_3}{2\Omega r^2}$$

Hence the degree of reaction increases from the hub to the casing. The mass flow rate through the annulus is given by

$$\dot{m} = 2\pi V_x \int_{r_h}^{r_c} \rho r \, dr$$

since V_x is constant. The integration could be carried out were the density variation with the radial position known. It is found by noting first that a free vortex design leads to equal work done on each blade element, and therefore the stagnation temperature will remain uniform in the annulus. The loss of stagnation pressure takes place in the wake as the flow leaves a blade row. There is also a loss in the endwall boundary layers and in the tip region of the blades. Secondary flow losses are more evenly distributed across the annulus. To make analytical progress, it is assumed that entropy and thus also stagnation pressure are uniform across the flow channel. This can be achieved by good lateral mixing of the flow. Even if such mixing cannot be justified, if this assumption is made, the stagnation density is also uniform. Of course, irreversibilities still cause the stagnation pressure loss in the flow direction. The density ratio may be written as

$$\frac{\rho}{\rho_0} = \left(\frac{T}{T_0}\right)^{1/(\gamma-1)} = \left(\frac{T_0 - V^2/2c_p}{T_0}\right)^{1/(\gamma-1)} = \left(1 - \frac{V^2}{2c_p T_0}\right)^{1/(\gamma-1)}$$

The velocity in this expression is seen to vary with radius according to

$$V^2 = V_x^2 + V_u^2 = V_x^2 + V_{um}^2 \frac{r_m^2}{r^2}$$

Here r_m is the mean radius, at which the tangential velocity is V_{um}. Defining $y = r/r_m$, the mass balance can be converted to the form

$$\dot{m} = 2\pi V_x r_m^2 \rho_0 \int_{2\kappa/(1+\kappa)}^{2/(1+\kappa)} \left(1 - \frac{V_x^2}{2c_p T_0} - \frac{V_{um}^2}{2c_p T_0} \frac{1}{y^2}\right)^{1/(\gamma-1)} y\, dy \qquad (6.27)$$

in which $r_c/r_m = 2/(1+\kappa)$, $r_h/r_m = 2\kappa/(1+\kappa)$, and $\kappa = r_h/r_c$. This integral can be evaluated in closed form, at least for values of γ for which $1/(\gamma-1)$ is an integer. But, for example, if $\gamma = \frac{4}{3}$, the result of the integration is sufficiently complicated, that it is better to proceed by numerical integration.

Cohen et al. [15] use the mean radius as a reference value and express at the inlet to the rotor, the free vortex velocity distribution in the form

$$r_{2m} V_{u2m} = r_2 V_{u2}$$

which can be recast as

$$r_{2m} V_{x2} \tan \alpha_{2m} = r_2 V_{x2} \tan \alpha_2$$

from which it follows that

$$\tan \alpha_2 = \frac{r_{2m}}{r_2} \tan \alpha_{2m} \qquad (6.28)$$

Here r_2 denotes an arbitrary radial position at the inlet of the rotor. Similarly, at the exit

$$\tan \alpha_3 = \frac{r_{3m}}{r_3} \tan \alpha_{3m} \qquad (6.29)$$

The relative velocity at the inlet is $W_{u2} = V_{u2} - U$, from which

$$\tan \beta_2 = \tan \alpha_2 - \frac{U}{V_{x2}}$$

or

$$\tan \beta_2 = \tan \alpha_2 - \frac{r_2}{r_{2m}} \frac{1}{\phi_2} \qquad (6.30)$$

Similarly

$$\tan \beta_3 = \tan \alpha_3 - \frac{r_3}{r_{3m}} \frac{1}{\phi_3} \qquad (6.31)$$

These equations are valid even if the axial velocity differs between the inlet and exit of the rotor.

■ **EXAMPLE 6.6**

Combustion gases with $\gamma = \frac{4}{3}$ and $R = 287 \text{ J/(kg} \cdot \text{K)}$ expand through a turbine stage. The inlet stagnation temperature is $T_{01} = 1100$ K and the stagnation pressure is $p_{01} = 420$ kPa. The mean radius is $r_m = 0.17$ cm, and it is constant across the stage. The turbine is flared so that the axial velocity remains constant. The hub-to-casing radius is $\kappa_2 = 0.7$ at the inlet and $\kappa_3 = 0.65$ at exit of the rotor. The flow angles at the mean radius at the inlet are $\alpha_{2m} = 60.08°$ and $\beta_{2m} = 25.98°$. At the exit the corresponding angles are $\alpha_{3m} = -21.20°$ and $\beta_{3m} = -58.59°$. The axial velocity is $V_x = 231$ m/s. (a) Plot the inlet and exit flow angles along the span of the blades. (b) Determine the reaction at the hub. (c) Calculate the mass flow rate using both numerical integration and the mean density.

Solution: (a) The reaction at the mean radius of this stage can be obtained by adding both parts of Eq. (6.16) together and doing the same for Eq. (6.12) and then dividing one by the other. This gives

$$1 - \frac{1}{R_m} = \frac{\tan \alpha_{2m} + \tan \alpha_{3m}}{\tan \beta_{2m} + \tan \beta_{3m}}$$

Substituting the values of the angles gives $R_m = 0.460$. The flow coefficient is then obtained from

$$\phi_m = \frac{1}{\tan \alpha_{2m} - \tan \beta_{2m}} = 0.800$$

and the blade-loading coefficient is

$$\psi = 2(1 - R_m - \phi_m \tan \alpha_{3m}) = 1.700$$

The flow angles are next calculated using Eqs. (6.28) – 6.31. At the hub

$$\tan \alpha_{2h} = \frac{1+\kappa_2}{2\kappa_2} \tan \alpha_{2m} = \frac{1.7}{1.4} \tan(60.08°) = 2.11 \qquad \alpha_{2h} = 64.64°$$

$$\tan \beta_{2h} = \frac{1+\kappa_2}{2\kappa_2} \tan \alpha_{2m} - \frac{2\kappa_2}{1+\kappa_2} \frac{1}{\phi} = 1.08 \qquad \beta_{2h} = 47.22°$$

At the casing

$$\tan \alpha_{2c} = \frac{1+\kappa_2}{2} \tan \alpha_{2m} = \frac{1.7}{2} \tan(60.08°) = 1.48 \qquad \alpha_{2c} = 55.90°$$

$$\tan \beta_{2c} = \frac{1+\kappa_2}{2} \tan \alpha_{2m} - \frac{2}{1+\kappa_2} \frac{1}{\phi} = 0.0064 \qquad \beta_{2c} = 0.37°$$

The exit angles are calculated similarly. They are

$$\alpha_{3h} = -26.72° \qquad \beta_{3h} = -55.90° \qquad \alpha_{3c} = -17.74° \qquad \beta_{3c} = -61.41°$$

The variation in flow angles along the span is obtained from Eqs. (6.28) – (6.31), which, together with the blade shapes, are shown in Figure 6.8.

(b) The reaction at the hub is

$$R_h = 1 - \frac{\phi_m}{2} \left(\frac{1+\kappa_2}{2\kappa_2} \right)^2 (\tan \alpha_{2m} + \tan \alpha_{3m})$$

$$= 1 - \frac{0.8}{2} \left(\frac{1.7}{1.4} \right)^2 (\tan(60.08°) + \tan(-21.2°)) = 0.204$$

(c) The mass flow rate is calculated from Eq. (6.27). The stagnation density is

$$\rho_{01} = \frac{p_{01}}{RT_{01}} = \frac{420,000}{287 \cdot 1100} = 1.330 \, \text{kg/m}^3$$

With hub radius $r_h = 0.14$ m, and casing radius, $r_c = 0.20$ m, numerical integration gives

$$\dot{m} = 15.068 \, \text{kg/s}$$

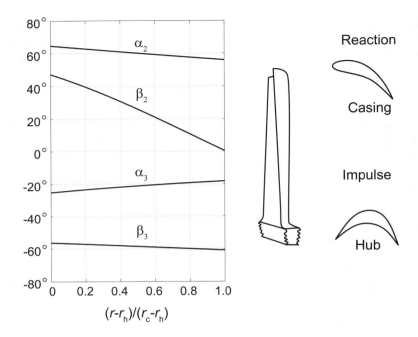

Figure 6.8 Variation of flow angles along the span for a gas with $\gamma = 1.4$.

Cohen et al. [15] suggested that a very good approximation may be obtained by ignoring the density variation along the span and using its value at the mean radius. Since

$$V_{u2m} = V_x \tan \alpha_{2m} = 231 \tan(60.08°) = 401.40 \, \text{m/s}$$

the mean temperature is

$$T_{2m} = T_{02} - \frac{V_x^2 + V_{u2m}^2}{2c_p} = 1100 - \frac{231^2 + 401.4^2}{2 \cdot 1148} = 1006.6 \, \text{K}$$

Hence

$$\rho_{2m} = \rho_{02} \left(\frac{T_{2m}}{T_{02}}\right)^{1/(\gamma-1)} = 1.33 \left(\frac{1006.6}{1100}\right)^3 = 1.019 \, \text{kg/m}^3$$

The flow rate is then

$$\dot{m} = \pi(r_c^2 - r_h^2)\rho_{2m} V_x = 2\pi r_m (r_c - r_h) \rho_{2m} V_x$$
$$= 2\pi \cdot 0.17 \cdot 0.06 \cdot 1.019 \cdot 231 = 15.086 \, \text{kg/s}$$

The approximation of using the density at the mean radius is seen to be excellent. ∎

6.5.2 Fixed blade angle

A design with a free vortex tangential velocity distribution has the attractive feature that each blade element delivers the same amount of work and that the axial velocity remains constant. On the other hand, the reaction varies quite strongly along the span of the blade. Other design possibilities exist. For example, if the nozzle angle is kept constant along the span, manufacturing cost of nozzles can be reduced. As the flow moves through the nozzles its stagnation enthalpy will not change, and, if the radial entropy gradients may be neglected, the equation for radial equilibrium

$$\frac{dh_0}{dr} - T\frac{ds}{dr} = V_x\frac{dV_x}{dr} + V_u\frac{dV_u}{dr} + \frac{V_u^2}{r}$$

reduces to

$$V_x\frac{dV_x}{dr} + V_u\frac{dV_u}{dr} + \frac{V_u^2}{r} = 0$$

as before. This can also be written as

$$\frac{d}{dr}\left(\frac{1}{2}V_x^2 + \frac{1}{2}V_u^2\right) + \frac{V_u^2}{r} = 0$$

Since $V_u = V \sin\alpha$ and $V_x^2 + V_u^2 = V^2$ this becomes

$$V\frac{dV}{dr} + \frac{V^2 \sin^2\alpha}{r} = 0$$

or

$$\frac{dV}{V} = -\sin^2\alpha\frac{dr}{r} \qquad (6.32)$$

If the angle α is constant, integrating gives

$$\frac{V(r)}{V_m} = \left(\frac{r_m}{r}\right)^{\sin^2\alpha}$$

In addition, $V_u = V\sin\alpha$ and $V_x = V\cos\alpha$, so that for constant α

$$r^{\sin^2\alpha}V_u = r_m^{\sin^2\alpha_m}V_{um}$$

For nozzles the exit flow angle is quite large (often between 60° and 70°). Therefore $\sin^2\alpha_2$ is in the range $0.75 - 0.88$ and the velocity distribution is nearly the same as for a free vortex. The rotor blades may then be twisted properly to give the free vortex distribution at their exit.

6.6 CONSTANT MASS FLUX

It has been seen that, if the tangential velocity varies inversely with radius, axial velocity is independent of radius. Since the density varies also, Horlock [35] suggested that a designer might decide to hold the axial *mass flux* ρV_x constant. This requires the blade angle to vary in such a way that

$$\rho V \cos\alpha = \rho_m V_m \cos\alpha_m$$

remains independent of radius. The flow angle in terms of the velocity and density ratios is therefore

$$\frac{\cos\alpha}{\cos\alpha_m} = \frac{V_m}{V}\frac{\rho_m}{\rho} \qquad (6.33)$$

From the definition of stagnation temperature

$$T_0 = T\left(1 + \frac{\gamma-1}{2}M^2\right) = T_m\left(1 + \frac{\gamma-1}{2}M_m^2\right)$$

the ratio of the static temperature to its value at the mean radius is

$$\frac{T}{T_m} = \frac{2+(\gamma-1)M_m^2}{2+(\gamma-1)M^2} \tag{6.34}$$

It then follows that

$$\frac{\rho}{\rho_m} = \left(\frac{2+(\gamma-1)M_m^2}{2+(\gamma-1)M^2}\right)^{1/(\gamma-1)} \tag{6.35}$$

and

$$\frac{p}{p_m} = \left(\frac{2+(\gamma-1)M_m^2}{2+(\gamma-1)M^2}\right)^{\gamma/(\gamma-1)} \tag{6.36}$$

The velocity ratio is obtained from the definition of Mach number:

$$\frac{V}{V_m} = \frac{M}{M_m}\sqrt{\frac{T}{T_m}} = \frac{M}{M_m}\left(\frac{2+(\gamma-1)M_m^2}{2+(\gamma-1)M^2}\right)^{1/2} \tag{6.37}$$

The ratio of cosines of the flow angles can now be written as

$$\frac{\cos\alpha}{\cos\alpha_m} = \frac{M_m}{M}\left(\frac{2+(\gamma-1)M^2}{2+(\gamma-1)M_m^2}\right)^{(\gamma+1)/2(\gamma-1)} \tag{6.38}$$

From the equation for radial equilibrium it follows that

$$\frac{dV}{V} = -\sin^2\alpha\,\frac{dr}{r} = (\cos^2\alpha - 1)\frac{dr}{r}$$

and by logarithmically differentiating Eq. (6.37) yields the equation

$$\frac{dM^2}{M^2(2+(\gamma-1)M^2)} = (\cos^2\alpha - 1)\frac{dr}{r} \tag{6.39}$$

Dividing through by $\cos^2\alpha - 1$ and integrating both sides gives

$$I = \int_{M_h}^{M_c} \frac{2\,dM}{M(2+(\gamma-1)M^2)(\cos^2\alpha - 1)} = \ln\frac{r_c}{r_h} = \ln\frac{1}{\kappa} \tag{6.40}$$

so that

$$\kappa = e^{-I}$$

Since the highest Mach number is at the hub, the way to proceed is to set it at an acceptable value. This may be slightly supersonic. Then, if the flow angle and Mach number are known at the average radius, Eq. (6.38) can be used to find the value of α_h. After that Eq. (6.38) is rewritten as

$$\frac{\cos\alpha}{\cos\alpha_h} = \frac{M_h}{M}\left(\frac{2+(\gamma-1)M^2}{2+(\gamma-1)M_h^2}\right)^{(\gamma+1)/2(\gamma-1)} \tag{6.41}$$

and this is substituted for $\cos \alpha$ in the integrand of Eq. (6.40). Finally, by trial, the value of M_c needs to be chosen so that the numerical integration gives the desired radius ratio κ. For a flow at the exit of the nozzle the stagnation temperature is known, and with the mean Mach number known, the mean temperature T_m can be determined. After that, the other ratios are calculated from Eqs. (6.34) – (6.37).

To obtain the radial locations that correspond to the calculated values of the thermodynamic properties, Eq. (6.39) is written as

$$\frac{2dM}{M(2+(\gamma-1)M^2)(\cos^2\alpha-1)} = \frac{dr}{r} \qquad (6.42)$$

and this is solved numerically using a fine grid.

Results from a sample calculation for $M_h = 1.15$ are shown in Figure 6.9. Panel (a) shows the variation of the thermodynamic variables, normalized with respect to their values at the mean radius. Since the flow resembles a free vortex type, the largest velocity is at

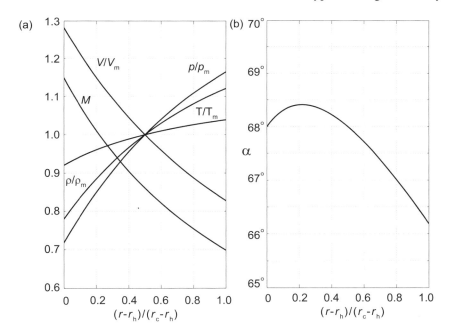

Figure 6.9 (a) Temperature, density, pressure, and velocity along the span for a gas with $\gamma = 1.4$; (b) flow angle leaving a nozzle as a function of the radial location.

the hub. The temperature there has the smallest value, since the stagnation temperature is constant across the span. Hence the Mach number is largest at the hub and drops to a value $M_c = 0.6985$ at the casing for a flow with radius ratio $\kappa = 0.6$. Four significant figures were used to make sure that κ was also accurate to the same number of significant figures. Radial equilibrium theory shows that pressure increases from the hub to the casing in a flow with a swirl component of velocity. The density follows the ideal gas law and it increases from the hub to the casing. The flow angles are shown in Figure 6.9b. The value at the mean radius was set at $\alpha_m = 68°$ and its value at the hub happens to come out to be the same. At the casing the flow angle drops to $\alpha_c = 66.18°$ and the entire variation is seen to be quite slight.

6.7 TURBINE EFFICIENCY AND LOSSES

Three methods are in common use for the calculation of losses in axial turbines. The correlation by Soderberg was introduced in Chapter 5. The other two methods are based on the original work of Ainley and Mathieson [2] and the studies of Craig and Cox [16]. The former is discussed below; the latter is presented by Wilson and Korakianitis [81] in their text on gas turbines. In this section analytical results are developed that relate the stage efficiency to the flow parameters. They enable the calculation of efficiency contours by methods introduced by Hawthorne [33] and Smith [71], and further developed by Lewis [50]. Horlock [35] gives a comprehensive review of the early work.

6.7.1 Soderberg loss coefficients

The loss correlation of Soderberg makes use of the static enthalpy loss coefficient

$$\zeta = \frac{h - h_s}{\frac{1}{2}V^2}$$

with V replaced by V_2 for the stator and by W_3 for the rotor. The nominal value of the loss coefficient is calculated from

$$\zeta^* = 0.04 + 0.06 \left(\frac{\varepsilon}{100}\right)^2$$

in which ε is the amount of turning, $\varepsilon_S = \alpha_2 - \alpha_3$ for the stator and $\varepsilon_R = \beta_2 - \beta_3$ for the rotor. The angles are in degrees. The nominal value, identified with superscript star, is for a blade height-to-axial chord ratio $b/c_x = 3.0$ and Reynolds number equal to 10^5. For different values of blade height to axial-chord-ratio, a new value for stator vanes is calculated from

$$\bar{\zeta} = (1 + \zeta^*)\left(0.993 + 0.021\frac{c_x}{b}\right) - 1$$

and for the rotor, from

$$\bar{\zeta} = (1 + \zeta^*)\left(0.975 + 0.075\frac{c_x}{b}\right) - 1$$

If Reynolds number differs from 10^5, the Reynolds number correction is obtained from

$$\zeta = \left(\frac{10^5}{\text{Re}}\right)^{1/4} \bar{\zeta}$$

The Reynolds number is based on the hydraulic diameter, which is given approximately by the expression

$$D_h = \frac{2sb\cos\alpha_2}{s\cos\alpha_2 + b}$$

for the stator and by

$$D_h = \frac{2sb\cos\beta_3}{s\cos\beta_3 + b}$$

for the rotor.

6.7.2 Stage efficiency

The process lines for a turbine stage are shown in Figure 6.10. The states of static enthalpy are drawn such that the enthalpy drop across the rotor is slightly larger than that across the stator. The reaction therefore is slightly larger than one-half. An isentropic expansion through the stator takes the process from state 1 to state $2s$, whereas the actual end state is at state 2. The stagnation enthalpy remains constant through the stator and its process line is horizontal. The stagnation pressure in the interblade gap is denoted by p_{02}.

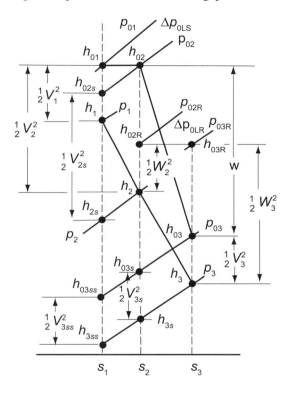

Figure 6.10 Thermodynamic states for expansion across a turbine stage.

The irreversible expansion across the rotor takes the process to state 3, with a corresponding stagnation stagnation enthalpy h_{03} and stagnation pressure p_{03}. The loss of stagnation pressure is discussed in the next subsection. The losses can be related to efficiency by first writing (as was done for steam turbines) the efficiency as

$$\eta_{tt} = \frac{h_{01} - h_{03}}{h_{01} - h_{03ss}}$$

and then manipulating it into the form

$$\frac{1}{\eta_{tt}} - 1 = \frac{h_{03} - h_{03ss}}{h_{01} - h_{03}} = \frac{h_3 - h_{3ss}}{w} + \frac{V_3^2 - V_{3ss}^2}{2w}$$

This can be further rearranged as

$$\frac{1}{\eta_{tt}} - 1 = \frac{(h_3 - h_{3s}) + (h_{3s} - h_{3ss})}{w} + \left(1 - \frac{V_{3ss}^2}{V_3^2}\right) \frac{V_3^2}{2w}$$

The first term in the numerator is simply

$$h_3 - h_{3s} = \frac{1}{2}\zeta_R W_3^2$$

Next, integrating the Gibbs equation along the constant-pressure line p_3 from state $3ss$ to $3s$ gives

$$s_2 - s_1 = c_p \ln \frac{T_{3s}}{T_{3ss}}$$

Similarly, integration along the constant pressure line p_2 yields

$$s_2 - s_1 = c_p \ln \frac{T_2}{T_{2s}}$$

Equating the RHSs gives

$$\frac{T_{3s}}{T_{3ss}} = \frac{T_2}{T_{2s}} \tag{6.43}$$

Subtracting one from each side, rearranging, and multiplying by c_p gives

$$h_{3s} - h_{3ss} = \frac{T_{3ss}}{T_{2s}}(h_2 - h_{2s})$$

Since $T_{3ss}/T_{2s} = T_{3s}/T_2$ and

$$h_2 - h_{2s} = \frac{1}{2}\zeta_S V_2^2$$

the expression for efficiency can be written as

$$\frac{1}{\eta_{tt}} - 1 = \frac{1}{2sw}\left[\zeta_R W_3^2 + \frac{T_{3s}}{T_2}\zeta_S V_2^2 + \left(1 - \frac{T_{3ss}}{T_3}\right)V_3^2\right]$$

In the last term the equality $V_{3ss}^2/V_3^2 = T_{3ss}/T_3$ was used, which follows from the fact that $M_3 = M_{3ss}$, as was shown in Chapter 5. Furthermore, since $V_x = W_3 \cos\beta_3 = V_2 \cos\alpha_2 = V_3 \cos\alpha_3$, the expression for efficiency can be recast as

$$\frac{1}{\eta_{tt}} - 1 = \frac{\phi^2}{2\psi}\left[\frac{\zeta_R}{\cos^2\beta_3} + \frac{T_{3s}}{T_2}\frac{\zeta_S}{\cos^2\alpha_2} + \left(1 - \frac{T_{3ss}}{T_3}\right)\frac{1}{\cos^2\alpha_3}\right] \tag{6.44}$$

Often this is approximated by

$$\frac{1}{\eta_{tt}} - 1 = \frac{\phi^2}{2\psi}\left(\frac{\zeta_R}{\cos^2\beta_3} + \frac{\zeta_S}{\cos^2\alpha_2}\right) \tag{6.45}$$

It will shown in an example that the error in using this approximation is very small.

6.7.3 Stagnation pressure losses

The stagnation pressure drop across the stator can be related to the static enthalpy loss coefficient by first integrating the Gibbs equation along the constant-stagnation-pressure line $h_{01} = h_{02}$:

$$s_2 - s_1 = R \ln\left(\frac{p_{01}}{p_{02}}\right)$$

Similarly, integrating the Gibbs equation along the constant-pressure line p_2 gives

$$s_2 - s_1 = \frac{\gamma R}{\gamma - 1} \ln\left(\frac{T_2}{T_{2s}}\right)$$

Equating the RHSs gives

$$\frac{p_{01}}{p_{02}} = \left(\frac{T_2}{T_{2s}}\right)^{\gamma/(\gamma-1)}$$

The definition for the static enthalpy loss coefficient

$$h_2 - h_{2s} = \frac{1}{2}\zeta_S V_2^2$$

can be written as

$$T_2 - T_{2s} = \frac{\gamma - 1}{2\gamma R} \zeta_S M_2^2 \gamma R T_2$$

from which

$$\frac{T_2}{T_{2s}} = \left(1 - \frac{\gamma - 1}{2}\zeta_S M_2^2\right)^{-1}$$

Since the second term involving the Mach number is small, this can be expanded as

$$\frac{T_2}{T_{2s}} = 1 + \frac{\gamma - 1}{2}\zeta_S M_2^2$$

The pressure ratio p_{01}/p_{02} is therefore

$$\frac{p_{01}}{p_{02}} = \left(1 + \frac{\gamma - 1}{2}\zeta_S M_2^2\right)^{\gamma/(\gamma-1)}$$

and expanding this gives

$$\frac{p_{01}}{p_{02}} = 1 + \frac{\gamma}{2}\zeta_S M_2^2$$

The expression for the stagnation pressure loss now takes the form

$$\Delta p_{0LS} = \frac{\gamma}{2}p_{02}\zeta_S M_2^2 = \frac{1}{2}\frac{T_{02}}{T_2}\rho_{02}\zeta_S V_2^2$$

or

$$\Delta p_{0LS} = \rho_{02}(1 + \frac{\gamma - 1}{2}M_2^2)\frac{1}{2}\zeta_S V_2^2 \tag{6.46}$$

Since the loss coefficient ζ_S is rather insensitive to Mach number, this shows how the stagnation pressure loss increases as compressibility becomes important.

The development of the stagnation pressure loss across the rotor is similar. It will be carried out in detail in order to highlight the use of stagnation properties on the basis of relative velocity. First, the stagnation enthalpy of relative motion is defined as

$$h_{03R} = h_3 + \frac{1}{2}W_3^2$$

and in using the Mach number in terms of W_3, defined as

$$M_{3R} = \frac{W_3}{\sqrt{\gamma R T_3}}$$

this can be rewritten in the form

$$\frac{T_{03R}}{T_3} = 1 + \frac{\gamma-1}{2}M_{3R}^2$$

The relative stagnation pressure is calculated from

$$\frac{p_{03R}}{p_3} = \left(1 + \frac{\gamma-1}{2}M_{3R}^2\right)^{\gamma/(\gamma-1)}$$

Since $h_{02R} = h_{03R}$ across the rotor, integrating the Gibbs equation along the line of constant relative stagnation enthalpy and also along the constant-pressure line p_3, gives

$$s_3 - s_2 = R\ln\frac{p_{02R}}{p_{03R}} \qquad s_3 - s_2 = c_p \ln\frac{T_3}{T_{3s}}$$

so that

$$\frac{p_{02R}}{p_{03R}} = \left(\frac{T_3}{T_{3s}}\right)^{\gamma/(\gamma-1)}$$

From the definition for static enthalpy loss coefficient for the rotor

$$h_3 - h_{3s} = \frac{1}{2}\zeta_R W_3^2$$

the temperature ratio

$$\frac{T_3}{T_{3s}} = \left(1 - \frac{\gamma-1}{2}\zeta_R M_{3R}^2\right)^{-1}$$

is obtained. Noting again that the term involving the Mach number is small and expanding the RHS gives

$$\frac{T_3}{T_{3s}} = 1 + \frac{\gamma-1}{2}\zeta_R M_{3R}^2$$

and the pressure ratio p_{02R}/p_{03R} is therefore

$$\frac{p_{02R}}{p_{03R}} = \left(1 + \frac{\gamma-1}{2}\zeta_R M_{3R}^2\right)^{\gamma/(\gamma-1)}$$

Expanding this gives

$$\frac{p_{02R}}{p_{03R}} = 1 + \frac{\gamma}{2}\zeta_R M_{3R}^2$$

The stagnation pressure loss across the rotor is therefore

$$\Delta p_{0LR} = \frac{\gamma}{2}p_{03R}\zeta_R M_{3R}^2 = \frac{1}{2}\frac{T_{03R}}{T_3}\rho_{03R}\zeta_R W_3^2$$

or

$$\Delta p_{0LR} = \rho_{03R}\left(1 + \frac{\gamma-1}{2}M_{3R}^2\right)\frac{1}{2}\zeta_R W_3^2 \qquad (6.47)$$

EXAMPLE 6.7

Combustion gases with $\gamma = \frac{4}{3}$ and $c_p = 1148\,\text{J}/(\text{kg}\cdot\text{K})$ flow through a normal turbine stage with $R = 0.60$. The flow enters the stator at $\alpha_1 = -33.0°$ and leaves at velocity $V_2 = 450\,\text{m/s}$. The inlet stagnation temperature is $1200\,\text{K}$, and the inlet stagnation pressure is $15\,\text{bar}$. The flow coefficient is $\phi = 0.7$, the blade height-to-axial chord ratio is $b/c_x = 3.5$, and the Reynolds number is 10^5. Find the efficiency of the stage.

Solution: The blade-loading coefficient is first determined from

$$\psi = 2(1 - R - \phi\tan\alpha_3) = 2(1 - 0.6 - 0.7\tan(-33°)) = 1.709$$

The flow angle leaving the stator is

$$\alpha_2 = \tan^{-1}\left(\frac{1 - R + \psi/2}{\phi}\right) = \tan^{-1}\left(\frac{1 - 0.6 + 1.709/2}{0.7}\right) = 60.84°$$

and the angle of the relative velocity leaving the stage is

$$\beta_3 = \tan^{-1}\left(\frac{-R - \psi/2}{\phi}\right) = \tan^{-1}\left(\frac{-0.6 - 1.709/2}{0.7}\right) = -64.30°$$

The angle of the relative velocity at the inlet of the rotor is

$$\beta_2 = \tan^{-1}\left(\frac{-R + \psi/2}{\phi}\right) = \tan^{-1}\left(\frac{-0.6 + 1.709/2}{0.7}\right) = 19.99°$$

The deflections are therefore

$$\varepsilon_S = \alpha_2 - \alpha_1 = 60.84 + 33.00 = 93.84°$$

$$\varepsilon_R = \beta_2 - \beta_3 = 19.99 + 64.30 = 84.29°$$

and the loss coefficients can now be calculated. First, the nominal values are

$$\bar{\zeta}_S = 0.04 + 0.06\left(\frac{\varepsilon_S}{100}\right)^2 = 0.04 + 0.06\left(\frac{93.84}{100}\right)^2 = 0.0928$$

and

$$\bar{\zeta}_R = 0.04 + 0.06\left(\frac{\varepsilon_R}{100}\right)^2 = 0.04 + 0.06\left(\frac{84.29}{100}\right)^2 = 0.0826$$

When corrected for $b/c_x = 3.5$ they are

$$\zeta_S = (1 + \bar{\zeta}_S)\left(0.993 + 0.021\frac{c_x}{b}\right) - 1 = (1 + 0.0928)\left(0.993 + \frac{0.021}{3.5}\right) = 0.0917$$

$$\zeta_R = (1 + \bar{\zeta}_R)\left(0.975 + 0.075\frac{c_x}{b}\right) - 1 = (1 + 0.0826)\left(0.975 + \frac{0.075}{3.5}\right) = 0.0788$$

The axial velocity is

$$V_x = V_2\cos\alpha_2 = 450\cos(60.84°) = 219.26\,\text{m/s}$$

and the tangential velocity leaving the rotor is

$$V_{u3} = V_x \tan\alpha_3 = 219.26\tan(-33°) = -142.39\,\text{m/s}$$

so that

$$V_3 = \sqrt{V_x^2 + V_{u3}^2} = \sqrt{219.26^2 + 124.39^2} = 261.44\,\text{m/s}$$

The blade speed is $U = V_x/\phi = 219.26/0.7 = 313.23\,\text{m/s}$. The tangential component of the relative velocity leaving the stage is

$$W_{u3} = V_{u3} - U = -142.39 - 313.23 = -455.62\,\text{m/s}$$

so that

$$W_3 = \sqrt{W_x^2 + W_{u3}^2} = \sqrt{219.26^2 + 455.62^2} = 505.63\,\text{m/s}$$

The work done by the stage is $w = \psi U^2 = 1.709 \cdot 313.23^2 = 167.69\,\text{kJ/kg}$. An approximate value for the stage efficiency may now be obtained by setting $T_{3ss} = T_{2s}$ and $T_{3ss} = T_3$ in

$$\frac{1}{\eta_{tt}} - 1 = \frac{\zeta_R W_3^2 + \dfrac{T_{3ss}}{T_{2s}}\zeta_S V_2^2 + \left(1 - \dfrac{T_{3ss}}{T_3}\right)V_3^2}{2w}$$

so that this expression evaluates to

$$\frac{1}{\eta_{tt}} - 1 = \frac{0.0788 \cdot 505.63^2 + 0.0917 \cdot 450^2}{2 \cdot 167690} = 0.124$$

with the result that $\eta_{tt} = 0.8965$.

To see the extent to which neglecting the temperature ratio T_{3s}/T_2 and the kinetic energy correction changes the efficiency, these terms are calculated next. The isentropic stage work is $w_s = w/\eta_{tt} = 167.69/0.8965 = 187.05\,\text{kJ/kg}$. With $T_{02} = T_{01} = 1200\,\text{K}$, the exit temperature is

$$T_{03} = T_{02} - \frac{w}{c_p} = 1200 - \frac{167{,}690}{1148} = 1053.9\,\text{K}$$

and for an isentropic process it is

$$T_{03ss} = T_{02} - \frac{w_s}{c_p} = 1200 - \frac{187{,}046}{1148} = 1037.1\,\text{K}$$

The static temperature at the exit is

$$T_3 = T_{03} - \frac{V_3^2}{2c_p} = 1053.9 - \frac{252.09^2}{2 \cdot 1148} = 1024.2\,\text{K}$$

and T_{3ss} can be calculated from

$$T_{3ss} = \frac{T_{03ss}}{T_{03}}T_3 = \frac{1037.1}{1053.9}1024.2 = 1007.8\,\text{K}$$

From the definition of static enthalpy loss coefficient across the rotor

$$T_{3s} = T_3 - \frac{\zeta_R W_3^2}{2c_p} = 999.0 - \frac{0.0788 \cdot 505.63^2}{2 \cdot 1148} = 1015.4\,\text{K}$$

In addition

$$T_2 = T_{02} - \frac{V_2^2}{2c_p} = 1200 - \frac{450^2}{2 \cdot 1148} = 1111.8\,\text{K}$$

so that

$$\frac{1}{\eta_{tt}} - 1 = \frac{0.0788 \cdot 505.63^2 + \frac{1015.4}{1111.8} 0.0917 \cdot 450^2 + (1 - \frac{1007.8}{1024.2})261.44^2}{2 \cdot 167690} = 0.1154$$

Therefore the total-to-total efficiency of the stage is $\eta_{tt} = 0.8965$, which to four significant figures is the same as that without the temperature and velocity correction. The temperature correction makes the loss through the rotor lower, but the velocity correction adds slightly to the losses, with the net result that these are compensating errors, and the shorter calculation gives an accurate result.

The stagnation pressure at the exit of the stator is obtained from

$$\frac{p_{01}}{p_{02}} = 1 + \frac{\gamma}{2}\zeta_S M_2^2$$

The Mach number is

$$M_2 = \frac{V_2}{\sqrt{\gamma R T_2}} = \frac{450}{\sqrt{1.333 \cdot 287 \cdot 1111.8}} = 0.690$$

so that

$$\frac{p_{01}}{p_{02}} = 1 + \frac{2 \cdot 0.0917 \cdot 0.690^2}{3} = 1.0291 \qquad p_{02} = \frac{1500}{1.0291} = 1457.6\,\text{kPa}$$

and the stagnation pressure loss across the stator is

$$\Delta p_{0\text{LS}} = 1500 - 1457.6 = 42.4\,\text{kPa}$$

To calculate the loss of stagnation pressure across the rotor, the relative and absolute Mach numbers at the exit are determined first:

$$M_3 = \frac{V_3}{\sqrt{\gamma R T_3}} = \frac{261.44}{\sqrt{1.333 \cdot 287 \cdot 1024.2}} = 0.4176$$

$$M_{3R} = \frac{W_3}{\sqrt{\gamma R T_3}} = \frac{505.63}{\sqrt{1.333 \cdot 287 \cdot 1024.2}} = 0.8077$$

Then stagnation pressure at the exit is

$$p_{03} = p_{01}\left(\frac{T_{03ss}}{T_{01}}\right)^{\gamma/(\gamma-1)} = 1500\left(\frac{1037.1}{1200}\right)^4 = 836.8\,\text{kPa}$$

and p_3 is

$$p_3 = p_{03}\left(1 + \frac{\gamma - 1}{2}M_3^2\right)^{-\gamma/(\gamma-1)} = 836.8\left(1 + \frac{0.4176^2}{6}\right)^{-4} = 746.14\,\text{kPa}$$

The stagnation pressure p_{03R} is obtained from

$$p_{03R} = p_3 \left(1 + \frac{\gamma - 1}{2} M_{3R}^2\right)^{\gamma/(\gamma-1)} = 746.14 \left(1 + \frac{0.8077^2}{6}\right)^4 = 1127.5 \, \text{kPa}$$

and the value of p_{02R} is then

$$p_{02R} = p_{03R} \left(1 + \frac{\gamma}{2} \zeta_R M_{3R}^2\right) = 1127.5 \left(1 + \frac{2}{3} 0.088 \cdot 0.8077^2\right) = 1166.1 \, \text{kPa}$$

so that the loss of stagnation pressure across the rotor is

$$\Delta p_{0LR} = 1166.1 - 1127.5 = 38.6 \, \text{kPa}$$

■

6.7.4 Performance charts

A useful collection of turbine performance characteristics was compiled by Smith [71] in 1965. His chart is shown in Figure 6.11. Each design is labeled in a small circle by the value of efficiency that may be achieved for a given choice of flow coefficient $\phi = V_x/U$ and a stage-loading coefficient $\psi = w/U^2 = \Delta h_0/U^2$. The curves of constant efficiency are based on a theory by Smith. He took blade losses other than tip losses into account, and for this reason the actual values of efficiency are expected to drop slightly.

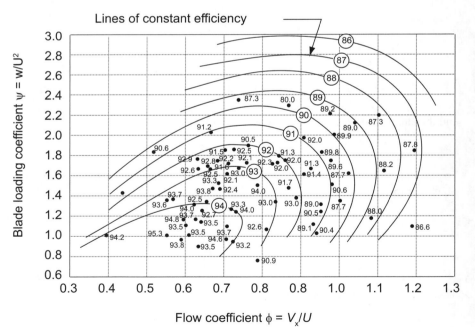

Figure 6.11 Variation of measured stage efficiency with stage loading coefficient and flow coefficient for axial-flow turbines. (Adapted from Smith [71].)

The flow coefficients in the range from $0.6 \leq \phi \leq 1.0$ give uncorrected efficiencies in a range from 90% to 94%, depending on how heavily loaded the blades are. Typical turbines have a blade loading coefficient in the range $1.5 < \psi < 2.2$, but there are designs outside this range.

Smith's method for calculating the main features of the performance characteristics has been extended by Lewis [50]. The present discussion follows Lewis, who suggests writing the efficiency in the form

$$\eta_{tt} = \frac{1}{1+f_L} \qquad (6.48)$$

The calculations in the foregoing example show that f_L can be approximated by

$$f_L = \frac{\zeta_R W_3^2 + \zeta_S V_2^2}{2w} = \frac{1}{2\psi}\left[\zeta_R\left(\frac{W_3}{U}\right)^2 + \zeta_S\left(\frac{V_2}{U}\right)^2\right] \qquad (6.49)$$

Since

$$\tan\alpha_2 = \frac{1-R+\psi/2}{\phi} \qquad \tan\beta_3 = \frac{-(R+\psi/2)}{\phi}$$

it follows that

$$\cos\alpha_2 = \frac{\phi}{\sqrt{\phi^2+(1-R+\psi/2)^2}} \qquad \cos\beta_3 = \frac{R+\psi/2}{\sqrt{\phi^2+(R+\psi/2)^2}}$$

and the velocity ratios may be written as

$$\left(\frac{V_2}{U}\right)^2 = \phi^2 + \left(1-R+\frac{\psi}{2}\right)^2$$

and

$$\left(\frac{W_3}{U}\right)^2 = \phi^2 + \left(R+\frac{\psi}{2}\right)^2$$

so that

$$f_L = \frac{1}{2\psi}\left[\zeta_R\left(\phi^2+\left(R+\frac{\psi}{2}\right)^2\right) + \zeta_S\left(\phi^2+\left(1-R+\frac{\psi}{2}\right)^2\right)\right]$$

This may be expressed in a more convenient form by defining $\nu = \zeta_S/\zeta_R$ and $F_L = f_L/\zeta_R$, so that

$$F_L = \frac{1}{2\psi}\left[\phi^2+\left(R+\frac{\psi}{2}\right)^2 + \nu\left(\phi^2+\left(1-R+\frac{\psi}{2}\right)^2\right)\right] \qquad (6.50)$$

and the efficiency can now be written as

$$\eta_{tt} = \frac{1}{1+F_L\zeta_R}$$

The maximum efficiency is obtained by minimizing F_L with respect to ψ with the value of ζ_R assumed to remain constant. Then, if the ratio ν is also assumed to remain constant, differentiating and setting the result to zero gives

$$\left(\frac{\partial F_L}{\partial \psi}\right)_{\nu,\phi,R} = \frac{1}{\psi}\left[\left(R+\frac{\psi}{2}\right)+\nu\left(1-R+\frac{\psi}{2}\right)\right]$$
$$-\frac{1}{\psi^2}\left[\phi^2+\left(R+\frac{\psi}{2}\right)^2+\nu\left(\phi^2+\left(1-R+\frac{\psi}{2}\right)^2\right)\right] = 0$$

which, when solved for ψ, leads to

$$\psi_{mm} = 2\sqrt{\frac{\phi^2 + R^2 + \nu(\phi^2 + (1-R)^2)}{1+\nu}} \qquad (6.51)$$

For $\nu = 1$ and $R = \frac{1}{2}$ this reduces to

$$\psi_m = \sqrt{4\phi^2 + 1} \qquad (6.52)$$

Contours of constant F_L are obtained by rearranging Eq. (6.50) first as

$$\psi^2 + \frac{4[\nu(1-R) + R - 2F_L]}{1+\nu}\psi + \frac{4[\phi^2 + R^2 + \nu(\phi^2 + (1-R)^2)]}{1+\nu} = 0$$

and solving it for ψ. This gives

$$\psi = \frac{2\left[2F_L - \nu(1-R) - R \pm \sqrt{4F_L^2 - 4F_L[\nu(1-R) + R] + 4\nu R(1-R) - \nu - (1+\nu)^2\phi^2}\right]}{1+\nu}$$

For $R = 0.5$ and $\zeta_S = \zeta_R = 0.9$, the two branches of each of the curves are shown Figure 6.12. The knee of the curves is where the discriminant is zero, namely, at

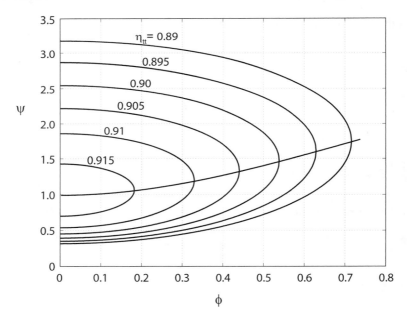

Figure 6.12 Contours of constant efficiency for an axial turbine stage with $R = 0.5$, $\zeta_S = 0.9$, and $\zeta_R = 0.9$; also shown is the curve of least losses.

$$\phi_m = \frac{\sqrt{4F_L^2 - 4F_L[\nu(1-R) + R] + \nu - 4\nu R(1-R)}}{1+\nu}$$

The locus of points of the blade-loading coefficient for which the losses are minimum, obtained from Eq. (6.52), is also shown. As stressed by Lewis, when the efficiency is written as

$$\eta_{tt} = \frac{1}{1 + F_L \zeta_R}$$

the factor F_L depends primarily on the shape of the velocity diagrams, which, in turn, are completely determined by ψ, ϕ, and R. The irreversibilities are taken into account by ζ_R and ζ_S. These depend on the amount of turning, but their influence on the shape of the efficiency contours is less than the influence of the flow angles.

The results in Figure 6.12 are qualitatively similar those in Figure 6.11, but they differ in important details. In the calculations the loss coefficients were assumed to be constant with values $\zeta_R = \zeta_S = 0.09$. They clearly depend on the amount of turning and thus on the values of ψ, ϕ, and R. As mentioned, Smith subtracted out the tip losses, and they are not included in Soderberg's correlation, either. The experiments on which the Smith plot is based were carried out on a test rig at low temperature, and therefore the results do not represent real operating conditions. The amount of turning today is approaching 90° or even higher [19]. This is achieved by using computational fluid dynamic analysis to design highly three-dimensional blades. In fact, the high efficiencies achieved today make further increases in efficiency more and more difficult to achieve [21].

The actual loss coefficients using the Soderberg correlation are easily included in the calculations, and the contours of constant efficiency and the deflection for a 50% reaction are shown in Figure 6.13. The results are based on a blade height-to-axial chord ratio of

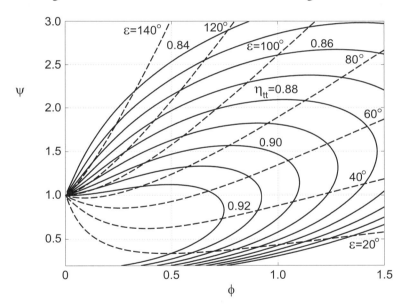

Figure 6.13 Contours of constant efficiency and deflection for a stage with $R = 0.5$, $b/c_x = 3$, and $\text{Re} = 10^5$.

$b/c_x = 3$, and the Reynolds number was set at $\text{Re} = 10^5$. From the expression

$$\psi = 2(1 - R - \phi \tan \alpha_3)$$

it is seen that when $\phi = 0$ and $R = \frac{1}{2}$, the blade-loading coefficient is $\psi = 1$. If this stage has an axial entry with $\alpha_3 = 0$, the loading coefficient is $\psi = 1$ for any value of ϕ. Examination of the figure shows that for an axial entry, as ϕ is increased to 0.5, the efficiency increases to slightly over 0.92 and decreases from there as ϕ is increased. Since the stage reaction is 50% the flow turns across the rotor and stator by an equal amount. At

$\phi = 0.5$ and $\psi = 1$, since $\alpha_3 = \alpha_1 = 0$, the flow angle α_2 is

$$\alpha_2 = \tan^{-1}\left(\frac{1-R+\psi/2}{\phi}\right) = \tan^{-1}(2) = 63.4°$$

On the other hand, if the entry angle is chosen to be $\alpha_3 = \alpha_1 = -45°$, then

$$\psi = 1 + 2\phi$$

and at $\phi = 0.5$ the blade-loading coefficient would be $\psi = 2$. The flow angle leaving the nozzle would then be $\alpha_2 = 71.56°$, and the flow angles for the rotor would have the values $\beta_2 = 45°$ and $\beta_3 = -71.56°$. Hence the deflection would reach $\varepsilon_R = 116.56°$. This is larger than that recommended, and the inlet angle is too steep. These calculations show that good designs are obtained for a range of flow coefficients of $0.5 < \phi < 1.5$ and the blade-loading coefficient in the range $0.8 < \psi < 2.7$, for then the deflection is less than $80°$. As the flow coefficient is increased from 0.5 to 1.5, the inlet flow angle may be changed from an axial entry to one with $\alpha_3 = -30°$.

Contours of constant-rotor-loss coefficients are shown in Figure 6.14. They are seen to follow the shape of the deflection lines in Figure 6.13. When there is no turning, the loss coefficient is $\zeta_R = 0.04$ and the line for $\zeta_R = 0.05$ corresponds to a turning of $\beta_2 - \beta_3 = 40°$.

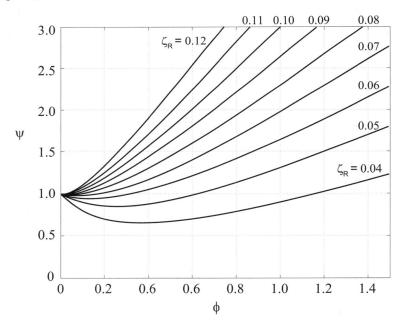

Figure 6.14 Lines of constant rotor loss coefficients for a stage with $R = 0.5$, $b/c_x = 3$, and $Re = 10^5$.

For a zero-reaction stage the contours of constant efficiency are given in Figure 6.15. Since $R = 0$, the equation

$$\psi = 2(1 - R - \phi \tan \alpha_3)$$

shows that at $\phi = 0$ the loading coefficient is $\psi = 2$. The lines of constant turning for the rotor are now straight lines, owing to the relationship $\beta_3 = -\beta_2$ and

$$\psi = \phi(\tan \beta_2 - \tan \beta_3) = 2\phi \tan \beta_2$$

If the deflection is kept at 70°, then $\beta_2 = 35°$, and at $\phi = 1.5$ the blade-loading coefficient is $\psi = 2.1$ and the efficiency is close to 0.87. At $\phi = 0.5$ the blade-loading coefficient has the value $\psi = 0.7$ and the efficiency is slightly over 0.92. For axial entry $\psi = 2$ independent of ϕ, and an efficiency of slightly under 0.89 may be maintained for the range of ϕ from 0.3 to 0.8.

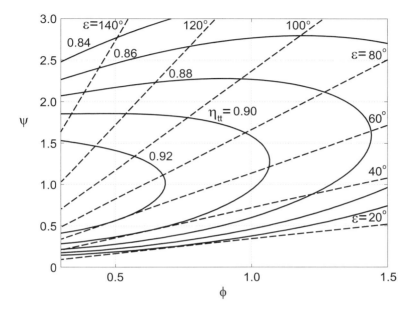

Figure 6.15 Contours of constant efficiency and deflection for a stage with $R = 0$, $b/c_x = 3$, and $\text{Re} = 10^5$.

6.7.5 Zweifel correlation

Zweifel [85] examined losses in turbines and developed a criterion for the space-to-chord ratio at which losses are the smallest. He put the loading of the blades into a nondimensional form by dividing the driving force by an ideal one defined as as the pressure difference $p_{02} - p_3$ times the axial chord. The stagnation pressure p_{02} is the maximum possible pressure encountered and p_3 is close to the minimum one as the flow accelerates through the passage. The ratio becomes

$$\psi_T = \frac{\rho V_x s (V_{u2} - V_{u3})}{(p_{02} - p_3) c_x}$$

in which c_x is the axial chord length. This can be written as

$$\psi_T = \frac{s}{c_x} \frac{\rho V_x^2 (\tan \alpha_2 - \tan \alpha_3)}{\frac{1}{2} \rho V_3^2}$$

or

$$\psi_T = 2 \frac{s}{c_x} \cos^2 \alpha_3 (\tan \alpha_2 - \tan \alpha_3) \tag{6.53}$$

In examining the performance of various turbines, Zweifel determined that the losses are minimized when $\psi_T = 0.8$. With this value for ψ_T, this equation is used to determine the spacing of the blades. With the spacing known, the cross-sectional area of the flow channel and its wetted area may be calculated and the Reynolds number determined. This may then be used to obtain the loss coefficients from the Soderberg correlation. For the rotor the flow angles of the absolute velocities are replaced by the relative flow angles.

6.7.6 Further discussion of losses

Ainley [1] carried out an experimental study of losses in turbines at about the same time as Soderberg. His loss estimates are shown in Figure 6.16 for a flow that turns only by about

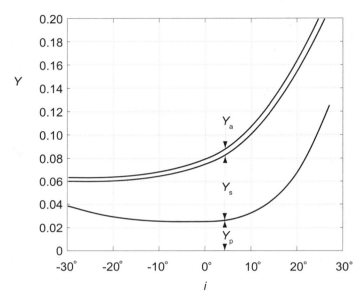

Figure 6.16 Stagnation pressure loss coefficients for turbine blades as a function of incidence, with $s/c = 0.77$ and $b/c = 2.7$ at Reynolds number $\mathrm{Re} = 2 \cdot 10^5$, from measurements by Ainley [1].

40° and thus has fairly small losses at the design condition. He presented the results in the form of a stagnation pressure loss coefficient, defined as

$$Y_\mathrm{p} = \frac{p_{01} - p_{02}}{p_{02} - p_2}$$

The denominator of this expression clearly depends on the Mach number and therefore, so does the value of the loss coefficient. The losses are separated into profile losses, secondary flow losses, and the losses in the annulus boundary layers. The profile losses, associated with the growth of the boundary layers along the blades, are lowest at a few degrees of negative incidence. The secondary flow losses dominate at all values of incidence. The losses in the annulus boundary layers represent a small addition to the secondary flow losses, and they are typically grouped together as it is difficult to separate them from each other.

The physical cause of the secondary flows is the curvature along the flow path. From study of fundamentals of fluid dynamics, it is known that pressure increases from concave to convex side of curved streamlines. This transverse pressure gradient gives rise to secondary flows for the following reason. In the inviscid stream far removed from the solid surfaces viscous forces are small and inertial forces are balanced by pressure forces. The transverse component of pressure force points from the pressure side of one blade to the suction side of the next one, and the transverse component of the inertial force is equal and opposite to this.

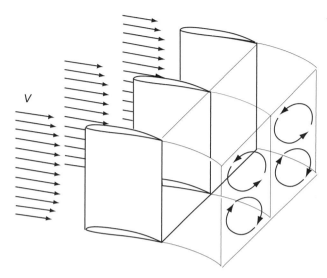

Figure 6.17 Secondary flows in a channel between two blades.

In the endwall boundary layer viscous forces in the main flow direction retard the flow, with the result that inertial forces in the boundary layers are smaller than those in the inviscid stream. However, owing to the thinness of the endwall boundary layers, pressure distribution in these layers is the same as in the inviscid stream. For this reason the unbalanced part of the pressure force causes a transverse flow in the endwall boundary layers toward the suction side of the blade. This is shown schematically in Figure 6.17. This secondary flow takes place in both endwall boundary layers. Continuity requires that there be a return flow across the inviscid stream. The return flow is more diffuse than that in the boundary layers as it occupies a large flow area. The effect is the development of two counterrotating secondary vortices, with axes in the direction of the main stream. Thus a secondary flow exists in these vortices. If this were all, the secondary flow would be easy to understand. But the flow in the boundary layer near the casing is also influenced by a vortex that develops at the tips of the rotor blades. This interaction increases the intensity of the secondary flow and the axis of the vortex migrates from the pressure side of the blade to the suction side as it traverses the flow passage. A sketch of this is shown in Figure 6.18. Further complication arises from the unsteadiness of the flow caused by a discrete set of rotor blades passing by the row of stator vanes.

6.7.7 Ainley–Mathieson correlation

Ainley and Mathieson [2] continued the work of Ainley [1] on turbine cascades. Their results are shown Figure 6.19. The flow enters the nozzles axially and leaves at the angle α_2 indicated. The profile loss coefficient Y_{pa} is seen to vary both with the space-to-chord ratio and the amount of turning of the flow. The two curves in the figure, for which the flow deflects $75°$ and $80°$, are marked with dashed line, for the flow is seldom turned this much.

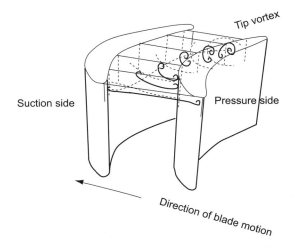

Figure 6.18 Distortion of tip vortex as it moves through the passage.

To facilitate the use of these results with hand calculators (or in short computer calculations), they have been fitted with two parabolas. The data in Figure 6.19 are correlated reasonably well by the biquadratic fit

$$
\begin{aligned}
Y_{\mathrm{pa}} &= \left[-0.627\left(\frac{\alpha_2}{100}\right)^2 + 0.821\left(\frac{\alpha_2}{100}\right) - 0.129\right]\left(\frac{s}{c}\right)^2 \\
&+ \left[1.489\left(\frac{\alpha_2}{100}\right)^2 - 1.676\left(\frac{\alpha_2}{100}\right) + 0.242\right]\left(\frac{s}{c}\right) \\
&-0.356\left(\frac{\alpha_2}{100}\right)^2 + 0.399\left(\frac{\alpha_2}{100}\right) + 0.0077
\end{aligned}
\tag{6.54}
$$

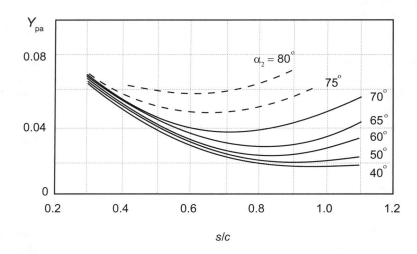

Figure 6.19 Stagnation pressure loss coefficients for nozzles, as measured by Ainley and Mathieson [2].

■ EXAMPLE 6.8

A nozzle row is tested with air. The air enters the row axially and leaves it at angle $60°$. The space-to-chord ratio is $s/c = 0.7$. The inlet pressure is $p_{01} = 200\,\text{kPa}$, stagnation temperature at the inlet is $T_{01} = 540\,\text{K}$, and the exit static pressure is $p_2 = 160\,\text{kPa}$. Find (a) the exit stagnation pressure and (b) the static enthalpy loss coefficient.

Solution: (a) For $s/c = 0.7$, the stagnation pressure loss coefficient is

$$Y_{\text{pa}} = \left[-0.627\left(\frac{60}{100}\right)^2 + 0.821\left(\frac{60}{100}\right) - 0.129\right]0.7^2$$

$$+ \left[1.489\left(\frac{60}{100}\right)^2 - 1.676\left(\frac{60}{100}\right) + 0.242\right]0.7$$

$$-0.356\left(\frac{60}{100}\right)^2 + 0.399\left(\frac{60}{100}\right) + 0.0077 = 0.0272$$

Examination of Figure 6.19 shows that this value is smaller than what can be read from the figure, which suggests that it should be increased to 0.032. If this error can be tolerated, solving next

$$Y_{\text{pa}} = \frac{p_{01} - p_{02}}{p_{02} - p_2}$$

for p_{02} gives

$$p_{02} = \frac{p_{01} + Y_{\text{pa}}p_2}{1 + Y_{\text{pa}}} = \frac{200 + 0.0272 \cdot 160}{1.0279} = 198.94\,\text{kPa}$$

so that $\Delta p_{0S} = 1.06\,\text{kPa}$. A more accurate value using the actual charts is $1.24\,\text{kPa}$.
(b) The exit static temperature is

$$T_2 = T_{02}\left(\frac{p_2}{p_{02}}\right)^{(\gamma-1)/\gamma} = 540\left(\frac{160{,}000}{198{,}913}\right)^{1/3.5} = 507.44\,\text{K}$$

and velocity is therefore

$$V_2 = \sqrt{2c_p(T_{02} - T_2)} = \sqrt{2 \cdot 1004.5(540 - 507.44)} = 255.77\,\text{m/s}$$

These give a Mach number value of $M_2 = V_2/\sqrt{\gamma RT_2} = 0.566$. The stagnation density at the exit is

$$\rho_{02} = \frac{p_{02}}{RT_{02}} = \frac{198{,}940}{287 \cdot 540} = 1.284\,\text{kg/m}^3$$

The static enthalpy loss coefficient is calculated from

$$\zeta_S = \frac{4\Delta p_{0S}/\rho_{02}}{V_2^2(2 + (\gamma-1)M_2^2)} == \frac{4 \cdot 1060/1.284}{255.77^2(2 + (1.4-1)0.566^2)} = 0.0237$$

■

The experiments of Ainley and Mathieson [2] also included the influence of the space-to-chord ratio for *impulse blades*. This is shown in Figure 6.20. Again, the approach velocity is at zero incidence. Since the rotor blades have the shape of impulse blades, the different curves are labeled by the relative flow angles. In a two-stage velocity compounded steam turbine, the stator between the two rotors has equiangular vanes as well, but this is an exceptional situation in turbines. Ainley and Mathieson showed that the results from the nozzles with axial entry and the impulse blades can be combined for use in other situations. This is discussed below.

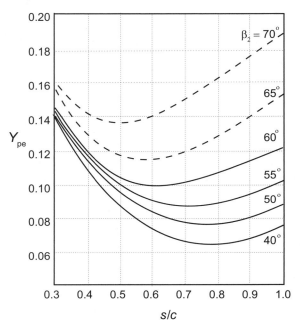

Figure 6.20 Loss coefficients for impulse blades, $\beta_3 = -\beta_2$, and $Re = 2 \cdot 10^5$, $M < 0.6$, as measured by Ainley and Mathieson [2].

The data shown in Figure 6.20 can be fitted to a biquadratic expression.

$$\begin{aligned} Y_{pe} &= \left[-1.56 \left(\frac{\alpha_2}{100} \right)^2 + 1.55 \left(\frac{\alpha_2}{100} \right) - 0.064 \right] \left(\frac{s}{c} \right)^2 \\ &+ \left[3.73 \left(\frac{\alpha_2}{100} \right)^2 - 3.43 \left(\frac{\alpha_2}{100} \right) + 0.290 \right] \left(\frac{s}{c} \right) \\ &- 0.83 \left(\frac{\alpha_2}{100} \right)^2 + 0.78 \left(\frac{\alpha_2}{100} \right) + 0.078 \end{aligned} \quad (6.55)$$

This is a reasonably good fit even if it was forced to a simple second-order polynomial form. On the basis of values obtained from Eqs. (6.54) and (6.55), Ainley and Mathieson recommend that for blades for which the inlet angle is between the axial entry of nozzles and that of the impulse blades, the stagnation pressure loss coefficient can be estimated from

$$Y_p = \left[Y_{pa} + \left(\frac{\alpha_1}{\alpha_2} \right)^2 (Y_{pe} - Y_{pa}) \right] \left(\frac{t/c}{0.2} \right)^{|\alpha_1/\alpha_2|} \quad (6.56)$$

The subscripts in this expression have the following meaning: Y_p is the *profile* loss at zero incidence; Y_{pa} is the profile loss coefficient for *axial* entry, and Y_{pe} is the profile loss coefficient for *equiangular* impulse blades. The absolute values are needed in the exponent because α_1 could be negative.

For a rotor the angle, α_1 is replaced by β_2 and α_2 is replaced by β_3. Thus Eq. (6.56) takes the form

$$Y_p = \left[Y_{pa} + \left(\frac{\beta_2}{\beta_3} \right)^2 (Y_{pe} - Y_{pa}) \right] \left(\frac{t/c}{0.2} \right)^{|\beta_2/\beta_3|} \tag{6.57}$$

In evaluating the loss coefficient from Eq. (6.55) for impulse blades, α_2 is replaced by β_2 for rotor blades, since β_2 will be positive and $\beta_3 = -\beta_2$. The ratio of maximum thickness to the length of the chord in these expressions is t/c, with a nominal value of 20%. Should it be greater than 25% the value $t/c = 0.2$ is used. If it is less than 15%, its value is set at $t/c = 0.15$.

6.7.8 Secondary loss

Secondary and tip losses require examination of lift and drag on the rotor blades. Unlike in the case for an airfoil, for which lift is the force component perpendicular to incoming flow direction, in turbomachinery flows the direction of the lift is defined as the force perpendicular to a *mean flow direction*. This, together with other important geometric parameters, is shown in Figure 6.21. The mean direction is obtained by first defining the mean tangential component of the relative velocity as

$$W_{um} = \frac{1}{2}(W_{u1} + W_{u2})$$

which can be rewritten in the form

$$W_{um} = \frac{1}{2} W_x (\tan \beta_2 + \tan \beta_3)$$

Next, writing $W_{um} = W_x \tan \beta_m$ gives

$$\tan \beta_m = \frac{1}{2} (\tan \beta_2 + \tan \beta_3)$$

and β_m defines the mean direction. For the situation shown in the figure $\tan \beta_m$ is negative and so is V_{um}.

Next, applying the momentum theorem to the rotor, the force in the direction of the wheel motion is given by

$$F_u = \rho s W_x (W_{u3} - W_{u2}) = \rho s W_x^2 (\tan \beta_3 - \tan \beta_2)$$

This is the force that the blade exerts on the fluid per unit height of the blade. The reaction force $R_u = -F_u$ is the force by the fluid on the blade, and it is given by

$$R_u = \rho s W_x^2 (\tan \beta_2 - \tan \beta_3) \tag{6.58}$$

The component of the reaction forces are related to lift L and drag D by

$$R_x = D \cos \beta_m - L \sin \beta_m \qquad R_u = D \sin \beta_m + L \cos \beta_m$$

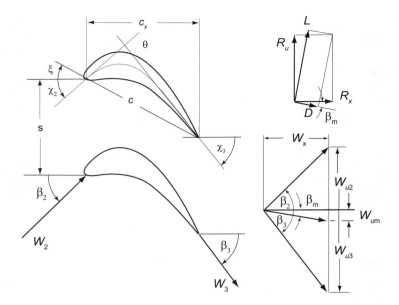

Figure 6.21 Illustration of flow angles, blade angles, mean direction, and lift and drag forces on a rotor.

The magnitude of the drag force has been drawn larger in the figure than its actual size, to make the sketch clearer. If it is neglected, the reaction R_u is related to lift by

$$R_u = L \cos \beta_m$$

and Eq. (6.58) can be written as

$$L \cos \beta_m = \rho s W_x^2 (\tan \beta_2 - \tan \beta_3)$$

Introducing the lift coefficient leads to the expression

$$C_L = \frac{L/c}{\frac{1}{2}\rho V_m^2} = 2 \left(\frac{s}{c}\right) \frac{W_x^2}{V_m^2} \frac{(\tan \beta_2 - \tan \beta_3)}{\cos \beta_m}$$

Since $W_x = W_m \cos \beta_m$ this reduces to

$$C_L = 2 \left(\frac{s}{c}\right) (\tan \beta_2 - \tan \beta_3) \cos \beta_m \qquad (6.59)$$

The secondary flow and tip losses according to Dunham and Came [24] are expressed as

$$Y_s + Y_k = \frac{c}{b} \left[0.0334 \frac{\cos \beta_3}{\cos \beta_2} + B \left(\frac{k}{c}\right)^{0.78} \right] \left(\frac{C_L}{s/c}\right)^2 \frac{\cos^2 \beta_3}{\cos^3 \beta_m} \qquad (6.60)$$

The second term, in which the gap width k appears, accounts for the tip losses. The parameter $B = 0.47$ for standard blades and $B = 0.37$ for shrouded blades. A shrouded blade is shown in Figure 3.11 in Chapter 3.

The loss coefficients are similar for the stator. The mean flow direction is given by

$$\tan \alpha_m = \frac{1}{2}(\tan \alpha_1 + \tan \alpha_2)$$

and the lift coefficient is

$$C_L = 2 \left(\frac{s}{c}\right) (\tan \alpha_2 - \tan \alpha_1) \cos \alpha_m$$

The loss coefficient of the secondary flows is

$$Y_s = \frac{c}{b} \left(0.0334 \frac{\cos \alpha_2}{\cos \alpha_1}\right) \left(\frac{C_L}{s/c}\right)^2 \frac{\cos^2 \alpha_2}{\cos^3 \alpha_m} \qquad (6.61)$$

as there are no tip losses.

■ **EXAMPLE 6.9**

Combustion gases, with $\gamma = \frac{4}{3}$ and $c_p = 1148 \text{ J/(kg} \cdot \text{K)}$, flow through a normal turbine stage with $R = 0.60$. The flow enters the stator at $\alpha_1 = -33°$ and leaves at velocity $V_2 = 450 \text{ m/s}$. The inlet stagnation temperature is $T_{01} = 1200 \text{ K}$, and the inlet stagnation pressure is 15 bar. The flow coefficient is $\phi = 0.7$, blade height-to-axial chord ratio is $b/c_x = 3.5$, and the Reynolds number is 10^5. The blades are unshrouded with $B = 0.4$, and the tip-gap-to-blade height ratio is $k/b = 0.02$. Find the stagnation pressure loss across the stator and the rotor based on Ainley–Matheison correlations.

Solution: The blade-loading coefficient is first determined from

$$\psi = 2(1 - R - \phi \tan \alpha_3) = 2(1 - 0.6 - 0.7\tan(-33°)) = 1.709$$

The flow angle leaving the stator is

$$\alpha_2 = \tan^{-1}\left(\frac{1 - R + \psi/2}{\phi}\right) = \tan^{-1}\left(\frac{1 - 0.6 + 1.709/2}{0.7}\right) = 60.84°$$

Hence the mean flow angle is

$$\alpha_m = \tan^{-1}\left[\frac{1}{2}(\tan \alpha_1 + \tan \alpha_2)\right] = 29.74°$$

The lift coefficient is

$$C_{LS} = 2\left(\frac{s}{c}\right)(\tan \alpha_2 - \tan \alpha_1)\cos \alpha_m$$
$$= 2 \cdot 0.9(\tan(60.84) - \tan(-33))\cos(29.74) = 3.82$$

The secondary loss coefficient for the nozzles is

$$Y_{sN} = 0.0334 \frac{c}{b} \frac{\cos^2 \alpha_2}{\cos^3 \alpha_m} = 0.0434$$

The value of profile losses obtained from Eqs. (6.54) and (6.55) are

$$Y_{paN} = 0.0269 \qquad Y_{peN} = 0.1151$$

so that for the particular nozzle row

$$Y_{pN} = \left[Y_{paN} + \left(\frac{\alpha_1}{\alpha_2}\right)^2 (Y_{peN} - Y_{paN})\right]\left(\frac{t/c}{0.2}\right)^{|\alpha_1/\alpha_2|}$$

comes out to be

$$Y_{\text{pN}} = \left[0.0269 + \left(\frac{-33}{60.84}\right)^2 (0.1151 - 0.0269)\right] = 0.0529$$

Thus the stagnation pressure loss coefficient is $Y_N = Y_{\text{pN}} + Y_{\text{sN}} = 0.0963$.
For the rotor, the angle of the relative velocity leaving the stage is

$$\beta_3 = \tan^{-1}\left(\frac{-R - \psi/2}{\phi}\right) = \tan^{-1}\left(\frac{-0.6 - 1.709/2}{0.7}\right) = -64.3°$$

and the angle of the relative velocity at the inlet of the rotor is

$$\beta_2 = \tan^{-1}\left(\frac{-R + \psi/2}{\phi}\right) = \tan^{-1}\left(\frac{-0.6 + 1.709/2}{0.7}\right) = 20.0°$$

The mean flow angle is

$$\beta_m = \tan^{-1}\left[\frac{1}{2}(\tan\beta_2 + \tan\beta_3)\right] = -40.6°$$

The lift coefficient is

$$C_{\text{LR}} = 2\left(\frac{s}{c}\right)(\tan\beta_2 - \tan\beta_3)\cos\beta_m$$

$$= 2 \cdot 0.9[\tan(20) - \tan(-64.3)]\cos(-40.6) = 3.34$$

The secondary flow and tip losses can then be determined from

$$Y_{\text{sR}} + Y_{\text{kR}} = \frac{c}{b}\left[0.0334\frac{\cos\beta_3}{\cos\beta_2} + B\left(\frac{k}{c}\right)^{0.78}\right]\left(\frac{C_L}{s/c}\right)^2 \frac{\cos^2\beta_3}{\cos^3\beta_m}$$

$$Y_{\text{sR}} + Y_{\text{kR}} = \frac{1}{3}\left[0.0334\frac{\cos(-64.3°)}{\cos(20.0°)} + 0.47 \cdot 0.02^{0.78}\right]\left(\frac{3.34}{0.9}\right)^2\frac{\cos^2(-64.3°)}{\cos^3|-40.6°|} = 0.075$$

From Eqs. (6.54) and (6.55) the profile losses coefficients are

$$Y_{\text{paR}} = 0.0322 \qquad Y_{\text{peR}} = 0.1334$$

so that

$$Y_{\text{pR}} = \left[Y_{\text{paR}} + \left(\frac{\beta_2}{\beta_3}\right)^2(Y_{\text{peR}} - Y_{\text{paR}})\right]\left(\frac{t/c}{0.2}\right)^{|\beta_2/\beta_3|}$$

is

$$Y_{\text{pR}} = \left[0.0322 + \left(\frac{-20.0°}{-64.3°}\right)^2(0.1334 - 0.0322)\right] = 0.0419$$

The stagnation pressure loss coefficient for the rotor is therefore $Y_R = Y_{\text{pR}} + Y_{\text{sR}} + Y_{\text{kR}} = 0.1169$.
The stagnation pressure loss across the nozzle row is obtained from

$$Y_N = \frac{p_{01} - p_{02}}{p_{02} - p_2} \qquad p_{01} - p_{02} = Y_N(p_{02} - p_2)$$

Dividing through by p_{02} gives

$$\frac{p_{01}}{p_{02}} = 1 + Y_N \left(1 - \frac{p_2}{p_{02}}\right) = 1 + Y_N \left[1 - \left(\frac{T_2}{T_{02}}\right)^{\gamma/(\gamma-1)}\right]$$

Since the temperature T_2 is

$$T_2 = T_{02} - \frac{V_2^2}{2c_p} = 1200 - \frac{450^2}{2 \cdot 1148} = 1111.8\,\text{K}$$

and $M_2 = V_2/\sqrt{\gamma R T_2} = 0.690$, the stagnation pressure ratio is

$$\frac{p_{01}}{p_{02}} = 1 + 0.0963 \left[1 - \left(\frac{1111.8}{1200}\right)^4\right] = 1.02534$$

Hence $p_{02} = 1500/1.02534 = 1462.9\,\text{kPa}$, and the stagnation pressure loss is $\Delta p_{0\text{LS}} = 37.2\,\text{kPa}$.

The static pressure at the exit is

$$p_2 = p_{02}\left(1 + \frac{\gamma-1}{2}M_2^2\right)^{-\gamma/(\gamma-1)} = 1462.9\left(1 + \frac{690^2}{6}\right)^{-4} = 1078.0\,\text{kPa}$$

Some preliminary calculations are necessary for the rotor. First

$$V_x = V_s \cos\alpha_2 = 450\cos(60.84) = 219.3\,\text{m/s}$$

and then

$$W_2 = \frac{V_x}{\cos\beta_2} = \frac{219.3}{\cos(20°)} = 233.4\,\text{m/s}$$

$$W_3 = \frac{V_x}{\cos\beta_3} = \frac{219.3}{\cos(-64.3°)} = 505.6\,\text{m/s}$$

$$V_3 = \frac{V_x}{\cos\alpha_3} = \frac{219.3}{\cos(-33°)} = 261.5\,\text{m/s}$$

Also

$$U = \frac{V_x}{\phi} = \frac{219.3}{0.7} = 312.2\,\text{m/s} \qquad w = \psi U^2 = 1.709 \cdot 313.2^2 = 167.69\,\text{kJ/kg}$$

so that

$$T_{03} = T_{02} - \frac{W}{c_p} = 1200 - \frac{167.69}{1148} = 1053.7\,\text{K}$$

and

$$T_3 = T_{03} - \frac{V_3^2}{2c_p} = 1053.7 - \frac{261.5^2}{2 \cdot 1148} = 1024.2\,\text{K}$$

Next

$$T_{03\text{R}} = T_3 + \frac{W_3^2}{2c_p} = 1024.2 + \frac{505.6^2}{2 \cdot 1148} = 1135.5\,\text{K}$$

and since $T_{02\text{R}} = T_{03\text{R}}$

$$p_{02\text{R}} = p_2 \left(\frac{T_{02\text{R}}}{T_2}\right)^{\gamma/(\gamma-1)} = 1078.0 \left(\frac{1135.5}{1111.8}\right)^4 = 1128.3\,\text{kPa}$$

The stagnation pressure loss coefficient for the rotor is

$$Y_R = \frac{p_{02R} - p_{03R}}{p_{03R} - p_3}$$

so that

$$\frac{p_{02R}}{p_{03R}} = 1 + Y_R\left(1 - \frac{p_3}{p_{03R}}\right) = 1 + Y_R\left[1 - \left(\frac{T_3}{T_{03R}}\right)^{\gamma/(\gamma-1)}\right]$$

The numerical value for the pressure ratio is

$$\frac{p_{02R}}{p_{03R}} = 1 + 0.1169\left[1 - \left(\frac{1024.2}{1135.5}\right)^4\right] = 1.040$$

and $p_{03R} = 1172.9/1.040 = 1128.28$ kPa. Hence the stagnation pressure loss across the rotor is $\Delta p_{0LR} = 44.6$ kPa. The value obtained the by Soderberg correlation was 38.6 kPa, but that correlation neglects the tip losses. For the nozzle row the loss is $\Delta p_{0LS} = 37.1$ kPa, and the Soderberg correlation gave 42.4 kPa. Ainley–Mathieson and Soderberg correlations are therefore in reasonable agreement. ∎

6.8 MULTISTAGE TURBINE

6.8.1 Reheat factor in a multistage turbine

Consider next a multistage turbine with process lines as shown in Figure 6.22. The isentropic work delivered by the jth stage is denoted as w_{js}, whereas the actual stage work is w_j. The stage efficiency is defined to be

$$\eta_s = \frac{w_j}{w_{js}}$$

and it is assumed to be the same for each stage. With

$$w_{js} = h_{0j} - h_{0,j+1,s}$$

the sum over all the stages gives

$$\sum_{j=1}^{N} w_{js} = \sum_{j=1}^{N}(h_{0j} - h_{0,j+1,s})$$

The isentropic work delivered by the turbine is $w_s = h_{01} - h_{0,N+1,ss}$. A *reheat factor* is defined as

$$RF = \frac{\sum_{j=1}^{N} w_{js}}{w_s} = \frac{\eta}{\eta_s}\frac{\sum_{j=1}^{N} w_j}{w}$$

in which $w = \sum_{j=1}^{N} w_j$, so that

$$RF = \frac{\eta}{\eta_s}$$

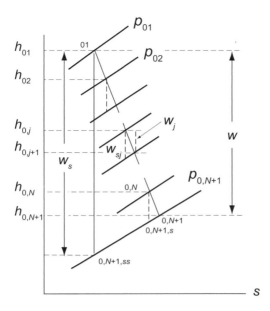

Figure 6.22 Processes in a multistage turbine.

Since the slope of constant-pressure line increases as temperature increases, it follows that

$$\sum_{j=1}^{N} w_{js} > w_s$$

and RF > 1. Hence $\eta > \eta_s$, and the overall efficiency of a turbine is greater than the stage efficiency. The reason is the internal heating, for the increase in static enthalpy by irreversibilities becomes *partly available* as the expansion proceeds over the next stage. The increase in the overall efficiency depends on the number of stages.

If the ideal gas model can be used, then the actual and ideal work delivered by the first stage are

$$w = c_p(T_{01} - T_{02}) \qquad w_s = c_p(T_{01} - T_{02s})$$

Since

$$\frac{T_{02s}}{T_{01}} = \left(\frac{p_{02}}{p_{01}}\right)^{(\gamma-1)/\gamma}$$

the temperature difference for the isentropic process becomes

$$T_{01} - T_{02s} = T_{01}\left[1 - \left(\frac{p_{02}}{p_{01}}\right)^{(\gamma-1)/\gamma}\right]$$

Letting

$$x = 1 - \left(\frac{p_{02}}{p_{01}}\right)^{(\gamma-1)/\gamma} \qquad \text{then} \qquad T_{01} - T_{02s} = xT_{01}$$

The actual temperature drop is then

$$T_{01} - T_{02} = \eta_s x T_{01} \qquad \text{so that} \qquad T_{02} = T_{01}(1 - \eta_s x)$$

Assuming that the pressure ratio and the efficiency are the same for each stage, similar analysis for the next stage gives

$$T_{02} - T_{03} = \eta x T_{02} \quad \text{so that} \quad T_{03} = T_{02}(1 - \eta_s x) = T_{01}(1 - \eta_s x)^2$$

and for the Nth stage

$$T_{0,N} - T_{0,N+1} = \eta_s x T_{0,N} \quad \text{so that} \quad T_{0,N+1} = T_{01}(1 - \eta_s x)^N$$

The work delivered by the entire turbine can then be expressed as

$$w = c_p(T_{01} - T_{02}) + c_p(T_{02} - T_{03}) + \cdots + c_p(T_{0,N} - T_{0,N+1}) = c_p(T_{01} - T_{0,N+1})$$

which can be written as

$$w = c_p T_{01}[1 - (1 - \eta_s x)^N]$$

The isentropic work by the entire turbine is given by this same equation when $\eta_s = 1$, or

$$w_s = c_p T_{01}[1 - (1 - x)^N]$$

The isentropic work can also be written as

$$w_s = c_p T_{01}\left[1 - \left(\frac{p_{0,N+1}}{p_{01}}\right)^{(\gamma-1)/\gamma}\right]$$

and therefore

$$\eta = \frac{w}{w_s} = \frac{1 - (1 - \eta_s x)^N}{1 - \left(\frac{p_{0,N+1}}{p_{01}}\right)^{(\gamma-1)/\gamma}}$$

so that the reheat factor becomes

$$\text{RF} = \frac{\eta}{\eta_s} = \frac{1 - (1 - \eta_s x)^N}{\eta_s\left[1 - \left(\frac{p_{0,N+1}}{p_{01}}\right)^{(\gamma-1)/\gamma}\right]}$$

6.8.2 Polytropic or small-stage efficiency

The polytropic process was introduced in Chapter 3. Here it is used for a small stage. If the stagnation enthalpy change is small across a stage, the stage efficiency approaches the polytropic efficiency. Consider the situation for which the ideal gas relation is valid and for which an incremental process is as shown in Figure 6.23. The temperature drops for actual and ideal processes are related and given by

$$\eta_p \, dT_{0s} = dT_0$$

For an isentropic expansion

$$c_p \frac{dT_{0s}}{T_0} = R \frac{dp_0}{p_0} \quad \text{or} \quad \frac{\gamma}{\eta_p(\gamma - 1)} \frac{dT_0}{T_0} = \frac{dp_0}{p_0}$$

Integrating this between the inlet and the exit gives

$$\frac{T_{0,N+1}}{T_{01}} = \left(\frac{p_{0,N+1}}{p_{01}}\right)^{(\gamma-1)\eta_p/\gamma}$$

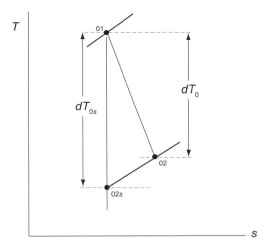

Figure 6.23 Processes across a small stage.

The reheat factor can then be written as

$$\text{RF} = \frac{\eta}{\eta_p} = \frac{1 - \left(\frac{p_{0,N+1}}{p_{01}}\right)^{(\gamma-1)\eta_p/\gamma}}{\eta_p\left[1 - \left(\frac{p_{0,N+1}}{p_{01}}\right)^{(\gamma-1)/\gamma}\right]}$$

The relationship between the turbine efficiency and polytropic, or small-stage efficiency is

$$\eta = \frac{1 - \left(\frac{1}{r}\right)^{(\gamma-1)\eta_p/\gamma}}{1 - \left(\frac{1}{r}\right)^{(\gamma-1)/\gamma}}$$

in which $r = p_{01}/p_{0e}$ is the overall pressure ratio of the turbine. This relationship is shown in Figure 6.24.

This completes the study of axial turbines, which began in the previous chapter on steam turbines. Wind turbines and some hydraulic turbines are also axial machines. They are discussed later. Many of the concepts introduced in this chapter are carried over to the next one, on axial compressors.

EXERCISES

6.1 At inlet to the rotor in a single-stage axial-flow turbine the magnitude of the absolute velocity of fluid is 610 m/s. Its direction is 61° as measured from the cascade front in the direction of the blade motion. At exit of this rotor the absolute velocity of the fluid is 305 m/s directed such that its tangential component is negative. The axial velocity is constant, the blade speed is 305 m/s, and the flow rate through the rotor is 5 kg/s. (a) Construct the rotor inlet and exit velocity diagrams showing the axial and tangential components of the absolute velocities. (b) Evaluate the change in total enthalpy across the rotor. (c) Evaluate the power delivered by the rotor. (d) Evaluate the average driving force

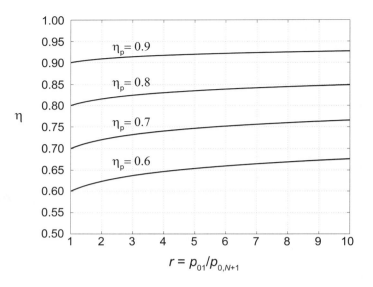

Figure 6.24 Turbine efficiency as a function of pressure ratio and polytropic efficiency for a gas with $\gamma = 1.4$.

exerted on the blades. (e) Evaluate the change in static and stagnation temperature of the fluid across the rotor, assuming the fluid to be a perfect gas with $c_p = 1148 \, \text{J}/(\text{kg} \cdot \text{K})$. (f) Calculate the flow coefficient and the blade-loading coefficient. Are they reasonable?

6.2 A small axial-flow turbine must have an output power of 37 kW when the mass flow rate of combustion gases is $0.5 \, \text{kg/s}$, and the inlet total temperature is 410 K. The value of the gas constant is $287 \, \text{J}/(\text{kg} \cdot \text{K})$ and $\gamma = 4/3$. The total-to-total efficiency of the turbine is 80%. The rotor operates at 50,000 rpm, and the mean blade diameter is 10 cm. Evaluate (a) the average driving force on the turbine blades, (b) the change in the tangential component of the absolute velocity across the rotor, and (c) the required total pressure ratio across the turbine.

6.3 A turbine stage of a multistage axial turbine is shown in Figure 6.3. The inlet gas angle to the stator is $\alpha_1 = -36.8°$, and the outlet angle from the stator is $\alpha_2 = 60.3°$. The flow angle of the relative velocity at the inlet to the rotor is $\beta_2 = 36.8°$ and the flow leaves at $\beta_3 = -60.3°$. The value of the gas constant is $287 \, \text{J}/(\text{kg} \cdot \text{K})$ and $\gamma = 4/3$. (a) Assuming that the blade speed is $U = 220$ m/s, find the axial velocity, which is assumed constant throughout the turbine. (b) Find the work done by the fluid on the rotor blades for one stage. (c) The inlet stagnation temperature to the turbine is 950 K, and the mass flow rate is $\dot{m} = 400 \, \text{kg/s}$. Assuming that this turbine produces a power output of 145 MW, find the number of stages. (d) Find the overall stagnation pressure ratio, given that its isentropic efficiency is $\eta_t t = 0.85$. (e) Why does the static pressure fall across the stator and the rotor?

6.4 A single-stage axial turbine has a total pressure ratio of 1.5 to 1, with an inlet total pressure 300 kPa and temperature of 600 K. The absolute velocity at the inlet to the stator row is in the axial direction. The adiabatic total-to-total efficiency is 80%. The relative velocity is at an angle of 30° at the inlet of the rotor and at the exit it is $-35°$. If the flow coefficient is $\phi = 0.9$, find the blade velocity. Use compressible flow analysis with $c_p = 1148\,\text{J}/(\text{kg} \cdot \text{K})$, $\gamma = \frac{4}{3}$, and $R = 287\,\text{J}/(\text{kg} \cdot \text{K})$.

6.5 An axial turbine has a total pressure ratio of 4 to 1, with an inlet total pressure 650 kPa and total temperature of 800 K. The combustion gases that pass through the turbine have $\gamma = \frac{4}{3}$, and $R = 287\,\text{J}/(\text{kg} \cdot \text{K})$. (a) Justify the choice of two stages for this turbine. Each stage is normal stage and they are designed the same way, with the blade-loading coefficient equal to 1.1 and the flow coefficient equal to 0.6. The absolute velocity at the inlet to the stator row is at angle 5° from the axial direction. The adiabatic total-to-total efficiency is 91.0%. Find, (b) the angle at which the the absolute velocity leaves the stator, (c) the angle of the relative velocity at the inlet of the rotor, (d) the angle at which the relative velocity leaves the rotor. (f) Draw the velocity diagrams at the inlet and outlet of the rotor. (g) What are the blade speed and the axial velocity? A consequence of the design is that each stage has the same work output and efficiency. Find, (h) the stage efficiency and (i) the pressure ratio for each stage.

6.6 For a steam turbine rotor the blade speed at the casing is $U = 300\,\text{m/s}$ and at the hub its speed is $240\,\text{m/s}$. The absolute velocity at the casing section at the inlet to the rotor is $V_{2c} = 540\,\text{m/s}$ and at the hub section it is $V_{2h} = 667\,\text{m/s}$. The angle of the absolute and relative velocities at the inlet and exit of the casing and hub sections are $\alpha_{2c} = 65°$, $\beta_{3c} = -60°$, $\alpha_{2h} = 70°$, and $\beta_{3h} = -50°$. The exit relative velocity at the casing is $W_{3c} = 456\,\text{m/s}$ and at the hub it is $W_{3h} = 355\,\text{m/s}$. For the tip section, evaluate (a) the axial velocity at the inlet and exit; (b) the change in total enthalpy of the steam across the rotor; and (c) the outlet total and static temperatures at the hub and casing sections, assuming that the inlet static temperature is 540°C and inlet total pressure is 7 MPa, and they are the same at all radii. Assume that the process is adiabatic and steam can be considered a perfect gas with $\gamma = 1.3$. The static pressure at the exit of the rotor is the same for all radii and is equal to the static pressure at inlet of the hub section. Repeat the calculations for the hub section. (d) Find the stagnation pressure at the outlet at the casing and the hub.

6.7 Combustion gases, with $\gamma = \frac{4}{3}$ and $R = 287\,\text{kJ}/(\text{kg} \cdot \text{K})$, flow through a turbine stage. The inlet flow angle for a normal stage is $\alpha_1 = 0°$. The flow coefficient is $\phi = 0.52$, and the blade-loading coefficient is $\psi = 1.4$. (a) Draw the velocity diagrams for the stage. (b) Determine the angle at which relative velocity leaves the rotor. (c) Find the flow angle at the exit of the stator. (d) A two-stage turbine has an inlet stagnation temperature of $T_{01} = 1250\,\text{K}$ and blade speed $U = 320\,\text{m/s}$. Assuming that the total-to-total efficiency of the turbine is $\eta_{tt} = 0.89$, find the stagnation temperature of the gas at the exit of the turbine and the stagnation pressure ratio for the turbine. (e) Assuming that the density ratio across the turbine based on static temperature and pressure ratios is the same as that based on the stagnation temperature and stagnation pressure ratios, find the ratio of the cross-sectional areas across the two-stage turbine.

6.8 Steam enters a 10-stage 50%-reaction turbine at the stagnation pressure 0.8 MPa and stagnation temperature 200°C and leaves at pressure 5 kPa and with quality equal to 0.86. (a) Assuming that the steam flow rate is 7 kg/s, find the power output and the overall efficiency of the turbine. (b) The steam enters each stator stage axially with velocity of 75 m/s. The mean rotor diameter for all stages is 1.4 m, and the axial velocity is constant through the machine. Find the rotational speed of the shaft. (c) Find the absolute and relative inlet and exit flow angles at the mean blade height assuming equal enthalpy drops for each stage.

6.9 Combustion gases enter axially into a normal stage at stagnation temperature $T_{01} = 1200$ K and stagnation pressure $p_{01} = 1500$ kPa. The flow coefficient is $\phi = 0.8$ and the reaction is $R = 0.4$. The inlet Mach number to the stator is $M_1 = 0.4$. Find, (a) the blade speed and (b) the Mach number leaving the stator and the relative Mach number leaving the rotor. (c) Using the Soderberg loss coefficients, find the efficiency of the stage. (d) Repeat the calculations with inlet Mach number $M_1 = 0.52$.

6.10 For a normal turbine stage fluid enters the stator at angle $10°$. The relative velocity has an angle $-40°$ as it leaves the rotor. The blade-loading factor is 1.6. (a) Determine the exit angle of the flow leaving the stator and the angle of the relative velocity as it enters the rotor. Determine the degree of reaction. (b) For the conditions of part (a), find the angle at which the flow leaves the stator and the angle of the relative velocity entering the rotor as well as the degree of reaction. (c) Calculate the total-to-total efficiency using Soderberg correlations. Take the blade height to axial chord ratio to be $b/c_x = 3.5$. (d) Determine the stagnation pressure loss across the stator, given that the inlet conditions are $T_{01} = 700$ K and $p_{01} = 380$ kPa, and the velocity after the stator is $V_2 = 420$ m/s.

6.11 For Example 6.6, write a computer program to calculate the mass flow rate and plot the variation of the reaction from the hub to the casing.

6.12 For a normal turbine stage the exit blade angle of the stator at $70°$ and relative velocity has angle $-60°$ as it leaves the rotor. For a range of flow coefficients $\phi = 0.2 - 0.8$ calculate and plot the gas exit angle from the rotor, the angle the relative velocity makes as it leaves the stator, the blade-loading coefficient, and the degree of reaction. Comment on what is a good operating range and what are the deleterious effects in flow over the blades if the mass flow rate is reduced too much or if it is increased far beyond this range.

6.13 For a normal turbine stage fluid enters the stator with the inlet conditions $T_{01} = 1100$ K and $p_{01} = 380$ kPa. The inlet flow angle is $10°$, and the velocity after the stator is $V_2 = 420$ m/s. The relative velocity has angle $-40°$ as it leaves the rotor. The blade loading factor is 1.6. (a) Determine the exit angle of the flow leaving the stator and the angle of the relative velocity as it enters the rotor. Determine the degree of reaction. (b) For the conditions in part (a), calculate the flow exit angle and the angle of the relative velocity entering the rotor. (c) Calculate the stagnation pressure losses across the stator and the rotor using Ainley–Mathiesen correlations. Take the space to axial chord ratio equal to $s/c_x = 0.75$ and assume that the maximum thickness-to-chord ratio is $t/c = 0.22$. (d) Determine the stagnation pressure loss across the stator, assuming the inlet conditions are $T_{01} = 1100$ K, and $p_{01} = 380$ kPa, and the velocity after the stator is $V_2 = 420$ m/s.

CHAPTER 7

AXIAL COMPRESSORS

In Chapter 4 it was pointed out that axial compressors are well suited for high flow rates and centrifugal machines are used when a large pressure rise is needed at a relatively low flow rate. To obtain the high flow rate, gas enters the compressor at a large radius. The gas is often atmospheric air and as it is compressed, it becomes denser and the area is reduced from stage to stage, often in such a way that the axial velocity remains constant. As in axial turbines, this may accomplished keeping the mean radius constant and by reducing the casing radius and increasing the hub radius. In jet engines high pressure ratios are obtained by pairing one multistage compressor with a turbine sufficiently powerful to turn that compressor. This arrangement of a turbine and compressor running on the same shaft is called a *spool*. A second spool consisting of intermediate-pressure (IP) compressor and turbine is then configured with a compressor in front of, and a turbine behind, the central high-pressure (HP) spool. The shaft of the IP spool is hollow and concentric with that of the HP spool. These high and intermediate pressure spools serve as *gas generators* to provide a flow to a low-pressure (LP) turbine that drives the fan in a turbofan jet aircraft or a generator for electricity production in a power plant [64].

The larger IP compressor turns at a lower speed than the HP spool in order to keep the blade speed sufficiently low to ensure that compressibility effects do not deteriorate the performance of the machine. In a typical spool one turbine stage drives six or seven compressor stages. The stage pressure ratio has increased with improved designs, reaching 1.3 and 1.4 in modern jet engine core compressors [64].

Pipeline compressors are often driven by a power turbine that uses a jet engine spool as the gas generator. Alternatively, a diesel engine may provide the power to the compressor.

An industrial compressor manufactured by MAN Diesel & Turbo SE in Germany is shown in Figure 7.1. It has 14 axial stages and one centrifugal compressor stage with shrouded blades.

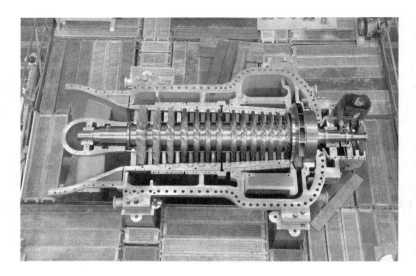

Figure 7.1 A 14-stage axial compressor with a single centrifugal stage. (Photo courtesy MAN Diesel & Turbo SE.)

The first section of this chapter is on the stage analysis of axial compressors. The theory follows closely that discussed in the previous chapter for axial turbines. Then empirical methods for calculation the flow deflection across the stator and rotor are introduced. After that a semiempirical method for allowable diffusion limit is discussed. Too much diffusion leads to separation of boundary layers, which may be catastrophic, with complete deterioration of the compressor performance. Next, the efficiency of a compressor stage is defined, followed by methods to calculate stagnation pressure losses. Three-dimensional effects will receive mention as well. Once the stagnation pressure losses have been related to flow angles, reaction, blade loading, and flow coefficient, an estimate of the stage efficiency can be obtained. The task of a compressor engineer is to use this information to design a well-performing multistage axial compressor.

7.1 COMPRESSOR STAGE ANALYSIS

A compressor stage consists of a rotor that is followed by a stator. In contrast to flow in turbines, in which pressure decreases in the direction of the flow, in compressors flow is against an adverse pressure gradient. The blade-loading coefficient is kept fairly low in order to prevent separation, with design range $0.35 \leq \psi \leq 0.5$. As a result, the amount of turning is about $20°$ and does not exceed $45°$ [19]. A typical range for the flow coefficient is $0.4 \leq \phi \leq 0.7$. If the flow is drawn into the compressor directly from an atmosphere, it enters the first stage axially. However, a set of inlet guidevanes may be used to change the flow angle α_1 to the first stage to a small positive value [18].

7.1.1 Stage temperature and pressure rise

Figure 7.2 shows a typical compressor stage. Since the rotor now precedes the stator, the inlet to the rotor is station 1 and its outlet is station 2. The outlet from the stator is station 3 and for a repeating stage the flow angles and velocity magnitudes there are equal to those at the inlet to the rotor.

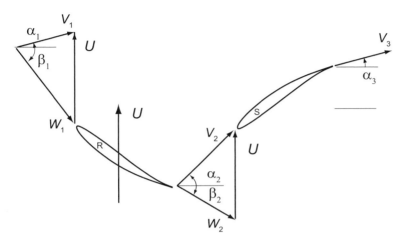

Figure 7.2 A typical axial compressor stage.

Work done by the blades is

$$w = U(V_{u2} - V_{u1}) \tag{7.1}$$

or

$$w = h_{02} - h_{01} = UV_x(\tan \alpha_2 - \tan \alpha_1) = UV_x(\tan \beta_2 - \tan \beta_1)$$

With $w = c_p(T_{03} - T_{01})$, this can also expressed in the form

$$\frac{w}{c_p T_{01}} = \frac{\Delta T_0}{T_{01}} = \frac{UV_x}{c_p T_{01}}(\tan \beta_2 - \tan \beta_1)$$

which gives the nondimensional stagnation temperature rise. From the definition of stage efficiency

$$\eta_{tt} = \frac{T_{03s} - T_{01}}{T_{03} - T_{01}}$$

the stage pressure ratio can be written as

$$\frac{p_{03}}{p_{01}} = \left(1 + \eta_{tt} \frac{\Delta T_0}{T_{01}}\right)^{\gamma/(\gamma-1)}$$

The stage temperature rise can also be written in the form

$$\frac{\Delta T_0}{T_{01}} = (\gamma - 1)\frac{U}{V_x}\frac{V_x^2}{c_{01}^2}(\tan \beta_2 - \tan \beta_1)$$

For axial entry, V_x is the inlet velocity. In terms of the flow coefficient and stagnation Mach number, defined as

$$\phi = \frac{U}{V_x} \qquad M_{01} = \frac{V_x}{c_{01}}$$

the temperature rise takes the form

$$\frac{\Delta T_0}{T_{01}} = (\gamma - 1)\frac{M_{01}^2}{\phi}(\tan\beta_2 - \tan\beta_1)$$

A typical inlet velocity is $V_x = 150$ m/s, and an inlet stagnation temperature is $T_{01} = 300$ K. The stagnation Mach number is therefore

$$M_{01} = \frac{V_x}{c_{01}} = \frac{150}{\sqrt{1.4 \cdot 287 \cdot 300}} = 0.432$$

The actual Mach number is obtained by noting that

$$M_1 = \frac{V_x}{c_{01}}\frac{c_{01}}{c_1} = M_{01}\sqrt{\frac{T_{01}}{T_1}} = M_{01}\left(1 + \frac{\gamma-1}{2}M_1^2\right)^{1/2}$$

so that

$$M_{01} = \frac{M_1}{\sqrt{1 + \frac{\gamma-1}{2}M_1^2}} \quad\text{and}\quad M_1 = \frac{M_{01}}{\sqrt{1 - \frac{\gamma-1}{2}M_{01}^2}}$$

Hence for this value of the stagnation Mach number, $M_1 = 0.440$. For a flow coefficient $\phi = 0.56$ (and axial entry), the relative velocity is at an angle

$$\beta_1 = \tan^{-1}\left(\frac{W_{u1}}{W_{x1}}\right) = \tan^{-1}\left(-\frac{U}{V_x}\right) = \tan^{-1}\left(-\frac{1}{\phi}\right) = \tan^{-1}\left(-\frac{1}{0.56}\right) = -60.75°$$

If the relative velocity is turned by 12°, the exit flow angle is $\beta_2 = -48.75°$. With these values, nondimensional stagnation temperature rise is

$$\begin{aligned}\frac{\Delta T_0}{T_{01}} &= (\gamma-1)\frac{M_{01}^2}{\phi}(\tan\beta_2 - \tan\beta_1) \\ &= \frac{0.4 \cdot 0.432^2}{0.56}(\tan(-48.75°) - \tan(-60.75°)) = 0.086\end{aligned}$$

so that the actual stagnation temperature rise is $\Delta T_0 = 25.8°$. Assuming a stage efficiency $\eta_s = 0.9$, the stage pressure rise is

$$\frac{p_{03}}{p_{01}} = (1 + 0.9 \cdot 0.086)^{3.5} = 1.30$$

This falls into the typical range of $1.3 - 1.4$ for core compressors.

The relative Mach number may be quite high at the casing, owing to the large magnitude of the relative velocity there. With $\kappa = r_h/r_C$, the blade speed at the casing is $U_c = 2U/(1+\kappa)$. Hence the relative flow angle, for $\kappa = 0.4$, at the casing is

$$\beta_{1c} = \tan^{-1}\left[\frac{2}{(1+\kappa)\phi}\right] = \tan^{-1}\left(\frac{2}{1.4 \cdot 0.56}\right) = 68.6°$$

and the relative mach number is

$$M_{1\text{Rc}} = \frac{M_1}{\cos\beta_{1c}} = \frac{0.44}{0.365} = 1.21$$

Hence the flow is supersonic (or transonic). As has been mentioned earlier, shock losses in transonic flows do not impose a heavy penalty on the performance of the machine and are therefore tolerable.

7.1.2 Analysis of a repeating stage

Equation (7.1) for work can be rewritten in a nondimensional form by dividing both sides by U^2, leading to

$$\psi = \phi(\tan\alpha_2 - \tan\alpha_1) = \phi(\tan\beta_2 - \tan\beta_1) \qquad (7.2)$$

The reaction R is the ratio of the enthalpy increase across the rotor to that over the stage

$$R = \frac{h_2 - h_1}{h_3 - h_1} = \frac{h_3 - h_1 + h_2 - h_3}{h_3 - h_1} = 1 - \frac{h_3 - h_2}{h_3 - h_1}$$

The work done by the blades causes the static enthalpy and kinetic energy (as seen from Figure 7.2) to increase across the rotor. In the stator stagnation enthalpy remains constant, and the static enthalpy and, therefore also a pressure increases, are obtained by decreasing the kinetic energy. In a design in which the areas are adjusted to keep the axial flow constant, the reduction in kinetic energy and increase in pressure result from turning the flow toward its axis. A similar argument holds for the rotor, but now it is the stagnation enthalpy of the relative flow that is constant. Hence pressure is increased by turning the relative velocity toward the axis. The velocity vectors in Figure 7.2 show this turning. The amount of turning of the flow through the rotor and stator are quite mild. A large deflection could lead to rapid diffusion and likelihood of stalled blades. Since compression is obtained in both the stator and the rotor, intuition suggests that the reaction ratio ought to be fixed to a value close to 50% in a good design. But in the first two stages, at least, where density is low and blades long, reaction increases from the hub to the casing and the average reaction is made sufficiently large to ensure that the reaction at the hub is not too low.

In the Figure 7.3 the thermodynamic states are displayed on a Mollier diagram. The distances that represent the absolute and relative kinetic energies are also shown. The relative stagnation enthalpy across the rotor remains constant at the value $h_{0R1} = h_1 + W_1^2/2 = h_2 + W_2^2/2$. The relative Mach number and the relative stagnation temperature are related by

$$\frac{T_{0R}}{T} = 1 + \frac{\gamma-1}{2}M_R^2$$

and the corresponding stagnation pressure and density are

$$\frac{p_{0R}}{p} = \left(\frac{T_{0R}}{T}\right)^{\gamma/(\gamma-1)} \qquad \frac{\rho_{0R}}{\rho} = \left(\frac{T_{0R}}{T}\right)^{1/(\gamma-1)}$$

The loss of stagnation pressure across the rotor is given by $p_{0R1} - p_{0R2}$. Across the stator the loss is $p_{02} - p_{03}$.

For the stator $h_{02} = h_{03}$ and therefore

$$h_2 + \frac{1}{2}V_2^2 = h_3 + \frac{1}{2}V_3^2 \quad \text{or} \quad h_3 - h_2 = \frac{1}{2}V_2^2 - \frac{1}{2}V_3^2$$

Since $V_2^2 = V_{x2}^2 + V_{u2}^2$ and $V_3^2 = V_{x3}^2 + V_{u3}^2$ and the axial velocity V_x is constant

$$h_3 - h_2 = \frac{1}{2}(V_{u2}^2 - V_{u3}^2) = \frac{1}{2}V_x^2(\tan^2\alpha_2 - \tan^2\alpha_3)$$

For *normal stage* $V_1 = V_3$ and $h_{03} - h_{01} = h_3 - h_1$. Since no work is done by the stator, across a stage

$$w = h_{03} - h_{01} = h_3 - h_1 = \psi U^2$$

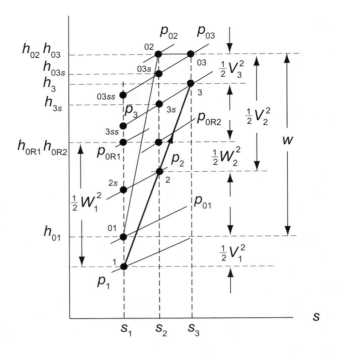

Figure 7.3 A Mollier diagram for an axial compressor stage.

Using these equations the reaction ratio, with $\alpha_1 = \alpha_3$, may be expressed as

$$R = 1 - \frac{V_x^2}{2} \frac{(\tan^2 \alpha_2 - \tan^2 \alpha_1)}{U^2 \psi} \quad \text{or} \quad R = 1 - \frac{\phi^2}{2\psi}(\tan^2 \alpha_2 - \tan^2 \alpha_1)$$

Substituting ψ from Eq. (7.2) into this gives

$$R = 1 - \frac{1}{2}\phi(\tan \alpha_2 + \tan \alpha_1) \tag{7.3}$$

Eliminating next α_2 from this, with the help of Eq. (7.2), yields

$$\psi = 2(1 - R - \phi \tan \alpha_1) \tag{7.4}$$

Equations (7.2) – (7.4) are identical to Eqs. (6.5), (6.8), and (6.9) for turbines, provided α is replaced by α_3. Hence the flow angles for the stator can be calculated from

$$\tan \alpha_1 = \frac{1 - R - \psi/2}{\phi} \qquad \tan \alpha_2 = \frac{1 - R + \psi/2}{\phi} \tag{7.5}$$

Similarly, for the rotor, the flow angles can be determined from

$$\tan \beta_1 = -\frac{R + \psi/2}{\phi} \qquad \tan \beta_2 = -\frac{R - \psi/2}{\phi} \tag{7.6}$$

In these four equations there are seven variables. Thus, once three have been specified, the other four can be determined. For example, specifying R, ψ, and ϕ, it is easy to obtain the flow angles. Suggested design ranges for these have already been mentioned. The situation changes if the axial velocity is not constant because then two flow coefficients need to be introduced. Equations (7.5) and (7.6) no longer hold, and the angles must be calculated using basic relations from velocity diagrams and definitions of blade-loading coefficient and reaction. Similarly, fundamental definitions need to be used if the flow angles entering and leaving are not the same.

There are two situations of particular interest. First, for a 50% normal reaction stage, these equations show that

$$\tan \alpha_1 = -\tan \beta_2 \qquad \alpha_1 = -\beta_2$$

$$\tan \alpha_2 = -\tan \beta_1 \qquad \alpha_2 = -\beta_1$$

and the velocity triangles are symmetric. Second, from Eq. (7.4) it is seen that for axial entry, with $\alpha_1 = 0$, the blade-loading coefficient is related to reaction by $\psi = 2(1 - R)$ and thus cannot be specified independently of the reaction. If the loading coefficient is to be in the range $0.35 < \psi < 0.5$, the stage would have to be designed for a reaction greater than 50%. But if the flow enters the stator at a small positive angle, then the blade loading can also be reduced by reducing the flow coefficient. The calculations for this situation are illustrated in the following example.

■ EXAMPLE 7.1

A normal compressor stage is designed for an inlet flow angle $\alpha_1 = 15.8°$, reaction $R = 0.63$, and the flow coefficient $\phi = 0.6$. (a) Find the blade-loading factor. (b) Determine the inlet and exit flow angles of the relative velocity to the rotor and the inlet flow angle to the stator.

Solution: (a) The value

$$\psi = 2(1 - R - \phi \tan \alpha_1) = 2(1 - 0.63 - 0.6 \tan(15.8°)) = 0.4$$

for a blade loading coefficient falls into a typical range.

(b) The flow angles are

$$\alpha_2 = \tan^{-1}\left(\frac{1 - R + \psi/2}{\phi}\right) = \tan^{-1}\left(\frac{1 - 0.63 + 0.2}{0.6}\right) = 43.54°$$

$$\beta_1 = \tan^{-1}\left(\frac{-R - \psi/2}{\phi}\right) = \tan^{-1}\left(\frac{-0.63 - 0.2}{0.6}\right) = -54.14°$$

$$\beta_2 = \tan^{-1}\left(\frac{-R + \psi/2)}{\phi}\right) = \tan^{-1}\left(\frac{-0.63 + 0.2}{0.6}\right) = -35.61°$$

The rotor turns the relative velocity by $\Delta\beta = \beta_2 - \beta_1 = 18.53°$, and the stator turns the flow by $\Delta\alpha = 27.74°$. These are also in the acceptable range. The velocity triangles in Figure 7.2 were drawn to have these angular values. A similar calculation shows that for a 50% reaction the blade-loading coefficient would increase to 0.66 and the amount of turning would be 38.34° in both the stator and the rotor. ■

In order to keep the diffusion low, a flow deflection of only 20° is typically used across the compressor blades [18]. A large deflection would lead to a steep pressure rise and possible separation of the boundary layer. A simple criterion, developed by de Haller, may be used to check whether the flow diffuses excessively [20]. He suggested that the ratios V_1/V_2 and W_2/W_1 should be kept above 0.72. These ratios can be expressed in terms of the flow angles, and for a normal stage they give the following conditions:

$$\frac{V_1}{V_2} = \frac{\cos\alpha_2}{\cos\alpha_1} > 0.72 \qquad \frac{W_2}{W_1} = \frac{\cos\beta_1}{\cos\beta_2} > 0.72$$

In the foregoing example when $R = 0.63$

$$\frac{V_1}{V_2} = \frac{\cos\alpha_2}{\cos\alpha_1} = \frac{\cos(43.54°)}{\cos(15.80°)} = 0.75 \qquad \frac{W_2}{W_1} = \frac{\cos\beta_1}{\cos\beta_2} = \frac{\cos(-54.14°)}{\cos(-35.63°)} = 0.72$$

so the de Haller criterion is satisfied. If the reaction is reduced to $R = 0.5$, then for both the rotor and the stator, these ratios are

$$\frac{\cos\alpha_2}{\cos\alpha_1} = \frac{\cos(54.14°)}{\cos(15.80°)} = 0.61$$

and now the de Haller criterion is violated. On the basis of this comparison the higher reaction keeps the diffusion within acceptable limits.

It has been mentioned that there is another reason why the reaction should be relatively large for the first two stages. Since the gas density there is low, to keep the axial velocity constant through the compressor, a large area and thus long blades are needed. This causes reaction to vary greatly from the blade root to its tip. To see this, consider again the equation

$$w = U(V_{u2} - V_{u1}) = \Omega(rV_{u2} - rV_{u1})$$

If the tangential velocity distribution is given by free vortex flow for which rV_u is constant then each blade section does the same amount of work. For a blading of this kind the equation for reaction

$$R = 1 - \frac{1}{2}\phi(\tan\alpha_2 + \tan\alpha_1)$$

may also be written as

$$R = 1 - \frac{V_{u2} + V_{u1}}{2U} = 1 - \frac{C_2 + C_1}{2rU}$$

where $C_1 = rV_{u1}$ and $C_2 = rV_{u2}$. Since $U = rU_m/r_m$, in which subscript m designates a condition at the mean radius, for this flow the reaction takes the form

$$R = 1 - \frac{A}{r^2}$$

which shows that the reaction is low at the hub and increases along the blades. Thus, if the reaction at the mean radius is to be 50%, the low reaction at the hub causes a large loading and greater deflection of the flow. This leads to greater diffusion.

If guidevanes are absent, the flow enters the stage axially. Hence $\alpha_1 = 0$ and Eq. (7.4) reduces to $\psi = 2(1 - R)$. When this is substituted into Eqs. (7.5) and (7.6) the following relations are obtained:

$$\tan\alpha_2 = \frac{2(1 - R)}{\phi} \qquad \tan\beta_1 = -\frac{1}{\phi} \qquad \tan\beta_2 = -\frac{2R - 1}{\phi}$$

Two of the parameters could now be assigned values and the rest calculated from these equations.

Another way to proceed is to use the de Haller criterion for the rotor flow angles and set

$$\frac{\cos\beta_1}{\cos\beta_2} = \sqrt{\frac{\phi^2 + (2R-1)^2}{\phi^2 + 1}} = D_R \quad (7.7)$$

in which $D_R = 0.72$, or slightly larger than this. This method of designing a stage is discussed in the next example.

■ EXAMPLE 7.2

A compressor stage is to be designed for axial entry and a reaction $R = 0.82$. Use the de Haller criterion to fix the flow angles for the stage design.

Solution: With $\alpha_1 = 0$ and $R = 0.82$, the blade loading coefficient is $\psi = 0.36$. Using the de Haller criterion, with $D_R = 0.72$, for limiting the amount of diffusion in the rotor, the flow coefficient can be solved from Eq. (7.7). This yields

$$\phi = \sqrt{\frac{D_R^2 - (2R-1)^2}{1 - D_R^2}} = \sqrt{\frac{0.72^2 - (2 \cdot 0.82 - 1)^2}{1 - 0.72^2}} = 0.4753$$

The remaining calculations give

$$\alpha_2 = \tan^{-1}\left(\frac{2-2R}{\phi}\right) = \tan^{-1}\left(\frac{0.36}{0.4753}\right) = 37.14°$$

$$\beta_1 = \tan^{-1}\left(-\frac{1}{\phi}\right) = \tan^{-1}\left(-\frac{1}{0.4753}\right) = -64.58°$$

$$\beta_2 = \tan^{-1}\left(\frac{1-2R}{\phi}\right) = \tan^{-1}\left(-\frac{0.64}{0.4753}\right) = -53.40°$$

Thus the de Haller criterion for the stator becomes

$$D_S = \frac{\cos\alpha_2}{\cos\alpha_1} = \frac{\cos(37.14°)}{\cos(0°)} = 0.797$$

and for the rotor it is

$$D_R = \frac{\cos\beta_1}{\cos\beta_2} = \frac{\cos(-64.58°)}{\cos(-53.40°)} = 0.720$$

in agreement with its specified value. The deflection across the stator is $\Delta\alpha = 37.14°$, and across the rotor it is $\Delta\beta = -53.40° + 64.58° = 11.18°$. ■

The velocity triangles of the foregoing example, drawn with U as a common side, are shown in Figure 7.4. The example shows that even if the flow turns by greater amount through the stator, it diffuses less than in the rotor. The reason is that the turning takes place at a low mean value of α. In fact, were the flow to turn from, say, $-10°$ to $10°$, there would be no diffusion at all, because the magnitude of the absolute velocity would be the same before and after the stator. Thus it is the higher stagger of the rotor that leads to large diffusion even at low deflection. For this reason, the de Haller criterion needs to be checked.

The deflection, represented by the change in the swirl velocity, is shown as the vertical distance in the top left of the diagram. Dividing it by the blade speed gives the loading

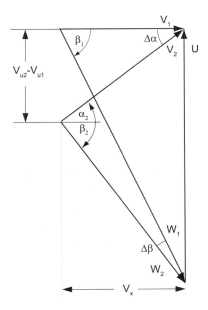

Figure 7.4 Velocity triangles on a common base for an axial compressor stage.

coefficient. The ratio of the horizontal V_x to blade speed is the flow coefficient. Thus a glance at the horizontal width of the triangles and comparison with the blade speed shows that the flow coefficient is slightly less than 0.5. The extents of turning across the stator and rotor are shown as angles $\Delta\alpha$ and $\Delta\beta$, respectively. The decrease in magnitude of the velocity across the stator is slightly larger than the length of the side opposite to the angle $\Delta\beta$ in the triangle with sides W_1 and W_2. Similarly, the length of the side opposite to the angle $\Delta\alpha$ in the triangle with V_1 and V_2 as its sides indicates the extent of reduction of the relative velocity. Hence inspection confirms that even slight turning, may lead to large diffusion when the blades are highly staggered.

7.2 DESIGN DEFLECTION

Figure 7.5 shows typical results from experiments carried out in a cascade tunnel [37]. It shows the deflection and losses from irreversibilities for a given blade as a function of incidence. The incidence is $i = \alpha_2 - \chi_2$, in which χ_2 is the metal angle. The losses increase with both positive and negative incidence, but there is a large range of incidence for which the losses are quite low. The deflection increases with incidence up to the stalling incidence ε_s, at which the maximum deflection is obtained. At this value losses have reached about twice their minimum value. This correspondence is not exact, but since the losses increase rapidly beyond this, a stage is designed for a nominal deflection of $\varepsilon^* = 0.8\varepsilon_s$, which also corresponds to an incidence at which the loss is near its minimum. As shown in the figure, at this condition the incidence is slightly negative. But for another cascade it may be zero, or slightly positive. The loss coefficients have been defined as

$$\omega_R = \frac{h_2 - h_{2s}}{\frac{1}{2}W_1^2} \qquad \omega_S = \frac{h_3 - h_{3s}}{\frac{1}{2}V_2^2}$$

The upstream velocity is now the reference velocity. The relationship between these and those based on the downstream velocity is

$$\omega_R = \zeta_R \frac{\cos^2 \beta_2}{\cos^2 \beta_1} \qquad \omega_S = \zeta_S \frac{\cos^2 \alpha_1}{\cos^2 \alpha_2}$$

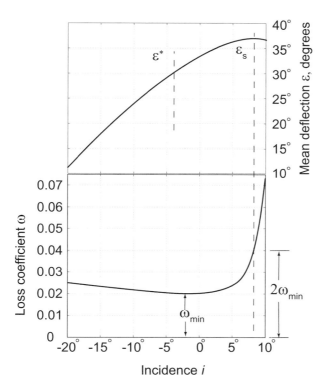

Figure 7.5 Mean deflection and stagnation loss coefficient as a function of incidence. (Drawn after Howell [37].)

The conclusion from a large number of experiments is that the nominal deflection is mainly a function the gas outlet angle and the space-chord ratio of the cascade. The camber and incidence are additive, and both are responsible for the amount of deflection of the flow. For the cascade shown in Figure 7.5, the nominal deflection is 30° with incidence at −4°. Thus the camber is quite low, making the blades rather flat. From such experiments a universal correlation, shown in Figure 7.6, that relates the deflection to the exit angle with solidity $\sigma = c/s$ as a parameter was developed by Howell [37]. Solidity is the ratio of the length of the blade chord to the spacing of the blades. It increases with reduced spacing, and the term suggests that in this case the solid blades fill the flow annulus more than the open passages. As the solidity increases the flow follows the blades better.

A curve fit for the nominal deflection, suitable for computer calculations, is

$$\varepsilon^* = (-3.68\sigma^2 + 17.2\sigma + 4.3)\left(\frac{\alpha_3^*}{100}\right)^2 + (12.6\sigma^2 - 54.3\sigma - 10.0)\frac{\alpha_3^*}{10} - 8.7\sigma^2 + 36.4\sigma + 6.1 \tag{7.8}$$

For example, for a cascade with $\sigma = 1.2$, if the relative velocity leaves the rotor at angle $\alpha_3 = 30°$, the nominal deflection is $\varepsilon^* = 21.92°$. For a cascade with solidity $\sigma = \frac{2}{3}$ and the flow leaving the stator at $\alpha_3 = -10°$, the nominal deflection is $\varepsilon^* = 30.7°$. For given ε^* and α_3, this equation is quadratic in σ. Its solution has an extraneous root with a value greater than 3 and it needs to be rejected.

This equation can also be used to calculate the deflection across a rotor by replacing α_3 with $-\beta_2$. The two examples worked out earlier in this chapter show that the deflection is larger across the stator than across rotor and that the angle at which the gas leaves the stator is not large. The flow over the rotor is turned less, and the absolute value of its exit angle is quite large. These are consistent with the results shown in Figure 7.6.

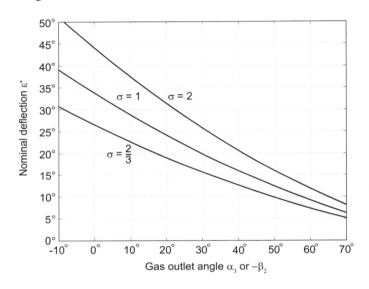

Figure 7.6 Nominal deflection as a function of gas outlet angle and solidity. Drawn after Howell [37].

An alternative to Eq. (7.8) is the *tangent difference formula*, which for the rotor is

$$\tan \beta_2^* - \tan \beta_1^* = \frac{1.55}{1 + 1.5/\sigma_R} \tag{7.9}$$

and a similar equation

$$\tan \alpha_2^* - \tan \alpha_3^* = \frac{1.55}{1 + 1.5/\sigma_S} \tag{7.10}$$

holds for a stator. They provide a quick way to calculate the nominal deflection and fit the data well over most of the outlet angles. At low deflections they underpredict the deflection, by about 3° at $\alpha_3^* = -10°$ and low solidity. Indeed, at $\alpha_3^* = -10°$ and $\sigma = \frac{2}{3}$ the tangent difference formula gives $\varepsilon^* = 26.7°$, whereas Eq. (7.8) yields 30.7°. Typically the flow does not leave the stator at a negative angle so that such exit angles are just outside the range of usual designs.

For a normal stage with constant axial velocity the nondimensional equation for work is

$$\psi = \phi(\tan \alpha_2 - \tan \alpha_1) = \phi(\tan \beta_2 - \tan \beta_1)$$

The tangent difference formulas now show that the solidity for the rotor is the same as that for the stator, as the design is based on nominal conditions. It is determined by solving

$$\frac{\psi^*}{\phi^*} = \frac{1.55}{1 + 1.5/\sigma}$$

for σ. In order to prevent resonance vibrations, the number of blades in the rotor is slightly different from the number of stator blades. Thus the spacing is changed, but the same solidity can still be achieved by changing the length of the chord.

■ **EXAMPLE 7.3**

A normal compressor stage is designed to have the flow leave the stator at the angle $\alpha_1 = 12.60°$. Assume that the optimum design condition for least losses is achieved when the reaction is $R = 0.68$, and the flow coefficient $\phi = 0.56$. For a normal stage, find the blade loading factor, and the optimum value for the solidity. Check also that the de Haller criterion is satisfied.

Solution: The loading coefficient is first determined from

$$\psi = 2(1 - R - \phi \tan \alpha_1) = 2(1 - 0.68 - 0.56 \tan(12.6°)) = 0.39$$

and the flow angles are

$$\alpha_2 = \tan^{-1}\left(\frac{1 - R + \psi/2}{\phi}\right) = \tan^{-1}\left(\frac{1 - 0.68 + 0.195}{0.56}\right) = 42.60°$$

$$\beta_1 = \tan^{-1}\left(\frac{-R - \psi/2}{\phi}\right) = \tan^{-1}\left(\frac{-0.68 - 0.195}{0.56}\right) = -57.38°$$

$$\beta_2 = \tan^{-1}\left(\frac{-R + \psi/2)}{\phi}\right) = \tan^{-1}\left(\frac{-0.68 + 0.195}{0.56}\right) = -40.90°$$

The deflections are therefore

$$\varepsilon_S^* = 30.00° \qquad \varepsilon_R^* = 16.48°$$

The diffusion factors are

$$\frac{\cos \alpha_2}{\cos \alpha_1} = \frac{\cos(42.60°)}{\cos(12.60°)} = 0.754 \qquad \frac{\cos \beta_1}{\cos \beta_2} = \frac{\cos(-57.38°)}{\cos(-40.90°)} = 0.713$$

so that the diffusion in the rotor is marginally too high. If the solidity is calculated from the tangent difference formula, it yields

$$\sigma = \frac{1.5 \psi}{1.55 \phi - \psi} = 1.22$$

■

In actual machines the values of ϕ, ψ, and R vary across the span, owing to the change in the blade velocity with the radius, and ψ tends to be high near the hub and low near the casing. The reaction is low near the hub and high near the casing, as R moves in opposite direction to the loading coefficient. The blade angles are adjusted to counteract the natural tendency that causes the values to change so that the loading can be kept more uniform. A 50% reaction ratio is common, and the blade-loading coefficient is typically in the range $0.3 < \psi < 0.45$.

7.2.1 Compressor performance map

Compressor blades tend to be quite thin, with maximum thickness-to-chord ratio of 5%. If the solidity is high, the blades guide the flow well. An operating condition in which the flow coefficient, ϕ_o, is larger than its design value, ϕ_d, is shown in Figure 7.7. It is seen that an increase in the flow rate causes a decrease in the blade-loading coefficient. This effect becomes amplified downstream as the density does not change according to design, and the difference cumulates from stage to stage. This subject is discussed in Cumpsty [18], who shows that the last stage is one that is likely to choke. Similarly, examination of Figure 7.7 shows that if the flow coefficient decreases, the blade loading coefficient increases. The blades now become susceptible to stall, and the last stage controls the stall margin.

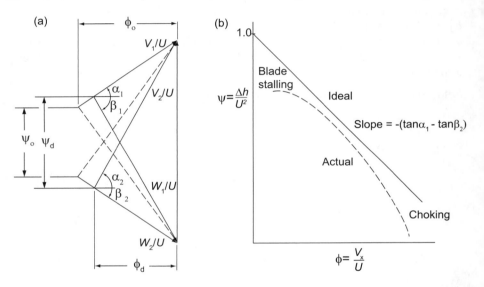

Figure 7.7 Normalized velocity triangles at design and off-design operation and their performance characteristics.

This reasoning can be carried out analytically be rewriting the Euler equation of turbomachinery

$$w = U(V_{u2} - V_{u1})$$

into a different form. Substituting

$$V_{u2} = U + W_{u2}$$

into the previous equation gives

$$w = U(U + W_{u2} - V_{u1}) = U[U - V_x(\tan\alpha_1 - \tan\beta_2)]$$

and dividing next each term by U^2 gives

$$\psi = 1 - \phi(\tan\alpha_1 - \tan\beta_2)$$

For a normal stage the exit angle from the stator is α_1, and the exit angle of the relative velocity from the rotor is β_2. These then tend to remain constant even at off-design conditions. Furthermore, the latter angle is usually negative so that the term in parentheses is positive. With the trigonometric factors constant, this means that increasing the flow coefficient decreases the loading.

This equation gives the compressor characteristic for an ideal compressor. It is a straight line with a negative slope when the blade-loading coefficient is plotted against the flow coefficient. This is shown in Figure 7.7b. The actual compressor characteristic is also shown, and the differences away from the design point are caused by irreversibilites. Far away from the design point to the left, the blades stall and at even lower flow rates (smaller ϕ) the compressor may experience surge. This means that the flow can actually reverse its direction and flow out the front of the compressor. For this reason compressors are operated some distance away from the stall line. To the right of the design point irreversibilities again cause deviation from the theoretical curve, and as the flow rate increases, the blade row will choke.

7.3 RADIAL EQUILIBRIUM

The mean line analysis on which most of the calculations in this text are based ignores the cross-stream variation in the flow. In axial-flow machines this means that only the influence of the blade speed, that increases with radius, is taken into account. Today it is possible to carry out calculations by CFD methods to resolve the three-dimensional aspects of the flow. However, as was seen in the discussion of axial turbines, the elementary radial equilibrium theory advances the understanding on how the important variables, such as the reaction and the loading, vary from the hub to the casing. There (and in Appendix A) it was shown that the principal equation to be solved is

$$\frac{dh_0}{dr} = T\frac{ds}{dr} + V_x\frac{dV_x}{dr} + V_u\frac{dV_u}{dr} + \frac{V_u^2}{r} \tag{7.11}$$

The first term on the right represents the entropy variation in the radial direction. Owing to the entropy production in the endwall boundary layers, tip vortices, and possible shocks, this is likely to be important near the walls. However, if the flow mixes well in the radial direction, entropy gradients diminish and it may be reasonable to neglect this term. Of course, irreversibilities would still cause entropy to increase in the downstream direction, even if it does not have radial gradients.

The stagnation enthalpy is uniform at the entrance to the first row of rotor blades, and if every blade section does an equal amount of work on the flow, it will remain uniform even if the stagnation enthalpy increases in the direction of the stream. As a consequence, the term on the left side of this equation may be neglected, and the equation reduces to

$$V_x\frac{dV_x}{dr} + V_u\frac{dV_u}{dr} + \frac{V_u^2}{r} = 0 \tag{7.12}$$

This is a relationship between the velocity components V_u and V_x. If the radial variation of one of them is assumed, the variation of the other is obtained by solving this equation. For example, as was seen in the discussion of axial turbines, if the axial component V_x is assumed to be uniform, then the radial component must satisfy the equation

$$\frac{dV_u}{V_u} = -\frac{dr}{r}$$

the solution of which is $rV_u = $ constant. This is the *free vortex* velocity distribution. Before and after a row of blades the tangential velocity is

$$V_{u1} = \frac{c_1}{r} \qquad V_{u2} = \frac{c_2}{r}$$

It was observed earlier that the expression for work is now

$$w = U(V_{u2} - V_{u1}) = r\Omega \left(\frac{c_2}{r} - \frac{c_1}{r}\right) = \Omega(c_2 - c_1)$$

so that the work is independent of radius. This justifies dropping the term dh_0/dr in Eq. (7.11).

7.3.1 Modified free vortex velocity distribution

The free vortex velocity distribution is a special case of the family of distributions

$$V_{u1} = cr^n - \frac{d}{r} \qquad V_{u2} = cr^n + \frac{d}{r} \qquad (7.13)$$

in which n is a parameter. Regardless of the value of n, each member of this family has a velocity distribution for which each blade section does an equal amount of work, as seen from

$$w = U(V_{u2} - V_{u1}) = \Omega r \left(\frac{d}{r} + \frac{d}{r}\right) = 2\Omega d$$

With the mean radius r_m the same at the inlet and the exit, a nondimensional radius may be introduced as $y = r/r_m$. The tangential velocity components for a free vortex distribution with $n = -1$ may now be written as

$$V_{u1} = \frac{a}{y} - \frac{b}{y} = \frac{a-b}{y} \qquad V_{u2} = \frac{a}{y} + \frac{b}{y} = \frac{a+b}{y}$$

When the Eq. (7.12) for radial equilibrium is recast into the form

$$V_x \frac{dV_x}{dy} = -\frac{V_u}{y} \frac{d}{dy}(yV_u)$$

it shows that for this velocity distribution RHS vanishes, as the substitution shows

$$V_x \frac{dV_x}{dy} = \frac{a-b}{y^2} \frac{d(a-b)}{dy} = 0 \qquad V_x = \text{constant}$$

Clearly, the same result is obtained for the outlet, as only the sign of b needs to be changed to describe the tangential velocity there. This reduced equation now shows that the axial velocity is constant along the span of the blade, but it does not follow that it has the same constant value before and after the blade row. The axial velocities can be made the same by proper taper of the flow channel.

The reaction is given by

$$R = 1 - \frac{1}{2}\frac{V_x}{U}(\tan\alpha_2 + \tan\alpha_2) = 1 - \frac{1}{2}\frac{V_{u2} + V_{u1}}{U_m y}$$

Here $U = r\Omega = rU_m/r_m$, or $U = U_m y$ was used. Since $V_{u2} + V_{u1} = 2a/y$, the reaction is

$$R = 1 - \frac{a}{U_m y^2}$$

If R_m is the reaction at the mean radius $y = 1$, then $a = U_m(1 - R_m)$ and the reaction can be written as

$$R(y) = 1 - (1 - R_m)\frac{1}{y^2}$$

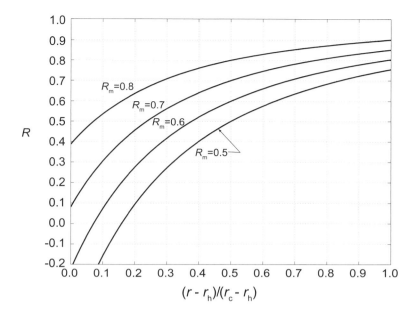

Figure 7.8 Reaction as a function of the radial position on the blade for a free vortex velocity distribution for $\kappa = 0.4$.

The parameter b can be related to the work done. From

$$w = U(V_{u2} - V_{u1}) = 2U_m b$$

and $b = w/2U_m = \frac{1}{2}\psi_m U_m$, in which $\psi_m = w/U_m^2$.

Another way to represent the data is to introduce a nondimensional radial coordinate $z = (r - r_h)/(r_c - r_h)$, so that

$$z = \frac{1+\kappa}{1-\kappa}\frac{y}{2} - \frac{\kappa}{1-\kappa}$$

in which $\kappa = r_h/r_c$. Clearly, the hub is now at $z = 0$ and the casing, at $z = 1$. The variation in the reaction as a function of this variable is shown in Figure 7.8, for the hub to tip radius $\kappa = 0.4$ and for values of the mean reaction in the range $0.5 \le R_m \le 0.8$. The graphs show that increasing the mean reaction lifts the negative reaction at the hub to a positive value.

Since the axial velocity is constant, the mean flow angles are given by Eqs. (7.5) and (7.6) for specified values of reaction, flow coefficient, and blade-loading coefficient. In this flow the flow coefficient and blade-loading coefficient vary along the blade only because the blade speed varies. Their local values, and that of the reaction, are given by

$$\phi = \frac{\phi_m}{y} \qquad \psi = \frac{\psi_m}{y^2} \qquad R = (1-R_m)\frac{1}{y^2}$$

The angles are then obtained from

$$\alpha_1 = \tan^{-1}\left(\frac{1-R-\psi/2}{\phi}\right) \qquad \beta_1 = \tan^{-1}\left(-\frac{R+\psi/2}{\phi}\right)$$

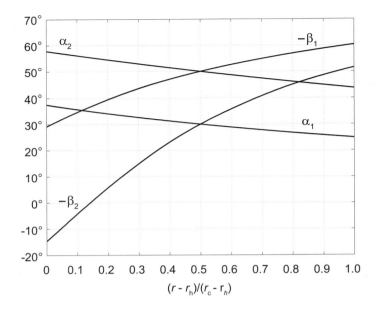

Figure 7.9 Flow angles as a function of the radial position on the blade for a free vortex velocity distribution for a stage with $\kappa = 0.6$. At the mean radius $R_m = 0.5$, and $\alpha_{1m} = 30°$, $\beta_{1m} = -50°$, $\alpha_{2m} = 50°$ and $\beta_{2m} = -50°$, with $\phi_m = 0.565$ and $\psi_m = 0.347$.

$$\alpha_2 = \tan^{-1}\left(\frac{1 - R + \psi/2}{\phi}\right) \qquad \beta_2 = \tan^{-1}\left(-\frac{R - \psi/2}{\phi}\right)$$

They are shown in Figure 7.9 for the situation for which $R_m = 0.5$, $\alpha_{1m} = 30°$, $\beta_{1m} = -50°$, $\alpha_{2m} = -50°$, and $\beta_{2m} = -30°$. This stage is a typical one deeper into the compressor, where blades are shorter ($\kappa = 0.6$) and reaction can be kept at 50% without becoming negative at the hub. In fact, the reaction at the hub is $R_h = 0.111$, and the blade-loading coefficient and flow coefficient there are $\psi_h = 0.617$ and $\phi_h = 0.754$. The relative flow undergoes a large deflection at the hub of the rotor, as the flow angle changes by 43.8°. Furthermore, as the hub is approached, β_2 changes sign, and at $(r - r_h)/(r_c - r_h) = 0.146$ the relative flow is exactly in the axial direction.

Since the axial velocity is independent of the radius in the free vortex blading,

$$V_1 \cos \alpha_1 = V_{1m} \cos \alpha_{1m} \qquad W_2 \cos \beta_2 = W_{2m} \cos \beta_{2m}$$

and dividing the first equation by c_{01} and the second by c_{02} gives

$$M_{01} = M_{01m} \frac{\cos \alpha_{1m}}{\cos \alpha_1} \qquad M_{02R} = M_{02mR} \frac{\cos \beta_{2m}}{\cos \beta_2}$$

The stagnation Mach numbers can be calculated from

$$M_{01m} = \frac{M_{1m}}{\sqrt{1 + \frac{\gamma-1}{2} M_{1m}^2}} \qquad M_{02mR} = \frac{M_{2mR}}{\sqrt{1 + \frac{\gamma-1}{2} M_{2mR}^2}}$$

The absolute and relative Mach numbers are then obtained from

$$M_1 = \frac{M_{01}}{\sqrt{1 - \frac{\gamma-1}{2} M_{01}^2}} \qquad M_{2R} = \frac{M_{02R}}{\sqrt{1 - \frac{\gamma-1}{2} M_{02R}^2}}$$

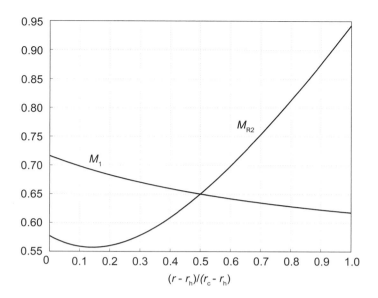

Figure 7.10 Radial variation of the Mach number as a function of radial position on the blade at the entrance to the stator and the relative Mach number at the entrance to the rotor. In both cases the respective Mach numbers were 0.65 at the mean radius. The other parameters are the same as in Figure 7.9.

Figure 7.10 shows that the relative Mach number has a minimum near the hub, where the relative velocity changes sign. It rises rapidly to a very high value at the shroud. For this reason its design value at the mean radius cannot be very large. The stagnation temperature increases across the rotor, and so does its speed of sound. This reduces the Mach number as the flow enters the stator. This was not taken into account in the graphs shown, for both Mach numbers were chosen to have the value 0.65 at the mean radius.

7.3.2 Velocity distribution with zero-power exponent

Another member of the velocity profiles given by Eq. (7.13) is

$$V_{u1} = a - \frac{b}{y} \qquad V_{u2} = a + \frac{b}{y}$$

for which $n = 0$. At the inlet, the radial equilibrium equation in this case reduces to

$$V_x \frac{dV_x}{dy} = -\left(\frac{a}{y} - \frac{b}{y^2}\right) \frac{d}{dy}(ay - b) = -\left(\frac{a^2}{y} - \frac{ab}{y^2}\right)$$

Integrating both sides from $y = 1$ to some arbitrary location gives

$$V_{x1}^2 - V_{x1m}^2 = -2\left[a^2 \ln y + ab\left(\frac{1}{y} - 1\right)\right]$$

Corresponding expression at the exit is obtained by changing the sign of b. It is

$$V_{x2}^2 - V_{x2m}^2 = -2\left[a^2 \ln y - ab\left(\frac{1}{y} - 1\right)\right]$$

The axial velocity now depends on the radial coordinate. For this reason the reaction is calculated from

$$R = 1 - \frac{h_3 - h_2}{h_3 - h_1} = 1 - \frac{V_2^2 - V_1^2}{2w} = 1 - \frac{V_{x2}^2 - V_{x1}^2 + V_{u2}^2 - V_{u1}^2}{2U(V_{u2} - V_{u1})}$$

If the taper is set such that the axial velocity at the mean radius is the same at the inlet and the outlet, then subtracting the two equations for the axial velocity gives

$$V_{x2}^2 - V_{x1}^2 = 4ab\left(\frac{1}{y} - 1\right)$$

and the sum and difference terms in $V_{u2}^2 - V_{u1}^2 = (V_{u2} - V_{u1})(V_{u2} + V_{u1})$ are

$$V_{u2} - V_{u1} = \frac{2b}{y} \qquad V_{u2} + V_{u1} = 2a$$

Hence, with $U = U_m y$, the reaction may be written as

$$R = 1 - \frac{a}{U_m}\left(1 - \frac{2}{y}\right)$$

and, with the reaction equal to R_m at $y = 1$, the parameter a is obtained from the expression $a = U_m(R_m - 1)$. The reaction depends on the radius as

$$R(y) = 1 + (1 - R_m)\left(1 - \frac{2}{y}\right)$$

As shown above, work done is uniform and $b = \frac{1}{2}\psi U_m$.

7.3.3 Velocity distribution with first-power exponent

A velocity distribution with $n = 1$ in Eq. (7.13) can be expressed as

$$V_{u1} = ay - \frac{b}{y} \qquad V_{u2} = ay + \frac{b}{y}$$

The radial equilibrium equation at the inlet now reduces to the form

$$V_x \frac{dV_x}{dy} = -\left(a - \frac{b}{y^2}\right)\frac{d}{dy}(ay^2 - b) = \frac{2ab}{y} - 2a^2 y$$

Integrating this from $y = 1$ to an arbitrary location gives

$$V_{x1}^2 - V_{x1m}^2 = 4ab \ln y - 2a^2(y^2 - 1)$$

and the same operations at the exit yield

$$V_{x2}^2 - V_{x2m}^2 = -4ab \ln y - 2a^2(y^2 - 1)$$

The reaction is calculated from

$$R = 1 - \frac{h_3 - h_2}{h_3 - h_1} = 1 - \frac{V_2^2 - V_1^2}{2w} = 1 - \frac{V_{x2}^2 - V_{x1}^2 + V_{u2}^2 - V_{u1}^2}{2U(V_{u2} - V_{u1})}$$

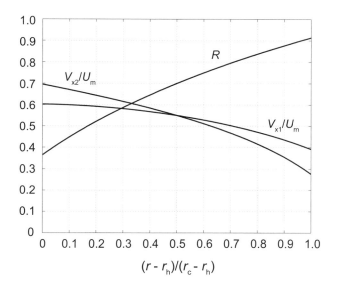

Figure 7.11 Normalized axial velocities before and after the rotor and reaction as a function of the radial position on the blade for a first power velocity distribution, for $\phi_m = 0.55$, $\psi = 0.18$, $R_m = 0.7$, and $\kappa = 0.4$.

Again, if the taper of the annulus is such that the velocities at the common mean radius are equal, then the difference of the squares of the axial velocities is

$$V_{x2}^2 - V_{x1}^2 = -8ab \ln y$$

and the difference and the sum of the swirl velocities are

$$V_{u2} - V_{u1} = \frac{2b}{y} \qquad V_{u2} + V_{u1} = 2ay$$

Hence the reaction may be written as

$$R = 1 + \frac{2a}{U_m} \ln y - \frac{a}{U_m}$$

With $R = R_m$ at $y = 1$, the parameter $a = U_m(1 - R_m)$. Similarly, $b = \frac{1}{2}\psi U_m$. The reaction, shown in Figure 7.11, can be written as

$$R(y) = 1 - (1 - R_m)(1 - 2\ln y)$$

Also shown are the axial velocities before and after the rotor. They are given by the expressions

$$\frac{V_{x1}}{U_m} = \sqrt{\phi_m^2 + 2(1 - R_m)\psi \ln y - 2(1 - R_m)^2(y^2 - 1)}$$

$$\frac{V_{x2}}{U_m} = \sqrt{\phi_m^2 - 2(1 - R_m)\psi \ln y - 2(1 - R_m)^2(y^2 - 1)}$$

in which $\phi_m = V_{xm}/U_m$ is the flow coefficient at the mean radius.

7.4 DIFFUSION FACTOR

Earlier in this chapter the de Haller criterion was used as a criterion to ensure that the diffusion in the flow passages would not be strong enough to cause separation of the boundary layers. The adverse pressure gradient associated with diffusion is seen in Figure 7.12, which shows a typical stator blade surface pressure measurement, plotted as a pressure coefficient $C_p = (p - p_2)/(p_{02} - p_2)$ against the fractional distance along the chord for three different values of solidity $\sigma = c/s$. The bottom branch of the curve gives the suction-side pressure distribution and the top branch, that for the pressure side of the blade. Under ideal conditions at low Mach number the spike in the pressure coefficient on the top branch should reach unity at the stagnation point located near the leading edge. From there the pressure falls as the flow accelerates to a location of maximum blade thickness followed by diffusion toward the trailing edge. Near the trailing edge the flow may accelerate slightly owing to its orientation and camber. On the suction side of the blade the negative spike is caused by the need for the flow to negotiate the blunt leading edge of the blade as it flows from the location of the stagnation point to the suction side. Pressure then increases sharply from the suction spike all the way to the trailing edge as the flow diffuses. This diffusion must be kept within acceptable limits. The difference in the value of pressure at any given chord location is a measure of the *local loading* on the blade and, as seen from Figure 7.12, the blade becomes *unloaded* as the trailing edge is approached.

Since the blade force is obtained by integrating the pressure acting over the blade surface, the area inscribed by the curves represents the blade force normal to the chord. As the solidity increases, this force is reduced, but now with reduced spacing, more blades can be fitted to the rotor wheel to carry the load. Increased solidity lifts the value of the minimum pressure, with the result that the pressure gradient decreases. Therefore, the flow can be turned by a greater amount without danger of boundary-layer separation.

A local *diffusion factor* is defined as

$$D_{\mathrm{loc}} = \frac{V_{\mathrm{max}} - V_3}{V_2}$$

in which V_{max} is the velocity at the location of minimum pressure. This definition would be useful, if the profile shape were known and V_{max} could be easily calculated. It can, in fact, be done quite readily with a computer code for inviscid flows. But before this became a routine task, an alternative was sought. Lieblein and Roudebush [51] suggested a diffusion factor of the form

$$DF = \frac{V_2 - V_3}{V_2} + \frac{1}{2\sigma}\frac{V_{u2} - V_{u3}}{V_2} \tag{7.14}$$

The first part is similar to the local diffusion factor, and the second part accounts for the amount of turning and the solidity of the blade row. Since reduced turning and higher solidity both contribute to lighter loading on the blades, diffusion is expected to diminish by both effects. The Lieblein diffusion factor for the stator can be expressed as

$$DF_{\mathrm{S}} = \frac{\cos \alpha_3 - \cos \alpha_2}{\cos \alpha_3} + \frac{1}{2\sigma_{\mathrm{S}}}(\tan \alpha_2 - \tan \alpha_3)\cos \alpha_2 \tag{7.15}$$

and for the rotor it takes the form

$$DF_{\mathrm{R}} = \frac{\cos \beta_2 - \cos \beta_1}{\cos \beta_2} + \frac{1}{2\sigma_{\mathrm{R}}}(\tan \beta_1 - \tan \beta_2)\cos \beta_1 \tag{7.16}$$

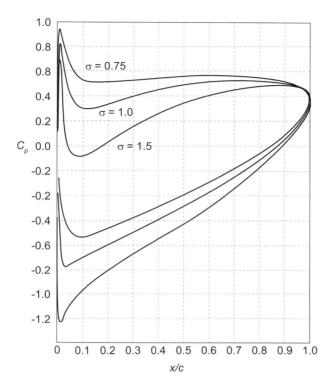

Figure 7.12 Pressure coefficient for a cascade with different solidities.

It is known that a boundary layers thicken faster in a flow with an adverse pressure gradient than when there is none. A relationship may be developed between diffusion and the boundary-layer thickness. Figure 7.13 is a plot of the diffusion factor as a function of the ratio of the *momentum thickness* of the boundary layer to the length of the chord θ/c. The

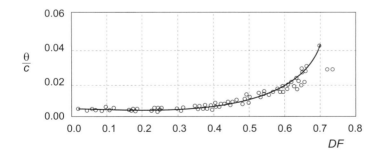

Figure 7.13 Dependence of the momentum thickness of the boundary layer on diffusion factor. (Drawn after Lieblein and Roudebush [51].)

curve begins its upward bend at the momentum thickness-to-chord ratio of about 0.007, and this corresponds to $DF = 0.45$. Thus values below these give good designs. For larger DF values the stagnation pressure losses grow appreciably.

Equations (7.15) and (7.16) are two additional equations among the design variables. In an earlier section it was shown that specifying the values for R, ψ, ϕ fixes the flow angles. With the tangent difference formulas and the two diffusion factor formulas, three more equations are added, but solidity is the only new unknown. Hence in principle the designer has the freedom to choose only one of the parameters, such as reaction, or else violate either the tangent difference equation, or one of the diffusion factor equations. However, if the design is such that either DF_S or DF_R, or both are less than 0.42, then the tangent difference formula fixes the solidity. Thus the choice for the acceptable range of values of reaction, blade loading, and the flow coefficient to achieve the mild diffusion restricts the design parameter space. The typical design ranges $0.35 \leq \psi \leq 0.5$ and $0.4 \leq \phi \leq 0.7$ reflect this.

7.4.1 Momentum thickness of a boundary layer

The *momentum thickness of a boundary layer* is related to the diffusion factor in Figure 7.13. It and the *displacement thickness* are now discussed further. The latter is denoted by δ^* and is introduced by the expression

$$\delta^* V_f = \int_0^\delta (V_f - V(y)) dy \tag{7.17}$$

in which V_f is the uniform velocity outside the boundary layer of thickness δ. Clearly, this equation can be rewritten as

$$\int_0^\delta V(y) dy = V_f(\delta - \delta^*) \tag{7.18}$$

and its interpretation is shown with the aid of Figure 7.14, in which the shadowed areas in parts (a) and (b) are equal according to Eq. (7.17). Figure 7.14b shows the *blocking effect* of the boundary layer; that is, the flow can be envisioned to have velocity V_f to the edge of δ^* and inside the layer to have zero velocity, as if the flow were completely blocked. Another way to say this is that if the wall were to be *displaced* by δ^*, the same mass flow rate would exist in the inviscid flow as in a real flow without a displacement of the wall.

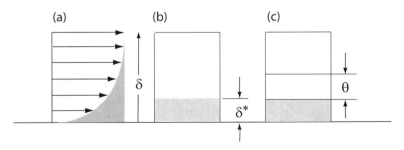

Figure 7.14 Illustration of displacement and momentum thickness.

The momentum thickness θ is defined by the equation

$$\theta V_f^2 = \int_0^\delta V(y)(V_f - V(y)) dy \tag{7.19}$$

This can be written as

$$\theta V_f^2 = V_f \int_0^\delta V(y)dy - \int_0^\delta V^2(y))dy = V_f^2(\delta - \delta^*) - \int_0^\delta V^2(y))dy$$

and reduced to the form

$$\int_0^\delta V^2(y))dy = V_f^2(\delta - \delta^* - \theta) \qquad (7.20)$$

This is also illustrated in the Figure 7.14, and it shows that the momentum thickness contributes a further blockage. Because the integrand in Eqs. (7.18) and (7.20) vanishes beyond the boundary layer thickness, δ can be replaced by the channel width L, in which case these become

$$\int_0^L V(y)dy = V_f(L - \delta^*) \qquad (7.21)$$

and

$$\int_0^L V^2(y))dy = V_f^2(L - \delta^* - \theta) \qquad (7.22)$$

Consider now the flow in a compressor cascade. Figure 7.15a shows two blades and boundary layers along them. As the boundary layers leave the blades, they form a wake behind the cascade. On the suction side the boundary layer is thicker than on the pressure side owing to the strongly decelerating flow there. Consider next the control volume in Figure 7.15b. The inflow boundary is at some location just before the trailing edge, and the outflow boundary is sufficiently far in the wake where mixing has made the velocity uniform. The side boundaries are streamlines that divide one flow channel from the next. Mass balance for this control volume gives

$$\dot{m}' = \int_0^L \rho V(y)dy = \rho L V_3$$

or

$$\dot{m}' = (L - \delta^*)\rho V_f = L\rho V_3$$

in which $\dot{m}'$ is the mass flow rate per unit width. The displacement thickness δ^* is that of the boundary layer on the suction side of the blade; the boundary layer along the pressure side has been ignored. Dividing through by L gives

$$\frac{\dot{m}'}{L} = \left(1 - \frac{\delta^*}{L}\right)\rho V_f = \rho V_3 \qquad (7.23)$$

The x component of the momentum equation gives

$$\int_0^L \rho V_3^2 \, dy - \int_0^L \rho V^2(y)dy = (p - p_3)L$$

in which the shear stress has been neglected, as it is very small along the dividing streamlines in the wake and also small along the blade surfaces, which now account for a small fraction of the control surface. This can be expressed as

$$\rho V_3^2 L - \rho V_f^2(L - \delta^* - \theta) = (p - p_3)L$$

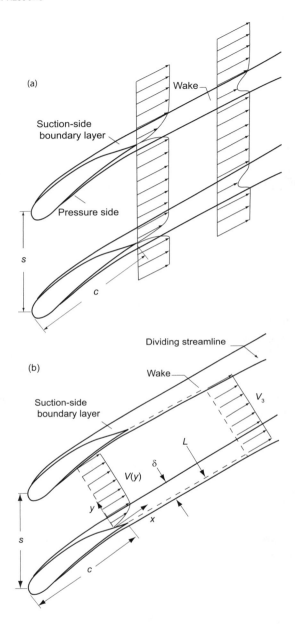

Figure 7.15 Flow in the wake of a compressor cascade.

Dividing through by L gives

$$p - p_3 = \rho V_3^2 - \left(1 - \frac{\delta^*}{L}\right)\rho V_{\mathrm{f}}^2 + \frac{\theta}{L}\rho V_{\mathrm{f}}^2$$

Adding the term $\rho V_{\mathrm{f}}^2/2$ to both sides, and subtracting $\rho V_3^2/2$ from both sides, gives

$$p_0 - p_{03} = p + \frac{1}{2}\rho V_{\mathrm{f}}^2 - \left(p_3 + \frac{1}{2}\rho V_3^2\right) = \frac{1}{2}\rho V_3^2 - \left(1 - \frac{\delta^*}{L}\right)\rho V_{\mathrm{f}}^2 + \frac{1}{2}\rho V_{\mathrm{f}}^2 + \frac{\theta}{L}\rho V_{\mathrm{f}}^2$$

Making use of Eq. (7.23) puts this into the form

$$p_0 - p_{03} = \frac{1}{2}\left(1 - \frac{\delta^*}{L}\right)^2 \rho V_f^2 - \left(\frac{1}{2} - \frac{\delta^*}{L}\right)\rho V_f^2 + \frac{\theta}{L}\rho V_f^2$$

Simplifying this gives

$$\frac{p_0 - p_{03}}{\frac{1}{2}\rho V_f^2} = \left(\frac{\delta^*}{L}\right)^2 + \frac{2\theta}{L}$$

The stagnation pressure loss takes place mainly in the decelerating boundary layer and the wake, so that it is reasonable to take $p_0 = p_{02}$. Also, the first term on the right is much smaller than the second one and can be neglected. Thus the stagnation pressure loss is related to the momentum boundary-layer thickness by

$$\frac{p_{02} - p_{03}}{\frac{1}{2}\rho V_f^2} = \frac{2\theta}{L}$$

Finally, V_f is approximately V_3 and $L = s\cos\alpha_3$, so that

$$\frac{p_{02} - p_{03}}{\frac{1}{2}\rho V_3^2} = \frac{2\theta}{s\cos\alpha_3} \qquad (7.24)$$

The boundary-layer thickness on the pressure side of the blade was neglected in the analysis. If it is included, as it should, the only change is that the momentum thickness is now the sum of the momentum boundary-layer thicknesses on both sides of the blade. It is, in fact, this thickness that is shown in the experimental data relating it to the diffusion factor. If $DF = 0.45$ is taken as the design value then at this value, the momentum boundary-layer thickness is given as $\theta = 0.007c$, so that the stagnation pressure losses for the stator and rotor are

$$\frac{p_{02} - p_{03}}{\frac{1}{2}\rho V_3^2} = \frac{0.014\,\sigma_S}{\cos\alpha_3} \qquad \frac{p_{01R} - p_{02R}}{\frac{1}{2}\rho W_2^2} = \frac{0.014\,\sigma_R}{\cos\beta_2} \qquad (7.25)$$

The stagnation pressure losses and and their relation to stage efficiency are discussed further in the next section.

7.5 EFFICIENCY AND LOSSES

The various stagnation pressure losses in a compressor stage are shown in Figure 7.16. It dates from 1945, and the estimates reflect the way in which compressors were designed then. Today the design flow coefficient has been lowered to the range $0.4 \leq \phi \leq 0.7$, and the losses are expected to be smaller as designs have improved. The data show that, owing to boundary-layer separation and blade stall, profile losses become a major loss component when a compressor is operated far from its design condition. *Profile losses* are associated with the boundary layers along the blades, and the *annular losses* come from the boundary layers along the endwalls of the flow annulus. *Secondary flow* arises from the interaction of the annulus boundary layers and the inviscid stream, which set up a circulation in the plane normal to the primary flow. Dissipation in the secondary flow accounts for these losses.

7.5.1 Efficiency

The stage efficiency is defined by

$$\eta_{tt} = \frac{h_{03ss} - h_{01}}{h_{03} - h_{01}} \qquad \text{so that} \qquad 1 - \eta_{tt} = \frac{h_{03} - h_{03ss}}{h_{03} - h_{01}}$$

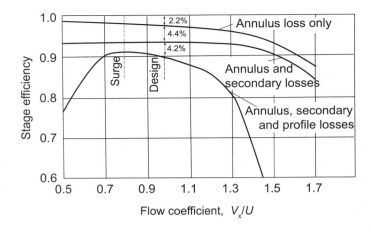

Figure 7.16 Losses in a typical compressor stage. (Drawn after Howell [37].)

where the states are as shown in Figure 7.3. This can be written as

$$1 - \eta_{\text{tt}} = \frac{h_3 - h_{3ss}}{w} + \frac{V_3^2 - V_{3ss}^2}{2w}$$

The numerator of the first term on the right is

$$h_3 - h_{3ss} = h_3 - h_{3s} + h_{3s} - h_{3ss}$$

From the definition of static enthalpy loss coefficient, the first of these can be written as

$$h_3 - h_{3s} = \frac{1}{2}\zeta_S V_3^2$$

and the second term is

$$h_{3s} - h_{3ss} = c_p T_{3s}\left(1 - \frac{T_{3ss}}{T_{3s}}\right)$$

Integrating the Gibbs equation along the constant-pressure lines p_2 and p_3 between states with entropies s_1 and s_2 gives

$$\frac{T_{2s}}{T_2} = \frac{T_{3ss}}{T_{3s}}$$

so that the previous equation can be recast as

$$h_{3s} - h_{3ss} = \frac{T_{3s}}{T_2}(h_2 - h_{2s})$$

When written in terms of the static enthalpy loss coefficient, this becomes

$$h_{3s} - h_{3ss} = \frac{1}{2}\zeta_R W_2^2 \frac{T_{3s}}{T_2}$$

The velocity term on the right in the expression for efficiency is

$$\frac{V_3^2 - V_{3ss}^2}{2w} = \left(1 - \frac{V_{3ss}^2}{V_3^2}\right)\frac{V_3^2}{2w}$$

But is has been shown in Chapter 5 that $M_{3ss} = M_3$. Therefore $V_{3ss}^2/V_3^2 = T_{3ss}/T_3$, and the expression for the efficiency now takes the form

$$1 - \eta_{tt} = \frac{\zeta_S V_3^2 + \zeta_R W_2^2 \dfrac{T_{3s}}{T_2} + \left(1 - \dfrac{T_{2s}}{T_2}\right) V_3^2}{2w}$$

The first and the last term have a common factor in V_3^2, and they could be combined. However, for a typical stage the temperature factor of the last term has a value of about 0.008. If this is neglected and the term T_{3s}/T_2 is set to unity, the right side is simplified somewhat. Therefore the reduced equation

$$1 - \eta_{tt} = \frac{\zeta_S V_3^2 + \zeta_R W_2^2}{2w}$$

although simple to use, underestimates the losses by about 1.5%.

The relationship to stagnation pressure loss can be developed by writing for the stator

$$h_3 - h_{3s} = c_p(T_3 - T_{3s}) = c_p T_3 \left(1 - \frac{T_{3s}}{T_3}\right) = c_p T_3 \left[1 - \left(\frac{p_{03}}{p_{02}}\right)^{(\gamma-1)/\gamma}\right]$$

which can be rewritten and then expanded for small stagnation pressure loss as

$$h_3 - h_{3s} = c_p T_3 \left[1 - \left(\frac{p_{03}}{p_{03} + \Delta p_{0LS}}\right)^{(\gamma-1)/\gamma}\right] = c_p T_3 \left[1 - \left(1 + \frac{\Delta p_{0LS}}{p_{03}}\right)^{-(\gamma-1)/\gamma}\right]$$

$$= c_p T_3 \frac{\gamma - 1}{\gamma} \frac{\Delta p_{0LS}}{p_{03}} = \frac{p_3}{p_{03}} \frac{\Delta p_{0LS}}{p_3}$$

Thus

$$h_3 - h_{3s} = \frac{\Delta p_{0LS}/\rho_3}{\left(1 + \dfrac{\gamma - 1}{2} M_3^2\right)^{\gamma/(\gamma-1)}}$$

A similar analysis for the rotor gives

$$h_2 - h_{2s} = c_p(T_2 - T_{2s}) = c_p T_2 \left(1 - \frac{T_{2s}}{T_2}\right) = c_p T_2 \left[1 - \left(\frac{p_{0R2}}{p_{0R1}}\right)^{(\gamma-1)/\gamma}\right]$$

which can be rewritten and then expanded for small stagnation pressure loss as

$$h_2 - h_{2s} = c_p T_2 \left[1 - \left(\frac{p_{0R2}}{p_{0R2} + \Delta p_{0LR}}\right)^{(\gamma-1)/\gamma}\right] = c_p T_2 \left[1 - \left(1 + \frac{\Delta p_{0LR}}{p_{0R2}}\right)^{-(\gamma-1)/\gamma}\right]$$

$$= c_p T_2 \frac{\gamma - 1}{\gamma} \frac{\Delta p_{0LR}}{p_{0R2}} = \frac{p_2}{p_{0R2}} \frac{\Delta p_{0LR}}{p_2}$$

Therefore

$$h_2 - h_{2s} = \frac{\Delta p_{0LR}/\rho_2}{\left(1 + \dfrac{\gamma - 1}{2} M_{R2}^2\right)^{\gamma/(\gamma-1)}}$$

The stagnation pressure losses can now be written as

$$\frac{\Delta p_{0\text{LS}}}{p_3} = \frac{1}{2}\zeta_\text{S} V_3^2 \left(1 + \frac{\gamma-1}{2}M_3^2\right)^{\gamma/(\gamma-1)} \tag{7.26}$$

and

$$\frac{\Delta p_{0\text{LR}}}{p_2} = \frac{1}{2}\zeta_\text{R} W_2^2 \left(1 + \frac{\gamma-1}{2}M_{\text{R}2}^2\right)^{\gamma/(\gamma-1)} \tag{7.27}$$

Since the rotor turns the flow toward the axis, the magnitude of the relative velocity diminishes and the relative Mach number leaving the rotor is fairly small. The same is true for the absolute Mach number leaving the stator. Hence, if the Mach number terms in Eqs. (7.26) and (7.27) are neglected, then with some loss of accuracy, substituting the these expressions for the stagnation pressure losses into Eqs. (7.25), gives

$$\zeta_\text{R} = \frac{0.014\sigma_\text{R}}{\cos\beta_2} \qquad \zeta_\text{S} = \frac{0.014\sigma_\text{S}}{\cos\alpha_3} \tag{7.28}$$

Since $V_x = V_3 \cos\alpha_3 = W_2 \cos\beta_2$ the efficiency may be rewritten in the form

$$1 - \eta_{tt} = \left(\frac{\zeta_\text{R}}{\cos^2\alpha_1} + \frac{\zeta_\text{S}}{\cos^2\beta_1}\right)\frac{\phi^2}{2\psi} \tag{7.29}$$

7.5.2 Parametric calculations

The development of the theory for axial compressors has been carried out in terms of assumed values for a flow coefficient ϕ, a blade-loading factor ψ, and a reaction ratio R. These three quantities are *nondimensional parameters* and are sufficient for calculating the flow angles. Solidity was obtained from experimental measurements of deflection for given flow outlet angle. It was used to establish the permissible diffusion, and the stagnation pressure coefficient for profile losses was established. The efficiency can then be cast in terms of the loss coefficients for the rotor and the stator. This is the most effective way to display results, and when the actual parameters for a machine are given, all the important results are quickly calculated from the nondimensional ones. Thus flow angles for a compressor stator and rotor have been shown to be

$$\tan\alpha_1 = \frac{1-R-\psi/2}{\phi} \qquad \tan\alpha_2 = \frac{1-R+\psi/2}{\phi}$$

$$-\tan\beta_1 = \frac{R+\psi/2}{\phi} \qquad -\tan\beta_2 = \frac{R-\psi/2}{\phi}$$

Hence with ϕ, ψ and R given, they can be determined. The rest of the calculations then follow directly. Results from such calculations are shown in Figure 7.17. The flow coefficient was taken to be $\phi = 0.6$ and the degree of reaction, $R = 0.4$. For Figures 7.17a – 7.17c, the Lieblein diffusion factor was taken to be equal to 0.45, and solidities of the rotor and stator were calculated in terms of this value. Figure 7.17a represents the diffusion in the flow. If the stagnation pressure losses are neglected and the flow is assumed incompressible, then

$$p_2 + \frac{1}{2}\rho V_2^2 = p_3 + \frac{1}{2}\rho V_3^2$$

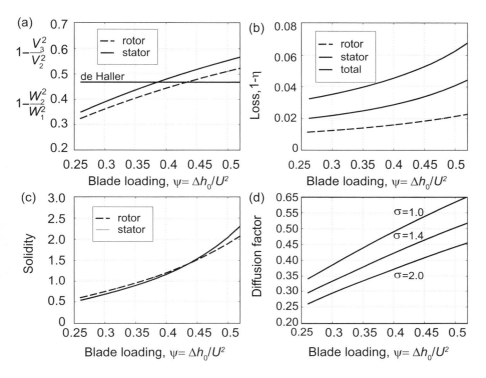

Figure 7.17 Design study for a compressor cascade with diffusion factor, $DF = 0.45$, flow coefficient $\phi = 0.6$, and reaction $R = 0.4$.

and

$$\frac{p_3 - p_2}{\frac{1}{2}\rho V_2^2} = 1 - \frac{V_3^2}{V_2^2}$$

Hence this also gives the pressure rise in a nondimensional form. A similar expression applies for the rotor. This nondimensional pressure rise should not be greater than 0.48, according to the recommendation by de Haller [20]. The de Haller criterion shows that the blade-loading factor must be kept below 0.39.

Figure 7.17b shows that if the de Haller criterion is obeyed, the stage efficiency is around 0.96. This efficiency calculation should not be considered conclusive, as it ignores the *annulus and secondary flow losses*, as well as *losses from the tip region*. The Figure 7.17c shows solidity to be near unity for both the rotor and the stator if the Lieblein diffusion factor is kept at 0.45. It turns out that Howell's criterion is also satisfied under these conditions. To be sure, the Howell's criterion shows that the solidity ought to be the same for the rotor and stator. So the system is overconstrained. The diffusion factor then becomes a check that the design parameters have been reasonably chosen. Figure 7.17d shows what happens to the diffusion factor as the solidity is varied independently. For a blade row of high solidity, diffusion is reduced as the load per blade is reduced.

7.6 CASCADE AERODYNAMICS

The development of axial compressor blade shapes has been carried out in extensive experiments in cascade tunnels. Such tunnels are designed to test in a two-dimensional arrangement, either a row of rotor blades, or a row of stator vanes. Both are shown in Figure 7.18 in which the flow moves from left to right. The blades are characterized by their chord, camber, thickness, and height. The axial chord c_x is the chord's projection along the x axis. In Figure 7.18b the blade (or metal) angle at the inlet is χ_2, and at the outlet it is χ_3. The flow *incidence* is defined as the difference between the flow angle and the metal angle. Thus $i = \alpha_2 - \chi_2$ at the inlet to the stator, and since the angles are negative, incidence is $i = |\beta_1 - \chi_1|$ at the inlet to the rotor. The figure is drawn to show a positive incidence for both. The angle $\delta = \alpha_3 - \chi_3$ at the exit of the stator is called *deviation*. Although the flow tends to leave at the exit metal angle, it turns slightly toward the suction surface.

The general orientation of the blades is given by the *stagger* angle ξ and their curvature, by the camber angle θ. The lines that form a triangle on the stator blade in the Figure 7.18b has the included angle $\chi_2 - \xi$ on the left, $\xi - \chi_3$ on the right, and $\pi - \theta$ at the top. These add up to π. Hence $\theta = \chi_2 - \chi_3$. These relationships are also true for the rotor in which negative angles are encountered. The exit metal angle χ_2 of the rotor has the same designation as the inlet metal angle to the stator. It will always be clear whether the rotor row or stator row is analyzed, so separate symbols are not needed.

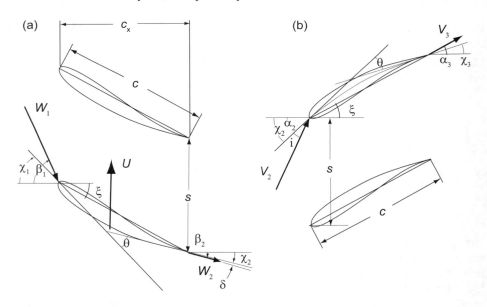

Figure 7.18 An axial compressor rotor cascade.

7.6.1 Blade shapes and terms

Today the designer is no longer constrained to the blade shapes that were in use during the early period of compressor development. Members from the NACA-65 series of blades in the United States were adopted as the base profiles. In the Great Britain blade designation, such as 6C7/25P40, carries with it the pertinent information about the maximum thickness

b and its location from the leading edge a. The notation means that the maximum thickness is 6% of the chord, so that $b/c = 0.06$. The next item C7 describes how the thickness is distributed over the blade. Next is the camber angle, which here is 25°. After that the letter P designates this to be a blade with a parabolic camber, with the location of maximum thickness at 40% from the leading edge; that is, $a/c = 0.4$. The circular arc profile 10C4/30C50 has 10% thickness and 30° camber. Its maximum thickness always occurs at the midchord point.

For parabolic arc profiles defining $\theta_1 = \chi_1 - \xi$ and $\theta_2 = \xi - \chi_2$, the angles θ_1 and θ_2 are given by

$$\tan \theta_1 = \frac{cb}{a^2} \qquad \tan \theta_2 = \frac{cb}{a(c-a)}$$

and, if the location of maximum camber a is close to the midchord position, these can be approximated by

$$\tan \theta_1 = \frac{4b}{4a - c} \qquad \tan \theta_2 = \frac{4b}{3c - 4a}$$

From these it follows that

$$\frac{b}{c} = \frac{1}{4\tan\theta} \left[\sqrt{(1 + 16\tan^2\theta) \left(\frac{a}{c} - \frac{a^2}{c^2} - \frac{3}{16} \right)} - 1 \right]$$

where $\theta = \theta_1 + \theta_2$. For blades for which the position of maximum camber is near the midchord point, useful approximations [18] to these are

$$\theta = 8\frac{b}{c} \qquad \theta_1 = \frac{1}{2}\theta\left[1 + 2\left(1 - 2\frac{a}{c}\right)\right] \qquad \theta_2 = \frac{1}{2}\theta\left[1 - 2\left(1 - 2\frac{a}{c}\right)\right]$$

7.6.2 Blade forces

Figure 7.19 shows a typical stator of an axial compressor. The x component of the momentum equation applied to a control volume consisting of the flow channel for the stator gives

$$\rho s V_x (V_{x3} - V_{x2}) = s(p_2 - p_3) + F_x$$

and if the axial velocity remains constant, this reduces to

$$F_x = s(p_3 - p_2)$$

Using the definition of stagnation pressure for an incompressible flow, this becomes

$$\frac{F_x}{s} = \frac{1}{2}(V_2^2 - V_3^2) - \Delta p_{0L} \qquad (7.30)$$

in which Δp_{0L} is the loss in stagnation pressure.

The force F_x is per unit height of the blade. Since the axial velocity is taken to be constant and $\alpha_3 < \alpha_2$, inspection of the magnitudes of V_2 and V_3 shows that $V_3 < V_2$ and therefore also $p_3 > p_2$, as it ought to be in a compressor. Hence the force F_x that blades exert on the fluid is positive. The y component gives

$$\rho s V_x (V_{u3} - V_{u2}) = -F_y$$

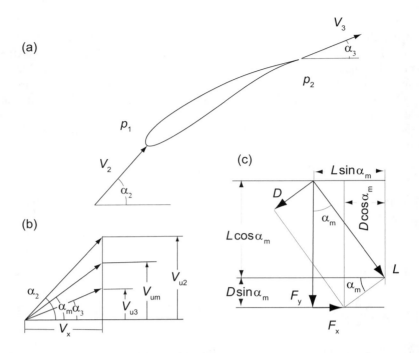

Figure 7.19 Compressor stator.

and the minus sign is inserted on the right side to render the numerical value of F_y positive, because $V_{u3} < V_{u2}$. This equation may also be written as

$$\frac{F_y}{s} = \rho V_x^2 (\tan \alpha_2 - \tan \alpha_3) \tag{7.31}$$

Since $V_{x2} = V_{x3}$, the kinetic energy difference in Eq. (7.30) may be expressed as

$$\begin{aligned} V_2^2 - V_3^2 &= (V_{x2}^2 + V_{u2}^2) - (V_{x3}^2 + V_{u3}^2) \\ &= V_{u3}^2 - V_{u3}^2 = (V_{u2} - V_{u3})(V_{u2} + V_{u3}) \end{aligned}$$

Next, let the mean tangential velocity be

$$V_{um} = \frac{1}{2}(V_{u2} + V_{u3})$$

then using this definition and inspection of Figure 7.19 gives

$$V_{um} = V_x \tan \alpha_m = \frac{V_x}{2}(\tan \alpha_2 + \tan \alpha_3)$$

in which the mean gas angle α_m is defined by the expression

$$\tan \alpha_m = \frac{1}{2}(\tan \alpha_2 + \tan \alpha_3) \tag{7.32}$$

The x component of the force may now be written as

$$F_x = \rho s V_x^2 (\tan \alpha_2 - \tan \alpha_3) \tan \alpha_m - s \Delta p_{0L} \tag{7.33}$$

In addition to the axial and tangential components of forces, the lift and the drag force are also shown in Figure 7.19. They are the forces that the blades exert on the fluid. Lift and drag are obtained by resolving the resultant along the direction of the mean flow angle and one perpendicular to it. With the resultant expressed as

$$\mathbf{F} = F_x \mathbf{i} - F_y \mathbf{j}$$

the component **D** parallel and **L** normal to the mean flow vector, as shown in Figure 7.19c, are

$$L = F_x \sin \alpha_m + F_y \cos \alpha_m \qquad (7.34)$$

$$D = -F_x \cos \alpha_m + F_y \sin \alpha_m \qquad (7.35)$$

Introducing Eqs. (7.33) and (7.31) into Eq. (7.34) gives

$$L = \rho s V_x^2 (\tan \alpha_2 - \tan \alpha_3) \sec \alpha_m - s \Delta p_{0L} \sin \alpha_m$$

The lift coefficient now becomes

$$C_L = \frac{L}{\frac{1}{2}\rho V_m^2 c} = \frac{2}{\sigma}(\tan \alpha_2 - \tan \alpha_3) \cos \alpha_m - \frac{\Delta p_{0L}}{\frac{1}{2}\rho V_m^2} \frac{\sin \alpha_m}{\sigma}$$

Substituting Eqs. (7.30) and (7.31) into Eq. (7.35) gives

$$D = s \Delta p_{0L} \cos \alpha_m \qquad (7.36)$$

The choice in defining the direction of drag and lift as parallel and perpendicular to the mean flow direction as given by the angle α_m is guided exactly by this result, since the stagnation pressure drop from profile losses should depend mainly on the drag and not the lift.

Defining next the drag coefficient as

$$C_D = \frac{D/c}{\frac{1}{2}\rho V_m^2} \qquad (7.37)$$

in which c is the chord, and making use of the preceding equation, gives

$$C_D = \frac{\Delta p_{0L} \cos \alpha_m}{\frac{1}{2}\rho V_m^2 \sigma}$$

For the stator the stagnation pressure loss can be written as

$$\Delta p_{0LS} = \frac{1}{2}\rho \zeta_S V_3^2 = \frac{1}{2}\rho V_m^2 \frac{\sigma_S C_D}{\cos \alpha_m}$$

so that

$$\zeta_S = \frac{\sigma_S \cos^2 \alpha_3}{\cos^2 \alpha_m} C_D \qquad (7.38)$$

in which $V_m/V_3 = \cos \alpha_3 / \cos \alpha_m$ has been used. For the rotor similarly

$$\zeta_R = \frac{\sigma_R \cos^2 \beta_2}{\cos^2 \beta_m} C_D \qquad (7.39)$$

The drag coefficients in these expressions are not the same, but are the appropriately calculated ones for the stator and the rotor. As will be seen below, the various losses are expressed in terms of the drag coefficients, and their sum gives the total. The two preceding equations can then be used to relate the drag coefficients to the static enthalpy loss coefficients.

The expression for the lift coefficient for the stator can now be written as

$$C_L = \frac{2}{\sigma_S}(\tan\alpha_2 - \tan\alpha_3)\cos\alpha_m - C_D \tan\alpha_m$$

In a typical unstalled blade $C_D \sim 0.025 C_L$, and, as will be shown shortly, it is advantageous to keep α_m less than $45°$. Under these conditions the drag term may be neglected. This gives the following expression for the lift-to-drag ratio:

$$\frac{C_L}{C_D} = \frac{L}{D} = \frac{2}{\zeta_S}(\tan\alpha_2 - \tan\alpha_3)\frac{\cos^2\alpha_3}{\cos^2\alpha_m}$$

To make use of these expressions for compressor rotors, the flow angle α_2 is replaced by β_1 and α_3 by β_2 and the rotor loss coefficient ζ_R is substituted for ζ_S.

7.6.3 Other losses

The drag coefficients for annulus losses are given by

$$C_{Da} = 0.02\frac{s}{b}$$

in which b is the blade height, and the annulus losses become a smaller fraction of the total losses as the blade height increases. Secondary losses are a complicated subject, but all these complications are sidestepped by calculating the loss coefficient from the expression

$$C_{Ds} = 0.018 C_L^2$$

The tip losses arise from a tip vortex and leakage loss through the tip clearance for the rotor. They are taken into account by the empirical relation

$$C_{Dt} = 0.29\frac{b_t}{b}C_L^{3/2}$$

where b_t is the tip clearance.

The total loss coefficient is then obtained by summing these

$$C_D = C_{Dp} + C_{Da} + C_{Ds} + C_{Dt}$$

The equivalent drag coefficients for secondary and tip losses are a way of expressing the total pressure loss on the same basis for profile and annulus losses. With the sum of drag coefficients determined, the total static enthalpy loss coefficient can be calculated from either Eq. (7.38) or (7.39). The calculation of losses using the method outlined above has been questioned over the years [18, 21], for it is clear that in the complicated flow structure with highly staggered blades, dividing the losses neatly into constituent parts and adding them is suspect. Modern CFD analysis gives a powerful alternative that can be used to calculate stagnation pressure losses. Even if the modern methods are an improvement over what has been done in the past, there are still hurdles to overcome. One important advancement has been to develop methods to include unsteady interactions between moving and stationary blade rows. The theory still hinges on accurate methods to calculate local entropy production of a turbulent velocity and temperature fields. New turbulence models and large-eddy-simulation methods show promise and for this reason industry has moved to rely more and more on CFD to resolve the flow fields in compressors and turbines.

7.6.4 Diffuser performance

The expression for the lift-to-drag ratio can be used to assess under which conditions a compressor cascade performs well as a diffuser. The diffuser efficiency is defined as

$$\eta_D = \frac{p_3 - p_2}{p_{3s} - p_2}$$

which is the ratio of the static pressure rise to the maximum possible. The labels are chosen to reflect that diffusion through the stator is considered. In a reversible adiabatic flow the stagnation pressure remains constant and

$$p_{02} - p_{03s} = p_2 - p_{3s} + \frac{1}{2}\rho(V_2^2 - V_{3s}^2) = 0$$

Hence

$$p_{3s} - p_2 = \frac{1}{2}\rho(V_2^2 - V_{3s}^2)$$

Assuming that $V_3 = V_{3s}$, the diffuser efficiency can thus be written as

$$\eta_D = \frac{p_3 - p_2}{\frac{1}{2}\rho(V_2^2 - V_3^2)}$$

Now

$$F_x = s(p_3 - p_2) \qquad F_x = F_y \tan \alpha_m - s\Delta p_{0S} \qquad F_y = \rho s V_x^2 (\tan \alpha_2 - \tan \alpha_3)$$

so that

$$p_3 - p_2 = \rho V_x^2 \tan \alpha_m (\tan \alpha_2 - \tan \alpha_3) - \Delta p_{0LS}$$

Thus

$$\eta_D = 1 - \frac{\Delta p_{0LS}}{\rho V_x^2 \tan \alpha_m (\tan \alpha_2 - \tan \alpha_3)}$$

or in terms of C_D and C_L the diffuser efficiency can be written as

$$\eta_D = 1 - \frac{2C_D}{C_L \sin 2\alpha_m}$$

The maximum efficiency is determined by differentiating. Thus

$$\frac{d\eta_D}{d\alpha_m} = \frac{4C_D \cos 2\alpha_m}{C_L \sin^2 2\alpha_m} = 0$$

Hence $\alpha_m = 45°$ at the condition of maximum diffuser efficiency and for $C_D/C_L = 0.05$ the efficiency is 90%. This result is due to Horlock [34].

7.6.5 Flow deviation and incidence

Before the blade angle can be set, the deviation at the exit from the blade row must be established. The flow deviation is given by $\delta = \alpha_2 - \chi_2$, in which χ_2 is the blade angle. Deviation is positive when the flow deflects from the pressure side toward the suction side. A positive deviation for the stator is shown in Figure 7.19, and for the rotor the flow is deflected in the opposite direction.

Deviation is an inviscid flow effect and not related to viscosity and therefore to losses. It can be regarded as incomplete turning of the flow. Hence, to obtain the desired turning, blades need to be curved more than in the absence of deviation.

Carter [14] recommends that deviation be calculated from

$$\delta^* = m\theta\sqrt{\frac{s}{c}}$$

at the design conditions, and Howell [37] proposed the formula

$$m = 0.92\frac{a^2}{c^2} + \frac{\alpha_2^*}{500}$$

for m, in which α_2^* is given in degrees and a/c is the ratio of maximum thickness to the chord.

■ EXAMPLE 7.4

A compressor stage with reaction ratio $R = \frac{1}{2}$, and stator outlet metal angle $\chi_3 = 3°$. The camber angle is $\theta = 34°$, pitch chord ratio is $s/c = 0.88$, and the position of maximum camber $a/c = 0.5$. The ratio of the blade height to the chord is $b/c = 2$. For 50% reaction the rotor flow angles have the same magnitudes as those for the stator, but they are of opposite sign. Find (a) the deviation of the flow leaving the stator, (b) the deflection, and (c) the flow coefficient and blade-loading coefficient, assuming that the inflow is at zero incidence.

Solution: (a) The flow deviation is calculated first. Since $\alpha_3^* = \chi_3 + \delta^*$, the flow angle can be calculated from

$$\alpha_3^* = \chi_3 + m\theta\sqrt{\frac{s}{c}} = \chi_3 + 0.92\frac{a^2}{c^2}\theta\sqrt{\frac{s}{c}} + \frac{\alpha_2^*}{500}\theta\sqrt{\frac{s}{c}}$$

Solving this for α_3^* gives

$$\alpha_3^* = \frac{\chi_3 + 0.92\frac{a^2}{c^2}\theta\sqrt{\frac{s}{c}}}{1 - \frac{\theta}{500}\sqrt{\frac{s}{c}}} = \frac{3 + 0.92 \cdot 0.25 \cdot 34\sqrt{0.88}}{1 - \frac{34}{500}\sqrt{0.88}} = 11.04°$$

so that $\delta^* = \alpha_3^* - \chi_3 = 11.04° - 3° = 8.04°$.

(b) The nominal deflection is calculated by first solving

$$\tan\alpha_2^* - \tan\alpha_3^* = \frac{1.55}{1 + 1.5s/c}$$

for α_2^*. It yields

$$\alpha_2^* = \tan^{-1}\left(\tan\alpha_3^* + \frac{1.55}{1 + 1.5s/c}\right) = \tan^{-1}\left(0.1927 + \frac{1.55}{1 + 1.5 \cdot 0.88}\right) = 40.8°$$

Thus the nominal deflection is

$$\epsilon^* = \alpha_2^* - \alpha_3^* = 40.8° - 11.0° = 29.8°$$

If the metal angle at the inlet to the stator is set at $\chi_2 = 40.8°$, then the flow enters the stator at *zero incidence*. If the metal angle is set, say, to $\chi_2 = 37°$, then the incidence angle is $i^* = \alpha_2^* - \chi_2 = 40.8 - 37 = 3.8°$.

(c) Since for a normal stage $\alpha_1^* = \alpha_3^*$, the flow coefficients is obtained from

$$\phi = \frac{1}{\tan \alpha_1^* + \tan \alpha_2^*} = \frac{1}{0.866 + 0.195} = 0.943$$

and the blade-loading coefficient, from

$$\psi = \phi(\tan \alpha_1^* - \tan \alpha_2^*) = 0.943(0.866 - 0.195) = 0.634$$

This value for the blade-loading coefficient is above the high end of the usual range $0.3 - 0.45$ of industrial practice. The flow coefficient is in the common range for axial compressors. For 50% reaction, the blading of the rotor is identical to that of the stator and the flow angle from the rotor would be $40.7°$.

■

7.6.6 Multistage compressor

The polytropic efficiency was introduced for turbines in the previous chapter. For a small change in the stagnation enthalpy across a stage, the stage efficiency approaches the polytropic efficiency. For an ideal gas the incremental process is as shown in Figure 7.20.

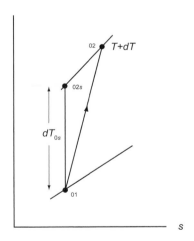

Figure 7.20 Processes across a small compressor stage.

The relation between the temperature increases for an actual and an ideal process is given by

$$dT_{0s} = \eta_p \, dT_0$$

For an isentropic expansion

$$c_p \frac{dT_{0s}}{T_0} = R \frac{dp_0}{p_0} \quad \text{or} \quad \frac{\gamma \eta_p}{(\gamma - 1)} \frac{dT_0}{T_0} = \frac{dp_0}{p_0}$$

Integrating this across an infinite number of infinitesimally small stages gives

$$\frac{T_{0,N+1}}{T_{01}} = \left(\frac{p_{0,N+1}}{p_{01}}\right)^{(\gamma-1)\eta_p/\gamma}$$

A reheat factor, defined as $\text{RF} = \eta_p/\eta$, can then be written as

$$\text{RF} = \frac{\eta_p}{\eta} = \frac{\eta_p \left[\left(\dfrac{p_{0,N+1}}{p_{01}}\right)^{(\gamma-1)/\eta_p\gamma} - 1\right]}{\left(\dfrac{p_{0,N+1}}{p_{01}}\right)^{(\gamma-1)/\gamma} - 1}$$

The relationship between the turbine efficiency and polytropic or small-stage efficiency is

$$\eta = \frac{r^{(\gamma-1)/\gamma} - 1}{r^{(\gamma-1)/\eta_p\gamma} - 1}$$

in which $r = p_{0e}/p_{01}$ is the overall pressure ratio of the turbine. This relationship is shown in Figure 7.21.

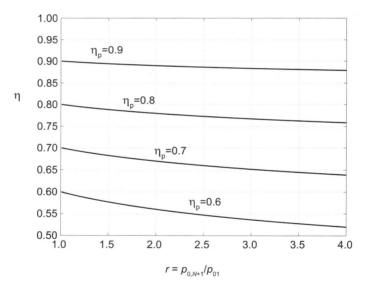

Figure 7.21 Compressor efficiency as a function of pressure ratio and polytropic efficiency for a gas with $\gamma = 1.4$.

In a multistage compressor the upstream stages influence those downstream. Smith [70] measured velocity and temperature profiles after each stage of a 12-stage compressor. These are shown in Figure 7.22, and examination shows that annulus boundary layers cause large decrease in the axial velocity and increase in total temperature. Although it has been assumed that average stagnation temperature does not change in adiabatic flow, the local values may change. These influences are taken into account by modifying the Euler equation for compressor work by introducing a *work-done factor* λ and expressing the work done as

$$h_{03} - h_{01} = \lambda U(V_{x2} - V_{x1})$$

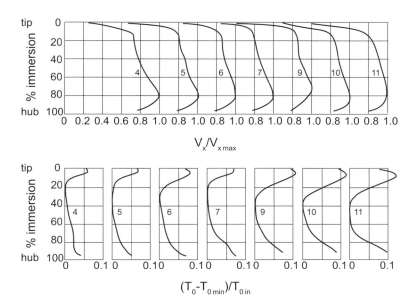

Figure 7.22 Velocity and temperature profiles in a 12-stage compressor. (From Smith [70].)

Comparing different compressors Howell and Bondham recommend a work-done factor in the range 0.86 to 0.96, the smaller corresponding to a compressor with 20 stages and the larger value is applicable to a compressor with only two stages [38]. Their results are shown in Table 7.1.

Table 7.1 Work-done factor λ in axial compressors with different number of stages.

stages	1	2	3	4	5	6	7	8	9
λ	0.982	0.952	0.929	0.910	0.895	0.883	0.875	0.868	0.863
stages	10	11	12	13	14	15	16	17	18
λ	0.860	0.857	0.855	0.853	0.851	0.850	0.849	0.848	0.847

7.6.7 Compressibility effects

The influence of Mach number on losses is substantial, and the loss coefficient increases rapidly with the incidence angle as Mach number at the inlet is increased from 0.4 to 0.8. A way in which a designer can help the situation is to choose a blade for which the location of maximum thickness is nearer to the leading edge.

EXERCISES

7.1 The inlet and exit total pressures of air flowing through a compressor are 100 and 1000 kPa. The inlet total temperature is 281 K. What is the work of compression if the adiabatic total-to-total efficiency is 0.75?

7.2 Air flows through an axial fan rotor at mean radius of 15 cm. The tangential component of the absolute velocity is increased by 15 m/s through the rotor. The rotational speed of the shaft is 3000 rpm. (a) Evaluate the torque exerted on the air by the rotor assuming that the flow rate is $0.471 \, \text{m}^3/\text{s}$ and the pressure and temperature of the air are 100 kPa and 300 K. (b) What is the rate of energy transfer to the air?

7.3 The blade speed of a compressor rotor is $U = 280 \, \text{m/s}$, and the total enthalpy change across a normal stage is 31.6 kJ/kg. If the flow coefficient $\phi = 0.5$ and the inlet to the rotor is axial, what are the absolute and relative gas angles leaving the rotor?

7.4 Air flows through an axial-flow fan, with an axial velocity of 40 m/s. The absolute velocities at the inlet and the outlet of the stator are at angles of 60° and 30°, respectively. The relative velocity at the outlet of the rotor is at an angle $-25°$. Assume reversible adiabatic flow and a normal stage. (a) Draw the velocity diagrams at the inlet and outlet of the rotor. (b) Determine the flow coefficient. (c) Determine the blade-loading coefficient. (d) Determine at what angle the relative velocity enters the rotor. (e) Determine the static pressure increase across the rotor in pascals. The inlet total temperature is 300 K and the inlet total pressure is 101.3 kPa. (f) Determine the degree of reaction.

7.5 Air flows through an axial-flow compressor. The axial velocity is 60% of the blade speed at the mean radius. The reaction ratio is 0.4. The absolute velocity enters the stator at an angle of 55° from the axial direction. Assume a normal stage. (a) Draw the velocity diagrams at the inlet and outlet of the rotor. (b) Determine the flow coefficient. (c) Determine the blade-loading coefficient. (d) Determine at what angle the relative velocity enters the rotor. (e) Determine at what angle the relative velocity leaves the rotor. (f) Determine at what angle the absolute velocity leaves the stator.

7.6 A single stage of a multistage axial compressor is shown in Figure 7.2. The angle at which the absolute velocity enters the rotor is $\alpha_1 = 40°$, and the relative velocity at the inlet of the rotor is $\beta_1 = -60°$. These angles at the inlet of the stator are $\alpha_2 = 60°$ and $\beta_2 = -40°$. The mean radius of the rotor is 30 cm, and the hub-to-tip radius is 0.8. The axial velocity is constant and has a value $V_x = 125 \, \text{m/s}$. The inlet air is atmospheric at pressure 101.325 kPa and temperature 293 K. (a) Find the mass flow rate. (b) What is the rotational speed of the shaft under these conditions? (c) What is the power requirement of the compressor?

7.7 A single stage of a multistage axial compressor is shown in Figure 7.2. The angle at which the absolute velocity enters the rotor is $\alpha_1 = 35°$, and the relative velocity makes an angle of $\beta_1 = -60°$. The corresponding angles at the inlet to the stator are $\alpha_2 = 60°$ and $\beta_2 = -35°$. The stage is normal, and the axial velocity is constant through the compressor. (a) Why does the static pressure rise across both the rotor and the stator? (b) Draw the velocity triangles. (c) If the blade speed is $U = 290 \, \text{m/s}$, what is the axial velocity? (d) Find the work done per unit mass flow for a stage and the increase in stagnation temperature across it. (e) The stagnation temperature at the inlet is 300 K. The overall adiabatic efficiency of the compressor is $\eta_c = 0.9$, and the overall stagnation pressure ratio is 17.5. Determine the number of stages in the compressor. (f) How many axial turbine stages will it take to power this compressor?

7.8 The blade speed of a rotor of an axial air compressor is $U = 150\,\text{m/s}$. The axial velocity is constant and equal to $V_x = 75\,\text{m/s}$. The tangential component of the relative velocity leaving the rotor is $W_{u2} = -30\,\text{m/s}$; the tangential component of the absolute velocity entering the rotor is $V_{u1} = 55\,\text{m/s}$. The stagnation temperature and pressure at the inlet to the rotor are 340 K and 185 kPa. The stage efficiency is 0.9, and one-half of the loss in stagnation pressure takes place through the rotor. (a) Draw the velocity diagrams at the inlet and exit of the rotor. (b) Find the work done per unit mass flow through the compressor. (c) Draw the states in an hs diagram. Find the (d) stagnation and static temperatures between rotor and stator, and (e) stagnation pressure between rotor and stator.

7.9 Air from ambient at 101.325 kPa and temperature 20°C enters into a blower axially with the velocity 61 m/s. The blade tip radius is 60 cm, and the hub radius is is 42 cm. The shaft speed is 1800 rpm. The air enters a stage axially and leaves it axially at the same speed. The rotor turns the relative velocity 18.7° toward the direction of the blade movement. The total-to-total stage efficiency is 0.87. (a) Draw the inlet and exit velocity diagrams for the rotor. (b) Draw the blade shapes. (c) Determine the flow coefficient and the blade-loading factor. (d) Determine the mass flow rate. (e) What is the specific work required? (f) What is the total pressure ratio for the stage? (g) Determine the degree of reaction.

7.10 Air from ambient at 101.325 kPa and temperature 300 K enters an axial-flow compressor stage axially with velocity 122 m/s. The blade casing radius is 35 cm, and the hub radius is 30 cm. The shaft speed is 6000 rpm. At the exit relative velocity is at an angle $-45°$. The total-to-total stage efficiency is 0.86. (a) Draw the inlet and exit velocity diagrams for the rotor. (b) Draw the blade shapes. (c) Determine the flow coefficient and the blade-loading factor. (d) Determine the mass flow rate. (e) What is the total pressure ratio for the stage? (f) Determine the degree of reaction.

7.11 Perform design calculations for a compressor stage with blade loading factor in the range $\psi = 0.25$ to $\psi = 0.55$, flow coefficient $\phi = 0.7$, and reaction $R = 0.6$. Keep the diffusion factor equal to 0.45. Calculate and plot $1 - \eta$, solidity, and the static pressure rise $1 - V_2^2/V_1^2$ for rotor and stator (including the de Haller criterion). What are the stagnation losses across stator and rotor?

7.12 A normal compressor stage has a reaction ratio $R = 0.54$, and stator outlet metal angle $\chi_3 = 14.5°$. The camber angle is $\theta = 32°$, pitch chord ratio is $s/c = 0.82$, and the position of maximum camber $a/c = 0.45$. The ratio of the blade height to the chord is $b/c = 1.7$. (a) Find the deviation of the flow leaving the stator. (b) Find the deflection. (c) Find the flow coefficient and blade-loading coefficient by assuming the inflow at zero incidence.

7.13 The circular arc blades of a compressor cascade have camber $\theta = 30°$ and maximum thickness at $a/c = 0.5$. The space-to-chord ratio is $s/c = 1.0$. The nominal outflow angle is $\alpha_3^* = 25°$. (a) Determine the nominal incidence. (b) Determine that lift coefficient at the nominal incidence given a drag coefficient of $C_D = 0.016$.

7.14 Air with density $1.21\,\text{kg/m}^3$ flows into a compressor stator with velocity $V_2 = 120\,\text{m/s}$ and leaves at angle $\alpha_3 = 30°$. If the Leiblein diffusion factor is to be held at

0.5 and the the stagnation pressure loss across a compressor stator is 0.165 kPa, assuming incompressible flow, what is the static pressure increase across the stage?

7.15 Air at temperature 288 K and at pressure 101.325 kPa flows into a compressor with 10 stages. The efficiency of the first stage is 0.920 and the second stage, 0.916. For the remaining stages, the stage efficiency drops by 0.004 successively so that the last stage has an efficiency of 0.884. The axial velocity is constant, and the flow angles are the same at the inlet and exit of each of the stages. Hence the work done by each stage is the same. (a) Find the overall efficiency of the compressor. (b) Find the overall efficiency by using the theory for a polytropic compression with small stage efficiency $\eta_p = 0.902$.

7.16 Air flows in a axial flow fan of free vortex design, with hub radius $r_h = 7.5$ cm and casing radius 17 cm. The fan operates at 2400 rpm. The volumetric flow rate $Q = 1.1 \text{ m}^3/\text{s}$ and the stagnation pressure rise is 3 cm H_2O. The fan efficiency is $\eta_{tt} = 0.86$. (a) Find the axial velocity. (b) Find the work done on the fluid. (c) Find the absolute and relative flow angles at the inlet and exit of the rotor when the inlet is axial.

7.17 Consider an axial-flow compressor in which flow leaves the stator with a tangential velocity distributed as a free vortex. The hub radius is 10 cm and the static pressure at the hub is 94 kPa, and the static temperature there is 292 K. The radius of the casing is 15 cm, and the static pressure at the casing is 97 kPa. The total pressure at this location is 101.3 kPa. Find the exit flow angles at the hub and the casing.

CHAPTER 8

CENTRIFUGAL COMPRESSORS AND PUMPS

This chapter is on centrifugal compressors and centrifugal pumps. Both achieve in a single stage a much higher pressure ratio than their axial counterparts. Large booster compressors are used in natural-gas transmission across continental pipelines and in offshore natural gas production. Multistage centrifugal compressors, also known as *barrel compressors*, are employed when very high delivery pressure is needed. When compressed air at high pressure is needed for industrial processes, a radial compressor fits this application well. Centrifugal compressors are often an integral part of a refrigeration plant to provide chilled water in HVAC systems. In contrast to these large compressors, small centrifugal compressors are found in a turbocharger and a supercharger in which rapidly changing operating conditions call for a machine of low inertia.

Centrifugal pumps operate on the same principles as compressors, but handle liquids in various industrial, agricultural, and sanitary applications. Small pumps perform a variety of tasks in households. The number of units of various kind of compressors and pumps is quite large, making their manufacture an important industry.

A sketch of a centrifugal compressor is shown in Figure 8.1. The axial part of the impeller at the inlet is called an *inducer*. The flow enters the impeller axially, or perhaps with a small amount of swirl, and leaves the wheel peripherally. Thus, in this machine the flow at the inlet has no radial component of velocity and at the outlet the axial component vanishes. In many blowers and fans the inducer section is left out. An example is the so-called *squirrel cage fan*, which provides large flow rates with modest pressure rise for certain industrial needs and which is also familiar as a small exhaust fan in a bathroom and a kitchen.

Principles of Turbomachinery. By Seppo A. Korpela
Copyright © 2011 John Wiley & Sons, Inc.

266 CENTRIFUGAL COMPRESSORS AND PUMPS

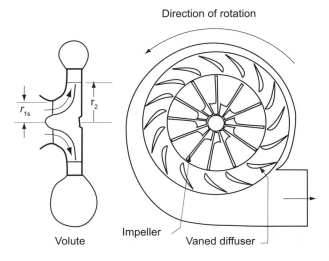

Figure 8.1 A sketch of a centrifugal compressor.

The blades, or vanes as they are called, in the impeller may be shrouded (as in the last stage of the compressor in Figure 1.6 in Chapter 1). Such impellers are used in multistage compressors, which, owing to the large load, require a large shaft, and the possible axial movement is better tolerated by a shrouded impeller than an unshrouded one. In an impeller with many vanes some of the vanes do not reach all the way into the inducer, for the finite thickness of the blades leaves less flow area and cause possible choking of the flow. The vanes that extend all the way to the inlet of the inducer are curved at the inlet sufficiently that the relative velocity enters the vanes tangentially when the absolute velocity is axial.

8.1 COMPRESSOR ANALYSIS

The velocity triangles at the inlet and outlet of a centrifugal compressor are shown in Figure 8.2. At the inlet the axial direction is to the right; at the outlet the radial direction is to the right and normal to U. When the work done is written in terms of kinetic energies, it is

$$w = \frac{1}{2}(V_2^2 - V_1^2) + \frac{1}{2}(U_2^2 - U_1^2) + \frac{1}{2}(W_1^2 - W_2^2)$$

It is clear that in order to obtain a high pressure increase in the compressor, the work transfer must be large. Only the first term on the right accounts for an actual increase in kinetic energy, and the other two represent increases in other thermodynamic properties. If the work done is written as

$$w = h_{02} - h_{01} = h_2 - h_1 + \frac{1}{2}(V_2^2 - V_1^2)$$

then equating the right sides of this and the previous equation gives

$$h_1 + \frac{1}{2}(W_1^2 - U_1^2) = h_2 + \frac{1}{2}(W_2^2 - U_2^2)$$

This is a statement that the *trothalpy* remains constant across the compressor wheel. Writing this as

$$h_2 - h_1 = u_2 - u_1 + p_2 v_2 - p_1 v_1 = \frac{1}{2}\left(U_2^2 - U_1^2 + W_1^2 - W_2^2\right) \qquad (8.1)$$

shows that the two kinetic energy terms on the right represent an increase in the internal energy of the gas and the difference in the flow work between the exit and the inlet.

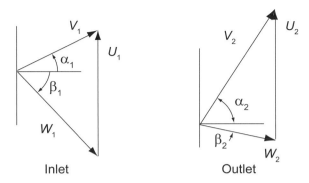

Figure 8.2 Velocity triangles for a centrifugal compressor.

Comparing Eq. (8.1) to the equation obtained for a rotor of an axial machine, it is seen to reduce to it when U at the inlet and the outlet are the same. In that case the increase in static enthalpy across the rotor was seen to arise from the decrease in relative velocity across the rotor. This effect is still there, but more importantly, enthalpy and thus pressure now rises owing to the centrifugal effect of a larger outlet radius. This increase in pressure is present even in absence of any flow, and thus does not have a loss associated with it. It is substantially larger than the diffusional effect of the relative velocity.

The expression for work in the form

$$w = U_2 V_{u2} - U_1 V_{u1}$$

shows that a large amount of work can be done on the air if both the blade speed U_2 and V_{u2} are large. However, a flow with a large V_{u2} and therefore a large V_2, may be difficult to diffuse to a slow speed in the volute. This must be taken into consideration in the design. The velocity diagrams in Figure 8.2 have been drawn to illustrate a general situation. Typically, in air compressors, air is drawn from the atmosphere and it is without a tangential component and therefore enters the compressor with $\alpha_1 = 0$. However, a maldistribution of the flow through the impeller and the vaned space at the exit may influence the upstream condition and lead to some swirl. In addition, in the second and the subsequent stages of a multistage compressor the inlet air might have some *prerotation*, and there are also single stage compressors with special vanes at the inlet that give the flow a small amount of swirl, with positive swirl decreasing the work and negative swirl increasing it.

8.1.1 Slip factor

At the exit of the blades of a centrifugal compressor the flow *deviates* in much the same way as was encountered in axial compressors. To account for it, a *slip factor* is introduced as

$$\sigma = 1 - \frac{V'_{u2} - V_{u2}}{U_2} = 1 - \frac{V_{us}}{U_2}$$

Here V_{u2} is the actual velocity, and V'_{u2} the velocity if there were no slip. The slip velocity is defined as $V_{us} = V'_{u2} - V_{u2}$. The flow angle of the relative velocity at the outlet is typically negative and greater in absolute value than the blade angle χ_2. The deviation is from the pressure side toward the suction side of the blades. From the equations

$$V'_{u2} = U_2 + V_{r2} \tan \chi_2 \qquad V_{u2} = U_2 + V_{r2} \tan \beta_2$$

the slip velocity can be determined, and then the slip coefficient can be expressed as

$$\sigma = 1 - \frac{V_{r2}}{U_2}(\tan \chi_2 - \tan \beta_2)$$

This can also be written as

$$\sigma U_2 = U_2 + W_{u2} - V_{r2} \tan \chi_2 = V_{u2} - V_{r2} \tan \chi_2$$

or in the more convenient form

$$V_{u2} = \sigma U_2 + V_{r2} \tan \chi_2 \qquad (8.2)$$

In addition, writing $V_{r2} = V_{u2}/\tan \alpha_2$, substituting and solving for V_{u2}, gives

$$V_{u2} = \frac{\sigma U_2}{1 - \dfrac{\tan \chi_2}{\tan \alpha_2}} \qquad (8.3)$$

If there is no inlet swirl, then $w = U_2 V_{u2}$, and this equation becomes

$$\psi = \frac{\sigma}{1 - \dfrac{\tan \chi_2}{\tan \alpha_2}} \qquad (8.4)$$

For radial blades $\chi_2 = 0$, and Eq. (8.3) reduces to

$$V_{u2} = \sigma U_2$$

The range of slip factor is $0.83 < \sigma < 0.95$, and the higher value is obtained if the number of blades is large and the flow is guided well.

The actual and ideal velocities are shown in Figure 8.3. Stodola gave the following

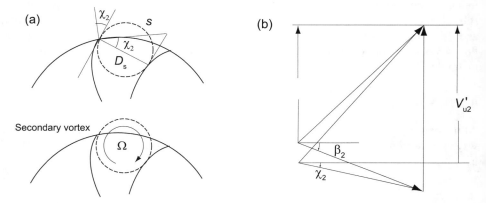

Figure 8.3 Illustration of slip.

argument to estimate the slip factor [75]. The fluid flow through the rotor is irrotational in the laboratory frame, except for that part of the flow that moves right next to the solid surfaces. Therefore, relative to the blade, the spin of the fluid particles (i. e. their vorticity) must be equal and opposite to that caused by the rotating coordinates. Hence a secondary forced vortex that rotates at the angular velocity Ω, as shown, can be assumed to exist in the flow channel. If s is the spacing between the blades along the peripheral circle, the diameter of the secondary vortex is roughly $D_s = s \cos \chi_2$, as the construction in Figure 8.3b shows. Hence the slip velocity can be taken to be

$$V_{us} = \frac{D_s}{2}\Omega = \frac{D_s U_2}{D} = \frac{U_2 s \cos \chi_2}{D}$$

If Z is the number of blades, then $\pi D = sZ$, so that $s/D = \pi/Z$ and the slip velocity takes the form

$$V_{us} = \frac{\pi U_2 \cos \chi_2}{Z}$$

and the slip factor is

$$\sigma = 1 - \frac{\pi \cos \chi_2}{Z} \qquad (8.5)$$

Stanitz [73] recommends

$$\sigma = 1 - \frac{0.63\pi}{Z} \qquad (8.6)$$

for the slip coefficient and assures it to be good in the range $-45° < \chi_2 < 45°$. Other efforts to establish the value for the slip velocity include those by Busemann for blades of logarithmic spiral shape. A synthesis of his and other results was carried out by Wiesner [80], who recommends the expression

$$\sigma = 1 - \frac{\sqrt{\cos \chi_2}}{Z^{0.7}} \qquad (8.7)$$

8.1.2 Pressure ratio

The Euler equation for turbomachinery is

$$w = h_{03} - h_{01} = U_2 V_{u2} - U_1 V_{u1}$$

and the ideal work is

$$w_s = h_{03ss} - h_{01}$$

The states for the compression process are shown in the Mollier diagram in Figure 8.4. With $\eta_{tt} = w_s/w$ from these equations, it follows that

$$\left(\frac{p_{03}}{p_{01}}\right)^{(\gamma-1)/\gamma} = 1 + \frac{\gamma - 1}{\gamma R T_{01}} \eta_{tt}(U_2 V_{u2} - U_1 V_{u1}) \qquad (8.8)$$

Since the inlet velocity is axial and therefore small, the stagnation enthalpy h_{01} is only slightly larger than h_1. The line of constant trothalpy

$$h_1 + \frac{1}{2}W_1^2 - \frac{1}{2}U_1^2 = h_2 + \frac{1}{2}W_2^2 - \frac{1}{2}U_2^2$$

is shown in Figure 8.4, as well as the magnitudes of the various kinetic energies. In particular, the blade velocity U_1 is smaller than U_2, and the relative velocity diffuses across

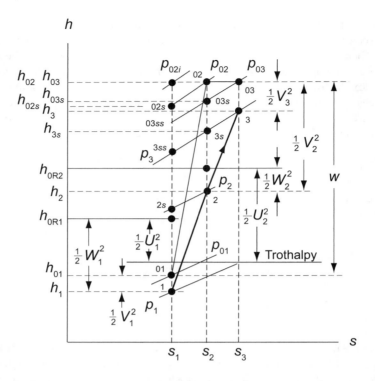

Figure 8.4 The thermodynamic states in a centrifugal compressor.

the rotor, so that $W_1 > W_2$. The kinetic energy leaving the rotor is quite large. The loss of stagnation pressure across the rotor is $\Delta p_{0LR} = p_{02i} - p_{02}$, and across the volute it is $\Delta p_{0LS} = p_{02} - p_{03}$. The pressure p_{02i} is the stagnation pressure for an isentropic compression process in which the same amount of work has been done as in the actual process.

The rotor efficiency is given by

$$\eta_R = \frac{h_{02s} - h_{01}}{h_{02} - h_{01}} \tag{8.9}$$

With η_R known, the stagnation temperature T_{02s} can be calculated from Eq. (8.9), and the stagnation pressure p_{02} can then be calculated from

$$p_{02} = p_{01} \left(\frac{T_{02s}}{T_{01}} \right)^{\gamma/(\gamma-1)}$$

In addition, integrating the Gibbs equation along the line of constant p_{02} and along the line of constant h_{02} between the states with entropies s_1 and s_2 gives

$$\frac{p_{02i}}{p_{02}} = \left(\frac{T_{02}}{T_{02s}} \right)^{\gamma/(\gamma-1)}$$

from which p_{02i} can be calculated.

For an axial inlet flow $V_{u1} = 0$ and introducing $V_{u2} = \psi U_2$ into Eq. (8.8), the pressure ratio can be written as

$$\frac{p_{03}}{p_{01}} = \left[1 + (\gamma - 1)\psi\eta_{tt} M_{0u}^2\right]^{\gamma/(\gamma-1)} \quad (8.10)$$

in which $M_{0u} = U_2/c_{01}$. A plot of this relation is shown in Figure 8.5. Similarly, the pressure ratio across the rotor alone is

$$\frac{p_{02}}{p_{01}} = \left[1 + (\gamma - 1)\psi\eta_R M_{0u}^2\right]^{\gamma/(\gamma-1)} \quad (8.11)$$

It was seen earlier that in the expression for constant trothalpy

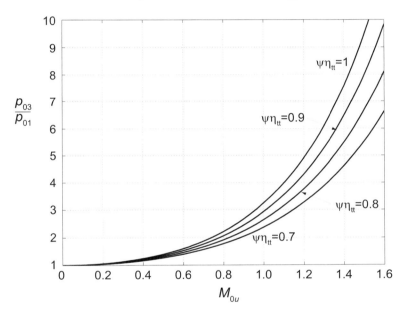

Figure 8.5 Pressure ratio as a function of blade stagnation Mach number, for various values of $\psi\eta_{tt}$ and for gas with $\gamma = 1.4$.

$$h_2 - h_1 = \frac{1}{2}(U_2^2 - U_1^2) + \frac{1}{2}(W_1^2 - W_2^2)$$

the first term on the right side is a kinematic effect and therefore represents a reversible process. If the left side is written as

$$h_2 - h_1 = h_2 - h_{2s} + h_{2s} - h_1$$

then the reversible enthalpy change may be written as

$$h_{2s} - h_1 = \frac{1}{2}(U_2^2 - U_1^2) + (1-f)\frac{1}{2}(W_1^2 - W_2^2)$$

and for the irreversible change

$$h_2 - h_{2s} = f\frac{1}{2}(W_1^2 - W_2^2)$$

in which f is the fraction of the change in relative kinetic energy lost to irreversibility. The static enthalpy loss coefficient ζ_R in

$$h_2 - h_{2s} = \frac{1}{2}\zeta_R W_2^2$$

is related to f by the equation

$$\zeta_R = f\left(\frac{W_1^2}{W_2^2} - 1\right)$$

The ratio $(1-\eta_R)/(1-\eta_{tt})$ of rotor losses to the total losses is between 0.5 and 0.6. Hence for $\eta_{tt} = 0.8$, the rotor efficiency is about $\eta_R = 0.89$.

Writing the rotor efficiency as

$$\eta_R = \frac{h_{2s} - h_1 + \frac{1}{2}(V_{2s}^2 - V_1^2)}{h_2 - h_1 + \frac{1}{2}(V_2^2 - V_1^2)}$$

and assuming that $V_{2s} = V_2$, this becomes

$$\eta_R = \frac{h_{2s} - h_2 + h_2 - h_1 + \frac{1}{2}(V_2^2 - V_1^2)}{h_2 - h_1 + \frac{1}{2}(V_2^2 - V_1^2)}$$

which can also be written as

$$\eta_R = 1 - \frac{h_2 - h_{2s}}{w} = 1 - \frac{\zeta_R W_2^2}{2w} = 1 - \frac{\zeta_R W_2^2}{2\psi U_2^2}$$

Squaring and adding the component equations

$$W_2 \sin \beta_2 = V_{u2} - U_2 \qquad W_2 \cos \beta_2 = V_{r2}$$

gives

$$W_2^2 = V_{u2}^2 - 2V_{u2}U_2 + U_2^2 + V_{r2}^2$$

In addition

$$V_{r2} = \frac{V_{u2}}{\tan \alpha_2}$$

so that

$$\frac{W_2^2}{U_2^2} = 1 - 2\psi + \frac{\psi^2}{\sin^2 \alpha_2} \tag{8.12}$$

and the rotor efficiency can be written as

$$\eta_R = 1 - \zeta_R \frac{(1 - 2\psi + \psi^2/\sin^2 \alpha_2)}{2\psi}$$

in which

$$\psi = \frac{\sigma}{1 - \dfrac{\tan \chi_2}{\tan \alpha_2}} \tag{8.13}$$

Solving the equation for rotor efficiency for ζ_R gives

$$\zeta_R = \frac{(1 - \eta_R)2\psi}{1 - 2\psi + \psi^2/\sin^2 \alpha_2}$$

For a typical case $\chi_2 = -40°$ and $\alpha_2 = 67°$. With $\sigma = 0.85$, these equations give $\psi = 0.627$, and for a rotor efficiency $\eta_R = 0.89$, the loss coefficient is $\zeta_R = 0.656$. This appears to be quite large, but when calculating the stagnation pressure losses this is multiplied by W_2^2, which is small because the flow will have been diffused through the rotor. In addition, a typical value for $W_1/W_2 = 1.6$. For these values the fraction of the relative kinetic energy lost to irreversibility is $f = 0.38$. This number accords with Cumpsty's [18] review of the field. Although this number appears to be large, in a typical compressor the centrifugal effect accounts for over three quarters of the pressure rise and the diffusion of the relative velocity the rest. Hence 38% loss keeps the rotor efficiency still over 90%.

■ **EXAMPLE 8.1**

Air flows from atmosphere at pressure 101.325 kPa and temperature 288 K into a centrifugal compressor with radial blades at the exit of the impeller. The inlet velocity is $V_1 = 93$ m/s, and there is no preswirl. The compressor wheel has 33 blades, and its tip speed is $U_2 = 398$ m/s. The total-to-total efficiency is $\eta_{tt} = 0.875$. Find (a) the stagnation pressure ratio and (b) the power required assuming that the mass flow rate is $\dot{m} = 2.2$ kg/s. Use the Stanitz formula to find the slip coefficient.

Solution: (a) Since the air comes from the atmosphere, where it is stagnant, the atmospheric pressure and temperature are the inlet stagnation properties. The stagnation speed of sound is

$$c_{01} = \sqrt{\gamma R T_{01}} = \sqrt{1.4 \cdot 287 \cdot 288} = 340.2 \text{ m/s}$$

and the stagnation blade Mach number is therefore

$$M_{0u} = \frac{U_2}{c_{01}} = \frac{398}{340.2} = 1.17$$

The Stanitz slip coefficient is

$$\sigma = 1 - \frac{0.63\pi}{Z} = 0.94$$

Since for radial blades Eq. (8.4) shows that $\psi = \sigma$, the pressure ratio can be calculated from

$$\frac{p_{03}}{p_{01}} = [1 + (\gamma - 1)\sigma\eta_{tt}M_{0u}^2]^{\gamma/(\gamma-1)} = (1 + 0.4 \cdot 0.94 \cdot 0.875 \cdot 1.17^2)^{3.5} = 3.674$$

(b) The stagnation temperature for an isentropic process is

$$T_{03s} = T_{01}\left(\frac{p_{03}}{p_{01}}\right)^{(\gamma-1)/\gamma} = 288 \cdot 3.674^{1/3.5} = 417.7 \text{ K}$$

and the ideal work is therefore

$$w_s = c_p(T_{03s} - T_{01}) = 1004.5(417.7 - 288) = 130.29 \text{ kJ/kg}$$

The actual work is $w = w_s/\eta_{tt} = 130.29/0.875 = 148.90$ kJ/kg, and the power required has the value $\dot{W} = \dot{m}w = 2.2 \cdot 148.90 = 327.6$ kW.

■

8.2 INLET DESIGN

Similar design limitations are encountered in the design of centrifugal compressors as in axial compressors. Namely, absolute and relative velocities must be kept sufficiently low to ensure that shock losses do not become excessive. The blade speed for high-performance compressors with a stainless-steel is kept below 450 m/s owing to limitations caused by high stresses.

Typically the inlet stagnation pressure and temperature are known, and when atmospheric air is compressed, they are the ambient values. If the compressor is not equipped with vanes to introduce preswirl into the flow, the absolute velocity is axial. This velocity is also uniform across the inlet. The inlet hub and shroud radii are r_{1h} and r_{1s}, and the inlet area is

$$A_1 = \pi(r_{1s}^2 - r_{1h}^2) = \pi r_{1s}^2(1 - \kappa^2) \qquad \text{where} \qquad \kappa = \frac{r_{1h}}{r_{1s}}$$

The flow rate

$$\dot{m} = \rho_1 A_1 V_1$$

may be expressed in a nondimensional form by diving it by the product $\rho_{01}\pi r_2^2 c_{01}$. Following Whitfield and Baines [79], this is written as

$$\Phi = \frac{\dot{m}}{\rho_{01}\pi r_2^2 c_{01}} = \frac{\rho}{\rho_{01}} \frac{r_{1s}^2}{r_2^2}(1-\kappa^2)\frac{W_{1s}}{c_{01}}\cos\beta_{1s} \qquad (8.14)$$

The inlet velocity V_1 is is related to the relative velocity at the shroud by $V_1 = W_{1s}\cos\beta_{1s}$ In addition, $W_{1s} = \sqrt{U_{1s}^2 + V_1^2}$ and since the blade speed is highest at the shroud, so is W_{1s} and β_{1s} has the largest absolute value there. With $r_{1s}/r_2 = U_{1s}/U_2$, the nondimensional mass flow rate may be written as

$$\Phi = \frac{\rho_1}{\rho_{01}} \frac{U_{1s}^2}{U_2^2}(1-\kappa^2)\frac{W_{1s}}{c_{01}}\cos\beta_{1s}$$

and it can be further manipulated into the form

$$\Phi = \frac{\rho_1}{\rho_{01}}\left(\frac{W_{1s}^2 - V_1^2}{c_1^2}\right)\frac{c_{01}^2}{U_2^2}\frac{c_1^2}{c_{01}^2}(1-\kappa^2)\frac{W_{1s}}{c_1}\frac{c_1}{c_{01}}\cos\beta_{1s} \qquad (8.15)$$

which can be written as

$$\Phi = \frac{\rho_1}{\rho_{01}}\left(\frac{T_1}{T_{01}}\right)^{3/2}\frac{M_{1Rs}^3}{M_{0u}^2}(1-\kappa^2)(1-\cos^2\beta_{1s})\cos\beta_{1s}$$

Here $M_{1Rs} = W_{1s}/c_1$ is the relative Mach number at the shroud. Defining next

$$\Phi_f = \frac{\Phi M_{0u}^2}{1 - \kappa^2}$$

the preceeding equation can be written as

$$\Phi_f = \frac{M_{1Rs}^3(1-\cos^2\beta_{1s})\cos\beta_{1s}}{\left(1 + \frac{\gamma-1}{2}M_{1Rs}^2\cos^2\beta_{1s}\right)^{(3\gamma-1)/(2\gamma-2)}} \qquad (8.16)$$

Graphs calculated from Eq. (8.16) are shown in Figure 8.6. The angle β_{1s} at which the flow rate reaches its maximum value is obtained differentiating this equation with respect to $\cos\beta_{1s}$ and setting the derivative to zero:

$$\cos^4\beta_{1s} - \frac{3+\gamma M_{1Rs}^2}{M_{1Rs}^2}\cos^2\beta_{1s} + \frac{1}{M_{1Rs}^2} = 0 \tag{8.17}$$

The solution of this equation is

$$\cos^2\beta_{1s} = \left(\frac{3+\gamma M_{1Rs}^2}{2M_{1Rs}^2}\right)\left(1 - \sqrt{1 - \frac{4M_{1Rs}^2}{(3+\gamma M_{1Rs}^2)^2}}\right) \tag{8.18}$$

Another way to plot the data is shown in Figure 8.7. The optimum angle from Eq. (8.18) is $\beta_{1s} = -57.15°$ for $M_{1Rs} = 0.6$, and if the relative Mach number is increased to $M_{1Rs} = 1.2$, the optimum angle is $\beta_{1s} = -62.56°$. In the incompressible limit the optimum angle is $\beta_{1s} = -54.74°$.

Equation (8.16) can be recast also in the form

$$\Phi_f = \frac{\rho_1}{\rho_{01}}\left(\frac{W_{1s}^2 - V_1^2}{c_1^2}\right)\frac{c_1^2}{c_{01}^2}\frac{V_1}{c_1}\frac{c_1}{c_{01}}$$

and further as

$$\Phi_f = \frac{(M_{1Rs}^2 - M_1^2)M_1}{\left(1 + \frac{\gamma-1}{2}M_1^2\right)^{(3\gamma-1)/(2\gamma-2)}}$$

Solving this for M_{1Rs}^2 gives

$$M_{1Rs}^2 = M_1^2 + \frac{\Phi_f}{M_1}\left(1 + \frac{\gamma-1}{2}M_1^2\right)^{(3\gamma-1)/(2\gamma-2)}$$

This result is plotted graphically in Figure 8.8. Since $M_1 = M_{1Rs}\cos\beta_{1s}$, the lines of constant $\cos\beta_{1s}$ are straight on this set of graphs and the optimum angle is at the minima of the curves.

■ **EXAMPLE 8.2**

In an air compressor the relative Mach number is $M_{1Rs} = 0.9$ at the shroud of the impeller. The inlet stagnation pressure and temperature are 101.325 kPa and 288 K. The mass flow rate is $\dot{m} = 1.2\,\text{kg/s}$. The hub-to-shroud radius ratio is 0.4, and the inlet flow is axial. Determine (a) the rotational speed of the impeller, (b) the axial velocity, and (c) the inducer shroud diameter, assuming that the inducer is designed for optimum relative flow angle, and that the compressor operates at the design point.

Solution: The stagnation density is

$$\rho_{01} = \frac{p_{01}}{RT_{01}} = \frac{101325}{287 \cdot 288} = 1.2259\,\text{kg/m}^3$$

and the stagnation speed of sound is

$$c_{01} = \sqrt{\gamma RT_{01}} = \sqrt{1.4 \cdot 287 \cdot 288} = 340.17\,\text{m/s}$$

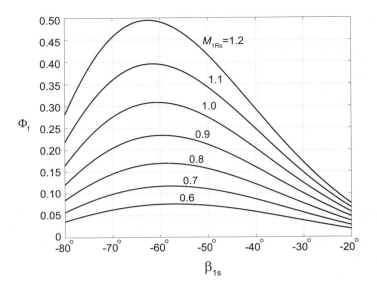

Figure 8.6 Nondimensional mass flow rate for given inlet angle of relative flow with relative Mach number as a parameter, for a gas with $\gamma = 1.4$.

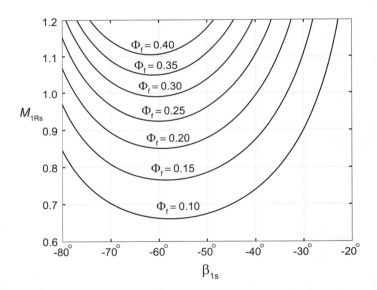

Figure 8.7 Relative Mach number for given inlet angle of relative flow with nondimensional flow rate as a parameter, for a gas with $\gamma = 1.4$.

The optimum relative flow angle can be found from

$$\cos^2 \beta_{1s} = \left(\frac{3 + \gamma M_{1Rs}^2}{2 M_{1Rs}^2} \right) \left(1 - \sqrt{1 - \frac{4 M_{1Rs}^2}{(3 + \gamma M_{1Rs}^2)^2}} \right)$$

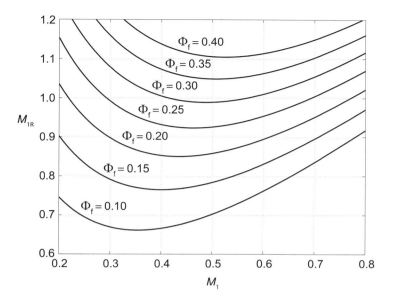

Figure 8.8 Relative Mach number for given inlet Mach number with nondimensional flow rate as a parameter and for a gas with $\gamma = 1.4$.

With $\gamma = 1.4$ and $M_{1\text{Rs}} = 0.9$ this gives

$$\cos^2 \beta_{1s} = \frac{3 + 1.4 \cdot 0.81}{2 \cdot 0.81}\left(1 - \sqrt{1 - \frac{4 \cdot 0.81}{(2 + 1.4 \cdot 0.81)^2}}\right) = 0.2546$$

and therefore $\beta_{1s} = -59.70°$. Next value of Φ_f is calculated from

$$\Phi_f = \frac{M_{1\text{Rs}}^3(1 - \cos^2 \beta_{1s})\cos \beta_{1s}}{\left(1 + \frac{\gamma-1}{2}M_{1\text{Rs}}^2 \cos^2 \beta_{1s}\right)^{(3\gamma-1)/(2\gamma-2)}}$$

and it yields the value

$$\Phi_f = \frac{0.9^3(1 - \cos^2(-59.7°))\cos(-59.7°)}{(1 + 0.2 \cdot 0.81 \cos^2(-59.7°))^4} = 0.2333$$

The equality

$$\Phi = \frac{\Phi_f M_{0U}^2}{1 - \kappa^2}$$

can next be rewritten as

$$\frac{\dot{m}}{\rho_{01} c_{01} \pi r_2^2} = \frac{\Phi_f(1 - \kappa^2)\rho_{01} c_{01}^3}{U_2^2}$$

After $U_2 = r_2 \Omega$ is substituted and r_2^2 canceled from both sides, this can be solved for Ω with the result:

$$\Omega = \sqrt{\frac{\Phi(1 - \kappa^2)c_{01}^3 \pi}{\dot{m}}} = \frac{\sqrt{0.2333 \cdot 0.84 \cdot 1.225 \cdot 340.17^3 \pi}}{1.2} = 47{,}510\,\text{rpm}$$

(b) The inlet Mach number is

$$M_1 = M_{1Rs} \cos \beta_{1s} = 0.9 \cos(-59.7°) = 0.454$$

and hence the inlet static temperature has the value

$$T_1 = \frac{T_{01}}{1 + \frac{\gamma-1}{2} M_2^2} = \frac{288}{1 + 0.2 \cdot 0.454^2} = 276.6 \text{ K}$$

The static density comes out to be

$$\rho_1 = \rho_{01} \left(\frac{T_1}{T_{01}}\right)^{1/(\gamma-1)} = 1.2259 \left(\frac{276.6}{288}\right)^{2.5} = 1.1081 \text{ kg/m}^3$$

and the axial velocity is

$$V_1 = M_1 \sqrt{\gamma R T_1} = 0.454\sqrt{1.4 \cdot 287 \cdot 276.6} = 151.4 \text{ m/s}$$

The mass flow rate,

$$\dot{m} = \rho_1 A_1 V_1 = \rho_1 \pi r_{1s}^2 (1 - \kappa^2) V_1$$

when solved for r_{1s} gives

$$r_{1s} = \sqrt{\frac{\dot{m}}{\rho_1 \pi (1 - \kappa^2) V_1}} = \sqrt{\frac{1.2}{1.1081 \cdot \pi \cdot 84 \cdot 151.4}} = 0.0521 \text{ m}$$

so that the inducer diameter of the impeller is $D_{1s} = 10.42$ cm. ∎

8.2.1 Choking of the inducer

The inducer is shown in Figure 8.9. The flow is axial at the inlet, and the relative velocity forms an angle β_1 at the mean radius and β_{1s} at the shroud. The corresponding blade velocities are U_1 and U_{1s} with $U_{1s} = r_{1s} U_1 / r_{1m}$. The blade angle is χ_1, and the stagger is ξ. The throat width is denoted by t. From the shape of the blade, its thickness distribution and the stagger, the width of the throat can be determined. As the sketch shows, a reasonable estimate is given by $t = s \cos \chi_1$. A typical value for incidence $i = \beta_1 - \chi_1$ is $-4°$ to $-6°$ [63].

Mass balance in the form

$$\dot{m} = \rho_1 A_1 W_1 \cos \beta_1 = \frac{p_1}{RT_1} A_1 M_{1R} \sqrt{\gamma R T_1} \cos \beta_1$$

can be written in terms of functions of the relative Mach number, by introducing the ratios

$$\frac{T_{0R}}{T_1} = 1 + \frac{\gamma - 1}{2} M_{1R}^2$$

and

$$\frac{p_{0R}}{p_1} = \left(1 + \frac{\gamma - 1}{2} M_{1R}^2\right)^{\gamma/(\gamma-1)}$$

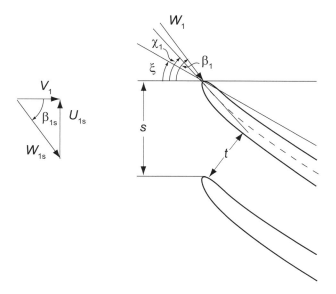

Figure 8.9 Detail of the inlet to the inducer.

with the result that the mass balance takes the form

$$\dot{m} = \frac{p_{0R} A_1 \cos \beta_1}{\sqrt{c_p T_{0R}}} \frac{\gamma}{\sqrt{\gamma - 1}} M_{1R} \left(1 + \frac{\gamma - 1}{2} M_{1R}^2\right)^{-(\gamma+1)/2(\gamma-1)}$$

At the throat the mass flow rate can be expressed as

$$\dot{m} = \rho_t A_t W_t = \frac{p_{0R} A_t}{\sqrt{c_p T_{0R}}} \frac{\gamma}{\sqrt{\gamma - 1}} M_{tR} \left(1 + \frac{\gamma - 1}{2} M_{tR}^2\right)^{-(\gamma+1)/2(\gamma-1)}$$

Equating the mass flow rates in the two preceding equations gives

$$\frac{A_1 M_{1R} \cos \beta_1}{\left(1 + \frac{\gamma-1}{2} M_{1R}^2\right)^{(\gamma+1)/2(\gamma-1)}} = \frac{A_t M_{tR}}{\left(1 + \frac{\gamma-1}{2} M_{tR}^2\right)^{(\gamma+1)/2(\gamma-1)}} \quad (8.19)$$

If the flow is choked, $M_{tR} = 1$, and this equation reduces to

$$\frac{A_t}{A_1 \cos \beta_1} = M_{1R} \left(\frac{2}{\gamma + 1} + \frac{\gamma - 1}{\gamma + 1} M_{1R}^2\right)^{-(\gamma+1)/2(\gamma-1)} \quad (8.20)$$

Figure 8.10 shows the graphs of the relative Mach number as a function of the effective area ratio for various values of the relative throat Mach number. If more accurate numerical values are needed than can be read from the graphs, they are easily calculated from Eq. (8.19).

Substituting $A_1 \cos \beta_1$ from Eq. (8.19) into

$$\dot{m} = \rho_1 A_1 W_1 \cos \beta_1$$

gives

$$\dot{m} = \rho_1 c_1 A_t M_{tR} \frac{\left(1 + \frac{\gamma-1}{2} M_{tR}^2\right)^{(\gamma+1)/2(\gamma-1)}}{\left(1 + \frac{\gamma-1}{2} M_{1R}^2\right)^{(\gamma+1)/2(\gamma-1)}}$$

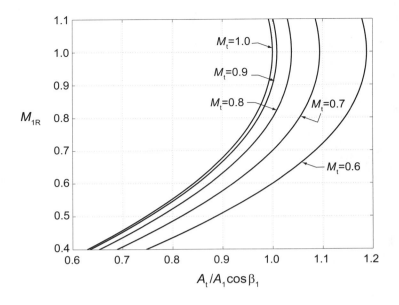

Figure 8.10 Throat area for a choked inducer for a gas with $\gamma = 1.4$.

which, when multiplied and divided by $\rho_{01} c_{01}$, can be written as

$$\dot{m} = \rho_{01} c_{01} A_t M_{tR} \frac{\left(1 + \frac{\gamma-1}{2} M_{1R}^2\right)^{(\gamma+1)/2(\gamma-1)}}{\left(1 + \frac{\gamma-1}{2} M_1^2\right)^{(\gamma+1)/2(\gamma-1)} \left(1 + \frac{\gamma-1}{2} M_{tR}^2\right)^{(\gamma+1)/2(\gamma-1)}}$$

In this equation the ratio of the two terms involving relative and absolute Mach numbers can be written as

$$\frac{1 + \frac{\gamma-1}{2} M_{1R}^2}{1 + \frac{\gamma-1}{2} M_1^2} = \frac{h_1 + \frac{1}{2} W_1^2}{h_1 + \frac{1}{2} V_1^2} = \frac{h_1 + \frac{1}{2} V_1^2 + \frac{1}{2} U_1^2}{h_1 + \frac{1}{2} V_1^2} = 1 + \frac{U_1^2}{2 h_{01}} = 1 + \frac{\gamma-1}{2} \frac{U_1^2}{c_{01}^2}$$

so that

$$\dot{m} = \rho_{01} c_{01} A_t M_{tR} \frac{\left(1 + \frac{\gamma-1}{2} \frac{U_1^2}{c_{01}^2}\right)^{(\gamma+1)/2(\gamma-1)}}{\left(1 + \frac{\gamma-1}{2} M_{tR}^2\right)^{(\gamma+1)/2(\gamma-1)}}$$

This shows that the mass flow rate increases as the blade speed increases. This happens even after the flow chokes, for then this equation becomes

$$\dot{m} = \rho_{01} c_{01} A_t \left(\frac{2}{\gamma+1} + \frac{\gamma-1}{\gamma+1} \frac{U_1^2}{c_{01}^2}\right)^{(\gamma+1)/2(\gamma-1)}$$

With increased blade speed more compression work is done on the gas, and its pressure temperature, and density, all increase in the flow channel. Thus at the choked throat the higher velocity, given by $V_t = \sqrt{\gamma R T_t}$, and higher density result in an increased flow rate

8.3 EXIT DESIGN

The characteristic design calculation in turbomachinery flows involves the relationship between the flow angles and the flow and blade loading coefficients. Density differences between the inlet and the exit of the impeller are considered only when the blade heights are determined. This was largely ignored in the axial compressor theory, as the annulus area could be reduced to keep the axial velocity constant. In radial compressor calculations the situation is somewhat more complicated, as the comparable criterion of constant axial velocity is absent. In this section the characteristics of the exit of the impeller are discussed.

8.3.1 Performance characteristics

If the inlet is axial, the work done is

$$w = U_2 V_{u2} \tag{8.21}$$

Since $V_{u2} = U_2 + W_2 \sin \beta_2$ and $V_{r2} = W_2 \cos \beta_2$, where V_{r2} is the radial velocity at the exit, this can be written as

$$w = U_2(U_2 + W_2 \sin \beta_2) = U_2^2 + V_{r2}U_2 \tan \beta_2$$

Defining the blade-loading coefficient for a centrifugal machine as $\psi = w/U_2^2$ and the flow coefficient as $\phi = V_{r2}/U_2$, this equation can be recast in the form

$$\psi = 1 + \phi \tan \beta_2 \tag{8.22}$$

This is a straight line in the performance plot shown in Figure 8.11. If $\beta_2 > 0$, the characteristic is rising and for $\beta_2 < 0$, it is falling. The outlet velocity diagrams in the same figure show what happens to the absolute velocity in these two cases. It is clearly much larger in a machine with a rising characteristic.

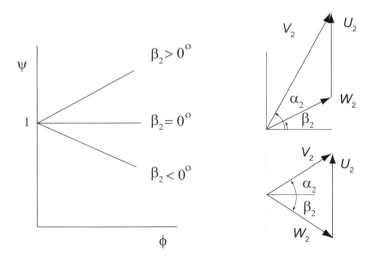

Figure 8.11 Idealized characteristic of a centrifugal machine.

Pressure rise in a centrifugal compressor takes place not only across the impeller but also in a vaned or vaneless diffuser and a volute. The latter are the stationary components of the machine and can be viewed as the stator. This is a useful way of thinking about the machine, and the degree of reaction can again be defined as the ratio of the static enthalpy rise across the rotor to the total.

If the outlet velocity from the impeller is large, then the flow needs to be diffused by a large amount in the stationary components of the machine. In order to prevent the flow separation and the irreversibilites associated with it, the diffusion in the volute must be kept sufficiently mild. This can be controlled by curving the blades backward, in order to reduce the absolute velocity at the outlet of the impeller.

Backward-swept blades have an another decisive advantage; namely, the operation of the machine is *stable*. If the flow rate is reduced by increasing the load, for example by partially closing a valve downstream, on a falling characteristic of backward-swept blades, the blade-loading coefficient increases leading to higher enthalpy and thus also pressure at the outlet of the impeller. Hence a higher pressure rise is obtained across the machine. This counteracts the increase in flow resistance, and a new stable operating point is established. On the other hand, for forward-swept blades a drop in the flow rate decreases the pressure across the machine and thus leads to further drop in the flow rate in the system. Thus the operation of a compressor with forward-swept blades is inherently unstable.

If the flow angle α_2 of the absolute velocity at the outlet is in the range $63° - 72°$ a good design is obtained, and the best designs are in the range $67° - 69°$. For radial blades, which are common, there is a wide range of fairly high efficiency. Radial blades have the added advantage that they experience no bending by centrifugal loads, and the state of stress tends to be dominated by tensile stresses.

The pressure ratio has been shown to be

$$\left(\frac{p_{03}}{p_{01}}\right)^{(\gamma-1)/\gamma} = 1 + (\gamma-1)\eta_{tt}\frac{U_2 V_{u2}}{c_{01}^2} = 1 + (\gamma-1)\eta_{tt}\psi M_{0u}^2$$

and the temperature ratio is

$$\frac{T_{03}}{T_{01}} = 1 + (\gamma-1)\psi M_{0u}^2$$

in which the blade-loading coefficient is given by Eq. (8.13). The exit Mach number can also be expressed in terms of the blade stagnation Mach number M_{0u}. The definition of exit Mach number can be rewritten as

$$M_2 = \frac{V_2}{c_2} = \frac{V_{u2}}{c_{01}\sin\alpha_2}\frac{c_{01}}{c_{02}}\frac{c_{02}}{c_2}$$

which can be rearranged to the form

$$M_2 = \frac{V_{u2}}{c_{01}\sin\alpha_2}\left(\frac{T_{01}}{T_{02}}\right)^{1/2}\left(\frac{T_{02}}{T_2}\right)^{1/2} = \frac{\psi M_{0u}}{\sin\alpha_2}\sqrt{\frac{1+\frac{\gamma-1}{2}M_2^2}{1+(\gamma-1)\psi M_{0u}^2}}$$

Solving this for M_2 gives

$$M_2 = \frac{\psi M_{0u}}{\sqrt{\sin^2\alpha_2 + (\gamma-1)\psi M_{0u}^2(\sin^2\alpha_2 - \frac{1}{2}\psi)}} \qquad (8.23)$$

The inlet Mach number can be written in terms of M_{0u} as well. First, the definition of M_1 can be written in the form

$$M_1^2 = \frac{V_1^2}{c_1^2} = \frac{U_{1s}^2}{c_{01}^2}\frac{c_{01}^2}{c_1^2}\frac{1}{\tan^2\beta_{1s}}$$

and then using $U_{1s} = r_{1s}U_2/r_2$ and $c_{01}^2/c_1^2 = T_{01}/T_1$ in this leads to

$$\tan^2 \beta_{1s} M_1^2 = M_{0u}^2 \frac{r_{1s}^2}{r_2^2}(1 + \frac{\gamma-1}{2}M_1^2)$$

Solving this for M_1 gives

$$M_1 = \frac{M_{0u}r_{1s}/r_2}{\sqrt{\tan^2 \beta_{1s} - \frac{\gamma-1}{2}M_{0u}^2 \frac{r_{1s}^2}{r_2^2}}} \quad (8.24)$$

The relative inlet Mach number at the shroud is $M_{1Rs} = M_1/\cos\beta_{1s}$. These relations are shown graphically in Figure 8.12 in the manner following Whitfield [78]. The blade angle varies from $\chi_2 = 0°$ for radial blades in increments of $-10°$ to $\chi_2 = -60°$. The outlet angle was assumed to be $\alpha_2 = 67°$ and the inlet flow angle, $\beta_{1s} = -60°$ at the shroud. The slip factor was assumed to be $\sigma = 0.85$, and the total-to-total efficiency was taken to be $\eta_{tt} = 0.8$. The radius ratio has the value $r_{1s}/r_2 = 0.6$. With these values fixed a typical

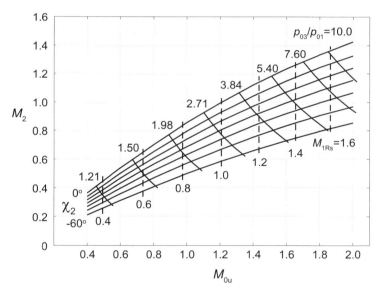

Figure 8.12 Exit Mach number, pressure ratio, and relative inlet Mach number as functions of blade stagnation Mach number for various blade angles χ_2. The exit flow angle is $\alpha_2 = 67°$ and at the inlet $\beta_{1s} = 60°$. The slip factor is $\sigma = 0.85$ and $\eta_{tt} = 0.8$. The radius ratio is $r_{1s}/r_2 = 0.6$, and $\gamma = 1.4$.

design would have $M_{0u} = 1.4$ and, say $\chi_2 = -30°$. These give a blade-loading coefficient of $\psi = 0.683$, an exit Mach number of $M_2 = 0.9$, a pressure ratio of $p_{03}/p_{01} = 3.84$, and a relative Mach number at the inlet shroud of $M_{1Rs} = 1.17$.

8.3.2 Diffusion ratio

The diffusion ratio for the relative velocity is discussed next. It is obtained by first writing at the inlet

$$W_{1s}\sin\beta_{1s} = U_{1s}$$

and then developing the relative velocity at the outlet into a convenient form. Starting with

$$W_2 \sin \beta_2 = V_{u2} - U_2 \qquad W_2 \cos \beta_2 = V_{r2}$$

and squaring and adding them gives

$$W_2^2 = V_{u2}^2 - 2V_{u2}U_2 + U_2^2 + V_{r2}^2$$

Noting that $V_{r2} = V_{u2}/\tan \alpha_2$, and dividing by U_2^2, gives

$$\frac{W_2^2}{U_2^2} = 1 - 2\psi + \frac{\psi^2}{\sin \alpha_2^2}$$

a result that was developed earlier in this chapter. Hence the diffusion ratio becomes

$$\frac{W_{1s}}{W_2} = -\frac{U_{1s}/U_2}{\sin \beta_{1s}\sqrt{1 - 2\psi + \dfrac{\psi^2}{\sin^2 \alpha_2}}}$$

which in terms of the radius ratio is clearly

$$\frac{W_{1s}}{W_2} = -\frac{r_{1s}/r_2}{\sin \beta_{1s}\sqrt{1 - 2\psi + \dfrac{\psi^2}{\sin^2 \alpha_2}}}$$

This ratio should be kept below $W_{1s}/W_2 < 1.9$. For radius ratio $r_{1s}/r_2 = 0.7$, graphs of the diffusion ratio as a function of χ_2 are plotted in Figure 8.13 for various values of the exit flow angle α_2. A typical design might have $\alpha_2 = 67°$ and if the diffusion ratio $W_{1s}/W_2 = 1.9$ is chosen, then $\chi_2 = -28.2°$.

8.3.3 Blade height

The blade height at the exit is determined by equating the mass flow rates at the inlet and the outlet of the rotor. Writing the mass flow rates in terms of the flow functions as

$$\frac{\dot{m}\sqrt{c_p T_{01}}}{p_{01} A_1} = F_1(M_1)$$

and

$$\frac{\dot{m}\sqrt{c_p T_{02}}}{p_{02} A_2} = F_2(M_2)$$

Diving the former by the latter gives

$$\frac{A_2}{A_1} = \sqrt{\frac{T_{02}}{T_{01}} \frac{p_{01}}{p_{02}}} \frac{F_1(M_1)}{F_2(M_2)} \tag{8.25}$$

The area ratio is seen to depend on the pressure ratio across the rotor, which is

$$\frac{p_{02}}{p_{01}} = \left(1 + (\gamma - 1)\eta_R \psi M_{0u}^2\right)^{\gamma/(\gamma-1)}$$

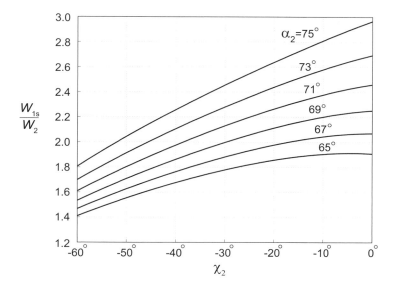

Figure 8.13 Diffusion ratio for $\sigma = 0.85$, inlet flow angle $\beta_{1s} = -60°$, $r_{1s}/r_2 = 0.7$, and $\gamma = 1.4$, as a function of χ_2 and for various values of α_2.

and the temperature ratio, which is

$$\frac{T_{02}}{T_{01}} = 1 + (\gamma - 1)\psi M_{0u}^2$$

The area ratio

$$\frac{A_2}{A_1} = \frac{2\pi b r_2}{\pi r_{1s}^2 (1 - \kappa^2)}$$

yields the blade height to radius ratio b_2/r_2, as

$$\frac{b}{r_2} = \frac{A_2}{A_1} \frac{r_{1s}^2}{r_2^2} \frac{(1 - \kappa^2)}{2}$$

To establish the area ratio in Eq. (8.25), the flow function $F_1(M_1)$ can be calculated in terms of M_{0u}, since Eq. (8.24) gives the relationship between M_1 and M_{0u}. Similarly, Eq. (8.23) is a relationship between M_2 and M_{0u}. Thus the blade width-to-radius ratio can now be calculated for various values of M_{0u}. The graphs shown in Figure 8.14 are again patterned after Whitfield [78].

8.4 VANELESS DIFFUSER

As the compressed gas leaves the rotor, it enters a vaneless space in which it diffuses to a lower velocity. The mass balance at the exit of the rotor can be written as

$$\dot{m} = \rho_2 2\pi r_2 b_2 V_2 \cos\alpha_2$$

A similar equation at the end of the diffusion process at r_{2e} is

$$\dot{m} = \rho_{2e} 2\pi r_{2e} b_{2e} V_{2e} \cos\alpha_{2e}$$

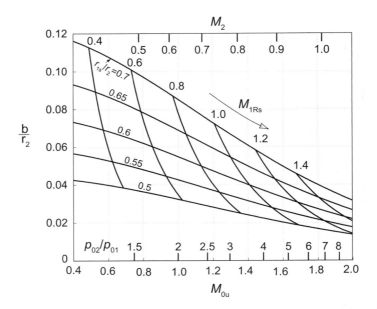

Figure 8.14 Blade width-to-radius ratio for $\sigma = 0.85$, $\eta_R = 0.9$ and $\eta_{tt} = 0.8$. At the inlet $\beta_{1s} = -60°$ and $\alpha_2 = 65°$. At the exit $\chi_2 = -40°$. The hub-to-shroud radius ratio is $\kappa = 0.4$, and the gas has $\gamma = 1.4$.

so that

$$\rho_2 r_2 b_2 V_2 \cos\alpha_2 = \rho_{2e} r_{2e} b_{2e} V_{2e} \cos\alpha_{2e} \qquad (8.26)$$

These mass balances can also be cast in the nondimensional forms:

$$\frac{\dot{m}\sqrt{c_p T_{02}}}{p_{02} 2\pi} = r_2 b_2 F_2 \cos\alpha_2$$

$$\frac{\dot{m}\sqrt{c_p T_{02e}}}{p_{02e} 2\pi} = r_{2e} b_{2e} F_{2e} \cos\alpha_{2e}$$

The stagnation temperature is constant so that $T_{02e} = T_{02}$. If it assumed that there are no losses, then since no work is done, $p_{02e} = p_{02}$. In addition, if the channel has a constant height, equating the foregoing two equations gives

$$r_{2e} F_{2e} \cos\alpha_{2e} = r_2 F_2 \cos\alpha_2 \qquad (8.27)$$

For a free vortex velocity distribution

$$r_{2e} V_{u2e} = r_2 V_{u2}$$

or

$$r_{2e} V_{2e} \sin\alpha_{2e} = r_2 V_2 \sin\alpha_2 \qquad (8.28)$$

For a channel of constant width $b_2 = b_{2e}$, so that dividing this equation by Eq. (8.26) yields

$$\frac{\tan\alpha_{2e}}{\rho_{2e}} = \frac{\tan\alpha_2}{\rho_2} \qquad (8.29)$$

Since both T_0 and p_0 are taken as constant, it then follows from the ideal gas relation that $\rho_{02} = \rho_{02e}$. Multiplying the left side of this equation by ρ_{02e} and the right side by ρ_{02} gives

$$\left(1 + \frac{\gamma-1}{2}M_{2e}^2\right)^{1/(\gamma-1)} \tan\alpha_{2e} = \left(1 + \frac{\gamma-1}{2}M_2^2\right)^{1/(\gamma-1)} \tan\alpha_2 \qquad (8.30)$$

When the conditions at the exit of the rotor are known, this equation and Eq. (8.27) are two equations for the two unknowns M_{2e} and α_{2e} for a specified location r_{2e}.

Mach number M_{2e} as a function of the radius ratio is shown in Figure 8.15 for a flow with $M_2 = 1.2$ at the exit of the rotor and flow angles in the range $62° < \alpha_2 < 70°$. The

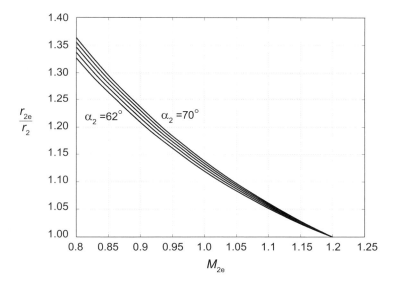

Figure 8.15 Mach number for a gas with $\gamma = 1.4$ as a function of radius ratio r_{2e}/r_2 for flow angles at the exit of the rotor in the range $62° < \alpha_2 < 70°$ and $M_2 = 1.2$.

value of M_{2e} at $r_{2e}/r_2 = 1.30$ is seen to be in the range $0.82 - 0.85$, as the flow angle is increased from $62°$ to $70°$. The diffusion of velocity is shown in Figure 8.16.

In the incompressible limit the mass balance and the irrotational flow condition give

$$r_2 V_2 \cos\alpha_2 = r_{2e} V_{2e} \cos\alpha_{2e} \qquad r_2 V_2 \sin\alpha_2 = r_{2e} V_{2e} \sin\alpha_{2e}$$

Taking ratio of these two equations gives $\tan\alpha_2 = \tan\alpha_{2e}$ indicating that the flow angle remains constant. In this case

$$\frac{V_{2e}}{V_2} = \frac{r_2}{r_{2e}}$$

and the diffusion varies inversely with the radius ratio. The compressible flow approaches this condition as the flow angle moves toward $90°$. Since the Mach number at any given radius is higher for the larger flow angles, density is lower, and Eq. (8.29) shows that the flow angle is now smaller. The change in flow angles is shown in Figure 8.17. That the flow angle α_{2e} increases with radius means that the flow remains longer in the vaneless space and therefore experiences losses as the boundary layers grow in a decelerating flow

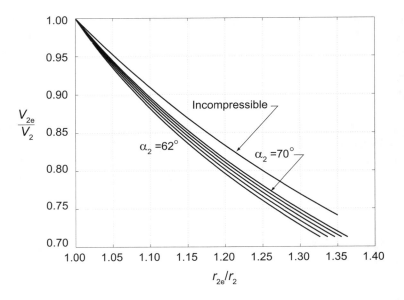

Figure 8.16 Diffusion ratio for a gas with $\gamma = 1.4$ as a function of radius ratio r_{2e}/r_2 for flow angles at the exit of the rotor in the range $62° < \alpha_2 < 70°$ and $M_2 = 1.2$; the incompressible flow case corresponds to constant flow angle.

in a narrow channel. For this reason, the vaneless space is kept small and vanes are inserted to guide the flow into the volute.

More information on losses can be found in Cumpsty [18] and in Whitfield and Baines [79]. The function of the vaned diffuser is to reduce the area over which the diffusion takes place. The vanes may be in the form of airfoils, triangular channels, or what are called *island diffusers*. For the triangular channels a rule of thumb is to keep the opening angle at less than $12°$. Again, Whitfield and Baines give more information, including results from a more advanced analysis.

■ **EXAMPLE 8.3**

Air from a centrifugal compressor leaves the blades at angle $\alpha_2 = 67.40°$ and $M_2 = 1.1$ at the radius $r_2 = 7.5\,\text{cm}$ as it enters a vaneless diffuser. (a) Find the radial location at which the flow reaches the sonic condition. (b) Find the Mach number at radius $r_{2e} = 0.10\,\text{cm}$.

Solution: (a) It is assumed that there are no losses, and denoting the sonic state by star, $\rho_{02}^* = \rho_{02}$, and

$$\frac{\rho_{02}}{\rho_2} \tan \alpha_2 = \frac{\rho_{02^*}}{\rho^*} \tan \alpha^*$$

The density ratio at the inlet to the vaneless diffuser is

$$\frac{\rho_{02}}{\rho_2} = \left(1 + \frac{\gamma - 1}{2} M_2^2\right)^{1/(\gamma-1)} = (1 + 0.2 \cdot 1.1^2)^{2.5} = 1.719$$

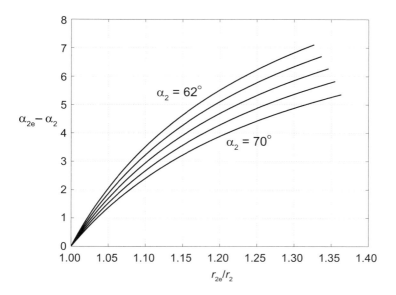

Figure 8.17 Difference in flow angles during the diffusion process, with angle at exit of the rotor in the range $62° < \alpha_2 < 70°$ and $M_2 = 1.2$, and the ratio of specific heats $\gamma = 1.4$.

and at the sonic state it is therefore

$$\frac{\rho_{02}^*}{\rho^*} = \left(\frac{\gamma+1}{2}\right)^{1/(\gamma-1)} = 1.2^{2.5} = 1.577$$

The flow angle at the sonic state is

$$\alpha^* = \tan^{-1}\left(\frac{\rho_{02}}{\rho_2}\frac{\rho^*}{\rho_{02}^*}\tan\alpha_2\right) = \tan^{-1}\left[\frac{1.719}{1.577}\tan(67.40°)\right] = 69.09°$$

The radial location where $M = 1$ is

$$r^* = r_2\frac{F_2\cos\alpha_2}{F^*\cos\alpha^*} = 0.075\frac{1.271\cos(67.40°)}{1.281\cos(69.09°)} = 0.0801\text{ m}$$

in which

$$F_2 = \frac{\gamma}{\sqrt{\gamma-1}}M_2\left(1+\frac{\gamma-1}{2}M_2^2\right)^{-(\gamma+1)/2(\gamma-1)} = \frac{1.4}{\sqrt{0.4}}1.1(1+0.2\cdot 1.1^2)^{-3} = 1.271$$

and

$$F^* = \frac{\gamma}{\sqrt{\gamma-1}}\left(\frac{\gamma+1}{2}\right)^{-(\gamma+1)/2(\gamma-1)} = \frac{1.4}{\sqrt{0.4}}(1.2)^{-3} = 1.281$$

(b) The two equations

$$r_{2e}\cos\alpha_{2e}F_{2e} = r_2\cos\alpha_2 F_2$$

and

$$\tan\alpha_{2e} = \frac{\left(1+\frac{\gamma-1}{2}M_2^2\right)^{1/(\gamma-1)}}{\left(1+\frac{\gamma-1}{2}M_{2e}^2\right)^{1/(\gamma-1)}}\tan\alpha_2$$

are to be solved simultaneously for α_{2e} and M_{2e}. The angle can be eliminated using

$$\cos\alpha_{2e} = \frac{1}{\sqrt{1+\tan^2\alpha_{2e}}}$$

Substituting, simplifying, and rearranging gives

$$\frac{r_{2e}\gamma M_{2e}}{\sqrt{\gamma-1}\sqrt{1+\frac{\gamma-1}{2}M_{2e}^2}} -$$

$$r\cos\alpha_2 F_2 \sqrt{\left(1+\frac{\gamma-1}{2}M_{2e}^2\right)^{1/(\gamma-1)} + \left(1+\frac{\gamma-1}{2}M_2^2\right)^{1/(\gamma-1)}\tan^2\alpha_2} = 0$$

Solving this by iteration gives $M_{2e} = 0.7571$. ∎

A well designed vaned diffuser improves the efficiency by 2% or 3% over a vaneless one. However, this comes at a cost, for at off-design operation the efficiency of a vaned diffuser will deteriorate; that is, a compressor with a vaned diffuser will have a narrow operating range at the peak efficiency, owing to stalling of the vanes. To widen the range, adjustable vanes can be implemented into the design at added complexity and initial cost. The payback is reduction in operating costs at higher efficiency. A low-cost option is to have a vaneless diffuser, which has a lower efficiency but a flatter operating range near peak efficiency. As the flow leaves the vaneless, or vaned, diffuser, it enters a volute. Its design with respect to pumps is discussed at the end of this chapter.

8.5 CENTRIFUGAL PUMPS

The operation and design of pumps follows principles similar to those of centrifugal compressors. Compressibility can be clearly ignored in pumping liquids, but it may also be neglected in fans in which the pressure rise is slight.

The first law of thermodynamics across a pump is

$$w = h_{02} - h_{01} = \left(u_2 + \frac{p_2}{\rho} + \frac{1}{2}V_1^2 + gz_1\right) - \left(u_1 + \frac{p_1}{\rho} + \frac{1}{2}V_2^2 + gz_2\right)$$

In incompressible fluids, as was discussed in Chapter 2, internal energy increases only as a result of irreversibilities in an adiabatic flow. Hence, if the flow through the pump is reversible and adiabatic, internal energy does not increase and $u_2 = u_1$. In this situation the preceding equation reduces to

$$w_s = \left(\frac{p_2}{\rho} + \frac{1}{2}V_2^2 + gz_2\right) - \left(\frac{p_1}{\rho} + \frac{1}{2}V_1^2 + gz_1\right)$$

The *total head* developed by a pump is defined as

$$H = \left(\frac{p_2}{\rho g} + \frac{1}{2g}V_2^2 + z_2\right) - \left(\frac{p_1}{\rho g} + \frac{1}{2g}V_1^2 + z_1\right)$$

so it represents the work done by a reversible pump per unit weight of the fluid. On the unit mass basis the reversible work is

$$w_s = gH$$

Since the total head is readily measurable, the pump industry reports it, as well as the overall efficiency, in the pump specifications.

The shaft power to the pump is given by

$$\dot{W}_o = \frac{\rho Q g H}{\eta}$$

The overall efficiency η can be expressed as the product

$$\eta = \eta_m \eta_v \eta_h$$

in which η_h is a hydraulic efficiency, η_v is a volumetric efficiency, and η_m is a mechanical efficiency. The hydraulic efficiency accounts for the irreversibilities in the flow through the pump. If the loss term is written as

$$gH_L = u_2 - u_1$$

then

$$w = w_s + gH_L = gH + gH_L$$

and the *hydraulic efficiency* is defined as

$$\eta_h = \frac{w_s}{w} = \frac{gH}{gH + gH_L}$$

The hydraulic efficiency may be calculated from the empirical equation

$$\eta_h = 1 - \frac{0.4}{Q^{1/4}} \quad (8.31)$$

where Q is in liters per second. If the volumetric flow rate is given in gallons per minute, as is still done in part of the pump industry today, the constant 0.4 has to be replaced by 0.8. It is in this form that this expression for hydraulic efficiency appears in the *Pump Handbook* [45].

A typical set of pump performance curves is given in Figure 8.18. (The quantity on the right ordinate axis, NPSH_R = required net positive suction head, and its significance is discussed later in conjunction with consideration of cavitation.) For an impeller diameter of 31 cm and flow rate of 18 L/s, the delivered head is 48 m at the shaft speed of 1750 rpm. The efficiency at this condition is about 0.63. The contours of constant efficiency and the power for pumping water are shown. Since the reversible work is $w_s = gH$, and $w = w_s/\eta_h$, the power calculated using $\dot{W} = \dot{m}w$, then, because of leakage flow through the clearances from the exit of the impeller back to the inlet, work will be redone on some of the fluid as it crosses the impeller, and the value obtained will be too low. To correct for this, the power into the impeller is obtained from

$$\dot{W}_R = (\dot{m} + \dot{m}_L)w$$

in which $\dot{m}_L$ is the leakage flow. The power transferred to the fluid is

$$\dot{W} = \dot{m}w$$

and the ratio of these two expression for power is defined as the volumetric efficiency. Hence it also equals the ratio of the mass flow rates and can be written as

$$\eta_v = \frac{\dot{m}}{\dot{m} + \dot{m}_L} = \frac{Q}{Q + Q_L} = \frac{Q}{Q_R} = \frac{\dot{W}}{\dot{W}_R}$$

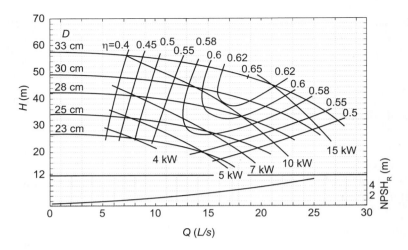

Figure 8.18 A typical set of performance curves for a centrifugal pump at 1750 rpm.

in which Q_R is the flow that passes over the blade passage. The volumetric efficiency for large pumps with flow rates of 600 L/s reaches 0.99 and for small pumps with flow rates of 3 L/s it drops to 0.86. Logan [52] correlated the volumetric efficiency according to

$$\eta_v = 1 - \frac{C}{Q^n} \qquad (8.32)$$

The constants are given in Table 8.1 as a function of the specific speed $\Omega_s = \Omega\sqrt{Q}/w_s^{3/4}$, with the flow rate in liters per second.

Table 8.1 Correlation for volumetric efficiency

Ω_s	C	n
0.20	0.250	0.500
0.37	0.122	0.380
0.73	0.047	0.240
1.10	0.023	0.128

Finally, there is bearing friction and disk drag that are not taken into account in the impeller losses. Hence, if the power needed from the prime mover to power the pump is $\dot{W}_o$ then the power delivered to the rotor $\dot{W}_R$ is less than this. Their ratio is defined as the mechanical efficiency

$$\eta_m = \frac{\dot{W}_R}{\dot{W}_o}$$

The power loss from mechanical friction can be estimated. But if the overall efficiency, hydraulic efficiency, and the volumetric efficiency are obtained from empirical relations, then the mechanical efficiency can be determined from the equation $\eta = \eta_h \eta_v \eta_m$.

As reported by Cooper in the *Pump Handbook* [45], the overall efficiency has been correlated by Anderson and is given by

$$\eta = 0.94 - 0.08955 \left[\left(\frac{1.660Q}{\Omega} \right) \left(\frac{3.56}{e_{rms}} \right)^2 \right]^{-0.21333} - 0.29 \left[\log_{10} \left(\frac{0.8364}{\Omega_s} \right) \right]^2 \quad (8.33)$$

The original correlation is in a mixed set of units, and even after it has been converted here to a form in which Q is given in liters per second and Ω in radians per second, it is not in a dimensionless form. Be as it may, according to Cooper, it gives satisfactory values for the overall efficiency, except at the upper half of the specific speed range, and he suggests that for large-capacity pumps the dashed line in Figure 8.19 be used. This figure gives a graphical representation of Eq. (8.33). The surface roughness of the flow passage is denoted by e_{rms} and its value is in micrometers, with $e_{rms} = 3.56\,\mu m$ for the graphs shown.

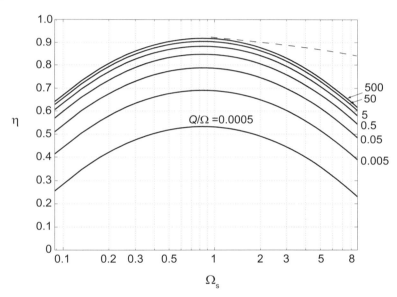

Figure 8.19 Efficiency of centrifugal pumps according to the correlation of Anderson, as quoted by Cooper [45].

The number of blades in the impeller is in the range $5 < Z < 12$, and the empirical equation of Pfleiderer and Petermann [59]

$$Z = 6.5 \left(\frac{r_2 + r_{1s}}{r_2 - r_{1s}} \right) \cos\left(\frac{1}{2}(\chi_2 + \beta_{1s}) \right) \quad (8.34)$$

can be used to calculate this number. It shows that Z increases as r_{1s}/r_2 increases.

Analysis of double-flow (double-suction) pumps, a sketch of which is shown in Figure 8.20, follows closely the analysis of single-flow pumps. The flow rate $Q/2$ is used to calculate the hydraulic and volumetric efficiencies as well as the specific speed. The mechanical efficiency is close to that for a single-flow pump. In the next section Cordier diagram is used to determine the size of a pump. When it is used for a double-flow pump, the entire flow rate Q is used in the definition of the specific diameter.

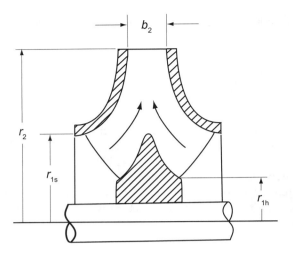

Figure 8.20 A double-flow pump.

8.5.1 Specific speed and specific diameter

A useful chart for pump selection was developed by Cordier. It is shown in Figure 8.21. The abscissa in the chart is the specific diameter, and the ordinate is the specific speed. These are defined as

$$D_s = \frac{D(gH)^{1/4}}{\sqrt{Q}} \qquad \Omega_s = \frac{\Omega\sqrt{Q}}{(gH)^{4/3}}$$

It was seen in Chapter 4 that specific speed is used to select a pump of certain shape. Once selected, the size of the pump can be obtained using the Cordier diagram. The curve has been constructed such that for the size selected, optimal efficiency is obtained.

Since the total head is reported, it is more convenient to define the blade loading coefficient in terms of the reversible work rather than the actual work. Therefore, the loading coefficient is defined as

$$\psi_s = \frac{w_s}{U_2^2} = \frac{gH}{U_2^2}$$

and the subscript s serves as a reminder that this definition differs from the conventional one. With $\psi = w/U_2^2$, the relation $\psi_s = \eta\psi$ relates the two definitions.

Another way to size pumps is given by Cooper in [45]. With a specified flow rate and head rise across a pump, the rotational speed is first chosen, with the understanding that the higher the speed, the more compact is the pump. Once the rotational speed is fixed, the flow coefficient $\phi = V_{m2}/U_2$ can be obtained from the correlation

$$\phi = 0.1715\sqrt{\Omega_s} \tag{8.35}$$

A correlation for the blade-loading coefficient is

$$\psi_s = \frac{0.386}{\Omega_s^{1/3}} \tag{8.36}$$

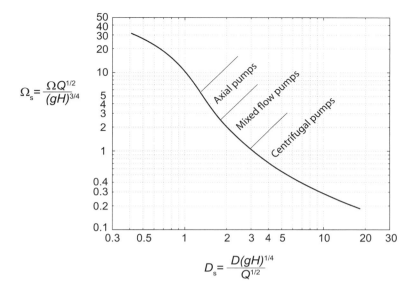

Figure 8.21 Cordier diagram for fans and pumps. (Adapted from Csanady [17].)

and after the loading coefficient is determined, the blade speed is obtained from

$$U_2 = \sqrt{\frac{gH}{\psi_s}}$$

After that, the impeller radius is calculated from $r_2 = U_2/\Omega$ and the flow meridional velocity determined from $V_{m2} = \phi U_2$. Finally the blade thickness b_2 is determined from

$$b_2 = \frac{Q_R}{2\pi r_2 V_{m2}}$$

where Q_R is the sum of the delivered flow Q and the leakage flow Q_L. Shapes of the velocity diagrams for low and high specific speeds are shown in Figure 8.22. Since

$$\psi_s = \frac{w_s}{U_2^2} = \frac{\eta w}{U_2^2} = \frac{\eta V_{u2} U_2}{U_2^2}$$

the relationship

$$\frac{V_{u2}}{U_2} = \frac{\psi_s}{\eta} = \frac{0.383}{\Omega_s^{1/3} \eta}$$

is obtained. When the specific speed is high, V_{u2} becomes much smaller than U_2 and vanes have a large backsweep. The backsweep reduces as the specific speed decreases. It also decreases because lowering the specific speed lowers the efficiency. A sufficient reduction in the specific speed leads to forward-swept vanes, and such pumps are prone to unstable operation if the load changes. When the specific speed becomes very low, the centrifugal pump is no longer suitable for the application and it should be replaced by a positive displacement pump, such as a screw pump or a rotary vane pump. In typical designs V_{u2} is slightly over 0.5 of U_2, and then the absolute values of both flow angles are quite large. For such pumps the exit relative flow angle ranges from $-65°$ to $-73°$.

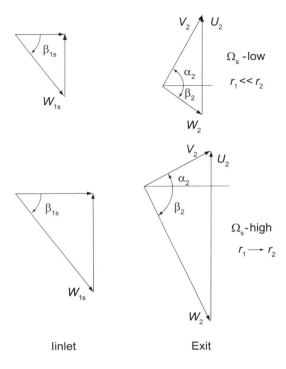

Figure 8.22 Velocity triangles for a low and a high specific speed centrifugal pump.

In the discussion of compressors an optimum inlet flow angle for the relative velocity was found, which gives the largest flow rate with a given relative Mach number. In the incompressible limit this gives $\beta_{1s} = -54.74°$. A range from $-65°$ to $-80°$ is typical for pumps, which means that a flow rate lower than the optimum is obtained for a fixed relative velocity.

For axial entry the volumetric flow rate can be written as

$$Q = A_1 V_1 = \pi r_{1s}^2 (1 - \kappa^2)\sqrt{W_{1s}^2 - r_{1s}^2 \Omega^2}$$

Solving this for W_{1s}^2 gives

$$W_{1s}^2 = r_{1s}^2 \Omega^2 + \frac{Q^2}{\pi^2 (1 - \kappa^2)^2 r_{1s}^4}$$

When r_{1s} is small, the second term causes W_{1s} to be large, and when r_{1s} is large, the first term increases the value of W_{1s}. The value of r_{1s} for which W_{1s} is minimum, is given by

$$r_{1s} = \left(\frac{\sqrt{2}Q}{\pi(1 - \kappa^2)\Omega}\right)^{1/3}$$

Typical values of κ are in the range from very small to about 0.5. The smallest value of the hub radius r_{1h} depends on the size of the shaft. The shaft diameter is easily determined from elementary torsion theory, once the torque is known. For double-suction pumps, in which the shaft penetrates the entire hub, κ is typically 0.5. These guidelines are illustrated next with examples.

EXAMPLE 8.4

A pump is to be selected to pump water at the rate of 50 L/s. The increase in total head across the pump is to be 35 m. An electric motor, connected with a direct drive and a rotational speed of 3450 rpm, provides the power to the pump. Water is drawn from a pool at atmospheric temperature and pressure. Its density is $\rho = 998 \text{ kg/m}^3$.
(a) Determine the type of pump for this application and its efficiency, assuming $e_{rms} = 3.56 \, \mu\text{m}$. (b) Calculate the pump diameter. (c) Estimate the pump efficiency and the power needed.

Solution: (a) The specific speed of this pump is

$$\Omega_s = \frac{\Omega\sqrt{Q}}{(gH)^{3/4}} = \frac{3450 \cdot \pi}{30} \frac{\sqrt{0.05}}{(9.81 \cdot 35)^{3/4}} = 1.013$$

From Figure 4.9 a pump with Francis-type impeller is chosen. The efficiency, calculated from Eq. (8.33), is $\eta = 0.815$.

(b) To determine the size of the pump, a Cordier diagram may be consulted. The specific diameter is estimated to be $D_s = 3.1$ so that the impeller diameter is

$$D = D_s \frac{Q}{(gH)^{1/4}} = \frac{3.1\sqrt{0.05}}{(9.81 \cdot 35)^{1/4}} = 16.1 \text{ cm}$$

(c) The power required is

$$\dot{W} = \frac{\rho Q g H}{\eta} = \frac{998 \cdot 0.05 \cdot 9.81 \cdot 35}{0.815} = 21.0 \text{ kW}$$

The specific speed of the pump in Example 8.4 is about the upper limit for centrifugal pumps. Beyond this value pumps fall into the category of mixed-flow type. In mixed-flow pumps the edge of the blade on the meridional plane is inclined with respect to the radial (or axial) direction. If the meridional velocity is perpendicular to the edge, then the effective

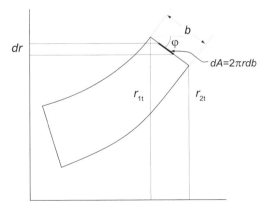

Figure 8.23 Sketch for calculation of blade width.

radius for calculating the volumetric flow rate is determined from the construction shown in Figure 8.23.

The differential area is $dA = 2\pi r\, db$ and $dr = \sin\varphi\, db$. Hence

$$A = \int_{r_{1t}}^{r_{2t}} \frac{2\pi r}{\sin\varphi} dr = \frac{\pi(r_{1t}^2 - r_{2t}^2)}{\sin\varphi} = \frac{\pi(r_{1t} - r_{2t})(r_{1t} + r_{2t})}{\sin\varphi} = \pi(r_{1t} + r_{2t})b$$

Thus the effective radius is the mean radius $r_m = \frac{1}{2}(r_{1t} + r_{2t})$ and $A_2 = 2\pi r_m b$.

■ EXAMPLE 8.5

A pump handles water at the rate of 10 L/s with a head of 100 m across the pump. The power is provided by an electric motor with shaft speed 3450 rpm. Water is at 20°C with density $\rho = 998\,\text{m}^3/\text{kg}$. (a) Calculate the specific speed of the pump. (b) Determine the flow coefficient and the blade-loading coefficient. (c) Find the directions of the absolute velocity and the relative velocity of water leaving the impeller. (d) Find the tip radius of the impeller. (d) Find the power needed.

Solution: (a) The specific speed of this pump is

$$\Omega_s = \frac{\Omega\sqrt{Q}}{(gH)^{3/4}} = \frac{3450 \cdot \pi}{30} \frac{\sqrt{0.01}}{(9.81 \cdot 100)^{3/4}} = 0.2061$$

(b) The flow coefficient is determined from

$$\phi = 0.1715\sqrt{\Omega_s} = 0.1715\sqrt{0.2061} = 0.0779$$

and the blade loading coefficient is obtained from

$$\psi_s = \frac{0.386}{\Omega_s^{1/3}} = \frac{0.386}{0.2061^{1/3}} = 0.6535$$

From $\psi_s = gH/U_2^2$ the blade tip speed is

$$U_2 = \sqrt{\frac{gH}{\psi_s}} = \sqrt{\frac{9.81 \cdot 100}{0.6535}} = 38.7\,\text{m/s}$$

(c) The hydraulic efficiency is

$$\eta_h = 1 - \frac{0.4}{Q^{1/4}} = 1 - \frac{0.4}{10^{1/4}} = 0.775$$

and the work done is therefore

$$w = \frac{w_s}{\eta_h} = \frac{9.81 \cdot 100}{0.775} = 1.266\,\text{kJ/kg}$$

The tangential and radial components of the velocity are

$$V_{u2} = \frac{w}{U_2} = \frac{1266}{38.7} = 32.7\,\text{m/s} \qquad V_{r2} = \phi U_2 = 0.0779 \cdot 38.7 = 3.01\,\text{m/s}$$

so that the flow angle is

$$\alpha_2 = \tan^{-1}\left(\frac{V_{u2}}{V_{r2}}\right) = \tan^{-1}\left(\frac{32.7}{3.01}\right) = 84.72°$$

The tangential and radial components of the relative velocity are

$$W_{u2} = V_{u2} - U_2 = 32.7 - 38.7 = -6.0 \, \text{m/s} \qquad W_{r2} = V_{r2} = 3.01 \, \text{m/s}$$

and therefore

$$\beta_2 = \tan^{-1}\left(\frac{W_{u2}}{W_{r2}}\right) = \tan^{-1}\left(\frac{-6.0}{3.01}\right) = -63.61°$$

(d) The impeller radius can be calculated to be

$$r_2 = \frac{U_2}{\Omega} = \frac{38.7 \cdot 30}{3450 \cdot \pi} = 0.1072 \, \text{m}$$

The volumetric efficiency is obtained by first finding the constants in Eq. (8.32) by interpolation. They are $C = 0.2454$ and $n = 0.4957$ for $\Omega_s = 0.2061$. The volumetric efficiency is then

$$\eta_v = 1 - \frac{C}{Q^n} = 1 - \frac{0.2454}{10^{0.4597}} = 0.9216$$

so that $Q_R = Q/\eta_v = 0.01/0.9216 = 0.01085 \, \text{m}^2/\text{s}$. The blade width is therefore

$$b_2 = \frac{Q_R}{2\pi \, r_2 \, V_{m2}} = \frac{0.01085}{2\pi \cdot 0.1077 \cdot 3.01} = 0.0053 \, \text{m} \qquad b_2 = 0.53 \, \text{cm}$$

(c) The overall efficiency is determined from Eq. (8.33) to be $\eta = 0.66$. Hence the power to the pump is

$$\dot{W} = \frac{\rho Q g H}{\eta} = \frac{998 \cdot 10 \cdot 9.81 \cdot 100}{1000 \cdot 0.66} = 14.8 \, \text{kW}$$

In the next example the number of vanes and their metal angle are also considered.

■ **EXAMPLE 8.6**

Water flows axially into a double-suction centrifugal pump at the rate of $0.120 \, \text{m}^3/\text{s}$. The pump delivers a head of $20 \, \text{m}$ while operating at 880 rpm. The hub-to-shroud ratio at the inlet is 0.50, and the relative velocity makes an angle $-73°$ at the inlet. (a) Find the reversible work done by the pump. (b) What is the work done by the impeller? (c) Find the radius of the impeller and the inlet radius of the shroud. (d) Determine the blade width at the exit of the impeller. (e) Assume a reasonable number of blades and calculate the blade angle at the exit. Use the Pfleiderer equation to determine more accurately the number of blades and recalculate the blade angle at the exit if needed.

Solution: (a) The reversible work is

$$w_s = gH = 9.81 \cdot 20 = 196.2 \, \text{J/kg}$$

(b) The hydraulic efficiency is

$$\eta_h = 1 - \frac{0.4}{Q^{1/4}} = 1 - \frac{0.4}{60^{1/4}} = 0.856$$

and the actual work by the impeller is

$$w = \frac{w_s}{\eta_h} = \frac{196.2}{0.856} = 229.1 \, \text{J/kg}$$

(c) The specific speed is

$$\Omega_s = \frac{\Omega\sqrt{Q}}{w_s^{3/4}} = \frac{880 \cdot \pi\sqrt{0.060}}{30 \cdot 196.2^{3/4}} = 0.431$$

and the loading coefficient is

$$\psi_s = \frac{0.383}{\Omega_s^{1/3}} = \frac{0.383}{0.431^{1/3}} = 0.507$$

Therefore the impeller tip speed is

$$U_2 = \sqrt{w_s \psi_s} = \frac{196.2}{0.507} = 21.25 \, \text{m/s}$$

and the impeller radius is

$$r_2 = \frac{U_2}{\Omega} = \frac{21.25 \cdot 20}{880 \cdot \pi} = 0.231 \, \text{m}$$

The volumetric flow rate can be written as

$$Q = A_1 V_1 = \frac{\pi r_{1s}(1 - \kappa^2) U_{1s}}{\tan(-\beta_{1s})} = \frac{\pi (1 - \kappa^2) \Omega r_{1s}^3}{\tan(\beta_{1s})}$$

Solving this for r_{1s} gives

$$r_{1s} = \left[\frac{Q \tan(-\beta_{1s})}{\pi(1 - \kappa^2)\Omega}\right]^{1/3}$$

so that

$$r_{1s} = \left[\frac{60 \cdot \tan(73°) \cdot 30}{1000 \cdot \pi^2 (1 - 0.50^2) \cdot 880}\right]^{1/3} = 0.0967 \, \text{m}$$

and

$$r_{1h} = \kappa r_{1s} = 0.5 \cdot 0.0967 = 0.0483 \, \text{m}$$

and thus the blade speed at the shroud is

$$U_{1s} = r_{1s} \Omega = \frac{0.0967 \cdot 880 \cdot \pi}{30} = 8.91 \, \text{m/s}$$

(d) The flow coefficient is

$$\phi = 0.1715\sqrt{\Omega_s} = 0.1715\sqrt{0.431} = 0.1125$$

and the radial velocity at the exit is then

$$V_{r2} = \phi U_2 = 0.1125 \cdot 21.25 = 2.39 \, \text{m/s}$$

To calculate the leakage flow, the coefficients for the expression of volumetric are interpolated to be

$$C = 0.1094 \qquad n = 0.3564$$

so that

$$\eta_v = 1 - \frac{C}{Q^n} = 1 - \frac{0.1094}{60^{0.3564}} = 0.975$$

and the flow through the exit is

$$Q_R = \frac{Q}{\eta_v} = \frac{120}{0.975} = 0.123 \, \text{m}^3/\text{s}$$

Hence the blade width has the value

$$b_2 = \frac{Q_R}{2\pi r_2 V_{r2}} = \frac{0.123}{2 \cdot \pi \cdot 0.231 \cdot 2.39} = 3.55 \, \text{cm}$$

(e) The tangential component of the exit velocity is calculated to be

$$V_{u2} = \frac{w}{U_2} = \frac{229.1}{21.25} = 10.78 \, \text{m/s}$$

and the flow angle at the exit is

$$\alpha_2 = \tan^{-1}\left(\frac{V_{u2}}{V_{r2}}\right) = \tan^{-1}\left(\frac{10.78}{2.39}\right) = 77.5°$$

The tangential component of the relative velocity becomes

$$W_{u2} = V_{u2} - U_2 = 10.78 - 21.25 = -10.47 \, \text{m/s}$$

so that the flow angle is

$$\beta_2 = \tan^{-1}\left(\frac{W_{u2}}{W_{r2}}\right) = \tan^{-1}\left(\frac{-10.47}{2.39}\right) = -77.1°$$

Next, the number of blades is assumed. Let $Z = 6$, and the blade angle is guessed to be, say, $\chi_2 = -60°$. Then the slip coefficient is calculated from

$$\sigma = 1 - \frac{\sqrt{\cos \chi_2}}{Z^{0.7}} = 1 - \frac{\sqrt{\cos(-60°)}}{6^{0.7}} = 0.798$$

and the equation

$$V_{u2} = \sigma U_2 + V_{r2} \tan \chi_2$$

is solved for χ_2, giving

$$\chi_2 = \tan^{-1}\left(\frac{V_{u2} - \sigma U_2}{V_{r2}}\right) = \tan^{-1}\left(\frac{10.78 - 0.798 \cdot 21.25}{2.39}\right) = -68.86°$$

Now a new value of σ is obtained from

$$\sigma = 1 - \frac{\sqrt{\cos \chi_2}}{Z^{0.7}} = 1 - \frac{\sqrt{\cos(-68.86°)}}{6^{0.7}} = 0.829$$

With this value for σ, repeating the calculation gives $\chi_2 = -70.7°$ and $\sigma = 0.836$. The number of blades can now be calculated from Pfleiderer's equation

$$Z = 6.5 \left(\frac{1 + r_{1s}/r_2}{1 - r_{1s}/r_2}\right) \cos\left(\frac{\beta_{1s} + \chi_2}{2}\right) = 6.00$$

so that the initial guess was correct. One more iteration gives $\chi_2 = -71.1°$.

8.6 FANS

An industrial fan with a wide impeller and blades in the shape of airfoils is shown in Figure 8.24. The impeller has no inducer, and the flow enters the fan axially. It then turns and enters the blade passage radially. The increase in radius between the inlet and outlet is quite modest, and for this reason such fans also have a low pressure rise and for this reason the flow can be considered incompressible. The blades can be made quite long, which gives a large flow area. Since the flow at the inlet is radial, at the inlet $W_{r1} = V_1$ and $W_{u1} = -U_1$.

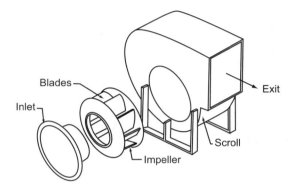

Figure 8.24 A centrifugal fan.

The blades are oriented such that the relative flow enters at the angle β_1 obtained by solving

$$\tan \beta_1 = \frac{W_{u1}}{W_{r1}}$$

Since the width of the flow areas at the inlet and exit are the same, and the density change is ignored, the radial velocities are related by $r_1 V_{r1} = r_2 V_{r2}$. With an inlet velocity without swirl, the work done is calculated in the same way as for centrifugal pumps.

8.7 CAVITATION

Common experience shows that water pressure increases with depth in a quiescent pool of water. Similarly, pressure decreases in a vertical pipe flow if the fluid moves to a higher elevation, not only because of this hydrostatic effect, but as a result of irreversibilities caused by turbulence and wall friction. If the inlet of the pipe is a short distance below a surface of a body of water, the pressure at the inlet of the pipe is the difference in the hydrostatic head and the drop caused by the dynamic head plus the frictional pressure loss. As a consequence, for an upward-moving flow the pressure in a sufficiently long pipe may drop enough to reach the saturation pressure corresponding to the prevailing temperature. The saturation pressure for water at 20° is 2.34 kPa.

After the saturation pressure has been crossed, vapor bubbles begin to form in the stream. When this happens in the blade passage of a turbomachine, the flow is said to undergo *cavitation*. The effects of cavitation are harmful, and the performance of the pump deteriorates. The work done by each element of the impeller vane increases the fluid pressure, and as the flow moves in the flow passages, it carries the bubbles into regions of higher pressure. There they collapse. The collapse is a consequence of an instability in

… the size and shape of the bubble. As the instability develops the bubble flattens out, and a liquid from the back accelerates toward the center, forming a jet that pierces through the bubble. These impinging jets from the bubbles located next to the impeller of the pump cause erosion. This kind of cavitation damage is seen also in marine propellers.

Any dissolved air tends to come out of the liquid at low pressures. These small air bubbles act as nucleation sites for bubble formation. They are aided in turbulent flow by local negative pressure spikes. The kinetics of nucleation, turbulence, and growth rates of bubbles are complicated subjects and make prediction of cavitation difficult. Hence pump manufacturers rely on experimentation to determine when the pump performance is significantly affected. A comprehensive review of the mechanisms of cavitation is given by Arakeri [4] and by Brennen [10, 9].

A pressure difference called *net positive suction pressure* (NPSP) is defined as

$$p_N = p + \frac{1}{2}\rho V^2 - p_v$$

in which p_v is the saturation pressure, p is the static pressure, and V is the velocity at the pump end of the suction pipe (which is the inlet to the pump). When expressed in units of a height of water, the net positive suction pressure is called the *net positive suction head* (NPSH).

The manufacturer tests the pump and gives a value for the *required* net positive suction head, NPSH_R. This increases with the flow rate as the accelerating flow into the inlet causes the pressure to drop. The application engineer can now determine what is the minimum total head at the pump end of the suction pipe and from this determine the *actual* net positive suction head, NPSH_A. In order to avoid cavitation, $\text{NPSH}_A > \text{NPSH}_R$. In the lower half of Figure 8.18 is a curve showing values (on the ordinate on the right) for the NPSH_R as a function of the flow rate.

A suction specific speed is defined as

$$\Omega_{ss} = \frac{\Omega\sqrt{Q}}{(g\,\text{NPSH})^{3/4}}$$

For a single-flow pump a rough rule is to keep the suction specific speed under $\Omega_{ss} = 0.3$ and for a double flow, under $\Omega_{ss} = 0.4$.

■ **EXAMPLE 8.7**

A pump draws water at the rate of 20 L/s from a large reservoir open to atmosphere with pressure 101.325 kPa. As shown in Figure 8.25, the pump is situated a height $z = 4$ m above the reservoir surface. The pipe diameter is 7.6 cm, and the suction pipe is 10 m in length. The entrance loss coefficient is $K_i = 0.8$, the loss coefficient of the elbow is $K_e = 0.6$, and the pipe roughness is 45 μ m. Find the suction specific speed given a shaft speed is of 1800 rpm. The viscosity of water is $1.08 \cdot 10^{-3}$ kg/(m · s).

Solution: A control volume containing the water in the reservoir and in the suction pipe is

$$\frac{p_a}{\rho} = \frac{p_1}{\rho} + \frac{1}{2}V_p^2 + gz_1 + \left(f\frac{L_p}{D_p} + K_i + K_e\right)\frac{1}{2}V_p^2$$

so that

$$p_1 + \frac{1}{2}V_p^2 - p_v = p_a - \rho gz_1 - \left(f\frac{L_p}{D_p} + K_i + K_e\right)\frac{1}{2}\rho V_p^2 - p_v$$

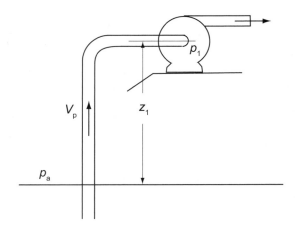

Figure 8.25 A pumping example illustrating possible cavitation.

and the positive suction head is

$$\text{NPSH} = \frac{p_a}{\rho g} - z - (f\frac{L_p}{D_p} + K_i + K_e)\frac{V_p^2}{2g} - \frac{p_v}{\rho g}$$

The velocity in the pipe is calculated as

$$V_p = \frac{Q}{A_p} = \frac{4Q}{\pi D_p^2} = \frac{4 \cdot 20}{1000 \cdot \pi \cdot 0.076^2} = 4.41 \text{ m/s}$$

and the Reynolds number has the value

$$\text{Re} = \frac{\rho V_p D_p}{\mu} = \frac{998 \cdot 4.41 \cdot 0.076}{1.08 \cdot 10^{-3}} = 309{,}620$$

The friction factor can now be calculated from Eq. (3.55) in Chapter 3. For a commercial steel pipe with roughness 0.045 mm, its value is $f = 0.0187$, and the net positive suction head is therefore

$$\text{NPSH} = \frac{101325}{998 \cdot 9.81} - 4 - \left(0.0187\frac{10}{0.076} + 0.6 + 0.8\right)\frac{4.41^2}{2 \cdot 9.81} - \frac{3782}{998 \cdot 9.81} = 2.14 \text{ m}$$

The value of the suction-specific speed becomes

$$\Omega_{ss} = \frac{\Omega\sqrt{Q}}{(g\text{NPSH})^{3/4}} = \frac{1800 \cdot \pi\sqrt{0.020}}{30(9.81 \cdot 2.14)^{3/4}} = 2.72$$

Since the suction specific speed is lower than the criterion $\Omega_{ss} = 3.0$, the pump on this basis will not experience cavitation. However, if the pump in Figure 8.18 is used, then this flow rate shows the value of $\text{NPSH}_R = 2.1$ m to be close to that calculated here, so that inception of the cavitation is close.

∎

8.8 DIFFUSER AND VOLUTE DESIGN

8.8.1 Vaneless diffuser

In the vaneless space in a flow without spin the tangential component of the velocity follows the free vortex distribution. This is a consequence of the law of conservation of angular momentum, if no moment is applied to the fluid particles. Thus

$$rV_u = r_2 V_{u2}$$

and, if the vaneless diffuser has a constant width, then, for an incompressible flow, the equation

$$\rho 2\pi r b V_r = \rho 2\pi r_2 b V_{r2}$$

reduces to

$$rV_r = r_2 V_{r2}$$

This and the condition for irrotationality $rV_u = $ constant then yields

$$rV = r_2 V_2$$

This is a special case of the general result discussed for centrifugal compressors. Since $V_u = V_r \tan \alpha$, the flow angle α remains constant.

From the flow trajectories constructed to Figure 8.26, it is easy to see that the flow angle

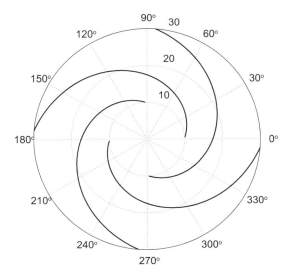

Figure 8.26 Logarithmic spiral with $\alpha = 70°$ and $10\,\text{cm} < r < 30\,\text{cm}$.

is given by

$$\tan \alpha = \frac{r d\theta}{dr}$$

which for a constant flow angle can be integrated to

$$\theta_2 - \theta_1 = \ln \frac{r_2}{r_1} \tan \alpha$$

The curve traced out is a *logarithmic spiral*. The incremental length of the path is

$$dL = \sqrt{dr^2 + r^2\, d\theta^2} = \sqrt{(1 + \tan^2 \alpha)}\, dr = \frac{dr}{\cos \alpha}$$

and integrating this gives $r_2 - r_1 = L \cos \alpha$. In the spirals shown in the Figure 8.26 $\alpha = 70°$. Therefore a spiral which starts at $r = 10$ cm and $\theta = 0$ and ends at $r = 30$ cm will have traversed and angular distance $\theta = 172.9°$. As the flow angle approaches $90°$, the length of the path increases greatly.

8.8.2 Volute design

In this section the calculations involved in the design of a volute are discussed. A schematic of a volute cross section for a centrifugal pump is shown Figure 8.27, in which the various radii are indicated. The diffuser includes a constant-width vaneless space, followed by a section with a linearly increasing gap, and then a circular volute. Although the principles for the calculation are straightforward, the details lead to complicated equations.

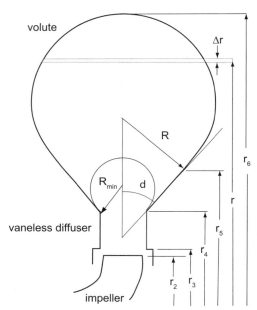

Figure 8.27 Sketch for volute design.

The side view of the pump is shown in Figure 8.28. The volute is a channel around the impeller in which the flow area increases slowly, leading to a decrease in velocity and thus an increase in pressure. The upstream section of the volute begins at a *tongue*, or *cutwater*, and the volute returns to the same location after turning $360°$. It then transitions into a conical diffuser that is connected to a high pressure delivery pipe.

The exit blade radius is labeled r_2, and the blade height is designated as b_2. The vaneless diffuser begins at radius r_3 and has a width b_3. In order to slide the impeller into the casing, radius r_3 is made slightly larger than r_2. For large pumps for which the casing is split in half, the impeller and the shaft can be lowered into place, and for a such a pump r_2 can be larger than r_3. The width of the vaneless diffuser b_3 is just a couple of millimeters larger

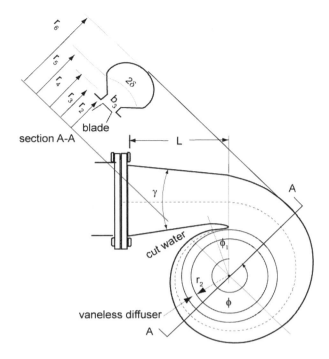

Figure 8.28 Centrifugal pump and its volute.

than the blade height b_2. For purposes of illustration, the radii r_2 and r_3 are assumed to be equal. With the radius r_3 decided, how the volute is developed depends on the design practice of each pump manufacturer. One possibility is to have the volute begin with a diffuser of trapezoidal cross section. The half-angle δ of the sidewalls and the height of the trapezoid is chosen such that a volute of circular cross section is fitted to the trapezoid in such a way that the slope of the circular section is the same as that of the sidewalls of the trapezoid at the point where they join. This is shown in Figure 8.27. The radii r_5 and r_6 increase in the flow direction in order to accommodate the increase in flow entering the volute. The calculations for this kind of design are illustrated in Wirzenius [83], and his analysis is partly repeated below.

The angle around the volute is ϕ, and it is convenient to measure this angle from the *tongue* or *lip* of the volute. The design of the tongue region requires special attention to make a smooth transition to the main part of the volute. The volute is to be designed in such a way that the pressure at the exit of the impeller is uniform and independent of ϕ. In such a situation the rate of flow into the volute is the same at every angular location, and, if Q_ϕ is the volumetric flow rate through the volute at the angle ϕ, then by simple proportionality

$$Q_\phi = \frac{\phi}{2\pi} Q$$

where Q is the total flow rate. Let V_u be the tangential component of the velocity in the volute. Then

$$Q_\phi = \int_{A_\phi} V_u \, dA = \int_{r_3}^{r_5} V_u b(r) \, dr$$

in which $b(r)$ is the volute width at the radius r.

The flow is assumed to be irrotational, and the tangential velocity therefore varies with r as
$$V_u = V_{u2}\frac{r_2}{r}$$
with V_{u2} the tangential velocity leaving the impeller. Hence
$$Q_\phi = \frac{\phi}{2\pi}Q = V_{u2}r_2 \int_{r_3}^{r_5} \frac{b(r)}{r} dr$$

Solving this for ϕ gives
$$\phi = \frac{2\pi V_{u2}r_2}{Q}\left[\int_{r_3}^{r_4} \frac{b(r)}{r} dr + \int_{r_4}^{r_5} \frac{b(r)}{r} dr + \int_{r_5}^{r_6} \frac{b(r)}{r} dr\right]$$

It remains to evaluate the integrals. To organize the work, let
$$I_1 = \int_{r_3}^{r_4} \frac{b(r)}{r} dr \qquad I_2 = \int_{r_4}^{r_5} \frac{b(r)}{r} dr \qquad I_3 = \int_{r_5}^{r_6} \frac{b(r)}{r} dr$$

For the first integral $b(r) = b_3$. Hence
$$I_1 = b_3 \int_{r_3}^{r_4} \frac{dr}{r} = b_3 \ln \frac{r_4}{r_3}$$

The channel width $b(r)$ for the linearly diverging part is given by
$$b(r) = b_3 + 2(r - r_4)\tan\delta$$

Hence
$$I_2 = \int_{r_4}^{r_5} \frac{b_3 + 2\tan\delta(r - r_4)}{r} dr = (b_3 - 2r_4 \tan\delta)\ln\frac{r_5}{r_4} + 2(r_5 - r_4)\tan\delta$$

With the aid of trigonometric relationships
$$r_5 = r_4 + R\frac{\cos\delta}{\tan\delta} - \frac{b_3}{2\tan\delta} \qquad (8.37)$$
the second integral therefore evaluates to
$$I_2 = (b_3 - 2r_4\tan\delta)\ln\left(1 + \frac{R\cos\delta}{r_4 \tan\delta} - \frac{b_3}{2r_4 \tan\delta}\right) + 2R\cos\delta - b_3$$

The radius that gives $I_2 = 0$ is the minimum radius $R_{\min} = b_3/2\cos\delta$; that is, the triangular section of the volute increases as the volute radius increases in the downstream direction. Therefore the radius r_5 also increases as R increases, whereas the radii r_4 and r_3 remain fixed, as does the angle δ.

The integral I_3 is the most complicated of the three integrals. The width of the volute at location r is given by $b(r) = 2R\sin\theta$. Also $r = r_6 - R + R\cos\theta$. Let
$$c = \frac{r_6}{R} - 1 = \frac{r_5}{R} + \sin\delta = \frac{r_4}{R} + \frac{\cos\delta}{\tan\delta} - \frac{b_3}{2R\tan\delta} + \sin\delta$$

so that
$$c = \frac{r_4}{R} + \frac{1}{\sin\delta} - \frac{b_3}{2R\tan\delta} \qquad (8.38)$$

The integral I_3 now can be written as

$$I_3 = \int_{r_5}^{r_6} \frac{b(r)}{r} dr = 2R \int_0^{\pi/2+\delta} \frac{\sin^2\theta}{\frac{r_5}{R} - 1 + \cos\theta} d\theta = 2R \int_0^{\pi/2+\delta} \frac{\sin^2\theta}{c + \cos\theta} d\theta$$

Evaluation of this leads to

$$I_3 = 2R \left[c \left(\frac{\pi}{2} + \delta \right) - \cos\delta - 2\sqrt{c^2 - 1} \tan^{-1}\left(\sqrt{\frac{c-1}{c+1}} \frac{1 + \tan(\delta/2)}{1 - \tan(\delta/2)} \right) \right]$$

Collecting the results gives the design formula

$$\phi = \frac{2\pi r_2 r_4 V_{u2}}{Q} \left\{ \frac{b_3}{r_4} \left(\ln \frac{r_4}{r_3} - \frac{1}{2} \right) \right.$$

$$+ \left(\frac{b_3}{r_4} - 2\tan\delta \right) \ln\left(1 + \frac{R\cos\delta}{r_4 \tan\delta} - \frac{b_3}{2r_4 \tan\delta} \right) + 2\frac{R}{r_4}\left[1 + c\left(\frac{\pi}{2} + \delta\right) \right.$$

$$\left. \left. - \cos\delta - 2\sqrt{c^2-1} \tan^{-1}\left(\sqrt{\frac{c-1}{c+1}} \frac{1+\tan(\delta/2)}{1-\tan(\delta/2)} \right) \right] \right\} \qquad (8.39)$$

The calculation now proceeds by starting with $R_{\min}$ and incrementing R by ΔR, so that the new value of R is $R_{\min} + \Delta R$. Next, the value of c is determined from Eq. (8.38) and r_5 is calculated from Eq. (8.37). Then r_6, the outer extent of the volute, is determined from $r_6 = (1+c)R$, but this is not needed in the calculation of angle ϕ. The value of angle ϕ is finally determined from Eq. (8.39). These values then trace out the volute and its trapezoidal base.

EXERCISES

8.1 An industrial air compressor has 29 backward-swept blades with blade angle $-21°$. The tip speed of the blades is $440\,\text{m/s}$, and the radial component of the velocity is $110\,\text{m/s}$. Air is inducted from atmospheric conditions at $101.3\,\text{kPa}$ and $298\,\text{K}$ with an axial velocity equal to $95\,\text{m/s}$. The hub-to-tip ratio at the inlet is 0.4. The total-to-total efficiency of the compressor is 0.83, and the mass flow rate is $2.4\,\text{kg/s}$. Find: (a) The total pressure ratio using the Stodola slip factor and (b) the tip radius of the impeller.

8.2 A centrifugal compressor has 23 radial vanes and an exit area equal to $0.12\,\text{m}^2$, where the radial velocity is $27\,\text{m/s}$, and the tip speed of the impeller is $350\,\text{m/s}$. The total-to-total efficiency is 0.83. (a) Find the mass flow rate of air, given that the total pressure and temperature are $101.3\,\text{kPa}$ and $298\,\text{K}$ at the inlet. (b) What is the exit Mach number? (c) If the blade height at the exit is $b = 3\,\text{cm}$ and there is no leakage flow, what is the tip radius of the impeller? (d) Find the rotational speed of the compressor wheel, and the required power neglecting mechanical losses.

8.3 A centrifugal compressor has an axial inlet and the outlet blades at an angle such that the tangential component of the exit velocity has a value equal to 0.9 times the blade speed. The outlet radius is $30\,\text{cm}$, and the desired pressure ratio is 3.5. The inlet stagnation temperature is $T_{01} = 298\,\text{K}$. If the total-to-total efficiency of the compressor is 0.8, at what angular speed does it need to be operated?

8.4 A small centrifugal compressor as a part of a turbocharger operates at $55{,}000$ rpm. It draws air from atmosphere at temperature $288\,\text{K}$ and pressure $101.325\,\text{kPa}$. The inlet

Mach number is $M_1 = 0.4$, and the flow angle of the relative velocity is $\beta_{1s} = -60°$ at the shroud. The radius ratio at the inlet is $\kappa = r_{1h}/r_{1s} = 0.43$. (a) Find the blade speed at the inlet shroud and (b) the mass flow rate. (c) If the inducer is choked, what is the throat area?

8.5 A centrifugal compressor in a turbocharger operates at 40,000 rpm and inlet Mach number $M_1 = 0.35$. It draws air from atmosphere at temperature 293 K and pressure 101.325 kPa. The radius ratio is $r_{1s}/r_2 = 0.71$, and the diffusion ratio is $W_{1s}/W_2 = 1.8$. The inlet angle of the relative velocity at the shroud is $\beta_{1s} = -63°$. The slip factor is $\sigma = 0.85$, and the flow angle at the exit is $\alpha_2 = 69°$. Find (a) the tip speed of the blade at the inlet, (b) the tip speed of the blade at the outlet, (c) the loading coefficient, and (d) the metal angle at the exit.

8.6 A small centrifugal compressor draws atmospheric air at 293 K and 101.325 kPa. At the inlet $r_{1h} = 3.2$ cm and $r_{1s} = 5$ cm. The rotor efficiency of the compressor is 0.88. The relative Mach number at the inlet shroud is 0.9 and the corresponding relative flow angle is $\beta_{1s} = -62°$. At the outlet the absolute velocity is at angle $\alpha_2 = 69°$. The diffusion ratio is $W_{1s}/W_2 = 1.8$ and the radius ratio is $r_{1s}/r_2 = 0.72$. Find, (a) the rotational speed of the shaft, (b) the blade-loading coefficient w/U_2^2, (c) the flow coefficient $\phi = V_{r2}/U_2$, and (d) the blade width at the exit.

8.7 Show that in the incompressible limit the angle of the relative velocity at the inlet is optimum at 54.7°.

8.8 Show that the expression for the dimensionless mass flow rate for a compressor with preswirl at angle α_1 is

$$\Phi_f = \frac{M_{1R}^3(\tan\alpha_1 - \tan\beta_{1s})^2 \cos^3\beta_{1s}}{\left(1 + \frac{\gamma-1}{2}M_{1R}^2 \frac{\cos^2\beta_{1s}}{\sin^2\alpha_1}\right)}$$

Plot the results for $\alpha_1 = 30°$, with β_{1s} on the abscissa and Φ_f on the ordinate, for relative Mach numbers 0.6, 0.7, 0.8. For a given mass flow rate, does the pre-swirl increase or decrease the allowable relative Mach number, and does the absolute value of the relative flow angle increase or decrease with preswirl?

8.9 Water with density 998 kg/m^3 flows through the inlet pipe of a centrifugal pump at a velocity of 6 m/s. The inlet shroud radius is 6.5 cm, and the hub radius is 5 cm. The entry is axial. The relative velocity at the exit of the impeller is 15 m/s and is directed by backward-curved impeller blades such that the exit angle of the absolute velocity is $\alpha_2 = 65°$. The impeller rotates at 1800 rpm and has a tip radius of 15 cm. Assume that the rotor efficiency of the pump is 75%. Evaluate (a) the power into the pump, (b) the increase in total pressure of the water across the impeller, and (c) the change in static pressure of the water between the inlet and outlet of the impeller. (d) What is the ratio of the change in kinetic energy of the water across the impeller to the total enthalpy of the water across the pump, the change in the relative kinetic energy, and the change in the kinetic energy owing to the centrifugal effect as a fraction of work done? (e) If the velocity at the exit of the volute is 6 m/s, what is the ratio of change in static pressure across the rotor to the change in static pressure across the entire pump?

8.10 A centrifugal-pump that handles water operates with backward-curving blades. The angle between the relative velocity and the tip section is 45°. The radial velocity at the tip section is 4.5 m/s, the flow at the inlet is axial, and the impeller rotational speed is 1800 rpm. Assume that there is no leakage and that the mechanical friction may be neglected, and that the total-to-total efficiency is 70%. (a) Construct the velocity diagram at the impeller exit and (b) evaluate the required tip radius for a water pressure rise of 600 kPa across the pump. (c) For the total pressure rise of 600 kPa, evaluate the difference between the total and static pressure of water at the impeller tip section.

8.11 A centrifugal water pump has an impeller diameter $D_2 = 27$ cm, and when its shaft speed is 1750 rpm, it produces a head $H = 33$ m. Find, (a) the volumetric flow rate, (b) the blade height at the exit of the impeller assuming that there are no leakage losses, and (c) the blade angle at the exit of the impeller, given that it has 11 blades.

8.12 A centrifugal water pump has an impeller diameter of $D_2 = 25$ cm, and when its shaft speed is 1750 rpm, it delivers 20 L/s of water. Find, (a) the head of water delivered by the pump, and (b) the power needed to drive the pump. (c) The impeller has nine blades. Use the Stanitz slip factor to find the blade angle at the exit of the impeller. (d) Use the Wiesner slip factor to find the blade exit blade angle.

8.13 A centrifugal pump delivers water at 0.075 m^3/s with a head of 20 m while operating at 880 rpm. The hub-to-shroud radius ratio at the inlet is 0.35, and the relative velocity makes an angle of $-52°$ at the inlet. (a) Find the reversible work done by the pump. (b) What is the work done by the impeller? (c) Find the impeller radius and the inlet radius of the shroud. (d) Determine the blade width at the exit of the impeller. (e) Assume a reasonable number of blades, and calculate the blade angle at the exit. Use the Pfleiderer equation to determine more accurately the number of blades and recalculate the blade angle at the exit if needed. (f) What is the power required to drive the pump?

8.14 A fan draws in atmospheric air at 0.4 m^3/s at pressure 101.32 kPa and temperature 288 K. The total pressure rise across the fan, which has 30 radial blades, is 2.8 cm of water. The inner radius is 14.8 cm, and outer radius is 17.0 cm. The rotational speed of the fan is 980 rpm and the hydraulic efficiency is the fan is 0.78. Use the Stanitz slip factor. (a) Assuming that velocity into the fan is radially outward, find the angle of the relative velocity at the shroud at the inlet. (b) Determine the power to the fan, given that 4% is lost to mechanical friction. (c) Find the angle blade angle at the exit.

8.15 A pump draws water at the rate of 75 L/s from a large tank with the air pressure above the free surface at 98.00 kPa. The pump is $z = 2$ m above the water level in the tank. The pipe diameter is 14.0 cm, and the suction pipe is 20 m in length. The entrance loss coefficient is $K_i = 0.2$, and the loss coefficient of the elbow is $K_e = 0.6$ and the pipe roughness is 55 μm. Find the suction specific speed given a shaft speed of 1800 rpm. The viscosity of water is $1.08 \cdot 10^{-3}$ kg/(m·s).

8.16 Consider a volute consisting only a of circular section in which the tangential velocity varies as $V_u = K/r$, and r is the location from the center of the volute to a location on the circular section. (a) Show that the value of K in terms of the volumetric flow rate Q, the radius of the circular section R, and the radius to the center of the of the circular section a, is given by

$$K = \frac{Q}{2\pi R(\lambda - \sqrt{\lambda^2 - 1})}$$

in which $\lambda = a/R$. (b) For $R = 0.5$ m, $a = 2$ m, and $Q = 1.5$ m^3/s, find the pressure difference $p_2 - p_1$ at the centerline, between the outside and inside edges of the section.

CHAPTER 9

RADIAL INFLOW TURBINES

The best known use of radial inflow turbines is in automobile turbochargers, but they also appear as auxiliary power turbines and, for example, in turboprop aircraft engines. They are used in processing industries (including refineries), natural-gas processing, air liquefaction, and geothermal energy production. In the automotive application burned gases from the engine exhaust manifold are directed into a radial inflow turbine of the turbocharger, which powers a centrifugal compressor on the same axis. The compressor, in turn, increases the pressure and density of the supply air to the engine. As the engine speed may change quite rapidly, turbochargers must respond to the changing operating conditions nimbly. Therefore they are made light in weight and low in inertia.

A sketch on the left side in Figure 9.1 shows a side view of a radial inflow turbine. It looks like a centrifugal compressor, but with a reversed flow direction. Hot gases enter through a volute and move into a vaned stator, which redirects them into a vaneless space and then into the rotor. On the right is a front view of the rotor. The velocity diagrams at the inlet and exit are shown in Figure 9.2. For best efficiency, the inflow angle β_2 is negative. But the blades at the inlet are typically radial, which means that the entering flow is at a negative incidence. Since the blades operate in a high-temperature environment and material strength diminishes as temperature increases, radial blades can withstand the imposed loads better than curved blades. It is for this reason that the blade angle at the inlet is set at $\chi_2 = 0$. The rotor turns the flow toward the axis as it passes through the flow channel, so that its radial velocity is zero at the exit and the absolute velocity is axial. For this reason these machines are also called 90° *inward-flow radial* (IFR) turbines.

Principles of Turbomachinery. By Seppo A. Korpela
Copyright © 2011 John Wiley & Sons, Inc.

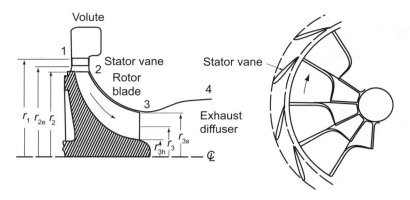

Figure 9.1 Radial inflow turbine.

9.1 TURBINE ANALYSIS

The velocity diagrams in Figure 9.2 are similar to those for centrifugal compressors, and at the exit the absolute velocity is axial, as it is at the inlet of a compressor. Work delivered

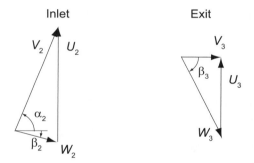

Figure 9.2 Velocity diagrams for a radial inflow turbine.

by the turbine, if written in terms of kinetic energies, is given by

$$w = \frac{1}{2}(V_2^2 - V_3^2) + \frac{1}{2}(U_2^2 - U_3^2) + \frac{1}{2}(W_3^2 - W_2^2) \tag{9.1}$$

Examination of this equation shows that increasing the inlet velocity V_2 increases the work. This is achieved by orienting the stator blades such that the flow enters the rotor at a large nozzle angle α_2. Similarly, a small value for W_2 increases the work, and this is obtained by directing the relative velocity radially inward at the inlet. The same reasoning leads to a design in which the exit velocity V_3 is axial and therefore as small as possible and in which the relative velocity W_3 is large. This can be obtained by making the magnitude of the flow angle of the relative velocity $|\beta_3|$ large. Finally, a large U_2 and a small U_3 increase the work delivered.

The usual expression for work

$$w = U_2 V_{u2} - U_3 V_{u3} \tag{9.2}$$

leads to same conclusions, namely, V_{u3} should be small and V_{u2} large. Similarly, U_2 should be large and U_3 small, which means that the ratio r_2/r_3 ought to be reasonably large. Work could be increased by making V_{u3} negative, but this would also increase the absolute value of β_3, and the exit relative Mach number might become so large as to cause the flow to choke. If the exit swirl is eliminated, the expression for work becomes

$$w = U_2 V_{u2}$$

When this equation is compared to Eq. (9.1) it is clear that the terms involving the exit station in that equation must cancel.

Since the blade speed $U_3 = r_3\Omega$ is smallest at the hub, if the exit velocity V_3 is designed to be uniform across the exit plane, then the flow angle β_{3s} and therefore also the blade angle must be turned more at the shroud than at the hub. This means that the relative velocity and the relative Mach number M_{3R} at the shroud will be the largest on the exit plane. They must be kept sufficiently small to prevent choking.

The equation for work delivered by a turbine

$$w = c_p(T_{01} - T_{03})$$

when written in the form

$$\frac{w}{c_p T_{01}} = 1 - \frac{T_{03}}{T_{01}} = s_\text{w}$$

defines a nondimensional specific work intensity $s_\text{w} = w/c_p T_{01} = \dot{W}/\dot{m} c_p T_{01}$, also known as a *power ratio*. If the exit kinetic energy is wasted (as is often the case), it is appropriate to use the total-to-static efficiency as the proper measure of efficiency. It is defined as

$$\eta_\text{ts} = \frac{T_{01} - T_{03}}{T_{01} - T_{3ss}} = \frac{1 - \dfrac{T_{03}}{T_{01}}}{1 - \dfrac{T_{3ss}}{T_{01}}}$$

which is clearly also

$$\eta_\text{ts} = \frac{1 - \dfrac{T_{03}}{T_{01}}}{1 - \left(\dfrac{p_3}{p_{01}}\right)^{(\gamma-1)/\gamma}}$$

Solving this for the pressure ratio gives

$$\frac{p_{01}}{p_3} = \left(1 - \frac{s_\text{w}}{\eta_\text{ts}}\right)^{-\gamma/(\gamma-1)} \tag{9.3}$$

Graphs of the pressure ratio as a function of s_w are shown in Figure 9.3. Power ratios in the range $0.15 < s_\text{w} < 0.25$ correspond to pressure ratios in the range of $2 < p_{01}/p_3 < 3$ for typical values of efficiency. From the pressure ratio, and an estimate of the efficiency, the power ratio can be calculated.

The expression for work can be written as

$$w = c_p T_{01} s_\text{w}$$

and with a typical power ratio $s_\text{w} = 0.2$ and inlet stagnation temperature $T_{01} = 1000\,\text{K}$, the specific work is $w = 229.60\,\text{kJ/kg}$ in expansion of combustion gases with $c_p = 1148\,\text{J/(kg}\cdot\text{K)}$. Thus the stagnation temperature drop is 200 K. If the relative flow is

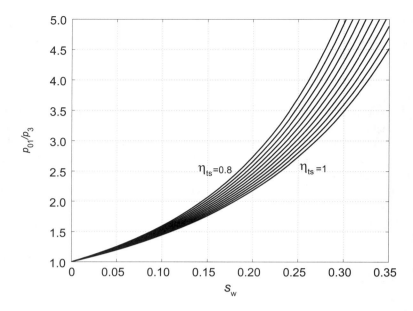

Figure 9.3 Pressure ratio for a gas with $\gamma = 1.4$ as a function of power ratio, or total-to-static efficiency $\eta_{ts} = 0.8$ to $\eta_{ts} = 1.0$ with adjacent graphs incremented by 0.02.

radially inward and there is no exit swirl, the Euler turbine equation shows that $w = U_2^2$, so that a typical blade speed is $479 \, \text{m/s}$. The size of the machine depends on the mass flow rate. Thus, as a rough estimate, a turbocharger operating at shaft speed of 40,000 rpm gives an inlet radius of about 10 cm, and one operating at 200,000 rpm gives a radius of 2 cm. A turbocharger from an automobile and with a rotor diameter of approximately 3 cm is shown in Figure 9.4.

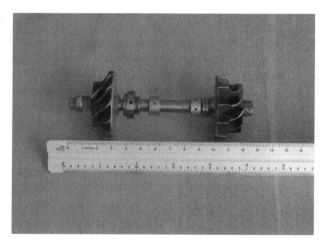

Figure 9.4 An automotive turbocharger.

At the exit, in addition to being uniform, the axial velocity V_3 should be small, so that the exit kinetic energy is small. Whatever exit kinetic energy is left in the exit stream may be recovered in an exit diffuser. Ideally the exit diffuser would reduce the velocity to zero as the flow enters the atmosphere. For isentropic flow the work delivered in this situation is

$$w_s = h_{01} - h_{3ss} = \frac{1}{2}V_0^2 \qquad (9.4)$$

and the quantity V_0 is called a *spouting velocity*. The kinetic energy associated with the spouting velocity is a convenient replacement for the maximum work that this turbine can deliver. This equation can also be interpreted as defining what velocity would be reached in a frictionless *nozzle* as the flow expands from pressure p_{01} to the exit static pressure p_3. The equation for work can now be written as

$$\frac{1}{2}V_0^2 = c_p T_{01}\left(1 - \frac{T_{3ss}}{T_{01}}\right)$$

or as

$$\frac{V_0^2}{2c_p T_{01}} = 1 - \left(\frac{p_3}{p_{01}}\right)^{(\gamma-1)/\gamma}$$

If the relative flow entering the turbine is radial and there is no exit swirl, then, the definition of total-to-static efficiency can be written as

$$\eta_{ts} = \frac{w}{w_s} = \frac{2U_2^2}{V_0^2} \qquad (9.5)$$

The largest value for U_2/V_0 according to this equation is 0.707, but since the highest efficiency is obtained with β_2 in the range from $-20°$ to $-40°$, and the nozzle angle is typically $\alpha_2 = 70°$, this result needs some modification and a typical range for this ratio is $0.55 < U_2/V_0 < 0.77$.

Since the isentropic work is given by Eq. (9.4), the value of the spouting velocity gives a way to calculate an initial estimate for the blade speed. With the value of blade speed U_2 known, the magnitude of the stresses can then be calculated. A compact design calls for a large shaft speed.

■ **EXAMPLE 9.1**

A radial inflow turbine with radial blades at the inlet operates at 62,000 rpm. Its inlet diameter is $D_2 = 12.6$ cm. The gases enter the blades radially and leave without exit swirl. The supply temperature is $T_{01} = 1150$ K, the pressure ratio of the turbine is $p_{01}/p_3 = 2$, and the mass flow rate is $\dot{m} = 0.31$ kg/s. The ratio of specific heats is $\gamma = 1.35$ and the gas constant is $R = 287$ J/(kg · K). Find (a) the ratio U_2/V_0, in which V_0 is the spouting velocity, (b) the total-to-static efficiency, and (c) the power delivered by the turbine.

Solution: (a) The specific heat is

$$c_p = \frac{\gamma R}{\gamma - 1} = \frac{1.35 \cdot 287}{0.35} = 1107 \text{ J/(kg · K)}$$

and the isentropic static temperature at the exit is

$$T_{3ss} = T_{01}\left(\frac{p_3}{p_{01}}\right)^{(\gamma-1)/\gamma} = 1150 \cdot 0.5^{0.35/1.35} = 960.84 \text{ K}$$

The isentropic work is therefore

$$w_s = c_p(T_{01} - T_{3ss}) = 1107(1150 - 960.84) = 209.40\,\text{kJ/kg}$$

and the spouting velocity is

$$V_0 = \sqrt{2w_s} = \sqrt{2 \cdot 209{,}400} = 647.1\,\text{m/s}$$

The blade speed at the inlet is

$$U_2 = r_2 \Omega = \frac{12.6 \cdot 62{,}000 \cdot \pi}{2 \cdot 100 \cdot 30} = 409.0\,\text{m/s}$$

so that the ratio U_2/V_0 becomes 0.632.

(b) Since there is no exit swirl, the work becomes

$$w = U_2^2 = 167.31\,\text{kJ/kg}$$

and the stagnation temperature drop amounts to 151°C. The total-to-static efficiency comes out to be

$$\eta_{ts} = \frac{w}{w_s} = \frac{167.31}{209.40} = 0.799$$

(c) The power delivered is

$$\dot{W} = \dot{m}w = 0.31 \cdot 167.31 = 51.87\,\text{kW}$$

■

The next example gives an analysis for a choked rotor passage.

■ **EXAMPLE 9.2**

Combustion gases with $\gamma = 1.35$, $c_p = 1107\,\text{J}/(\text{kg}\cdot\text{K})$ and $R = 287\,\text{J}/(\text{kg}\cdot\text{K})$ flow from conditions $p_{01} = 390\,\text{kPa}$ and $T_{01} = 1150\,\text{K}$ through a radial inflow turbine to an exit pressure $p_3 = 100\,\text{kPa}$. The total-to-static efficiency is $\eta_{ts} = 0.8$, and the flow leaving the stator is choked with $M_2 = 1$. (a) Find the work delivered by the turbine, given that the relative velocity at the inlet of the rotor is radial and that flow leaves without swirl. (b) Find the angle of the absolute velocity at the inlet of the rotor.

Solution: Since the stator is choked, the static temperature at the exit of the stator is

$$T_2 = \frac{2}{\gamma + 1}T_{02} = 978.7\,\text{K}$$

and the velocity is

$$V_2 = \sqrt{\gamma R T_2} = \sqrt{1.35 \cdot 287 \cdot 978.7} = 615.8\,\text{m/s}$$

Solving the definition of total-to-static efficiency

$$\eta_{ts} = \frac{T_{01} - T_{03}}{T_{01} - T_{3ss}}$$

for the exit stagnation temperature gives

$$\frac{T_{03}}{T_{01}} = 1 - \eta_{ts}\left[1 - \left(\frac{p_{01}}{p_3}\right)^{-(\gamma-1)/\gamma}\right]$$

so that
$$T_{03} = 1150\left[1 - 0.8\left(1 - 3.9^{-0.35/1.35}\right)\right] = 876.5\,\text{K}$$

Work delivered by the turbine is therefore
$$w = c_p(T_{01} - T_{03}) = 1107(1150 - 876.5) = 302.80\,\text{kJ/kg}$$

This is also
$$w = U_2^2$$

so that
$$U_2 = \sqrt{w} = \sqrt{302,800} = 550.27\,\text{m/s}$$

The flow angle at the inlet to the rotor is
$$\alpha_2 = \sin^{-1}\left(\frac{U_2}{V_2}\right) = \sin^{-1}(0.893) = 63.3°$$

■

The foregoing two examples assumed that the relative velocity at the inlet to the rotor is radial and that there is no exit swirl. In the next section a general development based on this assumption is carried out in order to relate the important flow parameters to the total-to-static efficiency.

9.2 EFFICIENCY

The total-to-static efficiency of a turbine is

$$\eta_{ts} = \frac{h_{01} - h_{03}}{h_{01} - h_{3ss}}$$

with the thermodynamic states as shown in Figure 9.5. This can be written as

$$\frac{1}{\eta_{ts}} - 1 = \frac{h_{03} - h_{3ss}}{h_{01} - h_{03}}$$

If the turbine is fitted with a diffuser and its exit pressure is p_4, then, if the flow exhausts to the atmosphere, the pressure p_4 is the atmospheric pressure and the small amount of residual kinetic energy is lost into the atmosphere. In this situation the pressure p_3 is below the atmospheric value. If the flow through this diffuser were reversible, then the pressure p_4 would correspond to the stagnation pressure p_{03}. In this case the appropriate efficiency to use is the total-to-total efficiency. On the other hand, if the flow is not diffused and the flow is exhausted to the atmosphere directly, then the exit pressure p_3 is the atmospheric pressure and the residual kinetic energy is lost. In this case the efficiency is the total-to-static efficiency. The numerator can be rewritten as

$$h_{03} - h_{3ss} = \frac{1}{2}V_3^2 + h_3 - h_{3s} + h_{3s} - h_{3ss}$$

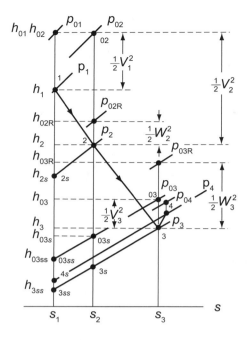

Figure 9.5 Velocity triangles and Mollier chart for a radial inflow turbine.

Next, integrating the Gibbs equation along the constant pressure lines p_2 and p_3 gives

$$\frac{T_2}{T_{2s}} = \frac{T_{3s}}{T_{3ss}} \quad \text{thus} \quad \frac{T_{3ss}}{T_{2s}} = \frac{T_{3s}}{T_2}$$

Adding minus one to both sides of the first of these equations, multiplying by c_p, rearranging, and using the second equation, gives

$$h_{3s} - h_{3ss} = \frac{T_{3ss}}{T_{2s}}(h_2 - h_{2s}) = \frac{T_{3s}}{T_2}(h_2 - h_{2s})$$

As a consequence the expression for efficiency is

$$\frac{1}{\eta_{ts}} - 1 = \frac{V_3^2}{2w} + \frac{\zeta_R W_3^2}{2w} + \frac{T_{3s}}{T_2}\frac{\zeta_S V_2^2}{2w}$$

in which loss coefficients

$$h_3 - h_{3s} = \frac{1}{2}\zeta_R W_3^2 \qquad h_2 - h_{2s} = \frac{1}{2}\zeta_S V_2^2$$

have been substituted for the internal heating terms. Making use of the relationships $w = U_2^2$, $U_2 = V_2 \sin\alpha_2$, $U_3 = -V_3 \tan\beta_3$, $U_3 = -W_3 \sin\beta_3$, and $U_3 = r_3 U_2/r_2$ turns this equation into

$$\frac{1}{\eta_{ts}} - 1 = \frac{1}{2}\left[\left(\frac{r_3}{r_2}\right)^2 (\cot^2\beta_3 + \zeta_R \csc^2\beta_3) + \frac{T_{3s}}{T_2}\zeta_S \csc^2\alpha_2\right] \qquad (9.6)$$

The temperature ratio in this expression can be written in a form from which it is easy to calculate. First, it is recast as

$$\frac{T_{3s}}{T_2} = 1 - \frac{1}{T_2}(T_2 - T_3 + T_3 - T_{3s})$$

in which temperature difference $T_2 - T_3$ is obtained from

$$w = U_2^2 = h_{02} - h_{03} = c_p(T_2 - T_3) + \frac{1}{2}V_2^2 - \frac{1}{2}V_3^2$$

Solving this for $T_2 - T_3$ gives

$$T_2 - T_3 = \frac{1}{c_p}\left(U_2^2 - \frac{1}{2}V_2^2 + \frac{1}{2}V_3^2\right)$$

But the from the velocity diagrams in Figure 9.2, since $\beta_2 = 0$ and $\alpha_3 = 0$, it is seen that

$$V_2^2 = W_2^2 + U_2^2 \qquad W_3^2 = V_3^2 + U_3^2$$

so that

$$T_2 - T_3 = \frac{1}{2c_p}(W_3^2 - W_2^2 + U_2^2 - U_3^2)$$

and therefore

$$\frac{T_{3s}}{T_2} = 1 - \frac{1}{2c_p T_2}(W_3^2 - W_2^2 + U_2^2 - U_3^2 + \zeta_R W_3^2)$$

in which $\zeta_R = 2c_p(T_3 - T_{3s})/W_3^2$ has been used. This can now be written as

$$\frac{T_{3s}}{T_2} = 1 - \frac{1}{2c_p T_2}(U_3^2 \csc^2\beta_3 - U_2^2 \cot^2\alpha_2 + U_2^2 - U_3^2 + \zeta_R U_3^2 \csc^2\beta_3)$$

and, with $U_3 = U_2 r_3/r_2$, as

$$\frac{T_{3s}}{T_2} = 1 - \frac{U_2^2}{2c_p T_2}\left[1 + \left(\frac{r_3}{r_2}\right)^2\left((1 + \zeta_R)\csc^2\beta_3 - 1\right) - \cot^2\alpha_2\right]$$

From

$$T_2 = T_{02} - \frac{1}{2c_p}V_2^2 \qquad 2c_p T_2 = 2c_p T_{02} - U_2^2 \csc^2\alpha_2$$

so that

$$\frac{U_2^2}{2c_p T_2} = \left(\frac{T_{02}}{T_2} - 1\right)\sin^2\alpha_2 = \frac{\gamma - 1}{2}M_2^2 \sin^2\alpha_2$$

and it follows that the temperature ratio can also be written as

$$\frac{T_{3s}}{T_2} = 1 - \frac{\gamma - 1}{2}M_2^2 \sin^2\alpha_2\left[1 + \left(\frac{r_3}{r_2}\right)^2\left((1 + \zeta_R)\csc^2\beta_3 - 1\right) - \cot^2\alpha_2\right]$$

The value of ζ_S is measured in a stationary test apparatus, and the total-to-static efficiency can be measured from the overall balance for the turbine [15]. The value of ζ_R can then be determined from the theory developed above. These calculations are illustrated in the following example.

EXAMPLE 9.3

In a radial-inflow turbine, combustion gases, with $\gamma = \frac{4}{3}$ and $c_p = 1148\,\text{J}/(\text{kg} \cdot \text{K})$, leave the stator at the angle $\alpha_2 = 67°$. The rotor blades at the inlet are radial, with a radius $r_2 = 5.8\,\text{cm}$. At the outlet the shroud radius of the blade is $r_{3s} = 4.56\,\text{cm}$, the hub-to-shroud radius ratio is $\kappa = 0.35$, and the relative flow makes an angle $\beta_3 = -38°$ with the axial direction at the exit. The relative velocity at the inlet to the rotor is radial. The power delivered by the turbine is $\dot{W} = 58.2\,\text{kW}$ when the mass flow rate is $\dot{m} = 0.34\,\text{kg/s}$ and the rotational speed is 64,000 rpm. The stagnation temperature and pressure at the inlet to the stator are $T_{01} = 1100\,\text{K}$ and $p_{01} = 2.5\,\text{bar}$. The static enthalpy loss coefficient for the flow across the stator is $\zeta_S = 0.08$. The outlet static pressure is $p_3 = 1\,\text{bar}$. Find (a) the Mach number at the inlet of the rotor, (b) the total-to-static efficiency, and (c) the static enthalpy loss coefficient of the rotor.

Solution: (a) The blade speed at the inlet is

$$U_2 = r_2 \Omega = \frac{0.058 \cdot 64000 \cdot \pi}{30} = 388.72\,\text{m/s}$$

and since the relative flow is radial, $V_{u2} = U_2$ and

$$V_2 = \frac{V_{u2}}{\sin \alpha_2} = \frac{388.72}{\sin(67°)} = 422.29\,\text{m/s}$$

Since $T_{02} = T_{01}$, the static temperature at the inlet is

$$T_2 = T_{02} - \frac{V_2^2}{2c_p} = 1100 - \frac{422.29^2}{2 \cdot 1148} = 1022.3\,\text{K}$$

and the Mach number becomes

$$M_2 = \frac{V_2}{\sqrt{\gamma R T_2}} = \frac{422.29}{\sqrt{1.333 \cdot 287 \cdot 1022.3}} = 0.675$$

(b) The specific work is given by

$$w = \frac{\dot{W}}{\dot{m}} = \frac{58200}{0.34} = 171.18\,\text{kJ/kg}$$

and the stagnation temperature at the outlet is therefore

$$T_{03} = T_{01} - \frac{w}{c_p} = 1100 - \frac{171.18}{1.148} = 950.9\,\text{K}$$

In an isentropic process to the exit pressure the static temperature at the exit is

$$T_{3ss} = T_{01} \left(\frac{p_3}{p_{01}}\right)^{(\gamma-1)/\gamma} = 1100 \left(\frac{1}{2.5}\right)^{0.25} = 874.8\,\text{K}$$

and the total-to-static efficiency is therefore

$$\eta_{ts} = \frac{T_{01} - T_{03}}{T_{01} - T_{3ss}} = \frac{1100 - 950.9}{1100 - 874.8} = 0.662$$

(c) If in the expression for efficiency the temperature ratio T_{3s}/T_2 is set to 1, then the static enthalpy loss coefficient ζ_R can be solved from

$$\frac{1}{\eta_{ts}} - 1 = \frac{1}{2\tan^2\beta_3}\left(\frac{r_3}{r_2}\right)^2 + \frac{\zeta_R}{2\sin^2\beta_3}\left(\frac{r_3}{r_2}\right)^2 + \frac{\zeta_S}{2\sin^2\alpha_2}$$

With $\zeta_S = 0.08$, $\beta_3 = -38°$, and $r_3 = (r_{3s} + r_{3h})/2 = 0.0308$ m, solving this equation gives $\zeta_R = 0.6256$. Now the temperature ratio is calculated as

$$\frac{T_{3s}}{T_2} = 1 - \frac{\gamma-1}{2}M_2^2\sin^2\alpha_2\left[1+\left(\frac{r_3}{r_2}\right)^2\left((1+\zeta_R)\csc^2\beta_3 - 1\right) - \cot^2\alpha_2\right] = 0.8876$$

Substituting this into Eq. (9.6) and repeating the calculation with the new temperature ratio gives $\zeta_R = 0.6399$. The static loss coefficient for the rotor is larger than that for the stator, not only because the flow path is long but also because the flow undergoes a great deal of turning through the rotor. ∎

9.3 SPECIFIC SPEED AND SPECIFIC DIAMETER

The specific speed, defined as

$$\Omega_s = \frac{\Omega\sqrt{Q_3}}{w_s^{3/4}}$$

characterizes the shape of a turbomachine, and a machine with a small specific speed has a low flow rate and a large specific work. A related quantity is the specific diameter given by

$$D_s = \frac{Dw_s^{1/4}}{\sqrt{Q_3}}$$

It is small in axial turbines because the flow rate is large and the work per stage is relatively small. For radial inflow turbines this quantity is large because the work per stage is large and the volumetric flow rate is rather small. The product of specific speed and specific diameter is

$$\Omega_s D_s = \frac{D\Omega}{\sqrt{w_s}} = 2^{3/2}\frac{U_2}{V_0}$$

When the work is given by $w = U_2^2$, this reduces to

$$\Omega_s D_s = 2\sqrt{\eta_{ts}} \tag{9.7}$$

Balje [5] has constructed a large number of diagrams with specific speed on the abscissa and specific diameter as the ordinate. A Balje diagram for radial inflow turbines is shown in Figure 9.6, and the region of highest efficiency falls into the range $0.2 < \Omega_s < 0.8$. The corresponding specific diameter can be calculated according to Eq. (9.7). Lines of constant r_3/r_2 are also drawn with an optimum value near 0.7. The recommended range is $0.53 < r_3/r_2 < 0.66$. The slight discrepancy in this range and the optimum value 0.7

arises from different loss model being used to calculate the results and scant experimental data to verify them. Here the radius

$$r_3 = \sqrt{\frac{1}{2}(r_{3h}^2 + r_{3s}^2)}$$

is used to define the mean exit radius. The line $\Omega_s D_s = 2$ corresponding to $\eta_{ts} = 1$ has been drawn into the Balje diagram as well.

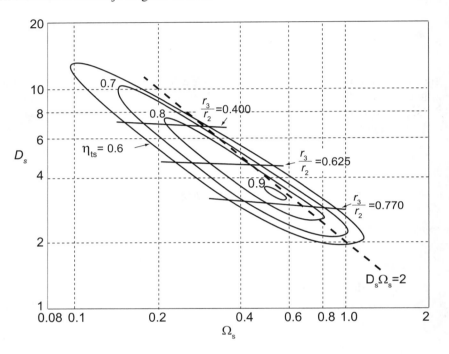

Figure 9.6 A Balje diagram for a radial inflow turbine. (Drawn after Balje [5].)

■ **EXAMPLE 9.4**

Combustion gases, with $\gamma = \frac{4}{3}$, $R = 287\,\mathrm{J/(kg \cdot K)}$, and $c_p = 1148\,\mathrm{J/kg\,K}$, flow through a radial inflow turbine. It is designed to deliver $152\,\mathrm{kW}$ of power. The exit pressure is atmospheric at $p_3 = 101.325\,\mathrm{kPa}$. The stagnation temperature to the turbine is $T_{01} = 960\,\mathrm{K}$, and the loss coefficient of the stator is $\zeta_s = 0.08$. The ratio of the blade radii at the exit is $\kappa = r_{3h}/r_{3s} = 0.35$. The rotor speed is 42,000 rpm, the blade speed is $U_2 = 440\,\mathrm{m/s}$, and the flow angle to the inlet of the rotor is $\alpha_2 = 72.1°$. Use a design point from the Balje diagram with $D_s = 2.6$, $\Omega_s = 0.71$ and $\eta_{ts} = 0.79$, and choose $r_{3s} = 0.75 r_2$. Find (a) the mass flow rate, (b) the inlet Mach number, and (c) the blade width at the inlet. Assuming that the absolute velocity is uniform at the exit, find (d) the relative Mach number at the exit at the shroud and the hub, and (e) the ratios W_{3s}/W_2 and W_{3h}/W_2.

Solution:
(a) The spouting velocity is first determined from the ratio

$$\frac{U_2}{V_0} = \frac{\Omega_s D_s}{2^{3/2}} = \frac{2.6 \cdot 0.71}{2^{3/2}} = 0.6527$$

so that

$$V_0 = \frac{440}{0.6527} = 674.2 \, \text{m/s}$$

The specific work is

$$w = U_2^2 = 440^2 = 193.60 \, \text{kJ/kg}$$

and the mass flow rate is therefore

$$\dot{m} = \frac{\dot{W}}{w} = \frac{152.00}{193.60} = 0.785 \, \text{kg/s}$$

(b) The inlet radius of the rotor has the value

$$r_2 = \frac{U_2}{\Omega} = \frac{440 \cdot 30}{42{,}000 \cdot \pi} = 0.100 \, \text{m}$$

and the relative and absolute velocities at the inlet are

$$W_2 = U_2 \cot \alpha_2 = 440 \cot(72.1°) = 142.1 \, \text{m/s}$$

$$V_2 = \frac{U_2}{\sin \alpha_2} = \frac{440}{\sin(72.1°)} = 462.38 \, \text{m/s}$$

The inlet Mach number is obtained by first calculating the static temperature at the inlet

$$T_2 = T_{02} - \frac{V_2^2}{2c_p} = 960 - \frac{462.38^2}{2 \cdot 1148} = 866.9 \, \text{K}$$

and then the Mach number at the inlet is

$$M_2 = \frac{V_2}{\sqrt{\gamma R T_2}} = \frac{462.38}{\sqrt{1.333 \cdot 287 \cdot 866.9}} = 0.803$$

(c) From the definition of the stator loss coefficient the isentropic inlet static temperature is

$$T_{2s} = T_2 - \zeta_S \frac{V_2^2}{2c_p} = 866.9 - 0.08 \frac{462.38^2}{2 \cdot 1148} = 859.4 \, \text{K}$$

and the isentropic static temperature at the exit is

$$T_{3ss} = T_{01} - \frac{V_0^2}{2c_p} = 960 - \frac{674.2^2}{2 \cdot 1148} = 762.0 \, \text{K}$$

hence the inlet stagnation pressure to the stator has the value

$$p_{01} = p_3 \left(\frac{T_{01}}{T_{3ss}} \right)^{\gamma/(\gamma-1)} = 101.325 \left(\frac{960}{762} \right)^4 = 255.2 \, \text{kPa}$$

At the inlet to the rotor the static pressure is

$$p_2 = p_{01}\left(\frac{T_{2s}}{T_{02}}\right)^{\gamma/(\gamma-1)} = 255.2\left(\frac{859.4}{960.0}\right)^4 = 163.9\,\text{kPa}$$

so that the density at the inlet becomes

$$\rho_2 = \frac{p_2}{RT_2} = \frac{163.9}{0.287 \cdot 866.9} = 0.659\,\text{kg/m}^3$$

The blade height can now be determined from the mass balance

$$b_2 = \frac{\dot{m}}{2\pi r_2 \rho_2 V_2 \cos\alpha_2} = \frac{0.785}{\pi \cdot 0.020 \cdot 0.659 \cdot 462.4 \cos(72.1°)} = 0.0133\,\text{m}$$

The ratio of the inlet blade height to the inlet radius is $b_2/r_2 = 0.133$.

(d) The rotor shroud and hub radii are

$$r_{3s} = 0.75 r_2 = 0.75 \cdot 0.1 = 0.075\,\text{m} \qquad r_{3h} = \kappa r_{3s} = 0.35 \cdot 0.075 = 0.0263\,\text{m}$$

When the Balje diagram is used, the average exit radius is its root mean square value

$$r_3 = \sqrt{\frac{1}{2}(r_{3h}^2 + r_{3s}^2)} = 0.0562\,\text{m}$$

The blade speed at the mean radius is therefore

$$U_3 = r_3 \Omega = \frac{0.0562 \cdot 42{,}000 \cdot \pi}{30} = 247.2\,\text{m/s}$$

At the shroud and the hub the blade speeds are

$$U_{3s} = r_{3s}\Omega = \frac{0.075 \cdot 42{,}000 \cdot \pi}{30} = 330.0\,\text{m/s}$$

and

$$U_{3h} = U_{3s}\frac{r_{3h}}{r_{3s}} = 330\frac{0.0263}{0.075} = 115.5\,\text{m/s}$$

To calculate the exit velocity, the exit Mach number is needed. First, the exit stagnation temperature is found to be

$$T_{03} = T_{02} - \frac{w}{c_p} = 960 - \frac{193{,}600}{1148} = 791.4\,\text{K}$$

Rearranging the mass balance

$$\dot{m} = \rho_3 V_3 A_3 = \frac{p_3 A_3}{RT_3} M_3 \sqrt{\gamma R T_3}$$

leads to

$$\frac{\dot{m}}{p_3 A_3}\sqrt{\frac{RT_{03}}{\gamma}} = M_3\left(1 + \frac{\gamma-1}{2}M_3^2\right)^{1/2}$$

which on squaring can be written as a quadratic equation in M_3^2

$$M_3^4 + \frac{2}{\gamma-1}M_3^2 - \frac{2RT_{03}\dot{m}^2}{\gamma(\gamma-1)p_3^2 A_3^2} = 0$$

and after numerical values are inserted this reduces to

$$M_3^4 + 6M_3^2 - 0.2548 = 0$$

Solution of this is $M_3 = 0.205$.

(e) The exit velocity can now be determined by first calculating the static temperature at the exit

$$T_3 = T_{03}\left(1 + \frac{\gamma-1}{2}M_3^2\right)^{-1} = 785.8\,\text{K}$$

which gives the exit velocity

$$V_3 = M_3\sqrt{\gamma R T_3} = 0.205\sqrt{1.333 \cdot 287 \cdot 785.8} = 112.6\,\text{m/s}$$

Since the absolute velocity is axial at the exit, the angles that the relative flow makes at the shroud and the hub at the exit are

$$\beta_{3s} = \tan^{-1}\left(\frac{U_{3s}}{V_3}\right) = \tan^{-1}\left(\frac{330.0}{112.6}\right) = 71.2°$$

$$\beta_{3h} = \tan^{-1}\left(\frac{U_{3h}}{V_3}\right) = \tan^{-1}\left(\frac{115.5}{112.6}\right) = 45.7°$$

The relative velocities at the shroud and hub are

$$W_{3s} = \sqrt{U_{3s}^2 + V_3^2} = \sqrt{330.0^2 + 112.6^2} = 348.7\,\text{m/s}$$

$$W_{3h} = \sqrt{U_{3h}^2 + V_3^2} = \sqrt{115.5^2 + 112.6^2} = 161.3\,\text{m/s}$$

and at the inlet the relative velocity is

$$W_2 = \sqrt{V_2^2 - U_2^2} = \sqrt{462.4^2 - 142.1^2} = 142.1\,\text{m/s}$$

Hence the relative velocity ratios are $W_{3s}/W_2 = 2.45$ and $W_{3h}/W_2 = 1.14$. The relative Mach numbers are $M_{3Rs} = 0.636$ and $M_{3Rh} = 0.294$. ■

In the foregoing example the mass balance was used to link the upstream and downstream states. Depending on what kind of information is known, this may become a somewhat involved calculation. The general approach is to express the mass balance in terms of the flow function. Upstream, this leads to

$$\frac{\dot{m}\sqrt{c_p T_{02}}}{p_{02} A_2} = F_2 \cos\alpha_2$$

and downstream the corresponding statement is

$$\frac{\dot{m}\sqrt{c_p T_{03}}}{p_{03} A_3} = F_3$$

in which the flow function is

$$F = \frac{\gamma}{\sqrt{\gamma-1}} M \left(1 + \frac{\gamma-1}{2} M^2\right)^{-(1/2)[(\gamma+1)/(\gamma-1)]}$$

The term $\cos\alpha_2$ appears in the upstream expression because it is the radial inflow velocity, $V_{r2} = V_2 \cos\alpha_2$, that enters the expression for the mass flow rate. Downstream the corresponding factor is missing because it is assumed that the absolute velocity is axial.

Dividing these gives

$$\frac{A_2}{A_3} = \frac{F_3}{F_2 \cos\alpha_2} \sqrt{\frac{T_{02}}{T_{03}} \frac{p_{03}}{p_{02}}}$$

The temperature ratio is

$$\frac{T_{02}}{T_{03}} = \frac{1}{1 - s_w}$$

and the pressure ratios may be written as

$$\frac{p_{03}}{p_{02}} = \frac{p_{03}}{p_3} \frac{p_3}{p_{01}} \frac{p_{01}}{p_{02}}$$

in which

$$\frac{p_{03}}{p_3} = \left(1 + \frac{\gamma-1}{2} M_3^2\right)^{\gamma/(\gamma-1)}$$

and

$$\frac{p_{01}}{p_3} = \left(1 - \frac{s_w}{\gamma-1}\right)^{\gamma/(\gamma-1)}$$

In addition

$$\frac{p_{01}}{p_{02}} = \left(\frac{T_2}{T_{2s}}\right)^{\gamma/(\gamma-1)} = \left(1 - \zeta_s \frac{\gamma-1}{2} M_2^2\right)^{-\gamma/(\gamma-1)}$$

Substituting gives

$$\frac{A_2}{A_3} = \frac{F_3}{F_2 \cos\alpha_2} \left(\frac{1 + \frac{\gamma-1}{2} M_3^2}{1 - \zeta_s \frac{\gamma-1}{2} M_2^2}\right)^{\gamma/(\gamma-1)} \left(1 - \frac{s_w}{\eta_{ts}}\right)^{\gamma/(\gamma-1)} \frac{1}{\sqrt{1 - s_w}}$$

From this it is clear that if the power ratio s_w can be determined and the total-to-static efficiency and the stator loss coefficient can be estimated, then the mass balance in this form involves only M_2, M_3, and the area ratio A_2/A_3 as unknowns. Hence, if two of these can be determined in other ways, this equation may be solved for the third. Clearly, if the area ratio is the only unknown, then it can be solved from this explicitly. However, if one of the Mach numbers is unknown, then an iterative solution of a nonlinear equation must be carried out. This can be readily performed with the aid of a computer. In hand calculations, it may be worthwhile to calculate the results directly from

$$\dot{m} = \rho_2 A_2 V_{r2} = \rho_3 A_3 V_3$$

by assuming one of the unknowns and then using the mass balance to check that the other converges by iterations to its correct value.

A collection of typical ranges of parameters for a well-designed radial inflow turbines has been compiled by Logan [52]. It is reproduced as Table 9.1.

Table 9.1 Design parameters for a radial inflow turbine. Source: Logan [52].

Parameter	Typical range
α_2	$68° - 76°$
β_3	$-50° - -70°$
r_{3h}/r_{3s}	< 0.4
r_{3h}/r_2	< 0.7
r_3/r_2	$0.53 - 0.66$
b_2/r_2	$0.1 - 0.3$
U_2/V_0	$0.55 - 0.8$
W_3/W_2	$2 - 2.5$
V_3/U_2	$0.15 - 0.5$
ζ_R	$0.4 - 0.8$
ζ_N	$0.06 - 0.24$

9.4 STATOR FLOW

The flow enters the stator row from a volute and leaves into a vaneless space. In the vaneless space, if losses are neglected, the flow may be assumed to be irrotational. The same assumption was made when the flow in the vaneless space in a centrifugal compressor was considered. In such an irrotational flow the tangential component of velocity varies inversely with radius, and thus rV_u is constant. Since mass flow rate is constant

$$\dot{m} = 2\pi r_{2e} b_{2e} \rho_{2e} V_{r2e} = 2\pi r_2 b_2 \rho_2 V_{r2}$$

and, as Figure 9.1 shows, the subscript in r_{2e} refers to a radius at the exit of the stator blades, and r_2 is the exit of the vaneless space and therefore also the inlet to the rotor. This equation can be recast as

$$V_{r2} = V_{r2e} \frac{b_{2e} \rho_{2e} r_{2e}}{b_2 \rho_2 r_2}$$

For a gap of constant width $b_{2e} = b_2$. The density ratio influences only the radial component of velocity, and since this component is much smaller than the tangential component only a small error is made if the density is assumed constant. If the density difference is ignored, then this equation and rV_u = constant can be written as

$$V_{r2} = V_{r2e} \frac{r_{2e}}{r_2} \qquad V_{u2} = V_{u2e} \frac{r_{2e}}{r_2}$$

Dividing gives

$$\frac{V_{u2}}{V_{r2}} = \frac{V_{u2e}}{V_{r2e}} = \tan \alpha_2$$

so the flow angle α_2 remains constant. The calculations are illustrated next.

■ **EXAMPLE 9.5**

A combustion gas mixture, with the ratio of specific heats $\gamma = \frac{4}{3}$ and gas constant $R = 287 \, \text{J}/(\text{kg} \cdot \text{K})$, has a specific heat $c_p = 1148 \, \text{J}/(\text{kg} \cdot \text{K})$. These gases flow through a radial inflow turbine with the inlet stagnation temperature $T_{01} = 1015 \, \text{K}$

and the stagnation pressure at the exit of the stator is $p_{02e} = 8.5\,\text{bar}$. The mass flow rate is $\dot{m} = 2.8\,\text{kg/s}$. The outlets of the stator blades are at the circle of radius $r_{2e} = 11.2\,\text{cm}$, where the flow angle is $\alpha_{2e} = 74.0°$. The stator blade height is $b = 1.7\,\text{cm}$. The inlet of the rotor blade is at $r_2 = 9.8\,\text{cm}$ and at the exit of the rotor the shroud radius is $r_{3s} = 7.7\,\text{cm}$ and the hub radius is $r_{3h} = 2.7\,\text{cm}$. The shaft rotates at 40,000 rpm and the turbine efficiency is $\eta_{tt} = 0.89$. At the exit the absolute velocity is axial. Find (a) the Mach number at the inlet to the rotor; (b) the power delivered by the turbine; (c) the Mach number at the exit of the turbine, assuming that the exit static pressure is 1 bar; (d) the flow angle of the relative velocity at the inlet to the rotor; and (e) the blade-loading coefficient $\psi = w/U_2^2$, the flow coefficient $\phi = V_{r2}/U_2$, and the specific speed of the turbine.

Solution: (a) If the thickness of the trailing edge of the stator blades is ignored, the flow area leaving the stator is

$$A_{2e} = 2\pi r_{2e} b = \frac{2\pi \cdot 11.2 \cdot 1.7}{100 \cdot 100} = 0.01196\,\text{m}^2$$

To determine the exit Mach number, the flow function is first calculated. It has the value

$$F_{2e} = \frac{\dot{m}\sqrt{c_p T_{02}}}{A_{2e} p_{02e} \cos\alpha_{2e}} = \frac{2.8\sqrt{1148 \cdot 1015}}{0.01196 \cdot 850{,}000 \cdot \cos(74°)} = 1.078$$

Now the exit Mach number is obtained by iteration from the nonlinear equation

$$F_{2e} = \frac{\gamma}{\sqrt{\gamma-1}} M_{2e}\left(1 + \frac{\gamma-1}{2} M_{2e}^2\right)^{-(1/2)[(\gamma+1)/(\gamma-1)]} = 1.078$$

With $\gamma = \frac{4}{3}$, its solution is $M_{2e} = 0.557$. The temperature can now be determined from

$$T_{2e} = T_{02}\left(1 + \frac{\gamma-1}{2} M_{2e}^2\right)^{-1} = 1015\left(1 + \frac{0.557^2}{6}\right)^{-1} = 965.1\,\text{K}$$

and the pressure is

$$p_{2e} = p_{02}\left(\frac{T_{2e}}{T_{02}}\right)^{\gamma/(\gamma-1)} = 8.5\left(\frac{965.1}{1015}\right)^4 = 6.948\,\text{bar}$$

The static density is therefore

$$\rho_{2e} = \frac{p_{2e}}{RT_{2e}} = \frac{6.948 \cdot 10^5}{287 \cdot 965.1} = 2.508\,\text{kg/m}^3$$

The exit velocity is next determined from

$$V_{2e} = M_{2e}\sqrt{\gamma R T_{2e}} = 0.557\sqrt{1.333 \cdot 287 \cdot 965.1} = 338.5\,\text{m/s}$$

and the radial velocity component is

$$V_{r2e} = V_{2e}\cos\alpha_{2e} = 338.5\cos(74°) = 93.3\,\text{m/s}$$

The tangential velocity component of the flow leaving the gap is

$$V_{u2} = \frac{r_{2e}}{r_2} V_{u2e} = \frac{11.2}{9.8} 325.4 = 371.9 \, \text{m/s}$$

and ignoring the density change in the radial term the corresponding value of the velocity component entering the blade is

$$V_{r2} = \frac{r_{2e}}{r_2} V_{r2e} = \frac{11.2}{9.8} 93.3 = 106.6 \, \text{m/s}$$

The magnitude of the absolute velocity is therefore

$$V_2 = \sqrt{V_{r2}^2 + V_{u2}^2} = \sqrt{106.6^2 + 371.9^2} = 386.9 \, \text{m/s}$$

The static temperature at this location is

$$T_2 = T_{02} - \frac{V_2^2}{2c_p} = 1015 - \frac{386.9^2}{2 \cdot 1148} = 949.8 \, \text{K}$$

and the Mach number is therefore

$$M_2 = \frac{V_2}{\sqrt{\gamma R T_2}} = \frac{386.9}{\sqrt{1.333 \cdot 287 \cdot 949.8}} = 0.642$$

If the density difference had been taken into account in the radial term, Mach number would have increased only to $M_2 = 0.644$.

(b) The tip speed of the rotor blade is

$$U_2 = r_2 \Omega = \frac{9.8 \cdot 40,000 \cdot \pi}{100 \cdot 30} = 410.5 \, \text{m/s}$$

and the tangential component of the relative velocity is

$$W_{u2} = V_{u2} - U_2 = 371.9 - 410.5 = -38.6 \, \text{m/s}$$

Since $W_{r2} = V_{r2}$, the flow angle of the relative velocity is

$$\beta_2 = \tan^{-1}\left(\frac{W_{u2}}{W_{r2}}\right) = \tan^{-1}\left(\frac{-38.6}{106.6}\right) = -19.9°$$

(c) Since there is no exit swirl, the work delivered becomes

$$w = U_2 V_{u2} = 410.5 \cdot 371.9 = 152.67 \, \text{kJ/kg}$$

and the isentropic work is

$$w_s = \frac{w}{\eta_{tt}} = \frac{152.67}{0.89} = 171.54 \, \text{kJ/kg}$$

The power delivered by the turbine is therefore

$$\dot{W} = \dot{m}w = 2.8 \cdot 152.67 = 427.5 \, \text{kW}$$

(d) The stagnation temperature at the exit is

$$T_{03} = T_{02} - \frac{w}{c_p} = 1015 - \frac{152,670}{1148} = 882.0\,\text{K}$$

and the corresponding temperature for an isentropic flow is

$$T_{03s} = T_{02} - \frac{w_c}{c_p} = 1015 - \frac{171,540}{1148} = 865.6\,\text{K}$$

The exit stagnation pressure is therefore

$$p_{03} = p_{02e}\left(\frac{T_{03s}}{T_{02}}\right)^{k/(k-1)} = 8.5\left(\frac{865.6}{1015}\right)^4 = 449.6\,\text{kPa}$$

With the exit area

$$A_3 = \pi(r_{3s}^2 - r_{3h}^2) = \pi(0.077^2 - 0.027^2) = 0.01634\,\text{m}^2$$

the exit flow function is equal to

$$F_3 = \frac{\dot{m}\sqrt{c_p T_{03}}}{A_3 p_{03}} = \frac{2.8\sqrt{1148 \cdot 882.0}}{0.01634 \cdot 449,550} = 0.384$$

and the Mach number can then be obtained by iteration from the nonlinear equation

$$F_3 = \frac{\gamma}{\sqrt{\gamma-1}} M_3 \left(1 + \frac{\gamma-1}{2} M_3^2\right)^{-(1/2)[(\gamma+1)/(\gamma-1)]} = 0.384$$

in which $\gamma = \frac{4}{3}$. The solution is $M_3 = 0.169$. The exit static temperature is then

$$T_3 = T_{03}\left(1 + \frac{\gamma-1}{2} M_3^2\right)^{-1} = 882.0\left(1 + \frac{0.169^2}{6}\right)^{-1} = 877.8\,\text{K}$$

so that the exit velocity becomes

$$V_3 = M_3\sqrt{\gamma R T_3} = 0.315\sqrt{1.333 \cdot 287 \cdot 877.8} = 97.9\,\text{m/s}$$

The volumetric flow rate is therefore

$$Q_3 = V_3 A_3 = 97.9 \cdot 0.01634 = 1.60\,\text{m}^3/\text{s}$$

(e) The blade-loading coefficient is

$$\psi = \frac{w}{U_2^2} = \frac{152,670}{410.5^2} = 0.906$$

and the flow coefficient is

$$\phi = \frac{V_{r2}}{U_2} = \frac{106.6}{410.5} = 0.260$$

The flow coefficient for radial inflow turbines is usually less than 0.5. The specific speed is

$$\Omega_s = \frac{\Omega\sqrt{Q_3}}{w_s^{0.75}} = \frac{40,000 \cdot \pi\sqrt{1.60}}{30 \cdot 171,540^{0.75}} = 0.628$$

and the specific diameter is

$$D_s = \frac{D_2 w_s^{0.25}}{\sqrt{Q_3}} = \frac{0.196 \cdot 171{,}540^{0.25}}{\sqrt{1.60}} = 3.15$$

These are in the range for good radial inflow turbines designs.

In the foregoing example losses in the stator and vaneless gap were not taken into account explicitly. In the next section methods to include them in the analysis are discussed.

9.4.1 Loss coefficients for stator flow

Baskharone [6] gives a correlation for the velocity coefficient for a flow through the stator:

$$c_v^2 = 1 - 1.8 \left(\frac{\theta}{s \cos \alpha_{2e} - t_e - \delta^*} \right) \left(1 + \frac{s \cos \alpha_m}{b} \right) \left(\frac{\text{Re}}{\text{Re}_f} \right)$$

in which $\text{Re}_f = 2.74 \cdot 10^6$ is a reference Reynolds number. It is based on the blade chord c and exit velocity V_{2e}. The angle α_m is the average flow angle between α_1 and α_{2e}. The spacing of the blades is s, and their height is b. The trailing-edge thickness of the blades is t_e. The sum of the pressure- and suction-side displacement thicknesses is $\delta^* = \delta_p^* + \delta_s^*$ and similarly for the total momentum thickness θ. It is assumed that the suction-side thicknesses are $\delta_s^* = 3.5\delta_p^*$ and $\theta_s = 3.5\theta_p$. The pressure-side values are obtained from

$$\frac{\delta_p^*}{c} = \frac{1.72}{\sqrt{\text{Re}}} \qquad \frac{\theta_p}{c} = \frac{0.664}{\sqrt{\text{Re}}}$$

which are valid for laminar flow when $\text{Re} < 2 \cdot 10^5$, and for turbulent boundary layers, when $\text{Re} > 2 \cdot 10^5$, they are

$$\frac{\delta_p^*}{c} = \frac{0.057}{\text{Re}^{1/6}} \qquad \frac{\theta_p}{c} = \frac{0.022}{\text{Re}^{1/6}}$$

Baskharone also recommends a model proposed by Khalil et al. [48] for taking into account the losses in the gap. It can be written as

$$Y_g = Y_r Y_\alpha Y_M Y_b$$

in which

$$Y_g = \frac{p_{02e} - p_{02}}{p_{02e} - p_2}$$

is the stagnation pressure loss coefficient and

$$Y_r = 0.193 \left(1 - \frac{r_2}{r_{2e}} \right)$$

$$Y_\alpha = 1 + 0.0641(\alpha_{2e} - \alpha_{2f}) + 0.0023(\alpha_{2e} - \alpha_{2f})^2$$
$$Y_M = 1 + 0.6932(M_{02e} - M_{02f}) + 0.4427(M_{02e} - M_{02f})^2$$
$$Y_b = 1 + 0.0923(\frac{r_{2e}}{b} - \frac{r_{2e}}{b_f}) + 0.0008(\frac{r_{2e}}{b} - \frac{r_{2e}}{b_f})^2$$

Here $M_{02e} = V_{2e}/\sqrt{\gamma R T_{02}}$ and the reference values are

$$\alpha_{2f} = 70° \qquad M_{02f} = 0.8 \qquad b_f = r_{2e}/10$$

The angle α_{2e} is measured in degrees.

EXAMPLE 9.6

A gas with $\gamma = \frac{4}{3}$ and $c_p = 1148 \, \text{J}/(\text{kg} \cdot \text{K})$ flows through a radial inflow turbine. The conditions at the inlet to the stator are $T_{01} = 1015 \, \text{K}$ and $p_{01} = 7.5 \, \text{bar}$. At the inlet the flow angle is $\alpha_1 = 20°$ and Mach number is $M_1 = 0.14$. The gas leaves the stator at angle $\alpha_{2e} = 71°$ and pressure $p_{2e} = 4.9 \, \text{bar}$. The number of stator vanes is $Z = 37$, their trailing edge thickness is $t_e = 1.5 \, \text{mm}$, their chord length is $c = 2.8 \, \text{cm}$, and their span is $b = 0.9 \, \text{cm}$. The radii are $r_1 = 12.0 \, \text{cm}$, $r_{2e} = 10.5 \, \text{cm}$, and $r_2 = 10.0 \, \text{cm}$. Assume the kinematic viscosity to be $\nu = 90 \cdot 10^{-6} \, \text{m}^2/\text{s}$. Find the stagnation pressure loss across the stator and the gap.

Solution: The static temperature at the inlet is

$$T_1 = T_{01}\left(1 + \frac{\gamma-1}{2}M_1^2\right)^{-1} = 1015\left(1 + \frac{0.14^2}{6}\right)^{-1} = 1011.7 \, \text{K}$$

and the static pressure is

$$p_1 = p_{01}\left(\frac{T_1}{T_{01}}\right)^{\gamma/(\gamma-1)} = 7.5\left(\frac{1011.7}{1015}\right)^4 = 7.40 \, \text{bar}$$

which give for the static density the value

$$\rho_1 = \frac{p_1}{RT_1} = \frac{7.4 \cdot 10^5}{287 \cdot 1011.7} = 2.55 \, \text{kg/m}^3$$

The velocity at the inlet is

$$V_1 = M_1\sqrt{\gamma RT_1} = 0.14\sqrt{1.333 \cdot 287 \cdot 1011.7} = 87.1 \, \text{m/s}$$

and its radial component is

$$V_{r1} = V_1 \cos\alpha_1 = 87.1 \cos(20°) = 81.86 \, \text{m/s}$$

Mass balance may be written as

$$\dot{m} = \rho_1 2\pi r_1 b V_{r1} = \rho_{2e} 2\pi r_{2e} b V_{r2e}$$

Introducing the flow function, the nondimensional mass flow rate may be written as

$$\frac{\dot{m}\sqrt{c_p T_{01}}}{A_1 p_{01}} = M_1 \cos\alpha_1 \left(1 + \frac{\gamma-1}{2}M_1^2\right)^{-(1/2)[(\gamma+1)/(\gamma-1)]} = F_1 \cos\alpha_1$$

at the inlet and at the exit it is

$$\frac{\dot{m}\sqrt{c_p T_{02}}}{A_{2e} p_{02e}} = M_{2e} \cos\alpha_2 \left(1 + \frac{\gamma-1}{2}M_{2e}^2\right)^{-(1/2)[(\gamma+1)/(\gamma-1)]} = F_{2e} \cos\alpha_2$$

Since the flow is adiabatic, $T_{01} = T_{02}$, and if it is assumed to be isentropic, then it follows that $p_{01} = p_{02e}$. With $A_1 = 2\pi r_1 b$ and $A_{2e} = 2\pi r_{2e} b$, the mass balance gives the relationship

$$F_{2e} = F_1 \frac{r_1 \cos\alpha_1}{r_{2e} \cos\alpha_{2e}}$$

The value of F_1 is

$$F_1 = \frac{\gamma M_1}{\sqrt{\gamma - 1}}\left(1 + \frac{\gamma - 1}{2}M_1^2\right)^{-(1/2)[(\gamma+1)/(\gamma-1)]} = \frac{4\sqrt{30.14}}{3}\left(1 + \frac{0.14^2}{6}\right)^{-3.5} = 0.3196$$

and therefore

$$F_{2e} = 0.3196 \frac{12.0 \cos(20°)}{10.5 \cos(71°)} = 1.054$$

Solving the nonlinear equation

$$F_{2e} = \frac{\gamma M_{2e}}{\sqrt{\gamma - 1}}\left(1 + \frac{\gamma - 1}{2}M_{2e}^2\right)^{-(1/2)(\gamma+1)/(\gamma-1)]}$$

with this value of F_{2e} and $\gamma = \frac{4}{3}$, yields $M_{2e} = 0.538$. Static temperature at the exit of the stator is

$$T_{2e} = T_{02}\left(1 + \frac{\gamma - 1}{2}M_{2e}^2\right)^{-1} = 1015\left(1 + \frac{0.539^2}{6}\right)^{-1} = 968.2\,\text{K}$$

and the velocity is

$$V_{2e} = M_{2e}\sqrt{\gamma R T_{2e}} = 0.538\sqrt{1.333 \cdot 287 \cdot 968.2} = 327.85\,\text{m/s}$$

Introduction of the flow function into the development assumes that the flow is isentropic. It is anticipated that no correction is needed, as V_{2e} is only used to calculate the Reynolds number in the loss correlation. The value of the Reynolds number is

$$\text{Re} = \frac{V_{2e}c}{\nu} = \frac{327.85 \cdot 2.8}{90 \cdot 10^{-6} \cdot 100} = 1.02 \cdot 10^5$$

The displacement thickness on the pressure side is therefore

$$\delta_p^* = \frac{1.72c}{\text{Re}^{1/2}} = \frac{1.72 \cdot 2.8}{100 \cdot (1.02 \cdot 10^5)^{1/2}} = 1.51 \cdot 10^{-4}\,\text{m}$$

Since the suction side displacement thickness is assumed to be $3.5\delta_p^*$, the sum becomes $\delta^* = 6.79 \cdot 10^{-4}$ m. The momentum thickness on the pressure side is

$$\theta_p = \frac{0.644c}{\text{Re}^{1/2}} = \frac{0.644 \cdot 2.8}{100 \cdot (1.02 \cdot 10^5)^{1/6}} = 5.65 \cdot 10^{-5}\,\text{m}$$

and with the suction side thickness assumed again to be $3.5\theta_p$, the sum is $\theta = 13.8 \cdot 10^{-4}$ m. Next, the spacing can be obtained from

$$s_2 = \frac{2\pi r_{2e}}{Z} = \frac{2\pi \cdot 10.5}{100 \cdot 37} = 0.0178\,\text{m}$$

The average flow angle is $\alpha_{2m} = 45.5°$, and the velocity coefficient is obtained from

$$c_v^2 = 1 - 1.8\left(\frac{\theta}{s_2 \cos\alpha_2 - t_e - \delta^*}\right)\left(1 + \frac{s_2 \cos\alpha_{2m}}{b}\right)\frac{\text{Re}}{\text{Re}_\text{f}}$$

with

$$\frac{\theta}{s_2 \cos\alpha_2 - t_e - \delta^*} = \frac{13.8 \cdot 10^{-4}}{0.0178 \cos(71°) - 1.5 \cdot 10^{-3} - 6.79 \cdot 10^{-4}} = 0.381$$

and
$$1 + \frac{s_2 \cos \alpha_{2m}}{b} = 1 + \frac{0.178 \cos(45.5°)}{0.009} = 2.389$$

the velocity coefficient becomes

$$c_v^2 = 1 - 1.8 \cdot 0.381 \cdot 2.389 \frac{102{,}000}{2{,}740{,}000} = 0.939 \quad \text{so that} \quad c_v = 0.969$$

The actual exit velocity is therefore

$$V_{2e} = c_v V_{2es} = 0.969 \cdot 327.8 = 317.7 \, \text{m/s}$$

As anticipated, the difference from the ideal is such that no correction is needed. The static temperature at the exit of the stator is now

$$T_{2e} = T_{02} - \frac{V_{2e}^2}{2c_p} = 1015 - \frac{317.7^2}{2 \cdot 1148} = 971.0 \, \text{K}$$

and the Mach number at the exit does not change significantly, for it is $M_{2e} = 0.521$. The stagnation pressure is now

$$p_{02e} = p_2 \left(\frac{T_{02}}{T_{2e}}\right)^{\gamma/(\gamma-1)} = 4.9 \left(\frac{1015}{971.0}\right)^4 = 5.849 \, \text{bar}$$

The loss of stagnation pressure across the stator is therefore

$$\Delta p_{0LS} = p_{01} - p_{02e} = 750.0 - 584.9 = 165.1 \, \text{kPa}$$

To determine the stagnation pressure loss across the gap, the stagnation Mach number at the inlet to the gap is needed. With $T_{01} = T_{02}$, it is

$$M_{02e} = \frac{V_{2e}}{\sqrt{\gamma R T_{02}}} = \frac{324.5}{\sqrt{1.333 \cdot 287 \cdot 1015}} = 0.5028$$

The stagnation loss coefficients for the gap are

$$Y_r = 0.193 \left(1 - \frac{10.0}{10.5}\right) = 0.00919$$
$$Y_\alpha = 1 + 0.0641(71 - 70) + 0.0023(71 - 70)^2 = 1.0664$$
$$Y_M = 1 + 0.6932(0.5028 - 0.8) + 0.4427(0.5028 - 0.8)^2 = 0.8410$$
$$Y_b = 1 + 0.0923\left(\frac{10.5}{0.9} - 10\right) + 0.0008\left(\frac{10.5}{0.9} - 10\right)^2 = 1.1561$$

and the overall loss coefficient comes out as

$$Y_g = 0.00919 \cdot 1.0664 \cdot 0.8410 \cdot 1.1561 = 0.00952$$

The stagnation pressure at the inlet to the rotor is therefore

$$p_{02} = p_{02e} - Y_g(p_{02e} - p_2) = 589.6 - 0.00952(589.6 - 490) = 588.6 \, \text{kPa}$$

and the stagnation pressure loss across the gap is

$$\Delta p_{0LG} = p_{02e} - p_{02} = 589.6 - 588.6 = 1.0 \, \text{kPa}$$

Hence this loss is relatively small, and upstream of the rotor the stagnation pressure loss is

$$\Delta p_{0LS} + \Delta p_{0LG} = 165.1 + 1.0 = 166.1 \, \text{kPa}$$

∎

Note that the Mach number increases across the stator from 0.14 to 0.52 owing to the acceleration of the flow as a result of the turning. There is a significant density change associated with this acceleration.

As has been shown above, across the gap the flow angle does not change for irrotational flow. Since the gap is also quite narrow, the area does not decrease greatly and the acceleration of the flow is rather slight. For this reason it is possible to assume the flow to be incompressible across the gap and calculate it by the formula

$$V_{r2} = V_{r2e} \frac{r_{2e}}{r_2}$$

This cannot be used for the stator flow, and using it in the foregoing example would have yielded for the radial component of the stator exit velocity the value

$$V_{r2e} = V_{r1} \frac{r_1}{r_{2e}} = 81.86 \frac{12.0}{10.5} = 93.55 \, \text{m/s}$$

and the exit velocity would have been

$$V_{2e} = \frac{V_{r2e}}{\cos \alpha_2} = \frac{93.55}{\cos(71°)} = 287.4 \, \text{m/s}$$

This would have been significantly in error as the correct value came out to be $324.5 \, \text{m/s}$.

9.5 DESIGN OF THE INLET OF A RADIAL INFLOW TURBINE

The foregoing developments and examples illustrate the basic methods for carrying out design calculations for radial inflow turbines. The loss calculations for the stator and interblade gap, together with the expression for the overall efficiency and its appropriate value obtained from the Balje diagram, are sufficient to determine the loss through the rotor. Table 9.1 gives the appropriate geometric length ratios for a design. Methods to obtain some of these values are developed from the theory discussed in this section. This work has been carried out, among others, by Rodgers and Geiser [63], Rohlik [65], and Whitfield [78]. The extensive study by Whitfield as been included in the book by Whitfield and Baines [79]. The following development follows their studies.

Since

$$s_w = \frac{w}{c_p T_{01}}$$

the blade loading coefficient is

$$\psi = \frac{w}{U_2^2} = \frac{s_w c_p T_{01}}{U_2^2} = \frac{1}{M_{0u}^2} \frac{s_w}{\gamma - 1}$$

in which the blade stagnation Mach number is

$$M_{0u} = \frac{U_2}{\sqrt{\gamma R T_{01}}}$$

This is based on the inlet stagnation temperature as a reference value.

If at the exit the flow has no swirl component and at the inlet the relative velocity is radially inward, then $w = U_2^2$, and the blade-loading coefficient is

$$\psi = \frac{w}{U_2^2} = 1$$

Experiments have shown that efficiency is increased if $-40° < \beta_2 < -20°$. This introduces a negative incidence into the flow since the blade angle χ_2 is invariably zero. The blade-loading coefficient is now given by

$$\psi = \frac{V_{u2}}{U_2} = \frac{U_2 + W_{u2}}{U_2} = 1 + \phi \tan \beta_2$$

in which the flow coefficient is defined as

$$\phi = \frac{V_{r2}}{U_2}$$

The flow angle of the relative velocity is the range $\beta_{2\,\text{min}} \le \beta_2 \le 0$, and $\beta_{2\,\text{min}}$ is chosen such that the blade loading coefficient is in the range $0 \le \psi \le 1$.

9.5.1 Minimum inlet Mach number

Whitfield [78] has shown how to optimize the inlet to the rotor by choosing (for a given power ratio) the absolute and relative flow angles that give the smallest inlet Mach number. Improper flow angles might lead to high inlet Mach numbers, and a possibility for choking. Following his development, the tangential velocity is first written as

$$V_{u2} = U_2 + W_{u2} = U_2 + V_{u2}\frac{\tan \beta_2}{\tan \alpha_2}$$

in which the second equation follows by using

$$V_{u2} = V_{r2} \tan \alpha_2 \qquad W_{u2} = W_{r2} \tan \beta_2$$

together with $V_{r2} = W_{r2}$. Multiplying through by V_{u2} and rearranging gives

$$V_{u2}^2 \left(1 - \frac{\tan \beta_2}{\tan \alpha_2}\right) = U_2 V_{u2}$$

The right side is the work delivered, and since this is related to the power ratio, it will be taken to be a fixed quantity. Dividing both sides by V_2^2 leads to

$$\sin^2 \alpha_2 \left(1 - \frac{\tan \beta_2}{\tan \alpha_2}\right) = u \qquad (9.8)$$

in which

$$u = \frac{U_2 V_{u2}}{V_2^2} = \frac{U_2 V_{u2}}{c_{02}^2}\frac{c_{02}^2}{V_2^2}$$

or
$$u = \frac{s_w}{(\gamma-1)}\frac{1}{M_{02}^2} = \frac{s}{m}$$

where the following notation has been introduced:
$$s = \frac{s_w}{\gamma-1} \qquad m = M_{02}^2$$

The inlet stagnation Mach number has been defined as
$$M_{02} = \frac{V_2}{\sqrt{\gamma R T_{02}}}$$

which can be written also as
$$M_{02}^2 = \frac{V_2^2}{c_{02}^2} = \frac{V_2^2}{c_2^2}\frac{c_2^2}{c_{02}^2} = M_2^2 \frac{T_2}{T_{02}}$$

so that
$$M_{02}^2 = \frac{M_2^2}{1+\frac{\gamma-1}{2}M_2^2} \qquad (9.9)$$

Dividing Eq. (9.8) next by $\cos^2 \alpha_2$, and noting that $1/\cos^2 \alpha_2 = \tan^2 \alpha_2 + 1$, gives
$$(1-u)\tan^2 \alpha_2 - \tan\beta_2 \tan\alpha_2 - u = 0$$

or
$$(m-s)\tan^2 \alpha_2 - m\tan\beta_2 \tan\alpha_2 - s = 0 \qquad (9.10)$$

For given values of s_w, M_2, and β_2, this quadratic equation can be solved for $\tan\alpha_2$. The solution is displayed in Figure 9.7 with β_2 and s_w as parameters.

For a given power ratio and relative flow angle, the minimum of each curve is sought next. To find it, let $a = \tan\alpha_2$ and $b = \tan\beta_2$. This puts Eq. (9.10) into the form
$$(m-s)a^2 - mba - s = 0 \qquad m = \frac{s(1+a^2)}{a(a-b)} \qquad (9.11)$$

The minimum for m, for a fixed s and b, is obtained by setting the derivative of this with respect to a to zero. Thus
$$\frac{dm}{da} = \frac{2s(a-b)a^2 - s(2a-b)(1+a^2)}{a^2(b-a)^2} = 0$$

and this reduces to
$$ba^2 + 2a - b = 0$$

Solving this quadratic equation gives
$$a = \frac{-1 \pm \sqrt{1+b^2}}{b} \qquad (9.12)$$

Substituting this value of a into Eq. (9.11) gives
$$m = \frac{2s}{1 \mp \sqrt{1+b^2}} \qquad (9.13)$$

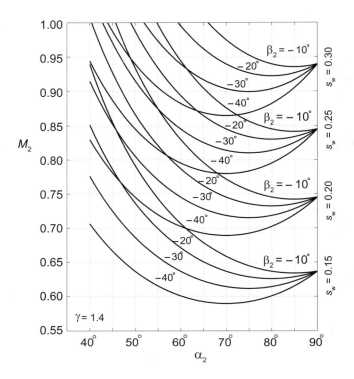

Figure 9.7 Inlet Mach number as a function of the nozzle angle for different relative flow angles and the power ratio with $\gamma = 1.4$.

for the minimum of m as a function of s and β_2. Since

$$\sqrt{1+b^2} = \sqrt{1+\tan^2\beta_2} = \frac{1}{\cos\beta_2}$$

Eq. (9.12) becomes

$$\tan\alpha_2 = \frac{-\cos\beta_2 \pm 1}{\sin\beta_2} \tag{9.14}$$

Next, making use of the identities

$$\cos\beta_2 = \cos^2\frac{\beta_2}{2} - \sin^2\frac{\beta_2}{2}$$

$$\sin\beta_2 = 2\sin\frac{\beta_2}{2}\cos\frac{\beta_2}{2}$$

in Eq. (9.14) leads to

$$\tan\alpha_2 = \frac{-\cos^2(\beta_2/2) + \sin^2(\beta_2/2) \pm 1}{2\sin(\beta_2/2)\cos(\beta_2/2)}$$

The positive sign gives

$$\tan\alpha_2 = \tan\frac{\beta_2}{2}$$

which is rejected because Figure 9.7 shows that $\alpha_2 > 0$ and $\beta_2 < 0$. The negative sign gives

$$\tan\alpha_2 = -\cot\frac{\beta_2}{2} \quad \text{or} \quad \tan\alpha_2 = \tan\left(\frac{\pi}{2} + \frac{\beta_2}{2}\right)$$

hence

$$\alpha_2 = \frac{\pi}{2} + \frac{\beta_2}{2}$$

when m is minimum. Substituting this into Eq. (9.13) gives

$$m = \frac{2s\cos\beta_2}{\cos\beta_2 \mp 1}$$

Since the minus sign gives a negative m, the positive sign is chosen. This then gives

$$M_{02\,\text{min}}^2 = \frac{2s\cos\beta_2}{1 + \cos\beta_2} \tag{9.15}$$

Equation 9.9 shows that

$$M_2^2 = \frac{M_{02}^2}{1 - \frac{\gamma-1}{2}M_{02}^2} \tag{9.16}$$

Substituting the value of $M_{02\,\text{min}}$ from Eq. (9.15) into this gives, after simplification

$$M_{2\,\text{min}} = \left[\left(\frac{2s_\text{w}}{\gamma-1}\right)\frac{\cos\beta_2}{1 + (1 - s_\text{w})\cos\beta_2}\right]^{1/2} \tag{9.17}$$

The minima for the curves, as seen from Figure 9.7, are at nozzle angles in their usual

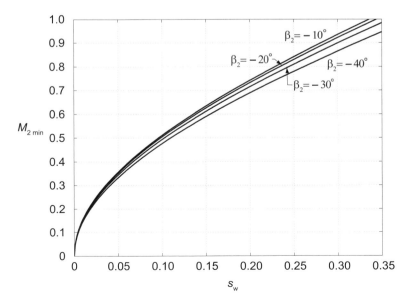

Figure 9.8 Minimum Mach number as a function of the power ratio for $\gamma = 1.4$ with angle β_2 as a parameter.

range between $60° < \alpha_2 < 80°$. For a given M_2, s_w, and β_2, there may be two angles α_2 that satisfy Eq. (9.10). The smaller angle is to be chosen. The larger angles put a limit on how large the inlet Mach number can be. As $\alpha_2 \to 90°$, $\tan \alpha_2 \to \infty$, and the second of Eqs. (9.11) shows that $m = s$. This means that at $\alpha_2 = 90°$ the inlet Mach number is

$$M_2 = \left[\frac{2s_w}{(\gamma - 1)(2 - s_w)} \right]^{1/2}$$

In particular, for $\gamma = 1.4$ and $s_w = 0.15$, $M_2 = 0.637$, as the graph shows. Figure 9.8 shows how the minimum Mach number depends on the power ratio and that its dependence on the angle β_2, and correspondingly on α_2, is weak.

■ **EXAMPLE 9.7**

Gas with $\gamma = 1.4$, and $R = 287 \text{ J/(kg} \cdot \text{K)}$ flows into a radial inflow turbine. The inlet stagnation temperature is $T_{01} = 1100 \text{ K}$, and the design-specific work is $w = 165.74 \text{ kJ/kg}$. (a) Find the value of the total-to-static efficiency that would give a pressure ratio of $p_{01}/p_3 = 2.0$. (b) At what angle should the flow leave the stator in order for $M_2 = 0.62$ be the minimum possible Mach number at the exit of the stator? (c) Find the blade speed at this condition. (d) Assume that the blade speed is increased to $U_2 = 460 \text{ m/s}$ and that the flow angle and the magnitude of the exit velocity from the stator remain the same. Find the new value for the relative flow angle β_2 entering the rotor.

Solution: (a) From the expression

$$w = s_w c_p T_{01}$$

the power ratio is

$$s_w = \frac{165,740}{1004.5 \cdot 1100} = 0.15$$

The pressure ratio is related to the power ratio and the total-to-static efficiency by Eq. (9.3), so that

$$\eta_{ts} = \frac{s_w}{1 - \left(\dfrac{p_3}{p_{01}} \right)^{(\gamma-1)/\gamma}} = \frac{0.15}{1 - 0.5^{1/3.5}} = 0.835$$

(b) The flow angle at the minimum inlet Mach number is obtained from the equation

$$M_2 = \left[\left(\frac{2s_w}{\gamma - 1} \right) \frac{\cos \beta_2}{1 + (1 - s_w) \cos \beta_2} \right]^{1/2}$$

which, when solved for $\cos \beta_2$, gives

$$\cos \beta_2 = \frac{M_2^2}{2s_w/(\gamma - 1) - (1 - s_w) M_2^2} \qquad \beta_2 = -24.74°$$

The minus sign must be chosen for the angle. At the minimum

$$\alpha_2 = 90° + \frac{\beta_2}{2} = 90° - 12.37° = 77.63°$$

DESIGN OF THE INLET OF A RADIAL INFLOW TURBINE 343

(c) To determine the blade speed, the velocity leaving the nozzle is needed, and therefore the stagnation Mach number is calculated first as:

$$M_{02} = \frac{M_2}{\sqrt{1 + \frac{\gamma - 1}{2} M_2^2}} = \frac{0.62}{\sqrt{1 + 0.2 \cdot 0.62^2}} = 0.5975$$

The exit velocity from the stator is therefore

$$V_2 = M_{02}\sqrt{\gamma R T_{02}} = 0.5975\sqrt{1.4 \cdot 287 \cdot 1100} = 397.2 \, \text{m/s}$$

and the tangential and radial components are

$$V_{u2} = V_2 \sin \alpha_2 = 397.2 \sin(77.63°) = 388.0 \, \text{m/s}$$

$$V_{r2} = V_2 \cos \alpha_2 = 397.2 \cos(77.63°) = 85.10 \, \text{m/s}$$

The blade speed can be obtained, for example, from

$$U_2 = \frac{w}{V_{u2}} = \frac{165,740}{388} = 427.2 \, \text{m/s}$$

(d) If the blade speed is increased to $U_2 = 460 \, \text{m/s}$ and the nozzle exit speed and direction are the same, then the new work done is

$$w = U_2 V_{u2} = 460 \cdot 388 = 178.48 \, \text{kJ/kg}$$

The new value for the tangential component of the relative velocity is

$$W_{u2} = V_{u2} - U_2 = 388.0 - 460.0 = -72.0 \, \text{m/s}$$

Since the radial component of the inlet velocity remains the same, the flow angle is

$$\beta_2 = \tan^{-1}\left(\frac{W_{u2}}{W_{r2}}\right) = \tan^{-1}\left(\frac{-72}{85.1}\right) = -40.2°$$

It is expected that the total-to-static efficiency will be somewhat lower at this angle. If the efficiency drops greatly, adjustable stator blades may be used to adjust the exit stator angle, which then influences the incidence. However, in a turbocharger an increase in blade speed comes from a larger cylinder pressure and a larger flow rate through the engine. Hence the radial component of the inlet velocity to the rotor also increases. The control of the stator blade angle must account for this as well. ■

9.5.2 Blade stagnation Mach number

The blade stagnation Mach number is yet another parameter of interest. It is obtained from the definition of the power ratio, which may be written as

$$\frac{U_2 \, V_{u2}}{c_{01} \, c_{01}} = \frac{s_w}{\gamma - 1}$$

so that

$$M_{0u} = \frac{U_2}{c_{01}} = \frac{s_w}{\gamma - 1} \frac{c_{01}}{c_2} \frac{c_2}{V_2 \sin \alpha_2}$$

Since $T_{01} = T_{02}$, this may be written as

$$M_{0u} = \frac{s_w}{\gamma - 1} \frac{1}{M_2 \sin \alpha_2} \sqrt{\frac{T_{02}}{T_2}}$$

and further as

$$M_{0u} = \frac{s_w}{\gamma - 1} \frac{\left(1 + \frac{\gamma - 1}{2} M_2^2\right)^{1/2}}{M_2 \sin \alpha_2} \quad (9.18)$$

For a given power factor $s_w/(\gamma - 1)$, the relative flow angle β_2, and Mach number M_2, Eq. (9.10) can be solved for α_2 and the nondimensional blade speed M_{0u} can be calculated from Eq. (9.18). These results are shown in Figure 9.9. Even if the blade speed is higher than the absolute velocity, the blade stagnation Mach number is based on the larger sonic speed and it therefore does not become as high as the inlet Mach number. This graph is useful when it is linked to the maximum attainable efficiency.

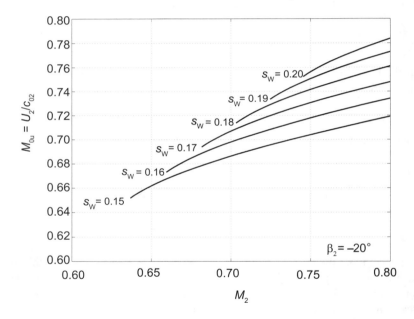

Figure 9.9 Blade stagnation Mach number as a function of M_2 for $\gamma = 1.4$ and for various power ratios and $\beta_2 = -20°$.

Experimental data for the total-to-static efficiency, measured by Rodgers and Geiser [63], are shown in Figure 9.10. The abscissa in this figure is the ratio of the blade speed to spouting velocity. Since

$$\frac{1}{2} V_0^2 = c_p T_{01} \left(1 - \frac{T_{3ss}}{T_{01}}\right) = c_p T_{01} \frac{s_w}{\eta_{ts}}$$

the ratio of the spouting velocity to the stagnation speed of sound is

$$\frac{V_0}{c_{01}} = \sqrt{\frac{2 s_w}{(\gamma - 1)\eta_{ts}}}$$

hence
$$M_{0u} = \frac{U_2}{V_0}\sqrt{\frac{2s_w}{(\gamma-1)\eta_{ts}}}$$

For $\gamma = \frac{4}{3}$, $\eta_{ts} = 0.9$, and $s_w = 0.15$, the factor involving the square root is unity. For values $\eta_{ts} = 0.8$ and $s_w = 0.2$, this equation yields $M_{0u} = 1.22 U_2/V_0$. Hence a turbine with a reasonably low power ratio and stagnation blade Mach number in the range $0.70 < M_{0u} < 0.75$ operates in the region of highest efficiency. Also, Figure 9.9 shows that under these conditions M_2 is quite high.

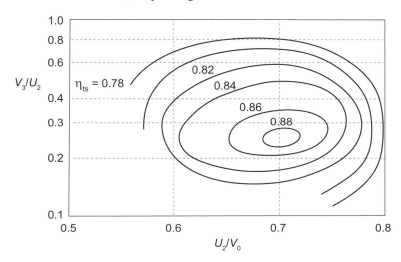

Figure 9.10 Total-to-static efficiency contours redrawn from the data of Rodgers and Geiser [63].

9.5.3 Inlet relative Mach number

The inlet relative Mach number can be calculated by writing the component equations
$$V_2 \sin\alpha_2 = U_2 + W_2 \sin\beta_2 \qquad V_2 \cos\alpha_2 = V_2 \cos\beta_2$$
and then squaring and adding them. This yields
$$V_2^2 = U_2^2 + W_2^2 + 2U_2 W_2 \sin\beta_2$$
and after each term is divided $c_2^2 = \gamma R T_2$, and the definition $M_u = U_2/c_2$ is substituted, then after rearrangement this reduces to
$$M_{2R}^2 + 2M_u \sin\beta_2 M_{2R} + M_u^2 - M_2^2 = 0$$
The solution of this is
$$M_{2R} = -M_u \sin\beta_2 + \sqrt{M_2^2 - M_u^2 \cos^2\beta_2}$$
in which the term M_u may be related to M_2 by
$$M_u = M_{0u}\sqrt{\frac{T_{02}}{T_2}} = M_{0u}\left(1 + \frac{\gamma-1}{2}M_2^2\right)^{1/2} = \left(\frac{s_w}{\gamma-1}\right)\frac{1+\frac{\gamma-1}{2}M_2^2}{M_2 \sin\alpha_2}$$

because the blade stagnation Mach number M_{0u} is given by Eq. (9.18). These results are plotted in Figure 9.11.

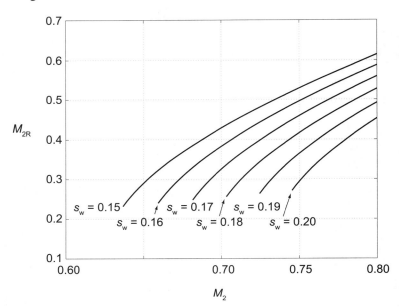

Figure 9.11 The relative inlet Mach number M_{2R} as a function of M_2 for various power ratios and for $\beta_2 = -20°$ and $\gamma = 1.4$.

9.6 DESIGN OF THE EXIT

The blade height at the exit is to be chosen sufficiently large to reduce the relative velocity low enough that the relative Mach number at the shroud does not reach unity. In this section its value is related to a nondimensional mass flow rate and the exit Mach number. In addition, the optimum angle of the relative flow at the exit is found.

9.6.1 Minimum exit Mach number

If the turbine operates under conditions such that the absolute velocity at the exit is axial, then the mass balance can be written as

$$\dot{m} = \rho_3 V_3 A_3 = \rho_3 V_3 \pi (r_{3s}^2 - r_{3h}^2) = \rho_3 V_3 \pi r_{3s}^2 (1 - \kappa^2)$$

in which $\kappa = r_{3h}/r_{3s}$. This can be changed to

$$\dot{m} = \frac{p_3}{RT_3} M_3 \sqrt{\gamma R T_3} \pi r_{3s}^2 (1 - \kappa^2)$$

Multiplying and dividing the right side by $p_{01} = \rho_{01} R T_{01}$ and introducing the inlet stagnation speed of sound $c_{01}^2 = \gamma R T_{01}$ converts this equation into

$$\dot{m} = \rho_{01} c_{01} \pi r_{3s}^2 (1 - \kappa^2) \frac{p_3}{p_{01}} \left(\frac{T_{01}}{T_3}\right)^{1/2} M_3$$

A form of a flow coefficient may now be defined as

$$\Phi = \frac{\dot{m}}{\rho_{01} c_{01} \pi r_2^2} = \frac{p_3}{p_{01}} \left(\frac{T_{01}}{T_3}\right)^{1/2} M_3 \frac{r_{3s}^2}{r_2^2} (1 - \kappa^2)$$

in which the denominator is large because it is a product of the stagnation density and the speed of sound at the upstream conditions (conditions at which both of these are large) and a fictitious large flow area πr_2^2. Substituting $U_{3s}/U_2 = r_{3s}/r_2$ into this gives

$$\Phi = \frac{p_3}{p_{01}} \left(\frac{T_{01}}{T_3}\right)^{1/2} M_3 \frac{U_{3s}^2}{U_2^2} (1 - \kappa^2) \tag{9.19}$$

Next, the absolute Mach number at the exit is related to the exit relative Mach number at the shroud of the blade. The relative velocity at the shroud is

$$W_{3s}^2 = V_3^2 + U_{3s}^2$$

Dividing through by U_2^2 leads to the following expression

$$\begin{aligned}
\frac{U_{3s}^2}{U_2^2} &= \frac{W_{3s}^2}{U_2^2} - \frac{V_3^2}{U_2^2} \\
&= \frac{W_{3s}^2}{c_3^2} \frac{c_3^2}{c_{01}^2} \frac{c_{01}^2}{U_2^2} - \frac{V_3^2}{c_3^2} \frac{c_3^2}{c_{01}^2} \frac{c_{01}^2}{U_2^2} \\
&= \frac{M_{3Rs}^2 - M_3^2}{M_{0u}^2} \frac{T_3}{T_{01}}
\end{aligned}$$

Substituting this into Eq. (9.19) leads to

$$\Phi = \frac{p_3}{p_{01}} \left(\frac{T_3}{T_{01}}\right)^{1/2} M_3 \frac{M_{3Rs}^2 - M_3^2}{M_{0u}^2} (1 - \kappa^2)$$

which when solved for M_{3Rs} gives

$$M_{3Rs}^2 = M_3^2 + \frac{\Phi M_{0u}^2}{1 - \kappa^2} \frac{1}{M_3} \left(1 + \frac{\gamma - 1}{2} M_3^2\right)^{1/2} \left(\frac{p_{01}}{p_3}\right) \left(\frac{T_{01}}{T_{03}}\right)^{1/2}$$

where the pressure ratio and the stagnation temperature ratio are given by

$$\frac{p_{01}}{p_3} = \left(1 - \frac{s_w}{\eta_{ts}}\right)^{-\gamma/(\gamma-1)} \qquad \frac{T_{01}}{T_{03}} = \frac{1}{1 - s_w}$$

Defining B as

$$B = \Phi_f \left(1 - \frac{s_w}{\eta_{ts}}\right)^{-\gamma/(\gamma-1)} \frac{1}{\sqrt{1 - s_w}}$$

in which

$$\Phi_f = \Phi_0 M_{0u}^2 / (1 - \kappa^2) \tag{9.20}$$

the relative Mach number at the exit has the form

$$M_{3Rs}^2 = M_3^2 + B \left(\frac{1}{M_3^2} + \frac{\gamma - 1}{2}\right)^{1/2}$$

The minimum value of $M_{3\text{Rs}}^2$ as a function of M_3^2, with other parameters held fixed, is obtained by differentiation:

$$\frac{dM_{3\text{Rs}}^2}{dM_3^2} = 1 - \frac{B}{2M_3^4}\left(\frac{1}{M_3^2} + \frac{\gamma-1}{2}\right)^{-(1/2)}$$

Setting this to zero gives the following equation for M_3,

$$M_3^8 + \frac{2}{\gamma-1}M_3^6 - \frac{B^2}{2(\gamma-1)} = 0$$

This fourth-order polynomial equation in M_3 is now numerically solved for M_3, and then $M_{3\text{Rs}}$ is determined from

$$M_{3\text{Rs}} = \left[M_3^2 + \frac{B}{M_3}\left(1 + \frac{\gamma-1}{2}M_3^3\right)^{1/2}\right]^{1/2}$$

The results are shown graphically in Figure 9.12, with the minima of $M_{3\text{Rs}}$ marked by small circles. Since the absolute velocity at the exit is axial, it follows that

$$\frac{M_3}{M_{3\text{Rs}}} = \frac{V_3}{W_{3\text{s}}} = \cos\beta_{3\text{s}}$$

This equation shows that lines of constant $\beta_{3\text{s}}$ are straight lines in Figure 9.12. The relative flow angle is plotted in Figure 9.13. It shows that the minimum relative Mach number at the exit occurs when $\beta_3 = -56°$ for $p_{01}/p_3 = 2$ and $\eta_{\text{ts}} = 0.85$. Clearly, since the absolute velocity is axial, the Mach number for the absolute flow is less than that for the relative flow.

9.6.2 Radius ratio $r_{3\text{s}}/r_2$

From the exit velocity diagram

$$\sin|\beta_3| = \frac{U_3}{W_3}$$

This equation applies at every radial location r_3 of the exit, and r_3 now denotes a radius that varies from the hub to the tip. It is assumed that the angle β_3 changes with radius such that the exit velocity V_3 is uniform. With $U_3 = U_2 r_3/r_2$, this may be written as

$$\sin|\beta_3| = \frac{U_2}{W_2}\frac{W_2}{W_3}\frac{r_3}{r_2}$$

therefore the radius ratio can be written as

$$\frac{r_3}{r_2} = \frac{W_3}{W_2}\frac{W_2}{U_2}\sin|\beta_3|$$

From the inlet velocity diagram the tangential components give

$$U_2 = V_2\sin\alpha_2 - W_2\sin\beta_2$$

or

$$\frac{U_2}{W_2} = \frac{V_2}{W_2}\sin\alpha_3 - \sin\beta_2$$

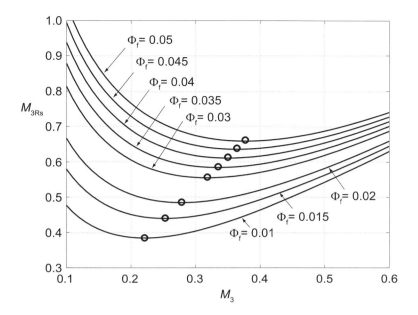

Figure 9.12 Relative Mach number as a function of Mach number, with Φ_f as a parameter. The pressure ratio is $p_{01}/p_3 = 2$, with $\eta_{ts} = 0.85$ and $\gamma = 1.4$. The locations of the minima of M_{3Rs} are marked by circles.

From the radial components

$$V_2 \cos \alpha_2 = W_2 \cos \beta_2 \qquad \frac{V_2}{W_2} = \frac{\cos \beta_2}{\cos \alpha_2}$$

Substituting this into the previous expression gives

$$\frac{U_2}{W_2} = \frac{\sin \alpha_2 \cos \beta_2 - \cos \alpha_2 \sin \beta_2}{\cos \alpha_2} = \frac{\sin(\alpha_2 - \beta_2)}{\cos \alpha_2}$$

which is just the law of sines.

The radius ratio may now be written as

$$\frac{r_3}{r_2} = \frac{W_3}{W_2} \frac{\cos \alpha_2 \sin |\beta_3|}{\sin(\alpha_2 - \beta_2)}$$

If the angle β_2 is chosen to be that for a minimum inlet Mach number, then $\beta_2 = 2\alpha_2 - 90$, and $\sin(\alpha_2 - \beta_2) = \sin \alpha_2$, and this expression reduces to

$$\frac{r_3}{r_2} = \frac{W_3}{W_2} \frac{\sin |\beta_3|}{\tan \alpha_2} \qquad (9.21)$$

At the shroud it is clearly

$$\frac{r_{3s}}{r_2} = \frac{W_{3s}}{W_2} \frac{\sin |\beta_{3s}|}{\tan \alpha_2}$$

Substituting $\alpha_2 = 90 + \beta_2/2$ gives the alternative form

$$\frac{r_{3s}}{r_2} = \frac{W_{3s}}{W_2} \tan \frac{\beta_2}{2} \sin \beta_{3s} \qquad (9.22)$$

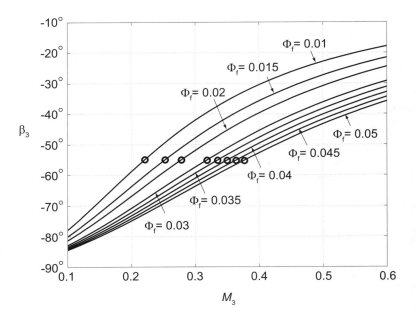

Figure 9.13 The flow angle of the relative velocity at the exit as a function of Mach number, with Φ_f as a parameter. The pressure ratio is $p_{01}/p_3 = 2$, with $\eta_{ts} = 0.85$ and $\gamma = 1.4$. The locations of the angle corresponding to minima of M_{3Rs} are marked by circles.

Rohlik [65] suggested that the relative velocity ratio $W_3/W_2 = 2$ gives a good design, or if the relative velocity at the shroud is used, then $W_{3s}/W_2 = 2.5$ is appropriate [79].

9.6.3 Blade height-to-radius ratio b_2/r_2

The final parameter to be determined is the blade height-to-radius ratio b_2/r_2 at the inlet. By casting the mass balance

$$\dot{m} = \rho_2 V_{r2} A_2 = \rho_3 V_3 A_3$$

in terms of the flow functions, the area ratio becomes

$$\frac{A_3}{A_2} = \frac{F_2 \cos\alpha_3}{F_3} \sqrt{\frac{T_{03}}{T_{02}} \frac{p_{02}}{p_{03}}} = \frac{F_2 \cos\alpha_3}{F_3} \sqrt{\frac{T_{03}}{T_{02}} \frac{p_{02}}{p_{01}} \frac{p_{01}}{p_3} \frac{p_3}{p_{03}}}$$

The stagnation pressure ratio p_{02}/p_{01} is related to the static temperature ratio T_2/T_{2s} by integrating the Gibbs equation along the constant-pressure line p_2 and along the line of constant stagnation temperature $T_{01} = T_{02}$. Equality of entropy changes then gives

$$\frac{p_{02}}{p_{01}} = \left(\frac{T_{2s}}{T_2}\right)^{\gamma/(\gamma-1)}$$

From the definition of the static enthalpy loss coefficient in the stator

$$\zeta_S = \frac{h_2 - h_{2s}}{\frac{1}{2}V_2^2}$$

the relationship

$$\frac{T_{2s}}{T_2} = 1 - \zeta_s \frac{\gamma - 1}{2} M_2^2$$

is obtained. In addition,

$$\frac{T_{03}}{T_{01}} = 1 - s_w \quad \text{and} \quad \frac{p_3}{p_{01}} = \left(1 - \frac{s_w}{\eta_{ts}}\right)^{\gamma/(\gamma-1)}$$

so the area ratio can now be written as

$$\frac{A_3}{A_2} = \frac{F_2 \sin \alpha_2}{F_3} \left(1 - \frac{s_w}{\eta_{ts}}\right)^{-\gamma/(\gamma-1)} \sqrt{1 - s_w} \left(\frac{1 - \zeta_s \frac{\gamma - 1}{2} M_2^2}{1 + \frac{\gamma - 1}{2} M_3^2}\right)^{\gamma/(\gamma-1)} \tag{9.23}$$

The blade width at the inlet is now obtained by writing the reciprocal of the area ratio as

$$\frac{A_2}{A_3} = \frac{2\pi r_2 b_2}{\pi(r_{3s}^2 - r_{3h}^2)} = 2\left(\frac{r_2}{r_{3s}}\right)^2 \frac{b_2}{r_2} \frac{1}{1 - \kappa^2}$$

in which $\kappa = r_{3h}/r_{3s}$. Solving for the blade height gives

$$\frac{b_2}{r_2} = \frac{1}{2}\left(\frac{r_{3s}}{r_2}\right)^2 (1 - \kappa^2) \frac{A_2}{A_3}$$

For a radial inflow turbine with a pressure ratio $p_{01}/p_3 = 2$, total-to-static efficiency $\eta_{ts} = 0.85$, the stator static enthalpy loss coefficient $\zeta_s = 0.15$, $\kappa = 0.2$, and $\gamma = 1.4$, the graphs for b_2/r_2 are shown in Figure 9.14. Turbines operated at low power factor have a blade height of about one fourth the inlet radius r_2.

9.6.4 Optimum incidence angle and the number of blades

For centrifugal compressors the Stanitz slip factor was given as

$$\sigma = \frac{V_{u2}}{V'_{u2}} \quad \text{with} \quad \sigma = 1 - \frac{0.63\pi}{Z}$$

in which Z is the number of blades and V'_{u2} is the tangential velocity component in the absence of slip. For radial blades $V'_{u2} = U_2$. Using this expression at the inlet to the rotor gives

$$\frac{V_{u2}}{U_2} = 1 - \frac{0.63\pi}{Z}$$

Since

$$V_{u2} = U_2 + W_{u2} = U_2 + W_{r2} \tan \beta_2 = U_2 + V_{r2} \tan \beta_2 = U_2 + V_{u2} \frac{\tan \beta_2}{\tan \alpha_2}$$

the ratio V_{u2}/U_2 becomes

$$\frac{V_{u2}}{U_2} = \frac{\tan \alpha_2}{\tan \alpha_2 - \tan \beta_2}$$

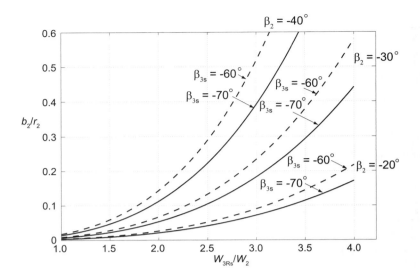

Figure 9.14 Blade width-to-radius ratio as a function of the relative velocity ratio W_{3s}/W_2 for relative flow angles corresponding to minimum Mach number at the inlet and at axial exit. The pressure ratio is $p_{01}/p_3 = 2$, with $\eta_{ts} = 0.85$, $\gamma = 1.4$, and the static enthalpy loss coefficient of the stator is $\zeta_S = 0.15$. The exit hub to shroud radius ratio is $\kappa = 0.2$.

At the condition of minimum inlet Mach number, at which $\alpha_2 = \pi/2 - \beta_2/2$, this reduces to
$$\frac{V_{u2}}{U_2} = \cos\beta_2$$
Hence the number of blades is related to the β_2 by
$$\cos\beta_2 = 1 - \frac{0.63\pi}{Z}$$
If this angle corresponds to the optimum nozzle angle for a minimum Mach number at the entry, then the substitution $\beta_2 = 2\alpha_2 - \pi$ gives
$$\cos\alpha_2 = \frac{1}{\sqrt{Z}}$$
This formula for the optimum nozzle angle was developed by Whitfield [78]. The second half of this chapter has been based on his original research on how the rotor blade design might proceed. The optimum angle is plotted in Figure 9.15 along with Glassman's suggestion
$$Z = \frac{\pi}{3}(110 - \alpha_2)\tan\alpha_2$$
These results agree when the number of blades is 13, but the incidence in the Glassman correlation decreases more rapidly as the number of blades increases.

■ **EXAMPLE 9.8**

Combustion gases with $\gamma = \frac{4}{3}$ and $R = 287\,\text{J}/(\text{kg}\cdot\text{K})$ enter the stator of an radial inflow turbine at $T_{01} = 1050\,\text{K}$ and $p_{01} = 250\,\text{kPa}$. The power produced by the

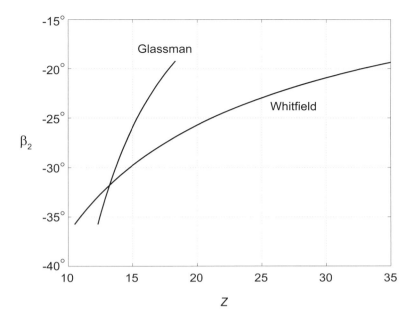

Figure 9.15 Optimum incidence angle β_2 as a function of the number of blades Z according to Glassman [27] and Whitfield [78].

turbine is $\dot{W} = 232\,\text{kW}$ at shaft speed of 35,000 rpm. The Mach number at the exit of the stator is $M_2 = 0.69$, and the flow angle there is $\alpha_2 = 67°$. The efficiency of the turbine is $\eta_{ts} = 0.89$, and the static enthalpy loss coefficient for the stator is $\zeta_S = 0.15$. There is no swirl in the exit flow, and the design seeks to have the incidence at the inlet to be at angle $\beta_2 = -18.9°$. Find (a) the exit blade radius entering the turbine and (b) the exit static pressure. (c) Given the exit Mach number $M_3 = 0.5$ and a ratio of the hub-to-shroud radius of blade of 0.3, find angle of the relative flow at the shroud radius of the exit. Find (d) the ratio W_{3s}/W_2 for the machine and (e) the blade height b_2 at the inlet.

Solution: (a) The stagnation speed of sound at the exit of the stator is

$$c_{01} = \sqrt{\gamma R T_{01}} = \sqrt{1.333 \cdot 287 \cdot 1050} = 633.9\,\text{m/s}$$

and the stagnation Mach number leaving the stator is therefore

$$M_{02} = \frac{M_2}{\sqrt{1 + \frac{\gamma-1}{2} M_2^2}} = \frac{0.69}{\sqrt{1 + \frac{0.69^2}{6}}} = 0.664$$

This gives the velocity leaving the stator the value

$$V_2 = M_{02} c_{01} = 0.664 \cdot 633.9 = 421.0\,\text{m/s}$$

The velocity components are then

$$V_{u2} = V_2 \sin\alpha_2 = 421.0 \sin(67°) = 387.5\,\text{m/s}$$

$$V_{r2} = V_2 \cos\alpha_2 = 421.0 \cos(67°) = 164.5\,\text{m/s}$$

With $W_{r2} = V_{r2}$, the relative velocity is

$$W_2 = \frac{W_{r2}}{\cos \beta_2} = \frac{164.49}{\cos(-18.9°)} = 173.9 \, \text{m/s}$$

and its tangential component is

$$W_{u2} = W_2 \sin \beta_2 = 173.9 \cdot \sin(-18.9°) = -56.3 \, \text{m/s}$$

The blade speed comes out to be

$$U_2 = V_{u2} - W_{u2} = 387.5 + 56.3 = 443.8 \, \text{m/s}$$

and the radius has the value

$$r_2 = \frac{U_2}{\Omega} = \frac{443.8 \cdot 30}{35,000 \cdot \pi} = 0.121 \, \text{m}$$

(b) The work delivered by the turbine is

$$w = U_2 V_{u2} = 443.8 \cdot 387.5 = 172.00 \, \text{kJ/kg}$$

and the stagnation temperature at the exit is therefore

$$T_{03} = T_{02} - \frac{w}{c_p} = 1050 - \frac{172.00}{1.148} = 900.2 \, \text{K}$$

With the total-to-static efficiency $\eta_{ts} = 0.89$ the isentropic work becomes

$$w_s = \frac{w}{\eta_{ts}} = \frac{172.0}{0.89} = 193.26 \, \text{kJ/kg}$$

so that the exit static temperature at the end of the isentropic process has the value

$$T_{3ss} = T_{01} - \frac{w_s}{c_p} = 1050 - \frac{193.26}{1.148} = 881.7 \, \text{K}$$

and the static pressure at the exit is

$$p_3 = p_{01} \left(\frac{T_{3ss}}{T_{01}} \right)^{\gamma/(\gamma-1)} = 250 \left(\frac{881.7}{1050} \right)^4 = 124.3 \, \text{kPa}$$

(c) To calculate the conditions at the shroud, first the mass flow rate is obtained from

$$\dot{m} = \frac{\dot{W}}{w} = \frac{232}{172} = 1.349 \, \text{kg/s}$$

and then the stagnation density at the inlet:

$$\rho_{01} = \frac{p_{01}}{RT_{01}} = \frac{250,000}{287 \cdot 1050} = 0.830 \, \text{kg/m}^3$$

Using these, the nondimensional mass flow rate becomes

$$\Phi = \frac{\dot{m}}{\rho_{01} c_{01} \pi r_2^2} = \frac{1.349}{0.830 \cdot 633.9 \cdot \pi \cdot 0.121^2} = 0.0557$$

DESIGN OF THE EXIT

The stagnation blade Mach number has the value

$$M_u = \frac{U_2}{c_{01}} = \frac{443.8}{633.9} = 0.70$$

which is used to calculate the modified nondimensional flow rate

$$\Phi_f = \frac{\Phi_0 M_u^2}{(1-\kappa^2)} = \frac{0.0557 \cdot 0.70^2}{1-0.09} = 0.030$$

With the exit Mach number $M_3 = 0.5$, the relative shroud Mach number can now be obtained from

$$M_{3Rs} = \left[M_3^2 + \frac{\Phi_f}{M_3}\left(1+\frac{\gamma-1}{2}M_3^2\right)^{1/2} \frac{p_{01}}{p_3}\left(\frac{T_{01}}{T_{03}}\right)^{1/2} \right]^{1/2}$$

$$= \left[0.5^2 + \frac{0.03}{0.5}\sqrt{1+\frac{0.5^2}{6}} \frac{250}{124.3}\sqrt{\frac{1050}{900.32}} \right]^{1/2} = 0.619$$

The flow angle is therefore

$$\beta_{3s} = -\cos^{-1}\left(\frac{M_3}{M_{3Rs}}\right) = -\cos^{-1}\left(\frac{0.50}{0.609}\right) = -36.1°$$

(d) To calculate the relative velocity at the shroud, the temperature T_3 is needed. It is given by

$$T_3 = \frac{T_{03}}{1+\frac{\gamma-1}{2}M_3^2} = \frac{900.2}{1+\frac{0.5^2}{6}} = 864.2\,\text{K}$$

The relative velocity at the shroud is therefore

$$W_{3s} = M_{3Rs}\sqrt{\gamma R T_3} = 0.619\sqrt{1.333 \cdot 287 \cdot 864.2} = 355.9\,\text{m/s}$$

and the ratio of relative velocities is

$$\frac{W_{3s}}{W_2} = \frac{355.9}{173.86} = 2.05$$

This is in the typical range for good designs.

(e) To determine the blade height at the inlet, the flow areas are calculated next. At the exit the shroud radius is obtained by first calculating the blade speed there. Its value is

$$U_{3s} = W_{3s}\sin|\beta_{3s}| = 355.9\sin|36.1°| = 209.7\,\text{m/s}$$

The shroud radius is now obtained as

$$r_{3s} = r_2 \frac{U_{3s}}{U_2} = 0.121\frac{209.7}{443.8} = 0.0572\,\text{m}$$

The exit area is of size

$$A_3 = \pi(r_{3s}^2 - r_{3h}^2) = \pi r_{3s}^2(1-\kappa^2) = \pi \cdot 0.0572^2(1-0.3^2) = 0.00936\,\text{m}^2$$

The flow functions at the inlet and exit of the turbine blade are

$$F_2 = \frac{\gamma M_2}{\sqrt{\gamma-1}} \left(1 + \frac{\gamma-1}{2} M_2^2 \right)^{-(1/2)[(\gamma+1)/(\gamma-1)]} = \frac{4 \cdot 0.69\sqrt{3}}{3} \left(1 + \frac{0.69^2}{6}\right)^{-3.5} = 1.202$$

$$F_3 = \frac{\gamma M_3}{\sqrt{\gamma-1}} \left(1 + \frac{\gamma-1}{2} M_3^2 \right)^{-(1/2)[(\gamma+1)/(\gamma-1)]} = \frac{4 \cdot 0.5\sqrt{3}}{3} \left(1 + \frac{0.5^2}{6}\right)^{-3.5} = 1.001$$

The area ratio is now

$$\frac{A_3}{A_2} = \frac{F_2 \cos \alpha_2}{F_3} \sqrt{\frac{T_{03}}{T_{01}} \frac{p_3}{p_{03}}} \left(\frac{1 - \zeta_S \frac{\gamma-1}{2} M_2^2}{1 + \frac{\gamma-1}{2} M_3^2}\right)^{\gamma/(\gamma-1)}$$

$$= \frac{1.202}{1.001} \sqrt{\frac{900.2}{1050} \frac{250}{124.3}} \left(\frac{1 - 0.15 \frac{0.69^2}{6}}{1 + \frac{0.5^2}{6}}\right)^4 = 0.718$$

so that

$$A_2 = \frac{A_2}{A_3} A_3 = \frac{0.00936}{0.718} = 0.1304 \, \text{m}^2$$

and the blade height is

$$b_2 = \frac{A_2}{2\pi r_2} = \frac{0.1304}{2\pi \cdot 0.121} = 0.0171 \, \text{m}$$

or $b_2 = 1.71$ cm.

∎

EXERCISES

9.1 Combustion gases with $\gamma = \frac{4}{3}$ and $c_p = 1148 \, \text{J}/(\text{kg} \cdot \text{K})$, at $T_{01} = 1050 \, \text{K}$ and $p_{01} = 310 \, \text{kPa}$ enter a radial inflow turbine. At the exit of the stator $M_2 = 0.9$. As the flow leaves the turbine, it is diffused to atmospheric pressure at $p_4 = 101.325 \, \text{kPa}$. The total-to-total efficiency of the turbine is $\eta_{tt} = 0.89$. Find the stator exit angle.

9.2 During a test air runs through a radial inflow turbine at the rate of $\dot{m} = 0.323 \, \text{kg/s}$ when the shaft speed is 55,000 rpm. The inlet stagnation temperature is $T_{01} = 1000 \, \text{K}$, and the pressure ratio is $p_{01}/p_3 = 2.1$. The blade radius at the inlet is $r_2 = 6.35 \, \text{cm}$. The relative velocity entering the blade is radial, and the flow leaves the blade without swirl. Find (a) the spouting velocity, (b) the total-to-static efficiency, and (c) the power delivered.

9.3 A radial turbine delivers $\dot{W} = 80 \, \text{kW}$ as its shaft turns at 44,000 rpm. Combustion gases with $\gamma = \frac{4}{3}$ and $c_p = 1148 \, \text{J}/(\text{kg} \cdot \text{K})$ enter the rotor with relative velocity radially inward at radius $r_2 = 8.10 \, \text{cm}$. At the exit the shroud radius is $r_{3s} = 6.00 \, \text{cm}$ and at this location $M_{3Rs} = 0.59$. The exit pressure is $p_3 = 101.325 \, \text{kPa}$ and exit temperature is $T_3 = 650 \, \text{K}$. The inlet Mach number is $M_2 = 0.9$. Find the hub to shroud ratio $\kappa = r_{3h}/r_{3s}$ at the exit.

9.4 A radial inflow turbine rotor, with rotor inlet radius $r_2 = 9.3 \, \text{cm}$ and blade height $b_2 = 1.8 \, \text{cm}$, turns at 42,000 rpm. Its working fluid is a gas mixture with $c_p = 1148 \, \text{J}/(\text{kg} \cdot \text{K})$ and $\gamma = \frac{4}{3}$. The exhaust pressure is $p_3 = 101.325 \, \text{kPa}$, and the total-to-static efficiency is $\eta_{ts} = 0.82$. The nozzle (stator) angle is $\alpha_2 = 67°$, and the velocity coefficient for the flow through the stator is $c_v = 0.96$. The Mach number at the exit of the stator is 0.8. Find (a) the inlet stagnation pressure to the stator, and (b) the stagnation pressure loss across the stator.

9.5 Gas with $\gamma = \frac{4}{3}$ and $c_p = 1148\,\text{K}/(\text{kg}\cdot\text{K})$ flows in a radial inflow turbine, in which the inlet stagnation temperature is $T_{01} = 980\,\text{K}$ and the inlet stagnation pressure is $p_{01} = 205.00\,\text{kPa}$. The exit pressure is $p_3 = 101.325\,\text{kPa}$, and the exit temperature is $T_3 = 831.5\,\text{K}$. The stagnation temperature at the exit is $T_{03} = 836.7\,\text{K}$. The pressure at the inlet to the rotor is $T_2 = 901.6\,\text{K}$, and the pressure is $p_2 = 142.340\,\text{kPa}$. The shaft speed is 160,000 rpm, and the radius ratio is $r_3/r_2 = 0.57$. Assume that the relative velocity is radial at the inlet and that there is no exit swirl. Find (a) the total-to-static efficiency, (b) the flow angles α_2 and β_3, and (c) η_S and η_R.

9.6 For an exit with no swirl, show that

$$\frac{W_3(r)}{W_2} = \frac{\sin(\alpha_2 - \beta_2)}{\cos\alpha_2}\frac{r_3}{r_2}\sqrt{\frac{r^2}{r_3^2} + \cot^2\beta_3}$$

in which r is the radius at an arbitrary location of the exit plane of the blade and r_3 and β_3 are the mean values of the exit radius and angle. Show further that at the mean radius

$$\frac{W_3}{W_2} = -\frac{\sin(\alpha_2 - \beta_2)}{\cos\alpha_2 \sin\beta_3}\frac{r_3}{r_2}$$

and plot the angle β_3 for the range $0.53 < r_3/r_2 < 0.65$ when $W_3/W_2 = 2$ and $\alpha_2 = 70°$ and $\beta_2 = -40°$.

9.7 An inexpensive radial inflow turbine has flat radial blades at both the inlet and the exit of the rotor. The shaft speed is 20,000 rpm. The radius of the inlet to the rotor is 10 cm and the mean radius at the exit is 6 cm. The ratio of blade widths is $b_3/b_2 = 1.8$ and the flow angle is $\alpha_2 = 75°$. The inlet stagnation temperature is $T_{01} = 420\,\text{K}$ and the exhaust flows into the atmospheric pressure 101.325 kPa. Assume that the gases that flow through the turbine have $\gamma = \frac{4}{3}$ and $c_p = 1148\,\text{J}/(\text{kg}\cdot\text{K})$ and the velocity coefficient of the nozzle is $c_N = 0.97$ and the rotor loss coefficient is $\zeta_R = 0.5$. (a) If the power output is 10 kW, what is the mass flow rate? Find (b) the static temperature at the exit of the stator, (c) the static temperature at the exit, (c) the blade height at the inlet and the exit, (d) the total-to-total efficiency, and (e) the total-to-static efficiency.

9.8 Combustion gases with $\gamma = \frac{4}{3}$ and $c_p = 1148\,\text{J}/(\text{kg}\cdot\text{K})$ enter a stator of a radial flow turbine with $T_{01} = 1150\,\text{K}$, $p_{01} = 1300\,\text{kPa}$, and $M_1 = 0.5$, and with a flow rate of $\dot{m} = 5.2\,\text{kg/s}$ and with the flow angle $\alpha_2 = 72°$. The radius of the inlet is $r_1 = 17.4\,\text{cm}$, the exit from the stator is at $r_{2e} = 15.8\,\text{cm}$, and the inlet to the rotor is at $r_2 = 15.2\,\text{cm}$. The chord of the stator is $c = 4.8\,\text{cm}$, and the width of the channel is $b = 1.2\,\text{cm}$. The rotational speed of the rotor is 31,000 rpm and the blade loading coefficient is $\psi = 1.3$. The exit static pressure is $p_3 = 320\,\text{kPa}$. The trailing-edge thickness of the 17 stator vanes can be ignored. Find (a) the total-to-static efficiency of the turbine, (b) the stagnation pressure loss across the stator, and (c) the stagnation pressure loss across the gap.

9.9 Combustion gases with $c_p = 1148\,\text{J/kg K}$ and gas constant $R = 287\,\text{J/kgK}$ enter the stator of a radial inflow turbine at the stagnation pressure $p_{01} = 346\,\text{kPa}$ and stagnation temperature $T_{01} = 980\,\text{K}$. They enter the rotor at the speed $V_2 = 481.4\,\text{m/s}$ with the relative flow making an angle $\beta_2 = -35°$ and exhaust into the atmosphere at 101.325 kPa.

The total-to-static efficiency of the turbine is $\eta_{ts} = 0.83$. Find (a) the angle at which the flow enters the rotor and (b) the relative Mach number at the inlet.

CHAPTER 10

HYDRAULIC TURBINES

Hydroelectric power plants are important providers of electricity in those countries in which there are large rivers or rainy high mountain regions, or both. In the United States large hydropower installations exist in the western states of Washington, Oregon, Idaho, and Nevada. The Tennessee Valley Authority, (TVA) has a large network of dams and hydropower stations in Tennessee and other southeastern states. Smaller plants in New England account for the rest of the installed capacity. In addition to the United States, large hydropower installations exist in Canada, Brazil, Russia, and China. Norway, Switzerland, Sweden, and Iceland obtain a large part of their power needs from hydroelectricity, and Norway in particular exports its excess generation. About 7% of primary energy production and 17% of total electricity generation in the world are obtained by hydropower. Hydropower plants exceeding 30 MW are designated as large plants. Plants with generation from 0.1 to 30 MW fall into a range of small plants and those with capacity below 0.1 MW are classified as microhydropower. Although hydropower is a clean form of energy, its negative aspects relate to blocking of fish migration paths and displacement of populations from valleys that have been cultivated for hundreds of years by local communities.

10.1 HYDROELECTRIC POWER PLANTS

The preliminary design of a power plant begins with the siting analysis, which consists of determining the available flow rate of water and its head. The rate of water flow depends on the season, and for large power plants with high capital costs, dams are used not only to

Principles of Turbomachinery. By Seppo A. Korpela
Copyright © 2011 John Wiley & Sons, Inc.

increase the head, but to store water for the dry season. When the water flow is sufficient, several turbines can be housed in the same plant.

A schematic of a hydroelectric power plant is shown in Figure 10.1. In small installations with high head a forebay is included and is made large enough that debris flowing into it will settle to the bottom. The floating debris is caught by an inlet screen. In a large installation the reservoir formed by an impound dam serves the same purpose. When the waterflow is regulated into the forebay, the head can be held constant and the turbine can be operated at its design condition. From the forebay the flow passes through a penstock into an inlet volute. After this the water is directed through a set of adjustable wicket or inlet gates to the runner. From the turbine the water is discharged into a draft tube and from there through a tailrace to the tailwater reservoir.

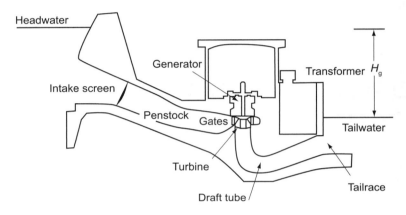

Figure 10.1 Hydroelectric power plant.

The elevation difference between the headwater and tailwater, denoted by H_g, is called the *gross head*. For a control volume with an inlet just below the free surface at the upper reservoir and an exit just below the free surface of the lower reservoir, the pressures at the inlet and exit are both atmospheric. Owing to the size of these reservoirs, the entering and leaving flow velocities are negligibly small. Hence the energy balance reduces to

$$gH_g = w_s + \sum_i gH_{Li}$$

in which $\sum_i H_{Li}$ is the head loss in the headwater, the penstock, the draft tube, and the exit to the tailwater. If the losses were absent $w_s = gH_g$ would be the reversible work delivered by the turbine. It is customary, however, to include the spiral volute and the draft tube as part of the turbine and call the difference $H_e = H_g - H_{Lp}$ the *effective head*, in which H_{Lp} represents the upstream loss. Then only the exit kinetic energy from the draft tube is a loss not attributed to the turbine. Efficiency now might be called the *total-to-static efficiency*, but this is not a customary practice. Clearly, the exit kinetic energy loss can be made small by making the draft tube long with a mild increase in area along its length.

In the same way as for centrifugal pumps the overall efficiency η is a product of three terms as,

$$\eta = \eta_m \eta_v \eta_h$$

in which η_m is the mechanical efficiency and η_v is the volumetric efficiency. The hydraulic efficiency η_h takes into account the various losses in the equipment downstream of the penstock. The overall efficiency can be written as

$$\eta = \frac{\dot{W}_o}{\dot{W}_R} \frac{\dot{W}_R}{\dot{W}} \frac{\dot{W}}{\rho Q g H_e}$$

in which $\dot{W}_o$ is the power delivered by the output shaft, $\dot{W}_R$ is the actual power delivered to the runner, and $\dot{W}$ is the rate at which the water does work on the turbine blades.

The hydraulic efficiency is defined as

$$\eta_h = \frac{\dot{W}}{\rho Q g H_e} = \frac{w}{g H_e}$$

If in the definition of effective head, the head losses in inlet spiral, the draft tube, and exit losses had also been subtracted out, the effective head would be smaller and the hydraulic efficiency larger.

There is some leakage flow that does no work on the turbine blades. For this reason the power delivered to the runner is

$$\dot{W}_R = \rho(Q - Q_L)w$$

in which Q_L is the volumetric flow rate of the leakage. The ratio

$$\frac{\dot{W}_R}{\dot{W}} = \frac{\rho(Q - Q_L)w}{\rho Q w} = \frac{Q - Q_L}{Q}$$

shows that the volumetric efficiency is also

$$\eta_v = \frac{Q - Q_L}{Q}$$

The mechanical efficiency is defined as the ratio

$$\eta_m = \frac{\dot{W}_o}{\dot{W}_R}$$

and the difference between $\dot{W}_R$ and $\dot{W}_o$ is caused by bearing friction and windage.

10.2 HYDRAULIC TURBINES AND THEIR SPECIFIC SPEED

The operating ranges for the three main types of hydraulic turbines are shown in Figure 10.2. In rivers with large flow rates Kaplan turbines and Francis turbines are used and Pelton wheels are the appropriate technology in very mountainous regions. A Pelton wheel is an impulse turbine in which a high-velocity jet impinges on buckets and the flow enters and leaves them at atmospheric pressure. Kaplan turbines are axial machines and handle large flow rates at relatively low head. They are mounted vertically with the electrical generators on the floor above the turbine bay. A bulb turbine is similar to a Kaplan turbine, but with a horizontal axis. Its handles even larger flow rates than does a Kaplan turbine at lower head. For Francis turbines the runner can be designed to accommodate a large variation in flow rates and elevation. A Deriaz turbine is similar to a Kaplan turbine, but with a conical runner. Turbines are classified according to their power-specific speed:

$$\Omega_{sp} = \frac{\Omega \sqrt{\dot{W}/\rho}}{(gH_e)^{5/4}}$$

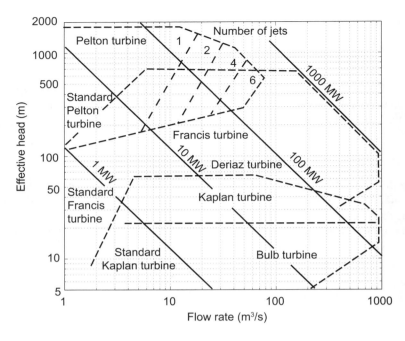

Figure 10.2 Types of hydraulic turbine and their operating ranges. (Drawn after charts by Sulzer Hydro Ltd. and Voith Hydro.)

Since

$$\dot{W} = \eta \rho Q g H_e$$

this reduces to

$$\Omega_{sp} = \frac{\Omega \sqrt{\eta Q}}{(g H_e)^{3/4}}$$

so that $\Omega_{sp} = \sqrt{\eta}\,\Omega_s$. For a well-designed hydropower station the overall efficiency at design condition is often greater than 0.9, and for this reason there is only slight difference between the two forms of specific speed.

Examination of Figure 10.2 shows that in the range where Pelton wheels are appropriate, the increased flow rate at a given head is accomplished by increasing the number of jets. In addition, if the flow rate is sufficient, multiple penstocks and Pelton wheels may be located at the same site.

The power-specific speeds of various machines are given in Table 10.1. It shows that the specific speed increases as the effective head decreases. Thus the high-specific-speed Francis runners operate at lower heads. For calculation of the specific speed, the shaft speed is needed. Since hydraulic turbines are used for electricity generation, the shaft speed must be synchronous with the frequency of the electric current. With some exceptions the frequency f is 50 Hz in Europe and Asia and 60 Hz in North and South America. To obtain the proper value for the line frequency, the shaft speed is to be

$$\Omega = \frac{120 f}{P}$$

in which the number of poles in the electric generator is P, which may vary from 2 to 48, and there is no fundamental reason why it could not be even higher. The number 120 is twice the 60 s in a minute, and the factor 2 arises from the fact that both ends of a magnet in an electric machine behave similarly. Thus a generator with 16 poles delivering energy at line frequency 50 Hz needs a turbine shaft speed of 375 rpm according to this formula. This is a typical shaft speed for large turbines.

Table 10.1 Power specific speed ranges of hydraulic turbines.

Type		Ω_{sp}	η %
Pelton wheel			
	Single jet	0.02 – 0.18	88 – 90
	Twin jet	0.09 – 0.26	89 – 92
	Three jet	0.10 – 0.30	89 – 92
	Four jet	0.12 – 0.36	86
Francis			
	Low-speed	0.39 – 0.65	90 – 92
	Medium-speed	0.65 – 1.2	93
	High-speed	1.2 – 1.9	93 – 96
	Extreme-speed	1.9 – 2.3	89 – 91
Kaplan turbine		1.55 – 5.17	87 – 94
Bulb turbine		3 – 8	

10.3 PELTON WHEEL

A Pelton wheel is shown in Figure 10.3. This machine provides an excellent way of producing power if the water reservoir is high. Water from the reservoir flows down a *penstock* and then through a set of nozzles that are directed against *buckets* fastened to a wheel. Penstocks are constructed from steel or reinforced concrete. Early ones were made of wood stave. Pelton wheels have been generally used if the total head is greater than 300 m. There is one installation in Switzerland in which the water head is 1700 m. An important advantage of Pelton wheels is that the machine can accommodate water laden with silt, for the erosion damage of its blades is easy to repair.

A Pelton wheel with six jets is shown in Figure 10.3. The number of buckets in the wheel can be calculated by an empirical formula of Tygun (cited by Nechleba [57]), namely,

$$Z = \frac{D}{2d} + 15$$

in which D is the wheel diameter and d is the jet diameter. In analysis of a Pelton wheel, the exit velocity is close to being axial and therefore is as small as possible. Generally the exit losses are ignored. The calculations are illustrated next.

■ **EXAMPLE 10.1**

A Pelton wheel generates $\dot{W}_o = 1\,\text{MW}$ of power as it operates under the effective head of $H_e = 410\,\text{m}$ and at 395 rpm. Its overall efficiency is $\eta = 0.84$, the nozzle velocity coefficient is $c_N = 0.98$, and the blade speed is $U = 37\,\text{m/s}$. Find (a) the specific speed and the recommended number of jets, (b) the wheel diameter,

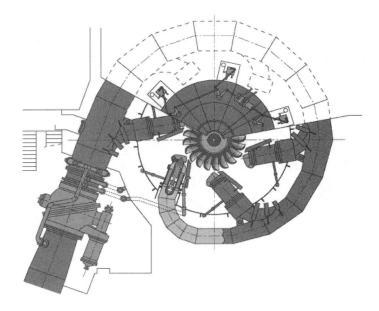

Figure 10.3 A six-jet Pelton wheel. (Drawing courtesy Voith Siemens Hydro.)

the diameter of the jet, as well as the recommended number of blades, and (c) the mechanical efficiency.

Solution: (a) The shaft power

$$\dot{W}_0 = \eta \rho Q g H_e$$

solved for the flow rate gives

$$Q = \frac{\dot{W}_o}{\eta \rho g H_e} = \frac{10^6}{0.84 \cdot 998 \cdot 9.81 \cdot 410} = 0.297 \, \text{kg/s}$$

and the specific speed comes out to as

$$\Omega_s = \frac{\Omega \sqrt{Q}}{(gH_e)^{3/4}} = \frac{395 \cdot \pi \sqrt{0.297}}{30 \cdot (9.81 \cdot 410)^{3/4}} = 0.0446$$

For this value of specific speed, one jet is sufficient.

(b) The jet velocity is given by

$$V_2 = c_N \sqrt{2gH_e} = 0.98\sqrt{2 \cdot 9.81 \cdot 410} = 87.90 \, \text{m/s}$$

so that the cross sectional area is

$$A = \frac{Q}{V_2} = \frac{0.297}{87.90} = 0.00337 \, \text{m}^2$$

and the jet diameter is

$$d = \sqrt{\frac{4A}{\pi}} = \sqrt{\frac{4 \cdot 0.00337}{\pi}} = 0.0655 \, \text{m} = 6.55 \, \text{cm}$$

The diameter of the wheel is

$$D = \frac{2U}{\Omega} = \frac{2 \cdot 37 \cdot 30}{395 \cdot \pi} = 1.79 \text{ m}$$

and the number of buckets comes out to be

$$Z = \frac{D}{2d} + 15 = \frac{179}{2 \cdot 6.55} + 15 = 28.6 \quad \text{or} \quad Z = 29$$

■

When using Tygun's formula for calculating the number of buckets, the typical range of diameter ratio $6 < D/d < 28$ gives the range $18 < Z < 29$ for a single wheel.

Consider a Pelton wheel in which the work done by the jet is

$$w = U(V_{u2} - V_{u3})$$

Owing to symmetry, only the stream deflected to the right needs to be analyzed. The inlet and outlet velocity diagrams are shown in Figure 10.4.

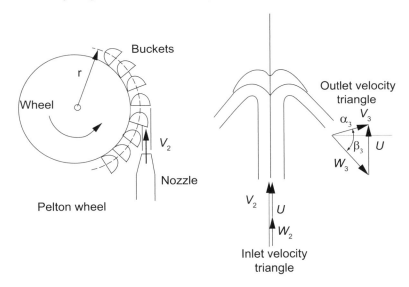

Figure 10.4 Setup for Pelton wheel analysis.

The fluid pressure at the inlet and exit is atmospheric, and frictional effects along the blades reduce the relative velocity to $W_3 = c_v W_2$, in which c_v is the velocity coefficient. At the inlet the flow is entirely tangential and thus $V_{u2} = V_2$ and $W_2 = V_2 - U$. The tangential component at the exit is

$$V_{u3} = U + W_3 \sin \beta_3 = U + c_v W_2 \sin \beta_3$$

Substituting $W_2 = V_2 - U$, gives

$$w = U(V_2 - U)(1 - c_v \sin \beta_3)$$

From this equation it is clear that the maximum power is obtained for $\beta_3 = -90°$, with other parameters held constant. It is not feasible to construct the buckets with such a large amount of turning, as the water would not clear the wheel. For this reason in actual installations β_3 is in the neighborhood of $-65°$. Any water that is partly spilled during the operation does no work on the blades. This spillage is taken into account by the volumetric efficiency.

In the expression for the overall efficiency

$$\eta = \eta_\text{h} \eta_\text{v} \eta_\text{m}$$

the hydraulic efficiency can be written as the sum of the nozzle efficiency and the rotor efficiency

$$\eta_\text{h} = \eta_\text{N} \eta_\text{R}$$

in which

$$\eta_\text{N} = \frac{V_2^2}{V_0^2}$$

and $V_0 = \sqrt{2gH_e}$ is the spouting velocity. The rotor efficiency is then

$$\eta_\text{R} = \frac{w}{\frac{1}{2} V_2^2}$$

The mechanical losses are taken to be proportional to U^2, with the proportionality constant denoted by K_u. The fluid dynamic losses in turbulent flows are often proportional to the square of the flow velocity, and therefore the windage losses should have this dependence on the blade speed. Then

$$\eta_\text{m} \eta_\text{v} = \frac{\dot{W}_o}{\dot{W}} = \frac{\dot{W}_\text{R} - \rho Q \frac{1}{2} K_u U^2}{\dot{W}} = \eta_\text{v} - \frac{K_u U^2}{2w}$$

Hence

$$\eta = \eta_\text{h} \left(\eta_\text{v} - \frac{K_u U^2}{2w} \right) = \left(\eta_\text{v} - \frac{K_u U^2}{2w} \right) \frac{w}{gH_e}$$

or

$$\eta = \frac{2\eta_\text{v} w}{2gH_e} - \frac{K_u U^2}{2gH_e} = \frac{1}{V_0^2} (2\eta_\text{v} w - K_u U^2)$$

Substituting the expression for work into this and replacing V_0 by $V_2/\sqrt{\eta_\text{N}}$ gives

$$\eta = \frac{\eta_\text{N}}{V_2^2} [2\eta_\text{v} U(V_2 - U)(1 - c_\text{v} \sin \beta_3) - K_u U^2]$$

This can be written is terms of the ratio of the blade speed to the nozzle velocity $\lambda = U/V_2$ as follows:

$$\eta = \eta_\text{N} (2\eta_\text{v} \lambda (1 - \lambda)(1 - c_\text{v} \sin \beta_3) - K_u \lambda^2)$$

To determine the rotational speed at which the power is maximum, this expression is differentiated with respect to λ while other parameters are held fixed. Setting the result to zero gives

$$\frac{d\eta}{d\lambda} = 2\eta_\text{v} (1 - 2\lambda)(1 - c_\text{v} \sin \beta_3) - 2K_u \lambda = 0$$

from which

$$\lambda_\text{m} = \frac{1 - c_\text{v} \sin \beta_3}{2\eta_\text{v} (1 - c_\text{v} \sin \beta_3) + K_u}$$

At this value the maximum efficiency is

$$\eta_{max} = \frac{\eta_N \eta_v^2 (1 - c_v \sin\beta_3)^2}{2\eta_v(1 - c_v \sin\beta_3) + K_u}$$

If $K_u = 0$ then $\eta_m = 1$ and this reduces to

$$\eta_{max} = \frac{1}{2}\eta_N \eta_v (1 - c_v \sin\beta_3) = \eta_N \eta_{Rmax} \eta_v$$

in which the maximum rotor efficiency is

$$\eta_{R\,max} = \frac{1}{2}(1 - c_v \sin\beta_3)$$

The efficiency of the Pelton wheel as a function of the speed ratio $\lambda = U/V_2$ is shown in Figure 10.5.

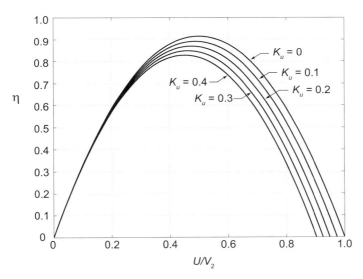

Figure 10.5 Efficiency of a Pelton wheel for various values of the windage loss coefficient.

■ EXAMPLE 10.2

A Pelton wheel operates with a gross head of 530 m and a flow rate of 9 m³/s. The penstock length is 800 m, its diameter is 1.2 m, and its RMS roughness is 0.1 mm. The minor losses increase the equivalent length of the penstock to 880 m. The hydraulic efficiency is $\eta_h = 0.84$, and the shaft speed is 650 rpm. Find (a) the effective head and the power delivered by the turbine and (b) the specific speed and from it the recommended number of jets and the number of buckets in the wheel. The nozzle coefficient is $c_N = 0.97$, and the ratio of the blade speed to the discharge velocity is $\lambda = U/V_2 = 0.45$.

Solution: (a) The velocity of water in the penstock is

$$V_p = \frac{Q}{A} = \frac{4Q}{\pi D_p^2} = \frac{4 \cdot 9}{\pi \cdot 1.22^2} = 7.96 \text{ m/s}$$

The Reynolds number is therefore

$$\mathrm{Re} = \frac{\rho V_{\mathrm{p}} D_{\mathrm{p}}}{\mu} = \frac{998 \cdot 7.96 \cdot 1.2}{1.01 \cdot 10^{-3}} = 9.44 \cdot 10^6$$

The Colebrook formula gives the friction factor the value $f = 0.0116$, so that the head loss is

$$H_{\mathrm{L}} = f \frac{L}{D_{\mathrm{p}}} \frac{V_{\mathrm{p}}^2}{2g} = \frac{0.116 \cdot 880 \cdot 7.96^2}{1.2 \cdot 2 \cdot 9.81} = 27.4 \,\mathrm{m}$$

The gross head is measured as the elevation difference between the headwater and the nozzle. Only the upstream losses are included in the calculation of the effective head, and whatever kinetic energy is left after the water is discharged from the wheel is taken into account in the turbine efficiency. Thus the turbine efficiency is the total-to-static efficiency. The effective head is $H_{\mathrm{e}} = 530 - 27.4 = 502.6 \,\mathrm{m}$.

The power delivered is

$$\dot{W}_{\mathrm{o}} = \eta \rho Q g H_{\mathrm{e}} = 0.84 \cdot 998 \cdot 9 \cdot 9.81 \cdot 502.6 = 37.20 \,\mathrm{MW}$$

(b) The specific speed comes out to be

$$\Omega_{\mathrm{s}} = \frac{\Omega \sqrt{Q}}{(gH_{\mathrm{e}})^{3/4}} = \frac{650 \cdot \pi \sqrt{9}}{30 \cdot (9.81 \cdot 502.6)^{3/4}} = 0.347$$

The recommended number of jets is four. Hence the discharge from each jet is $Q_{\mathrm{j}} = 2.25 \,\mathrm{m}^3/\mathrm{s}$. By calculating the discharge velocity as

$$V_2 = c_{\mathrm{N}} \sqrt{2gH_{\mathrm{e}}} = 0.97 \sqrt{2 \cdot 9.81 \cdot 502.6} = 96.32 \,\mathrm{m/s}$$

the nozzle diameter can be determined as

$$d = \sqrt{\frac{4Q_{\mathrm{j}}}{\pi V_2}} = \sqrt{\frac{4 \cdot 9}{4 \cdot \pi \cdot 96.32}} = 0.173 \,\mathrm{m}$$

and the blade speed is $U = 0.45 V_2 = 43.35 \,\mathrm{m/s}$. Therefore the diameter of the wheel is

$$D = \frac{2U}{\Omega} = \frac{2 \cdot 43.35 \cdot 30}{650 \cdot \pi} = 1.27 \,\mathrm{m}$$

and the number of buckets is

$$Z = \frac{D}{2d} + 15 = \frac{127}{2 \cdot 17.3} + 15 = 18.7 \quad \text{or} \quad Z = 19$$

■

■ **EXAMPLE 10.3**

A Pelton wheel operates with an effective head of 310 m producing 15 MW of power. The overall efficiency of the turbine is $\eta = 0.84$, the velocity coefficient of the nozzle is $c_{\mathrm{N}} = 0.98$, and the velocity coefficient of the rotor is $c_{\mathrm{v}} = 0.85$. The angle of the relative velocity leaving the rotor is $\beta_3 = -73°$. The wheel rotates at 500 rpm, and the ratio of the blade speed to nozzle velocity is 0.46. (a) Give a recommended number of jets to which the supply flow should be split and the number of blades on the wheel. (b) Find the mechanical and volumetric efficiencies given the value of $K_u = 0.05$. (c) Find also the optimum speed ratio.

Solution: (a) The volumetric flow rate can be calculated from

$$\dot{W}_o = \eta \rho Q g H_e$$

which gives

$$Q = \frac{\dot{W}_o}{\eta \rho g H_e} = \frac{15 \cdot 10^6}{0.84 \cdot 998 \cdot 9.81 \cdot 310} = 5.88 \, \text{m}^3/\text{s}$$

The spouting velocity is

$$V_0 = \sqrt{2gH_e} = \sqrt{2 \cdot 9.81 \cdot 310} = 78.0 \, \text{m/s}$$

so that the nozzle velocity comes out as

$$V_2 = c_N V_0 = 0.98 \cdot 78.0 = 76.43 \, \text{m/s}$$

The blade speed is therefore $U = \lambda V_2 = 0.46 \cdot 76.43 = 35.15 \, \text{m/s}$. With the specific speed qual to

$$\Omega_s = \frac{\Omega \sqrt{Q}}{(gH_e)^{3/4}} = \frac{500 \cdot \pi \sqrt{5.88}}{30 \cdot (9.81 \cdot 310)^{3/4}} = 0.310$$

the suitable number of jets is $N = 4$. The flow rate from each jet is therefore $Q_j = Q/N = 1.47 \, \text{m}^3/\text{s}$, and for this flow rate and velocity the nozzle diameter is

$$d = \sqrt{\frac{4Q_j}{\pi V_2}} = \sqrt{\frac{4 \cdot 1.47}{\pi \cdot 76.43}} = 0.156 \, \text{m} = 15.6 \, \text{cm}$$

The wheel diameter is

$$D = \frac{2U}{\Omega} = \frac{2 \cdot 35.16 \cdot 30}{500 \cdot \pi} = 1.343 \, \text{m}$$

so that the recommended number of buckets on the wheel is

$$Z = \frac{D}{2d} + 15 = \frac{134.3}{2 \cdot 15.6} + 15 = 19.3 \quad \text{or} \quad Z = 20$$

(b) The specific work done by the jet on the rotor is

$$\begin{aligned} w &= U(V_2 - U)(1 - c_v \sin \beta_3) \\ &= 35.16(76.43 - 35.16)(1 - 0.85 \sin(-73°)) = 2630.4 \, \text{J/kg} \end{aligned}$$

The nozzle efficiency and rotor efficiencies are

$$\eta_N = c_N^2 = 0.98^2 = 0.960 \qquad \eta_R = \frac{w}{\frac{1}{2}V_2^2} = \frac{2 \cdot 2630.6}{76.16^2} = 0.900$$

Hence the hydraulic efficiency is $\eta_h = \eta_N \eta_R = 0.960 \cdot 0.900 = 0.864$. The mechanical efficiency is obtained from

$$\eta_m = 1 - \frac{K_u U^2}{2\eta_v w}$$

Solving for η_v from

$$\eta = \eta_h \eta_v \eta_m \quad \text{gives} \quad \eta_v = \frac{\eta}{\eta_h \eta_m}$$

and substituting this into the previous equation leads to

$$\eta_m = 1 - \frac{K_u U^2 \eta_h \eta_m}{2\eta w}$$

which when solved for η_m results in

$$\eta_m = \left(1 + \frac{K_u U^2 \eta_h}{2\eta w}\right)^{-1} = \left(1 + \frac{0.05 \cdot 35.16^2 \cdot 0.864}{2 \cdot 0.84 \cdot 2630.6}\right)^{-1} = 0.988$$

The volumetric efficiency is therefore

$$\eta_v = \frac{\eta}{\eta_h \eta_m} = \frac{0.84}{0.864 \cdot 0.988} = 0.983$$

The optimum speed ratio is

$$\lambda_{opt} = \frac{\eta_v (1 - c_v \sin \beta_3)}{2\eta_v (1 - c_v \sin \beta_3) + K_u}$$
$$= \frac{0.984(1 - 0.85 \sin(-73°))}{2 \cdot 0.984(1 - 0.85 \sin(-73°)) + 0.05} = 0.493$$

■

10.4 FRANCIS TURBINE

A schematic of a Francis turbine is shown in Figure 10.6. It is a radial inward turbine similar in principle to the radial turbine discussed in Chapter 9. The flow enters the turbine through a set of adjustable inlet guidevanes, located downstream of a spiral volute and a penstock. The spiral volute is absent in Figure 10.1. Both the volute and the penstock may be absent if the head is low, but the flow capacity is large. In that case the turbine can be placed into a pit. The shape of the draft tube is conical with a straight axis. An alternative is to curve the axis as is shown in Figure 10.1. This way, the draft tube can be longer and the diffusion milder over its length.

The shape of the runner depends on the specific speed and thus on the flow rate and the effective head, with a low speed-runner having radial inlet and outlet. As the specific speed increases, the flow is turned such that it leaves in the axial direction. These modifications are illustrated in Figure 10.7.

The flow leaves a low-specific-speed runner in a radial direction, and it receives no further guidance from the blades as it turns toward the axial direction and then enters the draft tube. When the flow rate increases, the velocity in this turning region also increases, and in order to reduce it, the flow must be guided into the axial direction by properly shaped blades. The blade design aims to reduce the tangential component of the velocity to zero.

For a given flow rate and effective head Figure 10.2 is a guide to what kind of turbine is best suited to the proposed power plant site. There is some latitude in the choices

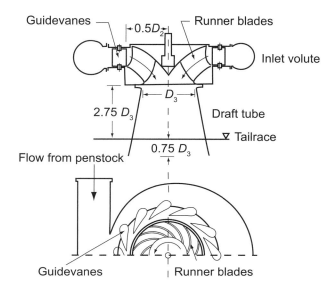

Figure 10.6 Francis turbine.

The axes in the figure are in terms of extensive properties and, as usual, an engineering design proceeds also in terms of not only intensive properties but, if possible, in terms of nondimensional parameters. The power-specific speed is one of them. The other is the specific diameter, defined by

$$D_s = \frac{D_2 (gH_e)^{1/4}}{\sqrt{Q}}$$

The loading coefficient, ψ and the flow coefficient ϕ, defined as

$$\psi = \frac{gH_e}{\Omega^2 D_2^2} \qquad \phi = \frac{Q}{\Omega D_2^3}$$

can be related to the specific speed and the specific diameter. It is readily shown that

$$\psi = \frac{1}{\Omega_s^2 D_s^2} \qquad \phi = \frac{1}{\Omega_s D_s^3}$$

The ratio of the blade speed to the spouting velocity $V_0 = \sqrt{2gH_e}$ is

$$\frac{U_2}{V_0} = \frac{\Omega D_2}{2\sqrt{2gH_e}} = \frac{1}{\sqrt{8}} \Omega_s D_s = \frac{1}{\sqrt{8\psi}}$$

Balje has shown that for hydraulic turbines the locus of maximum efficiency tends to be in the range $0.6 < U_2/V_0 < 1.1$. This is equivalent to $0.1 < \psi < 0.35$, with Pelton wheels and Francis turbines in the high end of this range for ψ and Kaplan turbines in the low end [5]. Figure 10.8 is a plot with specific speed on the abscissa and specific diameter on the ordinate, with some lines of constant loading and flow coefficient added. The figures of different runners are placed where their efficiency is the highest. Pelton wheels with multiple jets are not shown.

The flow angles shown for the different runners in Figure 10.7 can be calculated by noting that

$$\tan \alpha_2 = \frac{V_{u2}}{V_{r2}} = \frac{U_2 V_{u2}}{U_2 V_{r2}} = \frac{w}{U_2 V_{r2}}$$

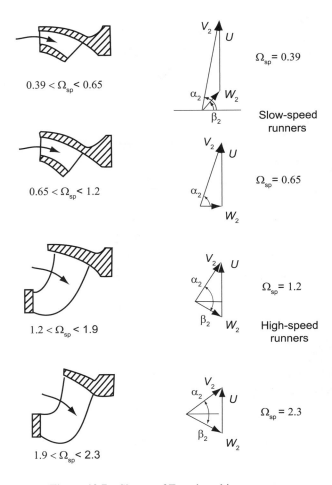

Figure 10.7 Shapes of Francis turbine runners.

and since $Q = \pi D_2 b_2 V_{r2}$, and $w = \eta_h g H_e$, this can be written as

$$\tan \alpha_2 = \frac{\eta_h g H_e \pi D_2 b_2}{U_2 Q}$$

From the definition of the specific speed the ratio

$$\frac{gH_e}{Q} = \frac{\Omega^2}{\Omega_s^2 \sqrt{gH_e}}$$

follows immediately, and with $\Omega = 2U_2/D_2$ and $V_0 = \sqrt{2gH_e}$, the formula for the tangent of α_2 can further be written as

$$\tan \alpha_2 = \frac{4\eta_h \sqrt{2}\pi b_r}{\Omega_s^2} \frac{U_2}{V_0}$$

in which the ratio $b_r = b/D_2$ appears. A graph of this ratio appears in the book by Nechleba [57]. A quadratic curve fit over the range of the power-specific speeds for Francis turbines

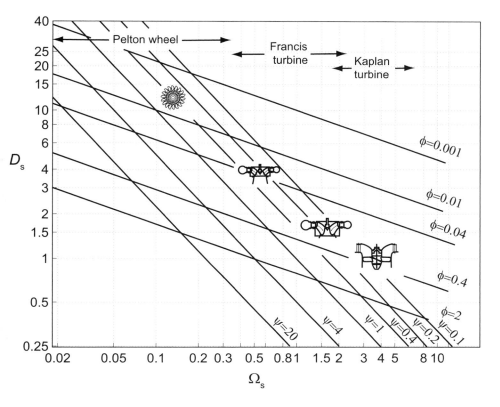

Figure 10.8 Specification map of hydraulic turbines.

gives
$$b_r = -0.0505\Omega_{sp}^2 + 0.26\Omega_{sp} + 0.018$$
In addition, the fit
$$\frac{U_2}{V_0} = 0.74\Omega_{sp}^{0.238}$$
over $0.39 < \Omega_{sp} < 2.3$ gives the range $0.59 < U_2/V_0 < 0.90$. The results of the calculations are shown in the Table 10.2. For these the hydraulic efficiency was taken to have the constant value $\eta_h = 0.9$ and for simplicity the mechanical and volumetric losses were neglected.

The angle β_2 can be determined from the exit velocity diagram, which yields
$$\tan\beta_2 = \frac{\sin\alpha_2 - U_2/V_2}{\cos\alpha_2}$$
and it is easy to show that the ratio U_2/V_2 is
$$\frac{U_2}{V_2} = \frac{\Omega_{sp}^4}{(4\pi b_r)^2 \eta_h^5} = \tan^2\alpha_2 \sin\alpha_2$$

For low-specific-speed runners the flow angle α_2 is quite large and the angle β_2 at first drops quite quickly as Ω_{sp} increases, but then it settles to a value around $-35°$ for the high-speed runners.

Table 10.2 Flow angles for various power-specific speeds for a Francis turbine of hydraulic efficiency $\eta_h = 0.9$.

Ω_{sp}	b_r	U/V_0	α_2	β_2
0.39	0.11	0.59	80.0°	49.5°
0.65	0.17	0.67	73.8°	−2.9°
1.20	0.26	0.77	61.1°	−33.3°
1.90	0.33	0.86	46.0°	−35.8°
2.30	0.35	0.90	38.0°	−34.0°

Neither the head, nor the volumetric flow rate, was needed to determine the shape of the velocity triangle, since the shape involves only angles and the ratio of two of its sides. These are nondimensional quantities. In a design for a given site, the head and flow rate are known, and there exist empirical relations that relate the specific speed to the effective head. One such equation is by Mozorov (cited by Nechleba [57]) can be written as

$$\Omega_{sp} = 0.82 \left(\frac{H_r}{H_e} \right)^{0.57}$$

in which $H_r = 100$ m has been chosen as a reference head. After that the diameter can be calculated from the calculated value of the specific diameter. Similarly by gathering data from a large number of installations, Lugaresi and Massa [53] carried out a statistical analysis and developed the correlation

$$\Omega_s = 1.14 \left(\frac{H_r}{H_e} \right)^{0.512}$$

The reference head is again $H_r = 100$ m. This correlation is based on the specific speed rather than power-specific speed, and is to be preferred over the proposal of Mozorov, as it is based on a more recent and larger set of turbines. It is important to note that these correlations are to be used as a guide and individual designs may differ from them. The use of the correlations is illustrated next.

■ **EXAMPLE 10.4**

A small Francis turbine is contemplated for a small power plant with an effective head of 220 m and capacity of 1.9 m³/s. The line frequency is 60 Hz and the anticipated overall efficiency is $\eta = 0.94$. Neglect mechanical and volumetric losses. (a) Determine the power delivered by the turbine. (b) What is your recommendation for the shaft speed if the electric line frequency is 60 Hz? (c) What should the diameter D_2 of the runner be, and what is the tip speed of the blade?

Solution: (a) The power can be calculated readily as

$$\dot{W}_o = \eta \rho Q g H_e = 0.94 \cdot 998 \cdot 1.9 \cdot 9.81 \cdot 220 = 3.85 \text{ MW}$$

(b) Examination of Figure 10.2 shows that a single-jet Pelton wheel and a Francis wheel are suitable for the site. It is therefore anticipated that a Francis turbine of low specific speed is a reasonable choice. Using the correlation for specific speed in terms of the effective head gives

$$\Omega_s = 1.14 \left(\frac{100}{220}\right)^{0.512} = 0.7614$$

and then from the definition of the specific speed the shaft speed can be determined to be

$$\Omega = \frac{\Omega_s(gH_e)^{3/4}}{\sqrt{Q}} = \frac{0.7614(9.81 \cdot 220)^{3/4}30}{\pi\sqrt{1.9}} = 1641 \text{ rpm}$$

If the shaft speed is taken to be 1800 rpm, then the number of poles in the generator is

$$P = \frac{120f}{\Omega} = \frac{120 \cdot 60}{1800} = 4$$

(c) With this shaft speed the specific speed is

$$\Omega_s = \frac{\Omega\sqrt{Q}}{(gH_e)^{3/4}} = \frac{1800 \cdot \pi\sqrt{1.9}}{30 \cdot (9.81 \cdot 220)^{3/4}} = 0.821$$

and the power-specific speed is $\Omega_{sp} = \sqrt{\eta}\,\Omega_s = \sqrt{0.94}\,0.821 = 0.796$. The velocity ratio is

$$\frac{U_2}{V_0} = 0.74\,\Omega_{sp}^{0.238} = 0.701$$

The specific diameter can then be calculated from

$$D_s = \frac{\sqrt{8}U_2/V_0}{\Omega_s} = \frac{0.701 \cdot \sqrt{8}}{0.821} = 2.415$$

and the diameter is

$$D_2 = \frac{D_s\sqrt{Q}}{(gH_e)^{1/4}} = \frac{2.415\sqrt{1.9}}{(9.81 \cdot 220)^{1/4}} = 0.488 \text{ m}$$

The blade speed is therefore $U_2 = D_2\Omega/2 = 46.0 \text{ m/s}$. ∎

To obtain more accurate results, the volumetric flow rate at the inlet to the runner can be calculated from

$$Q = \left(\pi D_2 - \frac{Zt}{\cos\beta_2}\right) b_2 W_{r2}$$

in which t is the blade thickness, Z is the number of blades, and W_{r2} is the radial component of the relative velocity with the flow angle β_2, at the inlet. The number of blades can be taken to be between $Z = 11V_0/U_2$ and $Z = 13V_0/U_2$, which for the typical range of U_2/V_0 comes close to 16 blades. The runner in Figure 10.9 has 17 blades, which is appropriately a prime number to prevent some resonance vibrations.

As the flow enters the draft tube, the axial velocity V_{x3} can be obtained from the expression for the volumetric flow rate, which is

$$Q = \frac{\pi D_3^2}{4} V_{x3} \qquad (10.1)$$

It is assumed that the flow enters without swirl so that $V_{x3} = V_3$. For the conical draft tube shown in Figure 10.6, the length of the tube is 2.5 – 3 times the diameter D_3 and its

Figure 10.9 Francis turbine viewed from downstream.

submerged length into the tailwater ranges from $0.5D_3$ to D_3. Since the function of the draft tube is to diffuse the flow, its flare must be kept small enough to prevent separation.

The diffuser efficiency is defined as

$$\eta_d = 1 - \frac{2gH_d}{V_3^2 - V_4^2} = 1 - \frac{2gH_d/V_3^2}{1 - A_3^2/A_4^2}$$

in which H_d is the head loss in the diffuser. For a conical diffuser the peak efficiency is at the cone angle $2\theta = 6°$ and has a value of about $\eta_{d\,max} = 0.9$. Defining the loss coefficient K_d via the equation

$$H_d = K_d \frac{V_3^2}{2g}$$

it can readily seen that it can be expressed in terms of the diffuser efficiency as

$$K_d = (1 - \eta_d)\left(1 - \frac{A_3^2}{A_4^2}\right)$$

Since all the kinetic energy leaving the draft tube is dissipated in the tail water pool, the exit loss is

$$H_e = \frac{V_4^2}{2g} = \frac{V_4^2}{V_3^2}\frac{V_3^2}{2g} = \frac{A_3^2}{A_4^2}\frac{V_3^2}{2g}$$

so that the loss coefficient based on the velocity V_3 is

$$K_e = \frac{A_3^2}{A_4^2}$$

The draft tube and exit loss coefficients and their sum are shown as a function of the ratio A_4/A_3 in Figure 10.10. If the inlet and exit diameters D_3 and D_4 are known, so is the area

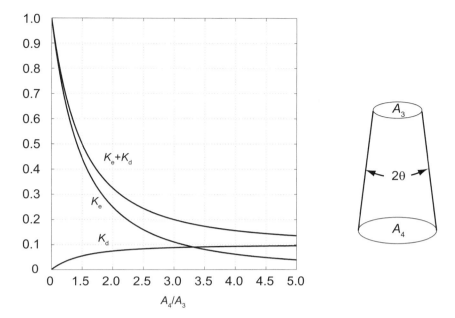

Figure 10.10 Loss coefficients for a draft tube.

ratio, and the inlet velocity V_3 to the draft tube is determined from Eq. (10.1). These then give a way to calculate the head loss in the draft tube. The true reversible work is given by

$$w_s = g\left(H_e - H_d - \frac{V_4^2}{2g}\right)$$

which can then be compared to the actual work so that the true efficiency of the turbine can be determined.

10.5 KAPLAN TURBINE

A Kaplan turbine is shown in Figure 10.11. It is an axial-flow turbine suitable for large installations and it generally operates under a head ranging from 20 to 70 m. The guidevanes are similarly adjustable as in a Francis turbine. Its runner is a propeller type, typically having 3 to 6 blades, but for smaller flow rates the number may reach 10. Each blade can be adjusted to the proper flow conditions by rotating it about its axis. This makes the power plant perform well also at off-design flow rates. In a typical installation the runner, turning at 120 rpm, might have a blade with 1 m hub radius and 3 m tip radius, with a volumetric flow rate of 200 m³/s through a turbine. Such a turbine produces around 74.5 MW of power from a single turbine.

A bulb turbine is similar to a Kaplan turbine but with a horizontal axis. A power plant based on a bulb turbine can be constructed at a lower cost than one with a Kaplan turbine. It can be used at low heads of 4 – 10 m when the speed of the flow is increased by proper design of the inlet channel. The design output power ranges between 100 kW and 50 MW for blade diameters between 0.8 and 8.4 m.

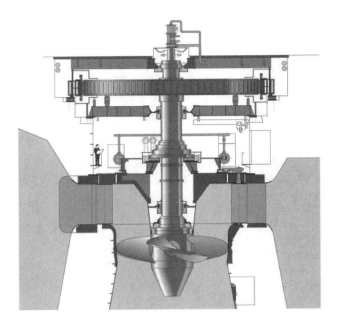

Figure 10.11 Kaplan turbine.

As the flow moves through the axial turbine its relative velocity is turned only moderately and by differing amounts at the hub and the tip of the blade. The blades extend over a considerable distance, and the local velocity of the blade $U = r\Omega$ increases outward. For this reason, for shockless entry, the absolute value of the blade angle must increase toward the tip. Thus the blade is oriented quite broadside to the flow at the tip. The flow, having started from stationary conditions at the headwater, remains free of vorticity as it flows into the turbine. Hence the tangential velocity has the *free vortex* distribution

$$K = rV_{u2}(r) = r_{2t}V_{u2t}$$

in which r_{2t} refers to the tip radius. If the exit swirl is absent then the work delivered is

$$w = U_2 V_{u2} = r\Omega V_{u2t}\frac{r_{2t}}{r} = \Omega V_{u2t} r_{2t} = K\Omega$$

which is independent of the radial location. This is the same result as was found when the axial turbines were discussed in Chapter 6.

An equation similar to that of Morozov for Francis turbines, can be obtained from the graphical result in Nechleba [57] for Kaplan turbines. However, by using data from more recent installations, Schweiger and Gregory [67] give the equation

$$\Omega_s = 2.76 \left(\frac{H_r}{H_e}\right)^{0.486} \tag{10.2}$$

in which the reference head has been chosen to be $H_r = 30\,\text{m}$. In addition, the hub-to-tip radii ratio may be correlated as

$$\frac{r_h}{r_t} = 0.8 - 0.1\,\Omega_{sp}$$

The number of blades increases with decreasing specific speed, with 3 blades when the power specific speed is 5.2 and and 10 blades when the power-specific speed drops to 1.7.

■ EXAMPLE 10.5

The Otari power plant in Japan delivers 100 MW of power when the flow rate is 220 m³/s and the effective head is 51 m. The diameter of the Kaplan turbine is $D_{2t} = 6.1$ m, and the hub-to-tip ratio is $\kappa = 0.6$. The generator has 48 poles and delivers power at a line frequency of 50 Hz. (a) Find the efficiency of the turbine. (b) Calculate the flow angles entering and leaving the rotor.

Solution: (a) The efficiency is

$$\eta = \frac{\dot{W}_0}{\rho Q g H_e} = \frac{10^8}{998 \cdot 220 \cdot 9.81 \cdot 51} = 0.91$$

(b) Since the line frequency is 50 Hz and the generator was chosen to have 48 poles, the shaft speed is $\Omega = 120 \cdot 50/48 = 125$ rpm. The tip speed of the runner blade is

$$U_{2t} = \frac{D_{2t}}{2}\Omega = \frac{6.1 \cdot 125 \cdot \pi}{2 \cdot 30} = 39.9 \text{ m/s}$$

The axial velocity is uniform and is given by

$$V_{x2} = \frac{4Q}{\pi D_{2t}^2(1 - \kappa^2)} = \frac{4 \cdot 220}{\pi \cdot 6.1^2(1 - 0.6^2)} = 17.76 \text{ m/s}$$

Since each blade element delivers the same amount of work, the tangential velocity at the tip can be determined from

$$V_{u2t} = \frac{\eta g H_e}{U_{2t}} = \frac{0.91 \cdot 9.81 \cdot 51}{39.9} = 11.41 \text{ m/s}$$

The flow angles for the absolute and relative velocities can now be determined. They are shown in Figure 10.12 across the span of the blades. To illustrate the calculations, at the mean radius $r_{2m} = 2.44$ m the absolute velocity makes an angle

$$\alpha_{2m} = \tan^{-1}\left(\frac{V_{u2t}r_t}{V_{x2}r_m}\right) = \tan^{-1}\left(\frac{11.41 \cdot 3.05}{17.76 \cdot 2.44}\right) = 50.5°$$

The tangential component of the relative velocity at this location is

$$W_{u2m} = V_{x2}\tan\alpha_{2m} - U_{2t}\frac{r_m}{r_t} = 11.76\tan(50.5°) - \frac{39.9 \cdot 2.44}{3.05} = -17.68 \text{ m/s}$$

and the flow angle is

$$\beta_{2m} = \tan^{-1}\left(\frac{W_{u2m}}{W_{x2}}\right) = \tan^{-1}\left(\frac{-17.68}{17.76}\right) = -56.4°$$

The flow leaves the runner axially. Therefore the tangential component of the relative velocity is $W_{u3m} = -U_{2t}r_m/r_t$ and with the axial velocity constant, the flow angle is

$$\beta_{3m} = \tan^{-1}\left(\frac{W_{u3m}}{W_{x3}}\right) = \tan^{-1}\left(\frac{-39.9 \cdot 2.33}{11.76 \cdot 3.05}\right) = -69.8°$$

■

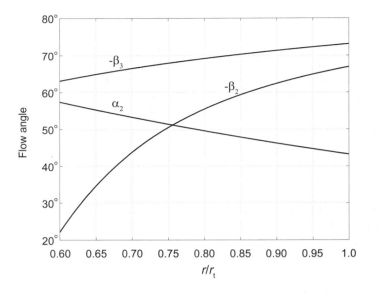

Figure 10.12 Flow angles for the turbine.

Small turbines have been excluded from Figure 10.2. The single-jet Pelton wheel with low flow rate extends past the left margin of the graph. Some who live far from the electricity grid and have access to streams with headwaters in high hills have built power plants using such small Pelton wheels. An alternative is a turgo turbine, shown in Figure 10.13a. It operates at the effective head of 50 – 250 m. The analysis is similar to those for single stage steam turbines and a Pelton wheel. The nozzle angle is about 70°.

A crossflow turbine, also called a *Banki–Mitchell turbine*, can be operated under heads between 5 and 200 m, but it is generally used in rural settings with ample flow rates and low head. It is shown in Figure 10.13b. It looks like a waterwheel, but the blades are designed is such a way that as water flows through the turbine, it does work on both the leading set of blades and on the trailing set. Because the axis is horizontal, the shaft and the blades can be made long to accommodate a large flow.

10.6 CAVITATION

Cavitation takes place in a hydraulic turbine if the minimum pressure drops below the vapor pressure of water at the prevailing temperature. For this reason, elevation of the plant in relation to the tailwater needs to be properly chosen. In addition, since the atmospheric pressure drops with increasing elevation, the margin between the minimum pressure and the vapor pressure diminishes, and this is to be taken into account. With large runners in hydroelectric power plants, the cavitation damage can become very expensive, not only in equipment repair but also in the unavailability of the plant during repairs.

The physical basis of cavitation is discussed in the review paper by Arakeri [4], who also proposes alternative criteria that factor in more precisely the location of the local minimum in pressure. In this section, however, the typical engineering criterion is used. The static

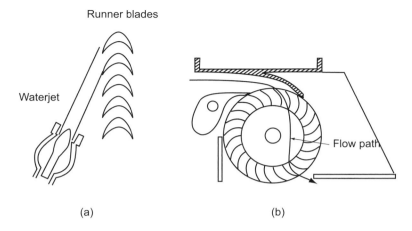

Figure 10.13 Illustrations of a turgo turbine (a) and a crossflow or a Banki–Mitchell turbine (b).

pressure at the exit of the turbine is given by

$$\frac{p_3}{\rho g} + \frac{V_3^2}{2g} + H_s = \frac{p_a}{\rho g} + \frac{V_4^2}{2g} + H_d$$

in which the last two terms represent the exit head loss and the head loss in the diffuser and H_s is the elevation of the turbine above the tailwater. The net positive suction pressure is defined as the sum of the static pressure head and the dynamic head minus the vapor pressure head of water at the ambient temperature

$$H_{sv} = \frac{p_3}{\rho g} + \frac{V_3^2}{2g} - \frac{p_v}{\rho g}$$

Substituting from the previous equation gives

$$H_{sv} = \frac{p_a - p_v}{\rho g} - H_s + \frac{V_4^2}{2g} + H_d$$

The head loss in the draft tube has been shown to be

$$H_d = (1 - \eta_d)\frac{(V_3^2 - V_4^2)}{2g}$$

Dividing through by the effective head leads to the definition of the Thoma cavitation parameter σ, as

$$\sigma = \frac{H_{sv}}{H_e} = \frac{p_a - p_v}{\rho g H_e} - \frac{H_s}{H_e} + \frac{V_4^2}{2g H_e} - \frac{H_d}{H_e}$$

Critical values for this parameter have been given by Moody and Zowski [56] for Francis turbines as

$$\sigma_c = 0.006 + 0.123 \Omega_{sp}^{1.8}$$

and for Kaplan turbines as

$$\sigma_c = 0.100 + 0.037 \Omega_{sp}^{2.5}$$

They are shown in Figure 10.14.

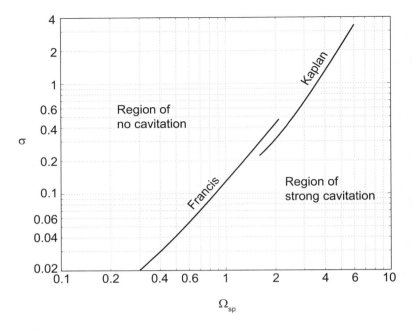

Figure 10.14 Critical Thoma parameter for Francis and Kaplan turbines.

EXERCISES

10.1 A Pelton wheel operates from an effective head of $H_e = 300$ m and at a flow rate of $4.2 \, \text{m}^3/\text{s}$. The wheel radius is $r_2 = 0.75$ m, and rotational speed is 450 rpm. The water that leaves the penstock is divided into 5 streams. The nozzle coefficient is $c_N = 0.98$ for each nozzle. Impulse blades turn the flow into the direction $\beta_3 = -65°$, and as a result of friction, the relative velocity reduces by an amount that gives a velocity coefficient $c_v = 0.90$. Find (a) the efficiency of the turbine, (b) the power-specific speed, and (c) the nozzle diameter and the number of buckets in the wheel.

10.2 The velocity diagrams for the runner of a small Francis turbine are shown in Figure 10.15. The discharge is $4.5 \, \text{m}^3/\text{s}$, the head is 150 m of water, and the rotational speed is 450 rpm. The inlet radius is $r_2 = 0.6$ m, and the water leaves the guidevanes at angle $\alpha_2 = 72°$ and velocity $V_2 = 53.3 \, \text{m/s}$. It leaves the turbine without swirl. (a) Find the velocity coefficient of the stator (inlet spiral and gates). (b) Find the inlet angle of the relative velocity β_2. (c) What is the output power? (d) What is the torque on the shaft? (d) Determine the power-specific speed and comment on whether the runner shape in the figure is appropriate.

10.3 The pressure at the entrance of a Francis turbine runner is 189.6 kPa, and at the exit it is 22.6 kPa. The shaft turns at 210 rpm. At the exit the flow leaves without swirl. The inlet radius is $r_2 = 910$ mm, and the exit radius is $r_3 = 760$ mm. The relative velocity entering the runner is $W_2 = 10.2 \, \text{m/s}$, and the flow angle of the relative velocity leaving the runner is $\beta_3 = -72°$. The blade height at the inlet is $b_2 = 600$ mm. (a) Compute the stagnation pressure loss in the runner. (b) Find the power delivered by the turbine.

10.4 A Francis turbine has an inlet radius of $r_2 = 1450$ mm and outlet radius of $r_3 = 1220$ mm. The blade width is constant at $b = 370$ mm. The shaft speed is 360 rpm and

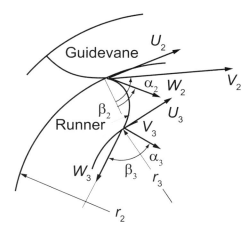

Figure 10.15 A runner of a Francis turbine.

the volumetric flow rate is $Q = 16.7 \, \text{m}^3/\text{s}$. The flow enters the runner at $\alpha_2 = 78°$. Water leaves the the turbine without swirl, and the outlet pressure $p_3 = 35 \, \text{kPa}$. The loss through the runner is $0.20 W_2^3/2g$ m. Find the pressure p_2 at the inlet and the head loss through the runner.

10.5 The relative velocities at the inlet and the exit of a Francis turbine are $W_2 = 10.0 \, \text{m/s}$ and $W_3 = 33.7 \, \text{m/s}$. The shaft speed is $\Omega = 200$ rpm. The inlet radius of the runner is $r_2 = 1880$ mm, and the outlet radius is $r_3 = 1500$ mm. The runner blade width is constant at $b = 855$ mm. Find (a) the flow rate through the turbine and (b) the torque, assuming that the flow leaves without exit swirl.

10.6 A Francis runner is to be designed for an effective head of $H_e = 140$ m and flow rate $Q = 20 \, \text{m}^3/\text{s}$. Assume that the efficiency is $\eta = 0.9$ and that there is no exit swirl. Use the formula of Lugaresi and Massa to calculate the specific speed. Use the other formulas in the text to obtain b_r and U_2/V_0 from Ω_{sp}. Assume that the mechanical and volumetric losses are negligible. Find, (a) the specific diameter on this basis, (b) the diameter at the inlet, (c) the blade speed at the inlet, and (d) the flow angles of the absolute and relative velocities at the inlet.

10.7 Water enters the runner of a Francis turbine with a relative velocity at angle $-12°$. The inlet radius is 2.29 m, and the mean radius at the exit is 1.37 m. The rotational speed is 200 rpm. The blade height at the inlet is $b_2 = 1.22$ m, and at the exit the inclined width of the blade is $b_3 = 1.55$ m. The radial velocity at the runner inlet is $10.0 \, \text{m/s}$, and the flow leaves the runner without swirl. Evaluate (a) the change in total enthalpy of the water across the runner, (b) the torque exerted by the water on the runner normalized to a metric ton per second, (c) the power developed, (d) the flow rate of water, and (e) the change in total pressure across the runner when the total-to-total efficiency is 95%, and the volumetric and mechanical losses can be neglected. (f) What is the change in static pressure across the runner?

10.8 The Otari number 2 power plant in Japan delivers 89.5 MW of power when the flow rate is $207 \, \text{m}^3/\text{s}$ and the head is 48.1 m. The diameter of the Kaplan turbine is $D_{2t} = 5.1$ m and the hub to tip ratio $\kappa = 0.56$. The generator has 36 poles and delivers power at a line frequency of 50 Hz. (a) Find the efficiency of the turbine. (b) Calculate and flow angles entering and leaving the rotor, and construct a graph to show their variation across the span.

CHAPTER 11

HYDRAULIC TRANSMISSION OF POWER

The subject of this chapter is *hydraulic transmission* of power by fluid couplings and torque converters. In these the *driver shaft* turns impeller blades of a centrifugal pump, and the *driven shaft* is powered by a radial turbine. The pump and its driver shaft are also called a *primary* and the turbine and the driven shaft, a *secondary*. The working fluid is oil, and the work done by the pump increases the oil pressure, which then drops as the oil flows through the turbine. The work done by the pump is greater than that delivered by the turbine; the difference is lost to irreversibilities.

Transmission of power by hydraulic means offers the advantages of quiet operation, damping of torsional vibrations, and low wear. As both the pump and the turbine are in a common casing, these machines are compact and safe to use in industrial settings. Such advantages come with some loss in efficiency when hydraulic transmissions are compared to rigid couplings.

11.1 FLUID COUPLINGS

A fluid coupling was invented by Hermann Föttinger while he was an engineer at AG Vulcan in Stettin, Germany.[5] Such a coupling is shown in Figure 11.1. It is toroidal in shape with a centrifugal pump on the primary side and a radial inflow turbine on the secondary side.

---

[5] H. Föttinger (1877–1945) received a PhD from TH München in 1904 and moved to work at AG Vulcan. In 1909 he left the company to assume a professorship in marine engineering at the Technical University of Danzig. From there he moved to the Chair of Turbomachinery of the Technical University of Berlin in 1924.

Principles of Turbomachinery. By Seppo A. Korpela
Copyright © 2011 John Wiley & Sons, Inc.

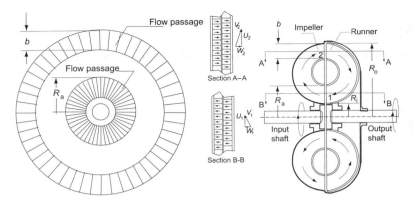

Figure 11.1 Sketch of a fluid coupling.

There are no guidevanes in the torus, and absence of a fixed member means that there is no restraining torque on the housing. Hence a free body diagram, cutting across both the driver and the driven shaft, shows that the torques in these shafts are equal. This result is independent of the shape of the blades.

11.1.1 Fundamental relations

The flow moves outward to a larger radius in the pump and inward in the turbine through a set of straight *radial blades* in both. Since the fluid flows from the primary to the secondary, the flow rate in each is the same. The torque *exerted by the impeller blades on the fluid* is

$$T_\mathrm{p} = \rho Q (r_2 V_{u2} - r_1 V_{u1})$$

Similarly, the torque that *the fluid exerts on the turbine blades* is

$$T_\mathrm{s} = \rho Q (r_2 V_{u2} - r_1 V_{u1})$$

Owing to irreversibilities in the coupling, the transmitted power is lower than the power inflow on the driver shaft. Therefore, with power given by $\dot W = T\Omega$, the output shaft must rotate at a lower speed than the input shaft. The difference in rotational speeds is called *slip*, denoted by s, and written in normalized form as:

$$s = \frac{\Omega_\mathrm{p} - \Omega_\mathrm{s}}{\Omega_\mathrm{p}} = 1 - \frac{\Omega_\mathrm{s}}{\Omega_\mathrm{p}}$$

From the equation for power

$$\dot W_\mathrm{p} = T\Omega_\mathrm{p} \qquad \dot W_\mathrm{s} = T\Omega_\mathrm{s}$$

the efficiency is given as

$$\eta = \frac{\dot W_\mathrm{s}}{\dot W_\mathrm{p}} = \frac{\Omega_\mathrm{s}}{\Omega_\mathrm{p}}$$

The dissipated power causes the temperature of the oil to increase, which, in turn, leads to heat transfer to the housing and a rise in its temperature. With the increase in temperature of the hardware, the dissipated energy is then transferred as heat to the surroundings. The fractional loss of power is given by

$$\frac{\dot{W}_p - \dot{W}_s}{\dot{W}_p} = 1 - \eta = 1 - \frac{\Omega_s}{\Omega_p} = s$$

and this gives an alternative meaning for slip. In steady operation efficiency of a fluid coupling reaches 96–98%. During transients slip increases, and so does the loss. This is of secondary importance, because the coupling was invented to transmit power smoothly precisely during such transients.

■ **EXAMPLE 11.1**

A prime mover delivers power to a fluid coupling at 140 kW, with a shaft rotating at 1800 rpm, and with a slip of 3%. (a) What is the shaft torque? (b) What is the power flow in the output shaft? (c) Find the rotational speed of the driven shaft. (d) At what rate is energy is dissipated as heat to the surroundings?

Solution: (a) Torque on the shaft is

$$T = \frac{\dot{W}_p}{\Omega_p} = \frac{140{,}000 \cdot 30}{1800\,\pi} = 742.7\,\text{N·m}$$

(b) The efficiency of the coupling at this operating point is

$$\eta = \frac{\Omega_s}{\Omega_p} = 1 - s = 1 - 0.03 = 0.97$$

and the power delivered is

$$\dot{W}_s = \eta\,\dot{W}_p = 0.97 \cdot 140 = 135.8\,\text{kW}$$

(c) The output shaft rotates at

$$\Omega_s = 0.97\,\Omega_p = 1746\,\text{rpm}$$

(d) The power dissipation rate is

$$\dot{W}_d = (1 - \eta)\,\dot{W}_p = 0.03 \cdot 140 = 4.2\,\text{kW}$$

■

The two half-toroids making up the fluid coupling are each fitted with a set of radial blades. A typical number is 30 blades, but in order to avoid resonance vibrations, each half has a slightly different number. Better guidance is obtained by increasing the number of blades, but this comes at the cost of increased frictional losses and thus a slower circulating flow rate for the oil in the coupling. The flow rate is an operating characteristic of a coupling and needs to be estimated. This is carried out next.

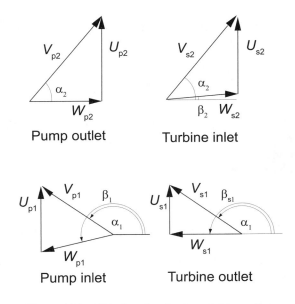

Figure 11.2 Velocity triangles for a fluid coupling.

11.1.2 Flow rate and hydrodynamic losses

Analysis of fluid couplings can be carried out by methods familiar from the analysis of centrifugal pumps and radial inflow turbines. The vector diagrams are shown in Figure 11.2. Since the relative velocity leaving the turbine and that leaving the pump are radial, the equation for the torque can be written as

$$T = \rho Q (\Omega_p r_2^2 - \Omega_s r_1^2) \qquad (11.1)$$

where

$$r_1 = \frac{1}{2}(R_i + R_a) \qquad r_2 = R_0 - \frac{1}{2}b$$

according to the labels in Figure 11.1. The power dissipation in the coupling is $T(\Omega_p - \Omega_s)$. This can be estimated by noting that the losses consist of the losses at the inlets to the pump and the turbine and the frictional losses along the flow passage.

Analysis of inlet losses is left as an exercise. It is based on destruction of the difference in the tangential component of the velocity [58]. When carried out, it shows that at the inlet to the turbine the tangential component of velocity undergoes a change equal to $(\Omega_p - \Omega_s)r_2$. Hence the kinetic energy change is equal to $(\Omega_p - \Omega_s)^2 r_2^2/2$ per unit mass. This kinetic energy is dissipated into internal energy of the fluid by turbulent mixing. Similarly, at the entrance to the pump the kinetic energy change is $(\Omega_p - \Omega_s)^2 r_1^2/2$. Hence the shock losses in the coupling amount to $(\Omega_p - \Omega_s)^2 (r_1^2 + r_2^2)/2$.

Inside the coupling, owing to skin friction, losses, as in pipe flow, are proportional to the square of the meridional velocity. If the effective friction factor is denoted by f the losses can be expressed as $fLQ^2/2D_h A^2$ in which L is the length of the flow path, A is the cross-sectional area, and D_h is the hydraulic diameter $D_h = 4A/C$. In this expression C is the circumference of the toroidal cross section. For fully turbulent flows, f is practically constant. The losses can now be expressed as

$$T(\Omega_p - \Omega_s) = \frac{1}{2}\rho Q \left[(\Omega_p - \Omega_s)^2(r_1^2 + r_2^2) + f\frac{L}{D_h}\frac{Q^2}{A^2}\right] \quad (11.2)$$

Eliminating the torque between Eqs. (11.1) and (11.2) gives for the flow rate the expression

$$Q = A\Omega_p r_2 \sqrt{\frac{D_h}{fL}}\sqrt{\left(1 - \frac{\Omega_s^2}{\Omega_p^2}\right)\left(1 - \frac{r_1^2}{r_2^2}\right)} \quad (11.3)$$

Substituting this into the equation for torque gives

$$T = \rho A \Omega_p^2 r_2^3 \sqrt{\frac{D_h}{fL}}\sqrt{\left(1 - \frac{\Omega_s^2}{\Omega_p^2}\right)\left(1 - \frac{r_1^2}{r_2^2}\right)}\left(1 - \frac{r_1^2}{r_2^2}\frac{\Omega_s}{\Omega_p}\right) \quad (11.4)$$

This shows that the torque varies as the square of the rotational speed of the driver shaft. Increasing the load on the driven shaft increases the torque as well, up to a maximum value, which corresponds to such a large load that the driven shaft stops turning. The maximum torque is given by

$$T_m = \rho A \Omega_p^2 r_2^3 \sqrt{\frac{D_h}{fL}}\sqrt{1 - \frac{r_1^2}{r_2^2}}$$

The equation for torque can now be expressed as

$$T = T_m \sqrt{1 - \frac{\Omega_s^2}{\Omega_p^2}}\left(1 - \frac{r_1^2}{r_2^2}\frac{\Omega_s}{\Omega_p}\right)$$

The power delivered is $\dot{W}_s = T\Omega_s$, so that it can be written as

$$\dot{W}_s = \Omega_s T_m \sqrt{1 - \frac{\Omega_s}{\Omega_p}}\left(1 - \frac{r_1^2}{r_2^2}\frac{\Omega_s}{\Omega_p}\right)$$

or in nondimensional form as

$$\frac{\dot{W}_s}{T_m \Omega_p} = \frac{\Omega_s}{\Omega_p}\sqrt{1 - \frac{\Omega_s^2}{\Omega_p^2}}\left(1 - \frac{r_1^2}{r_2^2}\frac{\Omega_s}{\Omega_p}\right)$$

The performance curves are shown in Figure 11.3. If the load on the driven shaft is very large, the coupling will experience a large slip and only a small amount of power will be transmitted to the driven shaft. Similarly, for light loads the torque is small and a small amount of power is needed to turn the shaft. Here again the power flow is low. Hence there is a maximum for some intermediate value, which depends on the dimensions of the coupling. For the characteristics shown $r_1/r_2 = 0.6$, and the maximum is at $\Omega_s/\Omega_p = 0.642$. The range $0.6 < r_1/r_2 < 0.8$ is typical.

By dimensional analysis, a torque coefficient can be defined and it depends on the slip, the Reynolds number, and geometric parameters, the most important of which is the ratio of the volume $\mathcal{V}$ of the space occupied by the working fluid and the cube of the outer diameter. These dependencies can be written as

$$C_T = \frac{T}{\rho \Omega_p^2 D^5} = f\left(s, \frac{\rho \Omega_p D^2}{\mu}, \frac{\mathcal{V}}{D^3}\right)$$

The flow through a coupling, as in other turbomachines, is dominated by inertial forces and the influence of the Reynolds number on the torque coefficient is much weaker than that caused by the slip. The only other parameter that needs to be investigated is the volume ratio.

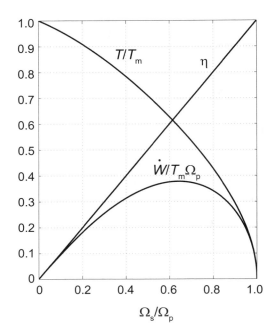

Figure 11.3 Characteristics of a fluid coupling with $r_1/r_2 = 0.6$.

11.1.3 Partially filled coupling

It was observed early in the development of fluid couplings that the torque could be reduced by filling the coupling only partially with fluid. Modern couplings have chambers into which some of the fluid can be drawn. Similarly, extra fluid can be added into the coupling. This provides a smooth start, and the amount of fluid can be adjusted to obtain the best efficiency during the nominal operating conditions.

The performance of a partially filled coupling is shown in Figure 11.4, where the parameter $\mathcal{V}/\mathcal{V}_m$ is the ratio of the actual volume of the fluid to that in a fully filled coupling. One way to see that torque must decrease in a partially filled coupling is to note that when some of the flow channels are starved for fluid, they are not effective in providing a full force into the turbine blades. Although such a fluid distribution is possible in principle, it is more likely that the fluid distribution is such that each channel is only partially filled and then the fluid is crowded to the pressure side of each blade in a pump, with the suction side lacking its share. Now the loading of the turbine blades is similar to that in an impulse turbine, but since the blades are flat, they do not perform well under this loading. As a consequence, an equally large pressure increase is not possible in the pump as in a completely filled coupling, and the torque is reduced. The graphs in this figure were drawn to match the results shown in Pfleiderer [58]. For couplings in use today, such performance data are obtained from manufacturers' catalogs.

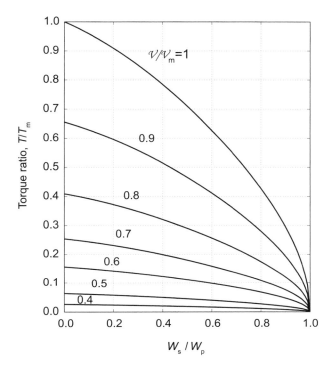

Figure 11.4 Partially filled coupling characteristics.

In addition to steady rotation, torsional vibrations may be present in the input shaft. Such torsional motions do not influence the flow in the coupling greatly, so they are not transmitted to the output shaft. The vibrations can be further suppressed by placing the coupling near the node of such a vibrational motion in the shaft, if it is not possible to change the natural frequency of the load. Fluid couplings vary in size, with small ones handling about one kilowatt. For those in the 2000 kW range, the outer diameter reaches over 1 m. They contain over 100 L of oil and have a mass approaching 2000 kg.

11.2 TORQUE CONVERTERS

The torque converter was also invented by H. Föttinger, while working at Vulcan Shipbuilding, where he was interested in finding an efficient way of connecting steam turbines to marine propellers. As the name suggests, *torque converter* changes and usually increases the input torque to a different output value. In order to achieve this, a set of guidevanes, fixed to the frame, direct the fluid into an impeller, which rotates with the input shaft. On the output shaft is a set of runner, or turbine, blades. To obtain the larger output torque, the blade shapes must be such that the fluid experiences a larger change in the tangential velocity through the turbine than through the pump. A torque converter with a single-stage radial outflow turbine is shown in Figure 11.5. More complicated designs having three turbine stages and two stationary guidevane stages exist. They have a wider operating range of high efficiency.

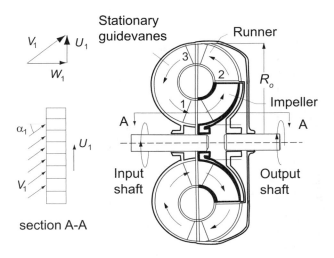

Figure 11.5 Sketch of a torque converter.

11.2.1 Fundamental relations

In order to keep the analysis general, the velocity triangles in Figure 11.6 are also drawn to depict the general situation. The equation for the output torque from the turbine is

$$T_s = \rho Q(r_2 V_{u2} - r_3 V_{u3})$$

and the torque in the input shaft is

$$T_p = \rho Q(r_2 V_{u2} - r_1 V_{u1})$$

For the fixed member the equation is

$$T_f = \rho Q(r_1 V_{u1} - r_3 V_{u3})$$

Adding the last two shows that

$$T_s = T_p + T_f$$

Examination of the velocity diagrams shows that at the inlet to the turbine the velocities are similar to those of an impulse turbine. Namely, the absolute velocity is at a steep angle to the axial direction. As the fluid flows through the turbine, its relative velocity is changed first toward the axis and then beyond it to a negative angle. Since at the exit of the turbine the axial velocity is in the direction exactly opposite to that at the inlet of the pump, the velocity diagram on the bottom part of the figure has been reflected about the vertical axis to reflect this. Also, at the inlet to the turbine, the absolute velocity is identical to the absolute velocity at the exit of the pump and the blade speed is lower than that of the pump.

A configuration of a torque converter that more closely reflects a centrifugal pump and a radial inflow turbine would have these two components as mirror images across the vertical centerline. The fixed member would then occupy only the lower part of the converter. In this case for a symmetric fixed member, the exit of the turbine would be at the same radius as the inlet to the pump.

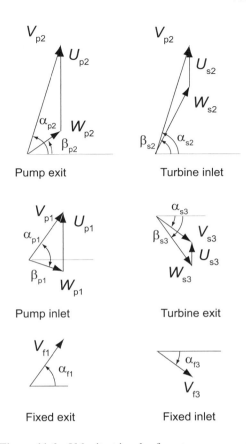

Figure 11.6 Velocity triangles for a torque converter.

The power in the input shaft is $\dot{W}_p = T_p \Omega_p$ and in the output shaft it is $\dot{W}_s = T_s \Omega_s$. The efficiency, given by

$$\eta = \frac{\dot{W}_s}{\dot{W}_p}$$

must be less than unity owing to irreversibilities. To obtain a large increase in torque, the output shaft must rotate substantially slower than the input shaft. For sufficiently large output load, the output shaft can be stopped. Under this situation the efficiency of the torque converter drops to zero and the irreversibilities cause its temperature to rise considerably. The irreversibilities are now mainly *shock losses*, although frictional loss is still substantial. A light load on the output shaft gives no torque multiplication at all, and as the shock losses vanish, the small losses are now almost entirely frictional losses.

■ **EXAMPLE 11.2**

A torque converter, when operating with a slip equal to 0.8, multiplies the torque in the ratio of 2.5 : 1. The circulatory flow rate in the converter is 140 kg/s. For the primary element, the ratio of inner to outer mean radii is 0.55 and the blade speed of the primary at the outer mean radius is 50 m/s. The axial velocity remains constant throughout the flow circuit. The absolute velocity at inlet to the primary and the relative velocity at the exit of the primary are both axially directed. Evaluate the power developed by the secondary element in this torque converter.

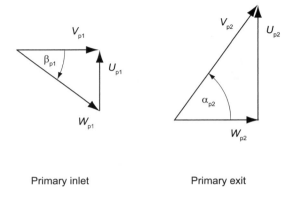

Figure 11.7 Velocity triangles for Example 11.2.

Solution: Let the inlet to the impeller be denoted by 1, the inlet to the runner by 2, and the inlet to the guidevanes as 3. The design and operation is such that the tangential velocity at the inlet to the impeller zero and the tangential component of the relative velocity to the runner vanishes also. For this reason the velocity diagrams are as shown in Figure 11.7.

Since the inlet angular momentum to the primary is zero, the torque is $T_\mathrm{p} = \rho Q r_2 V_{u2}$ and the input power is $\dot{W}_\mathrm{p} = T\Omega_\mathrm{p}$. Therefore

$$\dot{W}_p = \rho Q U_2 V_{u2} = \rho Q U_2^2 = 140 \cdot 50^2 = 350.0 \,\mathrm{kW}$$

Since the torque ratio is 2.5 and the slip is 0.8, it follows that

$$\frac{\dot{W}_\mathrm{s}}{\dot{W}_\mathrm{p}} = \frac{T_\mathrm{s}\Omega_\mathrm{s}}{T_\mathrm{p}\Omega_\mathrm{p}} = 2.5 \cdot 0.2 = 0.5$$

and $\dot{W}_\mathrm{s} = 0.5 \cdot 350.0 = 175.0\,\mathrm{kW}$. There is a large irreversiblity in the flow as the large torque multiplication requires heavily loaded blades. (Note that the ratio of the radii was not needed.)

∎

11.2.2 Performance

To illustrate how the general analysis may proceed, consider the torque converter shown in Figure 11.5. Assume that the blades are radial at the exit of the primary so that the angles at which the relative velocity leaves is zero and similarly for the flow at the exit of the secondary. Then, since the velocity does not change in magnitude or direction as the flow leaves the primary and enters the secondary, at the inlet of the secondary the angular momentum is $r_2 V_{su2} = r_2 V_{pu2} = r_2^2 \Omega_\mathrm{p}$. On the basis of what has been learned of the turbomachines studied so far, if the deviation is ignored, the exit angles of the relative velocity leaving the primary and the secondary do not change during the operation, and neither does the angle of the absolute velocity of at the exit of the fixed member. Then dimensional analysis of this particular torque converter, which also has fixed value for the diameter ratio, shows that the torque of the primary is related to the angular speed ratio and Reynolds number as

$$C_\mathrm{p} = \frac{T_\mathrm{p}}{\rho \Omega_\mathrm{p}^2 D^5} = f_1\left(\frac{\Omega_\mathrm{s}}{\Omega_\mathrm{p}}, \mathrm{Re}\right) \tag{11.5}$$

where the Reynolds number is $\mathrm{Re} = \rho \Omega_\mathrm{p} D^2/\mu$. Similarly for the secondary

$$C_\mathrm{s} = \frac{T_\mathrm{s}}{\rho \Omega_\mathrm{p}^2 D^5} = f_2\left(\frac{\Omega_\mathrm{s}}{\Omega_\mathrm{p}}, \mathrm{Re}\right) \tag{11.6}$$

Dividing these gives

$$\frac{T_\mathrm{s}}{T_\mathrm{p}} = f\left(\frac{\Omega_\mathrm{s}}{\Omega_\mathrm{p}}, \mathrm{Re}\right) \tag{11.7}$$

The efficiency is also a function of the same nondimensional groups

$$\eta = g\left(\frac{\Omega_\mathrm{s}}{\Omega_\mathrm{p}}, \mathrm{Re}\right) \tag{11.8}$$

Again the influence of the Reynolds number is much weaker than the angular velocity ratio [41].

Typical plots of the torque ratio and efficiency are shown in Figure 11.8. Experimental

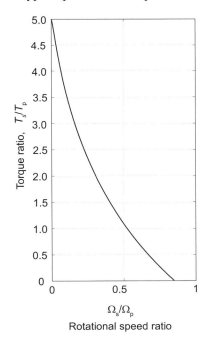
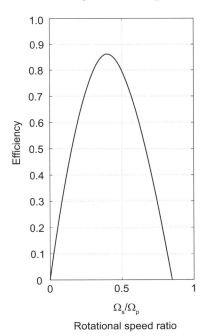

Figure 11.8 Performance curves for a torque converter.

evidence suggests that the torque coefficient of the primary C_p is independent of the speed ratio; that is, the amount of slip taking place in the torque converter influences mainly the downstream components. The primary torque is given by the angular momentum balance

$$T_\mathrm{p} = \rho Q(\Omega_\mathrm{p} r_2^2 - r_1 V_{u1})$$

With $V_{u1} = V_{x1}\tan\alpha_1 = Q\tan\alpha_1/A$, the torque coefficient of the primary is then

$$C_p = \frac{T_p}{\rho\Omega_p^2 r_3^5} = \frac{Q}{\Omega_p r_3^3}\left(\frac{r_2^2}{r_3^2} - \frac{Qr_1\tan\alpha_1}{Ar_3^3\Omega_p}\right)$$

The reaction torque of the secondary is given by

$$T_s = \rho Q(\Omega_p r_2^2 - \Omega_s r_3^2)$$

and the torque coefficient of the secondary is

$$C_s = \frac{T_s}{\rho\Omega_p^2 r_3^5} = \frac{Q}{\Omega_p r_3^3}\left(\frac{r_2^2}{r_3^2} - \frac{\Omega_s}{\Omega_p}\right)$$

The torque ratio becomes

$$\frac{T_s}{T_p} = \frac{r_2^2/r_3^2 - \Omega_s/\Omega_p}{r_2^2/r_3^2 - Qr_1\tan\alpha_1/A\Omega_p r_3^2} \tag{11.9}$$

The efficiency of the torque converter is then

$$\eta = \frac{T_s\Omega_s}{T_p\Omega_p} = \frac{(r_2^2/r_3^2 - \Omega_s/\Omega_p)\Omega_s/\Omega_p}{r_2^2/r_3^2 - Qr_1\tan\alpha_1/Ar_3^2\Omega_p} \tag{11.10}$$

Since C_p is assumed to be constant, it follows that Q/Ω_p is constant, and for a given torque converter the torque ratio is seen to decrease linearly with the speed ratio Ω_s/Ω_p. The efficiency then varies parabolically, being zero when the secondary shaft either is fixed or rotates at the value given by

$$\frac{\Omega_s}{\Omega_p} = \frac{r_2^2}{r_3^2}$$

At the point of maximum efficiency the speed ratio decreases to about one-half of this value. That the experimental curves of Figure 11.8 do not follow this theory exactly is caused by variation of C_p with the speed ratio.

Another design is shown in Figure 11.9. The output shaft is concentric with the input one, and the inlet mean radius of the fixed member is the same as its exit radius. This design is analyzed in the next example.

■ EXAMPLE 11.3

The mean radii of a torque converter of the type shown in Figure 11.9 are $r_1 = 10$ cm and $r_2 = 15$ cm. The primary operates with $\Omega_p = 3000$ rpm, and the secondary rotates at $\Omega_s = 1200$ rpm. The blades of the primary are oriented such that the angle of the relative flow at the exit is $\beta_{p2} = 35°$. The exit angle of the relative velocity of the secondary is $\beta_{s3} = -63°$. The fixed blades are shaped such that the exit velocity from them is at the angle $\alpha_{f1} = 55°$. The axial velocity is constant throughout the converter, and its value is $V_x = 15$ m/s. Determine the flow angles and calculate the torque ratio T_s/T_p and the efficiency of the torque converter.

Solution: The blade speeds are first determined as

$$U_{p1} = r_1\Omega_p = \frac{0.10 \cdot 2500 \cdot \pi}{30} = 26.78 \text{ m/s}$$

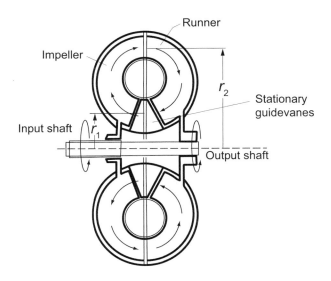

Figure 11.9 Torque converter with concentric shafts.

$$U_{p2} = r_3\Omega_p = \frac{0.15 \cdot 2500 \cdot \pi}{30} = 39.27 \text{ m/s}$$

$$U_{s2} = r_2\Omega_s = \frac{0.15 \cdot 1200 \cdot \pi}{30} = 18.85 \text{ m/s}$$

$$U_{s3} = r_3\Omega_p = \frac{0.10 \cdot 1200 \cdot \pi}{30} = 12.57 \text{ m/s}$$

Next, the tangential velocity at the exit of the primary is determined. Since the flow angle of the relative flow is given, the tangential component W_{pu2} is

$$W_{pu2} = V_x \tan \beta_{p2} = 15 \tan(35°) = 10.50 \text{ m/s}$$

and the tangential component of the absolute velocity is

$$V_{pu2} = U_{p2} + W_{pu2} = 39.27 + 10.50 = 49.77 \text{ m/s}$$

The flow angle is therefore

$$\alpha_{p2} = \tan^{-1}\left(\frac{V_{pu2}}{V_x}\right) = \tan^{-1}\left(\frac{49.77}{15.00}\right) = 73.23°$$

The absolute velocity and its flow angle at the inlet of the secondary are the same as those leaving the primary. Hence $V_{su2} = 49.77$ m/s. The relative velocity entering the secondary has the tangential component

$$W_{su2} = V_{su2} - U_{s2} = 49.77 - 18.85 = 30.92 \text{ m/s}$$

and its flow angle is

$$\beta_{s2} = \tan^{-1}\left(\frac{W_{su2}}{V_x}\right) = \tan^{-1}\left(\frac{30.92}{15.00}\right) = 64.12°$$

At the exit of the secondary the tangential velocity component of the relative velocity is

$$W_{su3} = V_x \tan \beta_{s3} = 15 \tan(-56°) = -22.24 \, \text{m/s}$$

so that the component of the absolute velocity and its flow angle are

$$V_{su3} = W_{su2} + U_{s3} = -22.24 + 12.57 = -9.67 \, \text{m/s}$$

$$\alpha_{s3} = \tan^{-1}\left(\frac{V_{su3}}{V_x}\right) = \tan^{-1}\left(\frac{-9.67}{15.00}\right) = -32.81°$$

The flow enters the fixed member at the same velocity as it leaves the secondary. Hence $V_{fu3} = -9.67 \, \text{m/s}$. At the exit of the fixed member

$$V_{fu1} = V_x \tan \alpha_{f1} = 15 \tan(55°) = 21.42 \, \text{m/s}$$

The flow enters the primary with this tangential velocity and angle, so that $V_{pu1} = 21.24 \, \text{m/s}$. The relative velocity at the inlet of the primary is then

$$W_{pu1} = V_{pu1} - U_{p1} = 21.42 - 26.18 = -4.76 \, \text{m/s}$$

The torques are

$$T_s = \rho Q (r_2 V_{su2} - r_3 V_{su3}) = 8.43 \rho Q$$

$$T_p = \rho Q (r_2 V_{pu2} - r_1 V_{pu1}) = 5.32 \rho Q$$

$$T_f = \rho Q (r_1 V_{fu1} - r_3 V_{fu3}) = 3.11 \rho Q$$

Hence

$$\frac{T_s}{T_p} = \frac{8.43}{5.32} = 1.58 \qquad \eta = \frac{T_s \Omega_s}{T_p \Omega_p} = 1.58 \frac{1200}{2500} = 0.761$$

■

The velocity triangles for the preceding example are close to what are shown in Figure 11.6.

EXERCISES

11.1 A fluid coupling operates with oil flowing in a closed circuit. The device consists of two elements, the primary and secondary, each making up one-half of a torus, as shown in Figure 11.1. The input power is 100 hp, and input rotational speed is 1800 rpm. The output rotational speed is 1200 rpm. (a) Evaluate both the efficiency and output power of this device. (b) At what rate must energy as heat be transferred to the cooling system, to prevent a temperature rise of the oil in the coupling?

11.2 (a) Carry out the algebraic details to show that the expression for the flow rate through a fluid coupling is given by Eq. (11.3) and assuming that for a low value of slip the friction factor is related to the flow rate by an expression

$$f = \frac{c}{\text{Re}} = \frac{c \mu A}{\rho Q D}$$

find the dependence of the flow rate on the slip for small values of s. (b) Carry out the algebraic details to show that the expression for the torque of a fluid coupling is given by Eq. (11.4). What is the appropriate form for this equation for low values of slip?

EXERCISES

11.3 A fluid coupling operates with an input power of 200 hp, 5% slip, and a circulatory flow rate of 1500 L/s. (a) What is the rate at which energy as heat must be transferred from the coupling in order for its temperature to remain constant? (b) What would be the temperature rise of the coupling over a period of 30 min, assuming that no heat is transferred from the device and that it has a mass of 45 kg, consisting of 70% metal with a specific heat 840 J/(kg · K), and 30% oil with a specific heat 2000 J/(kg · K)?

11.4 In the fluid coupling shown in Figure 11.1 fluid circulates in the direction indicated while the input and output shafts rotate at 2000 and 1800 rpm, respectively. The fluid is an oil having a specific gravity of 0.88 and viscosity 0.25 kg/(m · s). The outer mean radius of the torus is $r_2 = 15$ cm and the inner mean radius is $r_1 = 7.5$ cm. The radial height is $b = 2r_2/15$. The axial flow area around the torus is the same as the flow area at the outer clearance between the primary and secondary rotors. Given that the relative roughness of the flow conduit is 0.01, find the volumetric flow rate and the axial velocity.

11.5 Show that the kinetic energy loss model at the inlet to the turbine given by

$$\frac{1}{2}r_2^2(\Omega_p - \Omega_s)^2$$

is based on the conversion of the change in the one-half of the tangential component of the velocity squared, irreversibly into internal energy. To show this, note that the incidence of the relative velocity at the inlet to the turbine is β_2 since the blades are radial. This leads to a leading-edge separation, after which the flow reattaches to the blade. After this reattachement the radial component of the relative velocity is the same as in the flow incident on the blade.

11.6 For a fluid coupling for which $r_1/r_2 = 0.7$, develop an expression which from which by differentiation the value of the slip at which the power is maximum may be obtained.

11.7 A torque converter operates with oil flowing in a closed circuit. It consists of a torus consisting of a pump, a turbine, and a stator. The input and output rotational speeds are 4000 and 1200 rpm, respectively. At this operating condition the torque exerted on the stator is twice that exerted on the pump. Evaluate (a) the output to input torque ratio and (b) the efficiency.

11.8 A torque converter multiplies the torque by is designed to have to provide a torque multiplication ratio of 3.3 to 1. The circulating oil flow rate is 500 kg/s. The oil enters the fixed vanes in the axial direction at 10 m/s, and leaves at an angle 60° in the direction of the blade motion. The axial flow area is constant. Find the torque that the primary exerts on the fluid and the torque by the fluid on the blades of the secondary. The inlet and outlet radii of fixed vanes are 15 cm.

11.9 Develop the Eqs. (11.7) and (11.10). At what ratio of the rotational speeds is the efficiency maximum? From this and the experimental curves shown in the text, estimate (a) the ratio r_2/r_3 and (b) the value of $Qr_1 \tan \alpha_1 / Ar_3^2\Omega_p$.

CHAPTER 12

WIND TURBINES

A brief history of wind turbines was given in Chapter 1. The early uses for grinding grain and lifting water have been replaced by the need to generate electricity. For this reason the designation *windmill* has been dropped and it has been replaced by a *wind turbine*. Wind turbines, such as are shown in Figure 12.1b, are the most rapidly growing renewable energy technology, but as they provide for only 0.5% of the primary energy production in the world, it will take a long time before their contribution becomes significant. Since the installed base of wind turbines is still relatively small, even a large yearly percentage increase in their use does not result in a large increase in the net capacity. But the possibility of growing wind capacity is large. The most windy regions of the United States are in the North and South Dakotas. These states, as well as the mountain ridges of Wyoming, the high plains of Texas, and the mountain passes of California, have seen the early gains in the number of wind turbines.

In countries such as Denmark and Germany, the growth of wind turbine power has been quite rapid. The winds from the North Sea provide particularly good wind prospects both onshore and offshore in Denmark's Jutland region. In fact, during the year 2011, 22% of Denmark's electricity was generated from wind, and the entire power needs of western Denmark are provided by its windfarms on the windiest days. The installed capacity of 3800 MW in Denmark in year 2010 come from 5000 units. Because she adopted wind technology early, Denmark's old wind turbines are being replaced today with larger modern units. Denmark's electricity production from wind in 2010 was about 7810 GWh per year. The capacity factor is a modest 23% cent, owing to the intermittency of wind.

Figure 12.1 A Darrieus rotor (a) and a windfarm of modern wind turbines (b).

In Germany during the years 2001-2004 wind turbines were put into operation at the rate of two each day. The installed capacity was about 27,000 MW in 2010 and this was based on some 18,000 units. They provided about 36,500 GWh of electricity a year, giving a lower capacity factor than that in Denmark.

The cost of generating electricity from wind has dropped greatly since the 1980s. With the rising costs of fossil fuels and nuclear energy, it is now competitive with the plants using these fuels as sources of power.

12.1 HORIZONTAL-AXIS WIND TURBINE

Aerodynamic theory of wind turbines is similar to that of airplane propellers. Propeller theory, in turn, originated in efforts to explain the propulsive power of marine propellers. The first ideas for them were advanced by W. J. M. Rankine in 1865. They are based on what has come to be called the *momentum theory of propellers*. It ignores the blades completely and replaces them by an *actuator disk*. The flow through the disk is separated from the surrounding flow by a streamtube, which is called a *slipstream* downstream of the disk. For a wind turbine, as energy is drawn from the flow, the axial velocity in the slipstream is lower than that of the surrounding fluid. Some of the energy is also converted into the rotational motion of the wake.

The next advance was by W. Froude in 1878. He considered how a *screw propeller* imparts a torque and a thrust on the fluid that flows across an *element of a blade*. This *blade element theory* was developed further by S. Drzewieci at the beginning of the twentieth century. During the same period contributions were made by N. E. Joukowski in Russia, A. Betz and L. Prandtl in Germany, and F. W. Lancaster in Great Britain. These studies were compiled into a research monograph on *airscrews* by H. Glauert [28]. He also made important contributions to the theory at a time when aerodynamic research took on great urgency with the development of airplanes. In addition to marine propellers, aircraft propellers, and wind turbines, the theory of screw propellers can also be used in the study of helicopter rotors, hovercraft propulsion, unducted fans, axial pumps, and propellers in hydraulic turbines. The discussion below begins by following Glauert's presentation.

The aim of theoretical study of wind turbines is to determine what the length of blades should be for nominal wind conditions at a chosen site, and how the chord, angle of twist, and shape should vary along the span of the blade to give the blade the best aerodynamical performance. Thickness of the cross section of the blade is determined primarily by structural considerations, but a well-rounded leading edge performs better at variable wind conditions than a thin blade, so the structural calculations and aerodynamic analysis are complementary tasks.

12.2 MOMENTUM AND BLADE ELEMENT THEORY OF WIND TURBINES

In momentum theory airscrew is replaced by an *actuator disk*. When it functions as a propeller, it imparts energy to the flow; when it represents the blades of a wind turbine, it draws energy from the flow.

12.2.1 Momentum Theory

To analyze the performance of a windmill by momentum theory, consider the control volume shown in Figure 12.2. The lateral surface of this control volume is that of a streamtube that divides the flow into a part that flows through the actuator disk and an external stream. Assuming that the flow is incompressible, applying the mass balance to this control volume gives

$$A_a V = A V_d = A_b V_b$$

in which A_a is the inlet area and A_b is the outlet area. The approach velocity of the wind is V and the downstream velocity is V_b. The disk area is A, and the velocity at the disk is V_d.

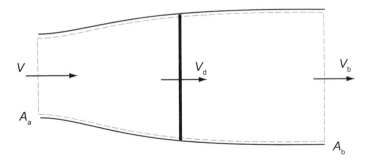

Figure 12.2 Control volume for application of the momentum theory for a wind turbine.

If there is *no rotation* in the slipstream and velocity and pressure at the inlet and exit are uniform, then an energy balance applied to the control volume gives

$$\dot{m}\left(\frac{p_a}{\rho} + \frac{1}{2}V^2\right) = \dot{W} + \dot{m}\left(\frac{p_a}{\rho} + \frac{1}{2}V_b^2\right) \qquad (12.1)$$

which gives for the specific work the expression

$$w = \frac{1}{2}(V^2 - V_b^2) \qquad (12.2)$$

Introducing the stagnation pressures

$$p_{0+} = p_a + \frac{1}{2}\rho V^2 \qquad p_{0-} = p_a + \frac{1}{2}\rho V_b^2$$

into Eq. (12.1), gives for the specific work the alternate form

$$w = \frac{p_{0+} - p_{0-}}{\rho} \tag{12.3}$$

and the specific work is evidently uniform across the disk. Making use of the fact that velocity V_d at the disk is the same on both sides, this can also be written as

$$w = \frac{p_+ - p_-}{\rho} \tag{12.4}$$

in which p_+ and p_- are the corresponding static pressures. Force balance across the disk gives

$$F_d = (p_+ - p_-)A \tag{12.5}$$

The force on the disk is also obtained by applying the momentum theorem to the control volume and assuming that the pressure along its lateral boundaries has a uniform value p_a. This yields

$$F_d = \int_A \rho V_d (V - V_b) dA = \rho A V_d (V - V_b) \tag{12.6}$$

and F_d here and in Eq. (12.5) is the force that the disk exerts on the fluid. It has been taken to be positive when it acts in the upstream direction. Equating Eq. (12.6) to Eq. (12.5) and making use of Eq. (12.3) gives

$$\rho V_d (V - V_b) = p_+ - p_- = \rho w$$

Substituting Eq. (12.2) for work gives

$$V_d(V - V_b) = \frac{1}{2}(V^2 - V_b^2) = \frac{1}{2}(V - V_b)(V + V_b)$$

from which it follows that

$$V_d = \frac{1}{2}(V + V_b)$$

Velocity at the disk is seen to be the arithmetic mean of the velocities in the free stream and in the far wake. Changes in velocity, total pressure, kinetic energy, and static pressure in the axial direction are shown in Figure 12.3.

A consequence of this analysis is that power delivered by the turbine is

$$\dot{W} = \rho A V_d w = A V_d (p_{0+} - p_{0-}) = A V_d (p_+ - p_-) = F_d V_d$$

This is a curious result, for in the previous chapters the work delivered by a turbine was always related to a change in the tangential velocity of the fluid, which produces a torque on a shaft. To reconcile this, one may imagine the actuator disk to consist of two sets of blades rotating in opposite directions such that the flow enters and leaves the set axially. Each rotor extracts energy from the flow with the total power delivered as given above. Also, since the velocity entering and leaving the disk is the same, it is seen that work extracted is obtained by the reduction in static pressure across the disk. This sudden drop in static pressure is shown schematically in Figure 12.3.

It is customary to introduce an *axial induction*, or *interference factor*, defined as

$$a = \frac{V - V_d}{V}$$

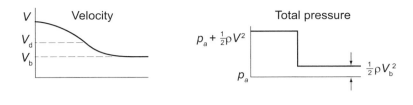
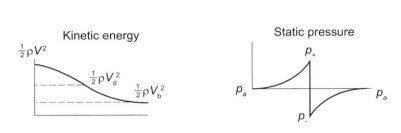

Figure 12.3 Variation of the different flow variables in the flow.

so that the velocities at the disk and far downstream are

$$V_d = (1-a)V \qquad V_b = (1-2a)V \qquad (12.7)$$

The second of these equations shows that if $a > \frac{1}{2}$, there will be reverse flow in the wake and simple momentum theory has broken down. The axial force on the blades is given by

$$F_d = \rho V_d (V_b - V) A = 2a(1-a)\rho A V^2 \qquad (12.8)$$

A force coefficient defined as

$$C_x = \frac{F_x}{\frac{1}{2}\rho V^2} = 4a(1-a) \qquad (12.9)$$

is seen to depend on the induction factor. The power delivered to the blades is

$$\dot{W} = \frac{1}{2}\rho V_d A (V^2 - V_b^2) = 2a(1-a)^2 \rho A V^3 \qquad (12.10)$$

from which the power coefficient, defined as

$$C_p = \frac{\dot{W}}{\frac{1}{2}\rho A V^3} = 4a(1-a)^2 \qquad (12.11)$$

is also a function of the induction factor only. Maximum power coefficient is obtained by differentiating this with respect to a, which gives

$$4(1-a)(1-3a) = 0 \qquad \text{so that} \qquad a = \frac{1}{3}$$

If the efficiency is defined as the ratio of power delivered to that in the stream moving with speed V over an area A, which is $\rho A V^3/2$, then the efficiency and power coefficient are defined by the same equation. The maximum efficiency is seen to be

$$\eta = \frac{16}{27} = 0.593$$

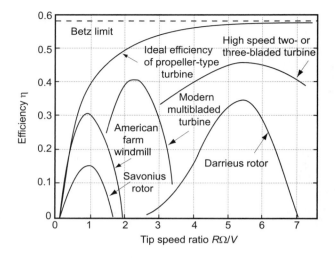

Figure 12.4 Efficiencies of various wind turbines.

This is called the *Betz limit*. Efficiencies of various windmills are shown in Figure 12.4. Depending on their design, modern wind turbines operate at tip speed ratios $R\Omega/V$ in the range 1–6. In the upper part of this range the number of blades is from one to three, and in the lower end wind turbines are constructed with up to two dozen blades, as is shown in the American wind turbine in Figure 1.4.

The flow through the actuator disk and its wake patterns have been studied for propellers, wind turbines, and helicopters. Helicopters, in particular, operate under a variety of conditions, for the flow through the rotor provides a thrust at climb and a brake during descent. The various flow patterns are summarized in the manner of Eggleston and Stoddard [25] in Figure 12.5. The representation of the axial force coefficient was extended by Wilson

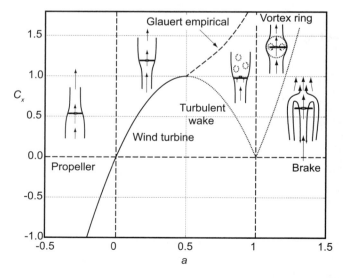

Figure 12.5 Operational characteristics of an airscrew. (After Eggleston and Stoddars [25].)

and Lissaman [82] to the brake range $a > 1$, by rewriting it as

$$C_x = 4a|1 - a|$$

This explains the diskontinuity in the slope at $a = 1$. For $a < 0$, the airscrew operates as a propeller. The limit of vanishing approach velocity corresponds to a tending to a large negative value at a rate that keeps V_d finite while V tends to zero. Under this condition the airscrew functions as an unducted axial fan. Since the flow leaves the fan as a jet, pressure is atmospheric short distance into the wake, and with a pressure increase across the fan, pressure will be lower than atmospheric at the inlet to the fan. This low pressure causes the ambient air to accelerate toward the front of the fan. At a given rotational speed, proper orientation of the blades gives a smooth approach. A slipstream forms downstream of the fan, separating the wake from the external flow. The wake is in angular rotation, and vorticity that is shed from the blades forms a cylindrical vortex sheet that constitutes the boundary of the slipstream. As a consequence, at the slipstream boundary velocity changes discontinuously from the wake to the surrounding fluid. This discontinuous change and rotation are absent in the flow upstream of the disk where the flow is axial, if the small radial component near the disk is neglected.

As the approach velocity increases to some small value V, the airscrew functions as a propeller, in which case V can be taken to be the velocity of an airplane flying through still air. The continuity equation now shows that the mass flow rate $\dot{m}$ across the propeller comes from an upstream cylindrical region of area A_a, given by $A_a = \dot{m}/\rho V$, and the area decreases as the velocity V increases.

A further increase in the approach velocity leads to a condition at which $a = 0$. At this state no energy is imparted to, or extracted from, the flow, and the slipstream neither expands nor contracts. As the velocity V increases from this condition, the angle of attack changes to transform what was the pressure side of the blade into the suction side, and the airscrew then extracts energy from the flow. The airscrew under this condition operates as a wind turbine. The blades are naturally redesigned so that they function optimally when they are used to extract energy from the wind. In a wind turbine pressure increases as it approaches the plane of the blades and drops across them. The diameter of the slipstream, in contrast to that of a propeller, increases in the downstream direction.

For $a > \frac{1}{2}$ the slipstream boundary becomes unstable and forms vortex structures that mix into the wake. This is shown in Figure 12.5 as the *turbulent wake* state. The theoretical curve in the figure no longer holds, and an empirical curve of Glauert gives the value of the force coefficient in this range. At the condition $a = 1$ the flow first enters a *vortex ring* state and for large values, a *brake state*. Flows in these regions are sufficiently complex that they cannot be analyzed by elementary methods.

The aim of wind turbine theory is to explain how the induction factors change as a result of design and operating conditions. When the theory is developed further, it will be seen that wind turbines operate in the range $0 < a < \frac{1}{2}$, which is consistent with the momentum theory.

12.2.2 Ducted wind turbine

Insight can be gained by repeating the analysis for a wind turbine placed in a duct and then considering the limit as the duct radius tends to infinity. Such an arrangement is shown in Figure 12.6. Applying the momentum equation to the flow through the control volume containing the slipstream gives

$$F_d - \int_{A_b} \rho V_b (V_b - V) dA_b = R + p_a A_a - p_b A_b \qquad (12.12)$$

in which R is the x component of the net pressure force that the fluid outside the streamtube (consisting of the slipstream and its upstream extension) exerts on the fluid inside. For the flow outside this streamtube, the momentum balance leads to

$$p_a(A_e - A_a) - p_b(A_e - A_b) - R = \rho V_c (V_c - V)(A_e - A_b)$$

which can be recast as

$$R = p_a(A_b - A_a) + (p_a - p_b)(A_e - A_b) + \rho V(V - V_c)(A_e - A_a)$$

The pressure difference $p_a - p_b$ can be eliminated by using the Bernoulli equation

$$p_a + \frac{1}{2}\rho V^2 = p_b + \frac{1}{2}\rho V_c^2$$

which transforms the expression for R into

$$R = p_a(A_b - A_a) - \frac{1}{2}(V^2 - V_c^2)(A_e - A_b) + \rho V(V - V_c)(A_e - A_a)$$

or

$$R = p_a(A_b - A_a) - \frac{1}{2}(V - V_c)\left[(V + V_c)(A_e - A_b) - 2V(A_e - A_a)\right]$$

The continuity equation for the flow outside the slipstream yields

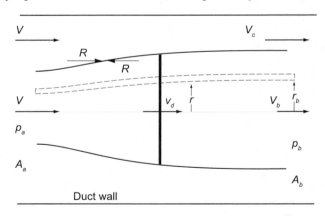

Figure 12.6 A ducted wind turbine.

$$V_c(A_e - A_b) = V(A_e - A_a)$$

so that

$$V - V_c = V\frac{A_a - A_b}{A_e - A_b} \qquad V + V_c = V\frac{2A_e - A_a - A_b}{A_e - A_b}$$

Substituting these into the equation for R leads to

$$R = p_a(A_b - A_a) - \frac{1}{2}\rho V^2 \frac{(A_b - A_a)^2}{A_e - A_b}$$

This shows that, as A_e becomes large in comparison to A_a and A_b, then

$$R = p_a(A_b - A_a)$$

The extra term then accounts for the variable pressure along the streamtube boundary. That the pressure is not exactly atmospheric along this boundary is also clear by noting that whenever streamlines are curved, pressure increases in the direction from the concave to the convex sides. Also, as the axial velocity decreases, pressure increases, but this is compensated partly by the flow acquiring a small radial velocity as the area of the streamtube increases downstream.

Equation 12.12 for the force on the blades now takes the form

$$\begin{aligned} F_d - \int_{A_b} \rho V_b(V_b - V) dA_b &= R + p_a A_a - p_b A_b \\ &= (p_a - p_b)A_b - \frac{1}{2}\rho V^2 \frac{(A_b - A_a)^2}{A_e - A_b} \\ &= \frac{1}{2}\rho(V_c^2 - V^2)A_b - \frac{1}{2}\rho V^2 \frac{(A_b - A_a)^2}{A_e - A_b} \end{aligned}$$

Since

$$V_c^2 - V^2 = (V_c - V)(V_c + V) = V^2 \left(\frac{A_b - A_a}{A_e - A_b}\right)\left(\frac{2A_e - A_a - A_b}{A_e - A_b}\right)$$

this can be written as

$$F_d - \int_{A_b} \rho V_b(V_b - V) dA_b = \frac{1}{2}\rho V^2 \frac{(A_b - A_a)}{(A_e - A_b)^2}[A_e(A_a + A_b) - 2A_a A_b]$$

For large A_e this reduces to

$$F_d = \int_{A_b} \rho V_b (V - V_b) dA_b \qquad (12.13)$$

This is equivalent to Eq. (12.6) obtained above. It is now assumed that this equation is also valid for each annular element of the streamtube shown in Figure 12.6. The differential form can be written as

$$dF_d = \rho V_b(V - V_b)dA_b = \rho V(V - V_b)dA$$

The analysis based on this equation is called *a blade element analysis* and is justified if no interactions take place between adjacent annular elements. This assumption has been criticized by Goorjian [30], but evidently in many applications of the theory the error is small.

12.2.3 Blade element theory and wake rotation

Wake rotation was included in the theory by Joukowski in 1918 and its presentation can be found in Glauert [28]. This theory is considered next.

The continuity equation for an annular section of the slipstream gives

$$V_{dx} dA = V_{bx} dA_b \qquad \text{or} \qquad V_{dx} r\, dr = V_{bx} r_b\, dr_b \qquad (12.14)$$

The axial component of velocity at the disk is denoted by V_{dx} and in the far wake is V_{bx}. The value of the radial location r_b far downstream depends on the radial location r at the disk. Since no torque is applied on the flow in the slipstream after it has passed through the disk, moment of momentum balance for the flow yields

$$rV_{du}\,V_{dx}\,2\pi r\,dr = r_b V_{bu}\,V_{bx}\,2\pi r_b\,dr_b$$

in which V_{du} is the tangential component of velocity just behind the disk and V_{bu} is its value in the distant wake. Using Eq. (12.14), this reduces to

$$rV_{du} = r_b V_{bu} \qquad (12.15)$$

This means that the angular momentum remains constant. If the rotation rate is defined by $\omega = V_{du}/r$, then this can be written as

$$r^2\omega = r_b^2 \omega_b \qquad (12.16)$$

Velocity triangles for the flow entering and leaving the blades are shown in Figure 12.7. Since trothalpy is constant across the blades, it follows that

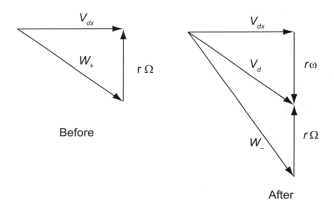

Figure 12.7 A velocity triangle for the flow leaving the blades of a wind turbine.

$$h_+ + \frac{1}{2}W_+^2 = h_- + \frac{1}{2}W_-^2$$

and for isentropic incompressible flow, this reduces to

$$p_+ + \frac{1}{2}\rho(V_{dx}^2 + r^2\Omega^2) = p_- + \frac{1}{2}\rho[V_{dx}^2 + (r\Omega + r\omega)^2]$$

which simplifies to

$$p_+ = p_- + \rho\left(\Omega + \frac{1}{2}\omega\right)r^2\omega \qquad (12.17)$$

The Bernoulli equation upstream of the disk yields

$$\frac{p_a}{\rho} + \frac{1}{2}V^2 = \frac{p_+}{\rho} + \frac{1}{2}V_{dx}^2$$

and downstream it yields

$$\frac{p_-}{\rho} + \frac{1}{2}V_{dx}^2 + \frac{1}{2}r^2\omega^2 = \frac{p_b}{\rho} + \frac{1}{2}V_{bx}^2 + \frac{1}{2}r_b^2\omega_b^2$$

Adding the last two equations and using Eq. (12.17) to eliminate the pressure difference $p_+ - p_-$ leads to

$$\frac{p_a}{\rho} - \frac{p_b}{\rho} = \frac{1}{2}(V_{bx}^2 - V^2) + \left(\Omega + \frac{1}{2}\omega_b\right)r_b^2\omega_b \qquad (12.18)$$

Differentiating this with respect to r_b gives

$$-\frac{1}{\rho}\frac{dp_b}{dr_b} = \frac{1}{2}\frac{d}{dr_b}(V_{bx}^2 - V^2) + \frac{d}{dr_b}\left[\left(\Omega + \frac{1}{2}\omega_b\right)r_b^2\omega_b\right]$$

Since the radial pressure variation in the downstream section is also given by

$$\frac{1}{\rho}\frac{dp_b}{dr_b} = \frac{V_{bu}^2}{r_b} = r_b\omega_b^2$$

equating these two pressure gradients produces

$$\frac{1}{2}\frac{d}{dr_b}(V_{bx}^2 - V^2) = -\frac{d}{dr_b}\left[\left(\Omega + \frac{1}{2}\omega_b\right)r_b^2\omega_b\right] - r_b\omega_b^2$$

or

$$\frac{1}{2}\frac{d}{dr_b}(V_{bx}^2 - V^2) = -\Omega\frac{dr_b^2\omega_b}{dr_b} - r_b\omega_b^2 - r_b^2\omega_b\frac{d\omega_b}{dr_b} - r_b\omega_b^2$$

$$= -\Omega\frac{dr_b^2\omega_b}{dr_b} - \omega_b\left(2r_b\omega_b + r_b^2\frac{d\omega_b}{dr_b}\right)$$

$$= -(\Omega + \omega_b)\frac{dr_b^2\omega_b}{dr_b} \qquad (12.19)$$

Momentum balance for the streamtube containing the flow through the disk yields

$$\int_{A_b}\rho V_{bx}(V_{bx} - V)dA_b = \int_{A_b}(p_a - p_b)dA_b - F_d$$

Assuming that this is also valid for an annular element yields

$$dF_d = \rho V_{bx}(V - V_{bx})dA_b + (p_a - p_b)dA_b$$

This elemental force is related to the pressures difference across the disk, which according to Eq. (12.17) can be related to rotation as

$$dF_d = (p_+ - p_-)dA = \rho\left(\Omega + \frac{1}{2}r^2\omega\right)$$

When Eq. (12.14) is used, this takes the form

$$dF_d = \rho\left(\Omega + \frac{1}{2}r^2\omega\right)\frac{V_{bx}}{V_{dx}}dA_b$$

Therefore the momentum balance for an elemental annulus becomes

$$\rho V_{bx}(V - V_{bx}) + p_a - p_b = \rho\left(\Omega + \frac{1}{2}\omega^2\right) r^2\omega \frac{V_{bx}}{V_{dx}}$$

Eliminating the pressure difference $p_a - p_b$ by making use of Eq. (12.18) gives

$$V_{bx}(V - V_{bx}) - \frac{1}{2}(V^2 - V_{bx}^2) + \left(\Omega + \frac{1}{2}\omega_b\right) r_b^2\omega_b^2 = \left(\Omega + \frac{1}{2}\omega^2\right) r^2\omega \frac{V_{bx}}{V_{dx}}$$

which simplifies to

$$\frac{(V - V_{bx})^2}{2V_{bx}} = \frac{(\Omega + \frac{1}{2}\omega_b)r_b^2\omega_b}{V_{bx}} - \frac{(\Omega + \frac{1}{2}\omega)r^2\omega}{V_{dx}} \quad (12.20)$$

This is the main result of Joukowski's analysis. If ω_b is now specified, then Eq. (12.19) can be solved for $V_{bx}(r_b)$. After this, from Eq. (12.16), rotation $\omega(r, r_b)$ at the disk can be obtained as a function of r and r_b. The axial velocity $V_{dx}(r, r_b)$ at the disk is then obtained from Eq. (12.20). Substituting these into the continuity equation [Eq. (12.14)] gives a differential equation that relates r_b to r. Appropriate boundary conditions fix the values for the integration constants.

12.2.4 Irrotational wake

The analysis of the previous section is valid for an arbitrary rotational velocity distribution in the wake. An important special case is to assume the wake to be irrotational. Then $r^2\omega = r_b^2\omega_b = k$ is constant. Equation (12.19) then shows that V_{bx} is constant across the wake, and Eq. (12.20) can be recast in the form

$$\frac{V_{dx}}{V_{bx}}\left[\frac{1}{2}(V - V_{bx})^2 - \left(\Omega + \frac{k}{2r_b^2}\right)k\right] = \left(\Omega + \frac{k}{2r^2}\right)k \quad (12.21)$$

Writing Eq. (12.14) as

$$\frac{V_{dx}}{V_{bx}} = \frac{r_b}{r}\frac{dr_b}{dr}$$

and making use of it in Eq. (12.21) leads to

$$\left[\frac{1}{2}(V - V_{bx})^2 - \Omega k\right] r_b\, dr_b - \frac{k^2}{2}\frac{dr_b}{r_b} = -\Omega kr\, dr - \frac{k^2}{2}\frac{dr}{r}$$

Integrating this gives

$$\left[\frac{1}{2}(V - V_{bx})^2 - \Omega k\right] \frac{r_b^2}{2} - \frac{k^2}{2}\ln\frac{r_b}{\epsilon R_b} = -\Omega k\frac{r^2}{2} - \frac{k^2}{2}\ln\frac{r}{\epsilon R}$$

where R is the disk radius and R_b the downstream slipstream radius. Owing to the logarithmic singularity, the lower limits of integration were taken to be ϵR and ϵR_b, so that as ϵ tends to zero, $\epsilon R/\epsilon R_b$ tends to R/R_b. Thus the solution takes the form

$$\frac{1}{2}(V - V_{bx})^2 - \Omega k\left(1 - \frac{r^2}{r_b^2}\right) = \frac{r^2}{r_b^2}\ln\left(\frac{r_b R}{r R_b}\right)$$

The most reasonable way to satisfy this equation is to choose the r dependence of r_b to be

$$r_b = \frac{R_b}{R} r$$

for the logarithmic term then vanishes and the second term is a constant, which balances the first constant term. Thus an equation is obtained that relates the far wake axial velocity to the slipstream radius R_b

$$\frac{1}{2}(V - V_{bx})^2 = \Omega k \left(1 - \frac{R^2}{R_b^2}\right) \quad (12.22)$$

Furthermore the following ratios are obtained:

$$\frac{r^2}{r_b^2} = \frac{\omega_b}{\omega} = \frac{V_{bx}}{V_{dx}} = \frac{R^2}{R_b^2} \quad (12.23)$$

Following Glauert's notation [28] this equation can be put into a nondimensional form by introducing the parameters

$$\lambda = \frac{V}{R\Omega} \quad \mu = \frac{V_{dx}}{R\Omega} \quad \mu_b = \frac{V_{bx}}{R\Omega} \quad \frac{q}{\mu} = \frac{k}{R^2\Omega}$$

In terms of these parameters, Eq. (12.22) can be written as

$$q = \frac{\mu^2}{2} \frac{(\lambda - \mu_b)^2}{(\mu - \mu_b)} \quad (12.24)$$

Equation (12.18), namely

$$\frac{p_a}{\rho} - \frac{p_b}{\rho} = \frac{1}{2}(V_{bx}^2 - V^2) + \left(\Omega + \frac{1}{2}\omega_b\right) r_b^2 \omega_b$$

evaluated at the edge of the slipstream, where $p_b = p_a$, gives

$$\frac{1}{2}(V^2 - V_{bx}^2) = \left(\Omega + \frac{k}{2R_b^2}\right)k = \left(\Omega + \frac{k}{2R^2}\frac{V_{bx}}{V_{dx}}\right)k$$

In terms of the nondimensional parameters this can be transformed into

$$\frac{1}{2}(\lambda^2 - \lambda_b^2) = \left(1 + \frac{q}{2\mu}\frac{\mu_b}{\mu}\right)\frac{q}{\mu} \quad (12.25)$$

Eliminating q between this and Eq. (12.24) gives

$$4(\mu - \mu_b)(2\mu - \mu_b - \lambda) = \mu(\lambda - \mu_b)^3 \quad (12.26)$$

Equations (12.25) and (12.26) determine μ and μ_b in terms of ratio λ and q. Equation (12.26) can be written as

$$\mu = \frac{1}{2}(\lambda + \mu_n) + \frac{\mu(\lambda - \mu_b)^3}{8(\mu - \mu_b)} \quad (12.27)$$

or in dimensional variables as

$$V_{dx} = \frac{1}{2}(V + V_{bx}) + \frac{V_{dx}(V - V_{bx})^3}{8(V_{dx} - V_{bx})\Omega^2 R^2}$$

This shows that
$$V_{dx} > \frac{1}{2}(V + V_{bx})$$
Since rotation of the flow takes place only in the wake, this means that the axial velocity in the wake V_{bx} is smaller than its value in a nonrotating wake.

Introducing the induction factors a and b by equations
$$\mu = \lambda(1-a) \qquad \mu_b = \lambda(1-b)$$
and substituting them into Eq. (12.27) gives
$$a = \frac{1}{2}b\left[1 - \lambda^2 \frac{(1-a)b^2}{4(b-a)}\right] \tag{12.28}$$
which, when solved for a, gives
$$a = \frac{3b}{4}\left[1 - \frac{1}{12}\lambda^2 b^2 - \frac{1}{3}\sqrt{[(1 - \frac{1}{4}\lambda^2 b^2)^2 + (2-b)\lambda^2 b]}\right]$$

It is customary in wind turbine analysis to replace the *advance ratio* λ by its reciprocal $X = R\Omega/V$, the tip speed ratio, and plot the values of b as a function of a with the tip speed ratio as a parameter. The curves, shown in Figure 12.8, clearly indicate that for X greater than 3, the approximation $b = 2a$ is quite accurate.

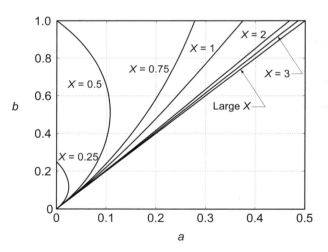

Figure 12.8 Induction factors for an irrotational wake.

For large values of X the induction factor a can be approximated by
$$a = \frac{b}{2} - \frac{b^2}{4}\left(1 - \frac{b}{2}\right)\frac{1}{X^2} + O\left(\frac{1}{X^4}\right)$$
or after replacing b by $2a$ in the higher-order terms, as
$$a = \frac{b}{2}\left[1 - \frac{a(1-a)}{X^2}\right] + O\left(\frac{1}{X^4}\right)$$

Similarly
$$b = 2a\left[1 + \frac{a(1-a)}{X^2}\right] + O\left(\frac{1}{X^4}\right)$$

This formula gives an estimate of the extent to which the axial velocity in the wake has been reduced as a result of wake rotation.

The expression for torque is

$$dT = 2\pi \rho V_{dx} r V_{du} r\, dr = 2\pi \rho V_{dx} kr\, dr$$

which when integrated over the span gives

$$T = \rho A V_{dx} k$$

Hence the torque coefficient

$$C_T = \frac{T}{\rho A \Omega^2 R^3} = \frac{V_{dx} k}{\Omega^2 R^3} = q$$

is given by the parameter q. With $\dot{W} = T\Omega$, the power coefficient becomes

$$C_p = \frac{T\Omega}{\rho V A \frac{1}{2} V^2} = \frac{2q}{\lambda^3} = \frac{b^2(1-a)^2}{b-a}$$

Since $r^2\omega$ is constant, the rate of rotation in an irrotational wake increases with decreasing radius and for small enough radius the rotation rate becomes unreasonably large. This leads to an infinite value for the power coefficient. To remedy this, the wake structure needs modification, and a reasonable model is obtained by making it a combination of solid body rotation at small radii and free vortex flow over large radii. The power coefficient for this, called the *Rankine combined vortex*, was stated by Wilson and Lissaman to be

$$C_p = \frac{b(1-a)^2}{b-a}[2aN + b(1-N)]$$

in which $N = \Omega/\omega_{\max}$. The details on how to develop this are not in their report, and it appears that in order to develop it, some assumptions need to be made [82].

12.3 BLADE FORCES

The blade forces can be calculated at each location of the span by the blade element analysis. This is carried out for an annular slice of thickness dr from the disk, as is shown in Figure 12.9. Across each blade element airflow is assumed to be the same as that for an isolated airfoil. The situation in which wake rotation is ignored is considered first. Then the general analysis including wake rotation and Prandtl's tip loss model is discussed.

12.3.1 Nonrotating wake

If the wake rotation is ignored, the velocity triangle at the midchord is as shown in Figure 12.10. The approach velocity at the disk is $(1-a)V$ and the blade element moves at the tangential speed $r\Omega$. The broken (dashed) line gives the direction of the chord and thus defines the blade pitch angle θ, which is measured from the plane of the disk. The angle

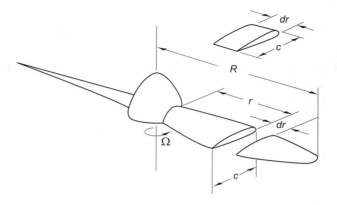

Figure 12.9 Illustration of a blade element.

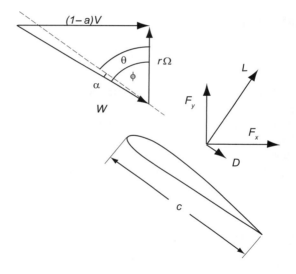

Figure 12.10 Illustration of the relative velocity across a blade.

of attack is α, and the flow angle of the relative velocity is $\phi = \alpha + \theta$, also measured from the plane of the disk.

The x and y components of the blade forces, expressed in terms of lift and drag, are

$$dF_x = (dL\cos\phi + dD\sin\phi)dr \qquad dF_y = (dL\sin\phi - dD\cos\phi)dr$$

For small angles of attack the lift coefficient is

$$C_L = 2\pi \sin\alpha = 2\pi \sin(\phi - \theta) = 2\pi(\sin\phi\cos\theta - \cos\phi\sin\theta)$$

From the velocity triangle

$$\sin\phi = \frac{(1-a)V}{W} \qquad \cos\phi = \frac{r\Omega}{W}$$

so the lift coefficient can be written as

$$C_L = 2\pi \left[\frac{(1-a)V}{W} \cos\theta - \frac{r\Omega}{W} \sin\theta \right]$$

On an element of the blade of width dr, if drag is neglected, the x component of the force on the blade is

$$dF_x = L\cos\phi\, dr = \frac{1}{2}\rho W^2 c\, C_L \cos\phi\, dr$$

or

$$dF_x = \pi\rho c[(1-a)V\cos\theta - r\Omega\sin\theta]r\Omega\, dr$$

so that the blade force coefficient is

$$C_x = \frac{dF_x}{\frac{1}{2}\rho V^2 2\pi r\, dr} = \frac{r\Omega}{V}\frac{c}{r}\left[(1-a)\cos\theta - \frac{r\Omega}{V}\sin\theta\right]$$

The ratio c/r is part of the definition of solidity, which is defined as $\sigma = c\,dr/2\pi r\,dr = c/2\pi r$. With $x = r\Omega/V$, the blade speed ratio, the blade force coefficient becomes

$$C_x = 2\pi\sigma x[(1-a)\cos\theta - x\sin\theta] \tag{12.29}$$

The blade force coefficient can also be written as

$$C_x = 4a(1-a) = 4a|1-a| \tag{12.30}$$

where the absolute value signs have been inserted so that the propeller brake mode, for which the induction factor is larger than unity, is also taken into account. Following Wilson and Lissaman [82], the values of C_x from both Eqs. (12.29) and (12.30) are plotted in Figure 12.11. The operating state of the wind turbine is at the intersection of these curves.

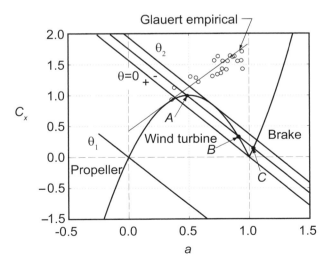

Figure 12.11 Blade force coefficient for different values of the interference factor. (Drawn after Wilson and Lissaman [82]).

The straight lines from Eq. (12.29) have a negative slope slope that depends mainly on the solidity times the blade speed ratio and weakly on angle θ. The blade force coefficient becomes zero at $a = 1$ and $a = 0$. At the former condition the blade pitch angle is $\theta = 0$, and at the latter condition it is $\theta_1 = \tan^{-1}(1/x)$. Increasing the blade angle further from this changes the operation from a windmill to a propeller. At $a = 0$, loading on the blades vanishes.

For $a = 1$, the blade angle becomes negative and the wind turbine operates as a propeller brake on the flow; that is, with a propeller turning in one direction and the wind attempting to turn it in an opposite direction, the result is a breaking action on the flow. This is best understood from the operation of helicopters. As they descend, their rotors attempt to push air into the direction in which they move and thereby cause a breaking action on their descending motion.

Since in the far wake the axial velocity is $(1 - 2a)V$, for states in the range $\frac{1}{2} < a < 1$ the simple momentum theory predicts a flow reversal somewhere between the disk and the far wake. Volkovitch [77] has shown that this anomaly disappears for yawed windmills and momentum theory is still valid. Blade angles for a working range can be obtained by equating the two expressions for the blade force coefficient and solving the resulting equation for the interference factor. This gives

$$a = \frac{1}{4}\left[2 + \pi x \sigma \cos\theta \pm \sqrt{(2 - \pi x \sigma \cos\theta)^2 + 8\pi\sigma x^2 \sin\theta}\right]$$

The two solutions of this equation are at points A and B, with point B corresponding to plus sign of the square root. The solution on this branch is unstable and can be ignored. The angle θ_2 is that for which the discriminant vanishes, and it can be determined as a root of

$$(2 - \pi x \sigma \cos\theta_2)^4 - 64\pi^2\sigma^2 x^2 (1 - \cos^2\theta_2) = 0 \qquad (12.31)$$

For a wind turbine at the span position where $\sigma = 0.03$ and blade speed ratio is $x = 4$, this comes out to be $\theta_2 = 12.76°$.

The experimental curve of Glauert is for free-running rotors, which thus have a power coefficient zero. These correspond to states of autorotation of a helicopter. A line drawn through the data gives $C_{xe} = C_1 - 4(\sqrt{C_1} - 1)(1 - a)$, in which $C_1 = 1.8$ is the value of the blade force coefficient at $a = 1$, as obtained from the experimental data. The straight line touches the parabola at $a^* = 1 - \frac{1}{2}\sqrt{C_1} = 0.33$.

■ **EXAMPLE 12.1**

Develop a Matlab script to establish the value of θ_2 for various values of solidity and tip speed ratio.

Solution: The discriminant given by Eq. (12.31) is a fourth order polynomial in $\cos\theta_s$; hence the Matlab procedure roots can be used to find the roots of this polynomial. Two of the roots are complex and can be disregarded. There is also an extraneous root, which arose when the expressions were squared. The script below gives the algorithm to calculate the roots:

```
sigma=0.03; x=4;
c(1)=(pi*x*sigma)^4;  c(2)=-8*(pi*x*sigma)^3;
c(3)=24*(pi*x*sigma)^2*(1+8*x^2/3);
c(4)=-32*pi*x*sigma;  c(5)=16-64*(pi*sigma*x^2)^2;
r=roots(c);
theta=min(acos(r)*180/pi);
```

For $\sigma = 0.03$ and $x = 4$ this gives $\theta_2 = 12.76°$.

■

12.3.2 Wake with rotation

Figure 12.12 shows a schematic of a flow through a set of blades. As the blades turn the flow, the tangential velocity increases. If its magnitude is taken to be $2a'r\Omega$ after the blade, at some location as flow crosses the blade it takes the value $a'r\Omega$. It is *assumed* that this coincides with the location at which relative velocity is parallel to the chord.[6]

When drag is included, the axial force is given by

$$dF_x = (L\cos\phi + D\sin\phi)dr$$

and the torque is $dT = r\,dF_y$, or

$$dT = r(L\sin\phi - D\cos\phi)dr$$

Denoting the number of blades by Z, these can be expressed as follows:

$$dF_x = \frac{1}{2}\rho W^2 Zc(C_L\cos\phi + C_D\sin\phi)dr \qquad (12.32)$$

$$dT = \frac{1}{2}\rho W^2 Zcr(C_L\sin\phi - C_D\cos\phi)dr \qquad (12.33)$$

If the influence of wake rotation on the value of the static pressure at the exit is neglected, the momentum equation applied to an annular streamtube gives

$$dF_x = (V - V_b)\rho 2\pi r V_d\,dr = 4a(1-a)V^2\rho\pi r\,dr$$

Equating this to dF_x in Eq. (12.32) gives

$$\frac{1}{2}W^2 Zc(C_L\cos\phi + C_D\sin\phi) = 4\pi ra(1-a)V^2 \qquad (12.34)$$

The velocity diagram at the *mean chord position* is shown on the upper part of Figure 12.12. The relationships between the flow angle, relative velocity, and the axial and tangential components of absolute velocity are

$$\sin\phi = \frac{(1-a)V}{W} \qquad \cos\phi = \frac{(1+a')r\Omega}{W} \qquad (12.35)$$

With solidity defined now as $\sigma = Zc/2\pi r$, Eq. (12.34) can be recast as

$$\frac{a}{1-a} = \frac{\sigma(C_L\cos\phi - C_D\sin\phi)}{4\sin^2\phi} \qquad (12.36)$$

Similarly, angular momentum balance applied to the elementary annulus gives

$$dT = rV_u\,d\dot{m} = rV_u\rho V_d 2\pi r\,dr = 4a'(1-a)\rho V\Omega r^3\pi\,dr$$

---

[6]This assumption is based on the examination of the induced velocity by the vortex structure in the wake.

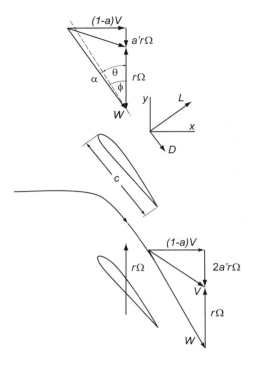

Figure 12.12 Illustration of the increase in tangential velocity across the disk.

and equating this to dT in Eq. (12.33) and making use of Eq. (12.35) yields

$$\frac{a'}{1+a'} = \frac{\sigma(C_L \sin\phi - C_D \cos\phi)}{4\cos\phi \sin\phi} \qquad (12.37)$$

Equations (12.36) and (12.37) can now be used to calculate the relationship between the parameters as follows. For a given blade element at location r/R, tip speed ratio $X = R\Omega/V$, and pitch angle θ, the solidity is $\sigma = Zc/2\pi r$ and the local blade speed ratio is $x = Xr/R = r\Omega/V$. With *assumed* values for a and a', the angle ϕ, as seen from Figure 12.12, is calculated from

$$\tan\phi = \frac{1-a}{1+a'}\frac{1}{x} \qquad (12.38)$$

and the angle of attack can then be determined from $\alpha = \phi - \theta$.

After this the lift coefficient can be calculated from the theoretical relation $C_L = 2\pi \sin\alpha$ and the value of the drag coefficient is found from experimental data. It is often given in the form $C_D = \varepsilon C_L$ with ε having a value in the range of $0.003 < \varepsilon < 0.015$ as C_L increases toward 1.2. After this the value of a is calculated from Eq. (12.36) and a' from Eq. (12.37). With these new estimates the calculations are repeated until the induction factors have converged. These calculations are demonstrated in the next example.

■ **EXAMPLE 12.2**

A three-bladed wind turbine with a rotor diameter of 40 m operates at the tip speed ratio 4. The blade has a constant cord of 1.2 m and a pitch angle of 12°. The drag-to-lift ratio is 0.005. Calculate the axial and angular induction factors at $r/R = 0.7$.

Solution: For given α the lift and drag coefficients are calculated from

$$C_{\text{L}} = 2\pi \sin \alpha \qquad C_{\text{D}} = \varepsilon C_{\text{L}}$$

The rest of the solution is carried out by iterations. First, for initial guesses for a and a'

$$\tan \phi = \frac{1-a}{1+a'} \frac{1}{x}$$

is solved for ϕ. Then the axial and tangential force coefficients are defined as

$$C_x = C_{\text{L}} \cos \phi + C_{\text{D}} \sin \phi \qquad C_y = C_{\text{L}} \sin \phi - C_{\text{D}} \cos \phi$$

and the axial and tangential induction factors are calculated from

$$\frac{a}{1-a} = \frac{\sigma C_x}{4 \sin^2 \phi} \qquad \frac{a'}{1+a'} = \frac{\sigma C_y}{4 \cos \phi \sin \phi}$$

or in explicit form, from

$$a = \frac{\sigma C_x}{4 \sin^2 \phi + \sigma C_x} \qquad a' = \frac{\sigma C_y}{4 \cos \phi \sin \phi - \sigma C_y}$$

Now a new value of ϕ can be determined, and the process is repeated. After the iterations have converged, the final value of the angle of attack is determined from

$$\alpha = \phi - \theta$$

A Matlab script gives the steps to carry out the calculations:

```
X=4; rR=0.7; R=20; theta=12*pi/180;
Z=3; c=1.2; ep=0.005; a(1)=0; ap(1)=0;
x=rR*X;
phi(1)=atan((1-a(1))/((1+ap(1))*x));
alpha(1)=phi(1)-theta;
CL(1)=2*pi*sin(alpha(1)); CD(1)=ep*CL(1);
Cx(1)=CL(1)*cos(phi(1))+CD(1)*sin(phi(1));
Cy(1)=CL(1)*sin(phi(1))-CD(1)*cos(phi(1));
sigma=Z*c/(2*pi*R*rR); imax=5;
for i=2:imax
    a(i)=sigma*Cx(i-1)/(4*sin(phi(i-1))^2+sigma*Cx(i-1));
    ap(i)=sigma*Cy(i-1)/(4*sin(phi(i-1))*cos(phi(i-1))-sigma*Cy(i-1));
    phi(i)=atan((1-a(i))/((1+ap(i))*x));
    alpha(i)=phi(i)-theta;
    CL(i)=2*pi*sin(alpha(i)); CD(i)=ep*CL(i);
    Cx(i)=CL(i)*cos(phi(i))+CD(i)*sin(phi(i));
    Cy(i)=CL(i)*sin(phi(i))-CD(i)*cos(phi(i));
end
fid=fopen('induction','w'); i=[1:imax];
fprintf(fid,'%12i%12.8f%12.8f\n',[i;a;ap]);
fclose(fid);
```

Convergence of the induction factors a and a' are given as follows:

i	a	a'
1	0.00000	0.000000
2	0.06665	0.009046
3	0.06366	0.007349
4	0.06389	0.007451
5	0.06387	0.007443

The induction factor for the wake rotation is seen to be one order of magnitude smaller than that for axial flow. Were the tip speed ratio smaller, the wake rotation would be more pronounced. The next example extends the analysis to the entire span.

■ EXAMPLE 12.3

A three-bladed wind turbine with a rotor diameter of 40 m operates at the tip speed ratio 4. The blade has a constant chord of 1.2 m and the blades are at $4°$ angle of attack. The drag-to-lift ratio is 0.005. Calculate the pitch angle along the span from $r/R = 0.2$ to 1.0.

Solution: The procedure for calculating the local parameters is the same as in the previous example. An outer loop needs to be inserted to extend the calculations to all the span locations. Results are shown in the table below.

r/R	ϕ	θ	a	a'
0.2000	50.2024	46.2024	0.0169	0.0237
0.3000	38.8601	34.8601	0.0204	0.0131
0.4000	31.1403	27.1403	0.0246	0.0089
0.5000	25.7391	21.7391	0.0292	0.0068
0.6000	21.8146	17.8146	0.0341	0.0055
0.7000	18.8599	14.8599	0.0392	0.0046
0.8000	16.5662	12.5662	0.0443	0.0039
0.9000	14.7392	10.7392	0.0497	0.0035
1.0000	13.2521	9.2521	0.0551	0.0031

Examination of Figure 12.12 shows that as the blade speed increases toward the tip, the pitch angle decreases as the tip is approached. This also causes the flow angle ϕ to decrease and the lift force to tilt toward the axial direction. This means that the tangential component of lift, which is the force that provides the torque on the shaft, diminishes. The same effect comes from the increase in induction factor along the span, for then the axial velocity across the blades at large radii is low, and this causes the lift force to tilt toward the axis. Another way to think about this is that an increase in the induction factor leads to lower mass flow rate and thus to lower work done on the blade elements near the tip.

The induction factor also varies in the angular direction, because in a two- or three-bladed wind turbine much of the flow does not come close to the blades at all, and hence this part of the stream is not expected to slow down substantially. Thus the induction factor in this part of the flow deviates from the factor that holds for an actuator disk.

The azimuthal variation of the induction factor has been calculated by Burton et al. [13] at four radial locations for a three-bladed wind turbine with the tip speed ratio of six. Their results are shown in Figure 12.13.

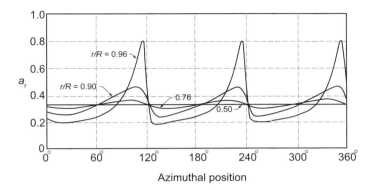

Figure 12.13 Azimuthal variation of induction factors for a three-bladed turbine at four radial locations. (Drawn after Burton et al. [13].)

The axial force and torque can now be calculated for the entire disk from

$$F_x = 4\rho\pi V^2 \int_0^R a(1-a)r\,dr \qquad (12.39)$$

$$T = 4\rho\pi\Omega V \int_0^R a'(1-a)r^3\,dr \qquad (12.40)$$

The power is then obtained as $\dot{W} = T\Omega$. A performance calculation for an actual wind turbine is given in the next example.

■ **EXAMPLE 12.4**

A three-bladed wind turbine with a rotor diameter of 40 m operates at the tip speed ratio 4. The blade has a constant chord of 1.2 m, and the blades are at 4° angle of attack. The drag-to-lift ratio is 0.005. The wind speed is $V = 10\,\text{m/s}$, and air density can be taken to be $\rho = 1.2\,\text{kg/m}^3$. Only the blade along $r/R = 0.2\text{--}1.0$ is to be used to calculate the total force and torque.

Solution: The calculation proceeds as in the previous example. After the results have been stored for various radial locations, the integrals can be evaluated by using the trapezoidal rule:

$$\int_a^b f(x)dx = \frac{\Delta r}{2}\sum_{i=1}^{n-1}(f(x_i)+f(x_{i+1})) = \Delta r\left[\sum_{i=1}^{n-1} f(x_i) + \frac{1}{2}(f(x_1)+f(x_n))\right]$$

The part of the blade close to the hub is ignored and in the numerical integrations of Eqs. (12.39) and (12.40) the lower limit of integration is replaced by $r/R = 0.2$. The result of the integration gives the values

$$F_x = 10{,}783\,\text{N} \qquad T = 51{,}403\,\text{N·m} \qquad \dot{W} = 102.8\,\text{kW}$$

for the blade force, torque, and power.

■

12.3.3 Ideal wind turbine

Glauert [28] developed relations from which the pitch angle can be calculated for a wind turbine with the highest value for the power coefficient. Simple relations are obtained if drag is neglected. Under this assumption Eq. (12.36) becomes

$$\frac{a}{1-a} = \frac{Zc\,C_L \cos\phi}{8\pi r \sin^2\phi}$$

and Eq. (12.37) reduces to

$$\frac{a'}{1+a'} = \frac{Zc\,C_L}{8\pi r \cos\phi}$$

Diving the second by the first gives

$$\tan^2\phi = \left(\frac{1-a}{1+a'}\right)\frac{a'}{a} \tag{12.41}$$

The velocity diagram at midchord shows that

$$\tan\phi = \frac{(1-a)}{(1+a')x}$$

in which the blade speed ratio is $x = r\Omega/V$. Making use of this expression reduces Eq. (12.41) to

$$a'(1+a')x^2 = a(1-a) \tag{12.42}$$

The power delivered by the turbine is

$$\dot{W} = \int_0^R \Omega\, dT = 4\pi\rho V\Omega^2 \int_0^R (1-a)ar^3\,dr$$

and the power coefficient can be written as

$$C_p = \frac{8}{X}\int_0^X (1-a)a'x^3\,dx \tag{12.43}$$

Since the integrand is positive, the maximum power coefficient can be obtained by maximizing the integrand, subject to the constraint Eq. (12.42).[7] Hence, differentiating the integrand and setting it to zero gives

$$(1-a)\frac{da'}{da} = a$$

and the same operation on the constraint leads to

$$(1+2a')x^2\frac{da'}{da} = 1 - 2a$$

---

[7] A problem of this kind is solved by methods of variational calculus and Lagrange multipliers. Owing to the form of the integrand, the direct method used above works.

Equating these gives
$$(1-a)(1-2a) = a'(1+2a')x^2 \tag{12.44}$$
Solving Eq. (12.42) for x^2 and substituting into this gives
$$\frac{1+2a'}{1+a'} = \frac{1-2a}{a}$$
from which
$$a' = \frac{1-3a}{4a-1} \tag{12.45}$$
This equation shows that for $\frac{1}{4} < a < \frac{1}{3}$, the value of a' is positive, and that as $a \to \frac{1}{4}$, then $a' \to \infty$. Also, as $a \to \frac{1}{3}$, then $a' \to 0$. Substituting $1 + a'$ from Eq. (12.45) into

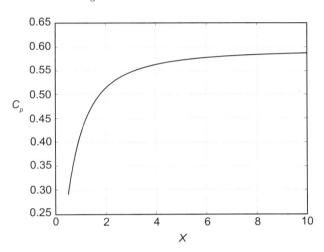

Figure 12.14 Power coefficient for optimum wind turbine.

Eq. (12.42) yields
$$a'x^2 = (1-a)(4a-1)$$
This shows that for $a = \frac{1}{4}$, the product $a'x^2 = 0$ and that for $a = \frac{1}{3}$, the value is $a'x^2 = \frac{2}{9}$. The power coefficient can now be determined by integrating
$$C_p = \frac{8}{X} \int_0^X (1-a)a'x^3 dx$$
for various tip speed ratios X. The results for this calculation are presented in the graph in Figure 12.14. For high tip speeds when $a = \frac{1}{3}$ and $a'x^2 = \frac{2}{9}$ the Betz limit $C_p = \frac{16}{27} = 0.593$ is approached.

With the coefficients a and a' known at various radial positions, both the relative velocity and the angle ϕ can be determined. From these the blade angle for a given angle of attack can be calculated.

12.3.4 Prandtl's tip correction

It has been seen that lift generated by a flow over an airfoil can be related to circulation around the airfoil. An important advance in the aerodynamic theory of flight was developed independently by Kutta and Joukowski, who showed that the magnitude of the circulation that develops around an airfoil is such that flow leaves the rear stagnation point move to the trailing edge. The action is a result of viscous forces, and the flow in the viscous boundary layers leaves the airfoil into the wake as a *free shear* layer with vorticity. The vorticity distribution in the wake is unstable and rolls into a *tip vortex*. In such a flow it is possible to replace the blade by *bound* vorticity along the span of the blade and dispense with the blade completely. For such a model, vortex filaments exist in an inviscid flow, and according to the vortex theorems of Helmholtz, they cannot end in the fluid, but must either extend to infinity or form closed loops. This is achieved by connecting the bound vortex into the tip vortex, which extends far downstream.

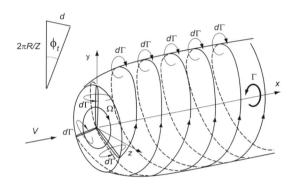

Figure 12.15 Wake vorticity. Drawn after Burton et al. [13].

For a wind turbine with an infinite number of blades, the vortex system in the wake consists of a cylindrical sheet of vorticity downstream from the edge of an actuator disk, as shown in Figure 12.15, and bound vorticity along radial lines starting from the center. From there an axial vortex filament is directed downstream. The angular velocity in the wake in this model is a result of the induced velocity of this *hub vortex*, and it explains why an *irrotational vortex* flow in the wake is a reasonable model.

When the number of blades is finite, vorticity leaves the blades of a wind turbine in a twisted helical sheet owing to rotation of the blades. One model is to assume that it organizes itself into a cylindrical vortex sheet. The twisted sheets are separated since the blades are discrete. Prandtl [60] and Goldstein [29] developed ways to take into account the influence of this vorticity, which now is assumed to be shed from the blade tips.

Prandtl's model is based on flow over the edges of a set of vortex sheets distance d apart, as shown in Figure 12.16. As the sheets move downstream, they induce a periodic flow near their edges. This leads to lower transfer of energy to the blades.

Prandtl's analysis results in the introduction of factor F, given by

$$F = \frac{2}{\pi} \cos^{-1}\left[\exp\left(-\frac{\pi(R-r)}{d}\right)\right]$$

into the equation for the blade force of an elemental annulus. The distance d between helical sheets is given by

$$d = \frac{2\pi R}{Z}\sin\phi_t$$

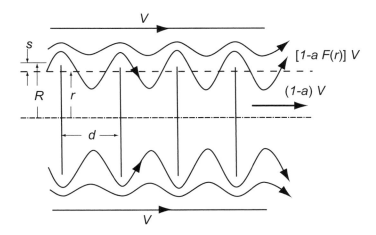

Figure 12.16 Prandtl's tip loss factor.

as the diagram in Figure 12.16 shows. The sine of the flow angle at the tip is

$$\sin \phi_t = \frac{(1-a)V}{W_t}$$

so that

$$\frac{\pi(R-r)}{d} = \frac{1}{2} Z \left(\frac{R-r}{R}\right) \frac{W_t}{(1-a)V}$$

Following Glauert and assuming that

$$\frac{W_t}{R} = \frac{W}{r}$$

yields

$$\frac{\pi(R-r)}{d} = \frac{1}{2} Z \left(\frac{R-r}{r}\right) \frac{W}{(1-a)V}$$

The relative velocity is

$$W = \sqrt{(1-a)^2 V^2 + (r\Omega)^2}$$

so that

$$\frac{W}{(1-a)V} = \sqrt{1 + \frac{(r\Omega)^2}{(1-a)^2 V^2}} = \sqrt{1 + \frac{X^2}{(1-a)^2} \frac{r^2}{R^2}}$$

which gives

$$\frac{\pi(R-r)}{d} = \frac{1}{2} Z \left(\frac{R}{r} - 1\right) \sqrt{1 + \frac{X^2}{(1-a)^2} \frac{r^2}{R^2}}$$

The tip loss correction factor is then given by

$$F = \frac{2}{\pi} \cos^{-1} \left[\exp\left(-\frac{1}{2} Z \left(\frac{R}{r} - 1\right) \sqrt{1 + \frac{X^2}{(1-a)^2} \frac{r^2}{R^2}}\right) \right] \qquad (12.46)$$

The tip loss factor calculated for four tip speed ratios is shown in Figure 12.17 for a two-bladed wind turbine. As the expression in Eq. (12.46) shows, decreasing the number of blades moves the curves in the same direction as decreasing the tip speed ratio. Both of these increase the spacing d and make the winding of the helix looser. Conversely, an actuator disk with an infinite number of blades results in $F = 1$. The blade force is thus reduced by factor F for a wind turbine with a discrete number of blades. The axial force

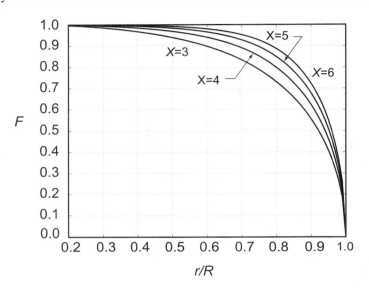

Figure 12.17 Prandtl's tip loss factor for two-bladed wind turbine with tip speed ratios $X = 3, 4, 5, 6$ at various radial positions r/R.

and torque are given by

$$dF_x = 4\rho\pi V^2 a(1-a)Fr\,dr \qquad dT = 4\rho\pi\Omega V a'(1-a)Fr^3\,dr$$

and in terms of the axial and tangential force coefficients, these are

$$dF_x = \frac{1}{2}\rho W^2 Zc\, C_x\, dr \qquad dT = \frac{1}{2}\rho W^2 Zc\, C_y\, r\,dr$$

Equating these gives

$$8\pi a(1-a)V^2 F = Zc\, W^2 C_x \qquad 8\pi a'(1-a)V\Omega r^2 F = Zc\, W^2 C_y$$

Making use of

$$\cos\phi = \frac{(1-a)V}{W} \qquad \sin\phi = \frac{(1+a')r\Omega}{W}$$

and solving for a and a' then gives

$$a = \frac{\sigma C_x}{4F\sin^2\phi + \sigma C_x} \qquad a' = \frac{\sigma C_y}{4F\cos\phi\sin\phi - \sigma C_y} \qquad (12.47)$$

These can now be used to calculate the wind turbine design and performance parameters. For the data given in the previous examples, if the lift coefficient is kept at $C_L = 0.3$ along the entire span (corresponding to an angle of attach of 2.74°), then the pitch angle and induction factors are as shown in table below. The results differ slightly from those in which the tip correction was neglected.

r/R	ϕ	θ	a	a'
0.2000	44.0145	41.2778	0.0176	0.0168
0.3000	32.8470	30.1103	0.0224	0.0095
0.4000	25.7736	23.0369	0.0279	0.0066
0.5000	21.0284	18.2917	0.0340	0.0051
0.6000	17.6605	14.9237	0.0408	0.0043
0.7000	15.1463	12.4095	0.0491	0.0037
0.8000	13.1644	10.4277	0.0612	0.0035
0.9000	11.4195	8.6828	0.0876	0.0038

∎

12.4 TURBOMACHINERY AND FUTURE PROSPECTS FOR ENERGY

In some sections of this book the reader's attention has been drawn to how the present primary energy resources of the world are used and how important turbomachines are in converting the energy in fossil fuels into shaft work. Turbomachines powered by hydroelectric energy and wind energy are the main two machines that contribute to significant forms of renewable energy in the mix. It is unfortunate that they account for such a small fraction of the world's energy supply, because the fossil energy resources will be mostly exhausted during this century. The recent high prices for oil are a signal that soon the world's aging oil fields will be unable to keep up with the demand of the world's growing population (which in 2011 reached 7 billion). The production curves for oil, natural gas, and coal are bell-shaped, and the significant event on the road to their exhaustion is when each production peak will be passed. After that, the world will need to get by with less of these nonrenewable fuels. The peak production for world's oil is likely to occur before the year 2015, and the production peaks for natural gas and coal will not be far behind. Coal, which is thought to be the most abundant of the three, has seen its reserve-to-production ratio drop from 227 years to 119 years during the first decade of the present century. A 119-life for the existing reserve might seem like a long time, but the 108-year drop during the last decade indicates that the rate at which the reserves decrease is quite rapid. In addition, the reserve-to-production ratio gives a misleading number, as it assumes that production stays flat until the resource is completely exhausted. A 40-year reserve-to-production ratio is, in fact, an indication of a production peak [49].

In the United States coal became the primary source of energy in the late 1800s overtaking biomass, which met the energy needs of the early settlers. It took some 40 years for coal's share to rise from 1% to 10% of the mix and the historical record shows similar timespans for the rise of oil and natural gas as well. Nuclear energy's share, after 50 years of effort, is still less than 10% of the energy supply in the United States. The prospects for renewable forms to offset the growing population and its growing demand for energy are exceedingly slim. Decisions to launch a comprehensive strategy for coping with the exhaustion of fossil fuels have been delayed many times despite warnings throughout the last century [11, 39]. Still installed wind energy capacity is increasing rapidly (even if not rapidly enough), and wind power offers the best possibility of becoming a significant source of energy in the

United States. Denmark in particular has demonstrated that wind power can make up a sizeable fraction of a country's energy supply with a continuous and sustained effort.

EXERCISES

12.1 Reconsider the ducted wind turbine, but now let the duct be a cylindrical control volume A_e in cross sectional area. Show that the axial force on the blades is

$$F_d = \rho A_b V_b (V - V_b)$$

which is in agreement with Eq. (12.6).

12.2 The nondimensional pressure difference can be expressed as

$$\frac{p_+ - p_a}{\frac{1}{2}\rho V^2}$$

in which the pressure difference is that just upstream the disk and the free stream. (a) Use the momentum theory for a wind turbine to express this relationship in terms of the interference factor a. For which value of a is this maximum? Interpret this physically. (b) Develop a similar relationship for the expression

$$\frac{p_a - p_-}{\frac{1}{2}\rho V^2}$$

and determine for which value of a is this the maximum?

12.3 A wind turbine operates at wind speed of $V = 12\,\text{m/s}$. Its blade radius is $R = 20\,\text{m}$ and its tip speed radius $R\Omega/V = 4$. It operates at the condition $C_p = 0.3$. Find (a) the rate of rotation of the blades, (b) the power developed by the turbine, (c) the value of the interference factor a using the momentum theory, and (d) the pressure on the front of the actuator disk, assuming that the free stream pressure is $101.30\,\text{kPa}$.

12.4 Using the axial momentum theory, calculate the ratio of the slipstream radius to that of the disk radius in terms of the interference factor a. If the wind turbine blades are $80\,\text{m}$ long, what is the radius of the slipstream far downstream? What is the radius of the streamtube far upstream?

12.5 In a wind with speed $V = 8.7\,\text{m/s}$ and air density $\rho = 1.2\,\text{kg/m}^3$, a wind turbine operates at a condition with $C_p = 0.31$. Find the blade length, assuming that the power delivered to the turbine is to be $\dot{W} = 250\,\text{kW}$.

12.6 Consider a three-bladed wind turbine with blade radius of $R = 35\,\text{m}$ and constant chord of $c = 80\,\text{cm}$, which operates with a rotational speed of $\Omega = 10$ rpm. The wind speed is $V = 12\,\text{m/s}$. (a) Find the axial and tangential induction factors at $r = 10\,\text{m}$, assuming that the angle of attack is $6°$ and $C_D = 0.01 C_L$. (b) Find the axial force and the torque, assuming the air density is $\rho = 0.12\,\text{kg/m}^3$. (c) Calculate the axial force and the torque by assuming that the induction factors are uniform and equal to their values at $r/R = 0.6$.

Appendix A
Streamline curvature and radial equilibrium

A.1 STREAMLINE CURVATURE METHOD

In this appendix the governing equations for the streamline curvature method are developed. The derivation of the acceleration terms follows that of Cumpsty [18]. The blade surfaces may be highly curved, but it is useful to visualize locally flat surface patches in them.

A.1.1 Fundamental equations

The acceleration of a fluid particle is given by its substantial derivative

$$\mathbf{a} = \frac{\partial \mathbf{V}}{\partial t} + \mathbf{V} \cdot \nabla \mathbf{V}$$

in which the unsteady term vanishes in steady flow. The second term represents the spatial acceleration of the flow. In cylindrical coordinates the gradient operator can be written as

$$\nabla = \mathbf{e}_r \frac{\partial}{\partial r} + \frac{\mathbf{e}_\theta}{r} \frac{\partial}{\partial \theta} + \mathbf{e}_z \frac{\partial}{\partial z}$$

By defining the meridional component as

$$\mathbf{V}_m = V_m \mathbf{e}_m = V_r \mathbf{e}_r + V_z \mathbf{e}_z$$

Principles of Turbomachinery. By Seppo A. Korpela
Copyright © 2011 John Wiley & Sons, Inc.

the velocity vector can also be written as

$$\mathbf{V} = V_m \mathbf{e}_m + V_\theta \mathbf{e}_\theta$$

The scalar product of the unit vector in the direction of $\mathbf{e}_m$ on the meridional plane and the gradient operator gives the directional derivative in the direction of the unit vector:

$$\mathbf{e}_m \cdot \nabla = \frac{\partial}{\partial m} = (\mathbf{e}_m \cdot \mathbf{e}_r)\frac{\partial}{\partial r} + (\mathbf{e}_m \cdot \mathbf{e}_z)\frac{\partial}{\partial z}$$

The term in the tangential direction has dropped out because the vector $\mathbf{e}_m$ is orthogonal to $\mathbf{e}_\theta$.

The angle between the directions of $\mathbf{e}_m$ and $\mathbf{e}_z$ is denoted by ϕ. In terms of this angle the partial derivatives may be written as

$$\frac{\partial}{\partial m} = \sin\phi \frac{\partial}{\partial r} + \cos\phi \frac{\partial}{\partial z}$$

and the gradient operator takes the form

$$\nabla = \mathbf{e}_m \frac{\partial}{\partial m} + \frac{\mathbf{e}_\theta}{r}\frac{\partial}{\partial \theta}$$

The acceleration of a fluid particle can now be expressed as

$$\mathbf{a} = (V_m \mathbf{e}_m + V_\theta \mathbf{e}_\theta) \cdot \left(\mathbf{e}_m \frac{\partial}{\partial m} + \mathbf{e}_\theta \frac{1}{r}\frac{\partial}{\partial \theta}\right)(V_m \mathbf{e}_m + V_\theta \mathbf{e}_\theta)$$

which leads to

$$\mathbf{a} = V_m \frac{\partial}{\partial m}(V_m \mathbf{e}_m + V_\theta \mathbf{e}_\theta) + \frac{V_\theta}{r}\frac{\partial}{\partial \theta}(V_m \mathbf{e}_m + V_\theta \mathbf{e}_\theta)$$

When this is expanded, it reduces to

$$\mathbf{a} = \mathbf{e}_m V_m \frac{\partial V_m}{\partial m} - \mathbf{e}_n \frac{V_m^2}{R} + \mathbf{e}_\theta V_m \frac{\partial V_\theta}{\partial m} - \mathbf{e}_r \frac{V_\theta^2}{r}$$

To arrive at these relations, the formulas

$$\frac{\partial \mathbf{e}_m}{\partial m} = -\frac{\mathbf{e}_n}{R} \qquad \frac{\partial \mathbf{e}_m}{\partial \theta} = 0 \qquad \frac{\partial \mathbf{e}_\theta}{\partial m} = 0 \qquad \frac{\partial \mathbf{e}_\theta}{\partial \theta} = -\mathbf{e}_r$$

were used. The radius of curvature of a streamline on the meridional plane is denoted by R and is taken as positive when the streamline is concave away from the z axis. The direction of the unit vector $\mathbf{e}_n$ is perpendicular to the direction of vector $\mathbf{e}_m$ on the meridional plane in such a way that a right-handed triple $(\mathbf{e}_n, \mathbf{e}_\theta, \mathbf{e}_m)$ is retained. Since this was obtained by rotation by the angle ϕ about the axis of $\mathbf{e}_\theta$, the n direction coincides with the radial direction and the m direction coincides with z direction, when the angle ϕ is zero.

Consider next a direction specified by the unit vector $\mathbf{e}_q$ on the meridional plane. It is inclined to the radial direction by angle γ. This and other angles are shown in Figure A.1. The angle γ is the *sweep* angle of the blade at its leading edge. It is positive for a sweep toward the positive z direction. The angle between direction $\mathbf{e}_q$ and the m axis is $\pi/2 - (\gamma + \phi)$. Hence the component of acceleration in the direction $\mathbf{e}_q$ is

$$a_q = (\mathbf{e}_q \cdot \mathbf{e}_m) V_m \frac{\partial V_m}{\partial m} - (\mathbf{e}_q \cdot \mathbf{e}_n) \frac{V_m^2}{R} + (\mathbf{e}_q \cdot \mathbf{e}_\theta) V_m \frac{\partial V_\theta}{\partial m} - (\mathbf{e}_q \cdot \mathbf{e}_r) \frac{V_\theta^2}{r}$$

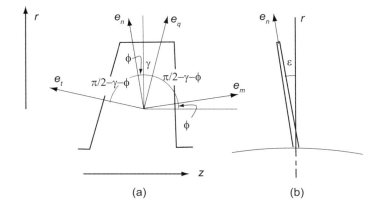

Figure A.1 Unit vectors on the meridional plane (a) and the angle of lean (b).

This reduces to

$$a_q = \sin(\gamma + \phi) V_m \frac{\partial V_m}{\partial m} - \cos(\gamma + \phi) \frac{V_m^2}{R} - \cos\gamma \frac{V_\theta^2}{r} \qquad (A.1)$$

Next, let the unit vector e_t denote a direction normal to e_q on the meridional plane. The component of acceleration in this direction is given by

$$a_t = (\mathbf{e}_t \cdot \mathbf{e}_m) V_m \frac{\partial V_m}{\partial m} - (\mathbf{e}_t \cdot \mathbf{e}_n) \frac{V_m^2}{R} + (\mathbf{e}_t \cdot \mathbf{e}_\theta) V_m \frac{\partial V_\theta}{\partial m} - (\mathbf{e}_t \cdot \mathbf{e}_r) \frac{V_\theta^2}{r}$$

which reduces to

$$a_t = \cos(\gamma + \phi) V_m \frac{\partial V_m}{\partial m} + \sin(\gamma + \phi) \frac{V_m^2}{R} - \sin\gamma \frac{V_\theta^2}{r}$$

Finally, the acceleration component in the tangential direction can be written as

$$a_\theta = V_m \frac{\partial V_\theta}{\partial m}$$

If the blades lean at an angle ε from the meridional plane, in the direction opposite to θ, the acceleration components $\mathbf{a}_q$ and $\mathbf{a}_\theta$ can be used to construct new components that lie on the plane containing a blade with this lean angle. The transformation to this plane is equivalent to rotation of the surface about the axis containing the unit vector $\mathbf{e}_t$. With $\mathbf{e}_e$ denoting the unit vector obtained by rotating $\mathbf{e}_q$ by angle ε counterclockwise, the acceleration component in the direction of $\mathbf{e}_e$ is given by

$$a_e = (\mathbf{e}_e \cdot \mathbf{e}_q) a_q + (\mathbf{e}_e \cdot \mathbf{e}_\theta) a_\theta + (\mathbf{e}_e \cdot \mathbf{e}_t) a_t$$

which reduces to

$$a_e = \cos\varepsilon\, a_q - \sin\varepsilon\, a_\theta$$

The Euler equation for an inviscid flow in vector notation is

$$\mathbf{a} = -\frac{1}{\rho} \nabla p + \frac{\mathbf{F}}{\rho} \qquad (A.2)$$

The components of this equation in any direction can now be obtained by taking its scalar product with a unit vector in the chosen direction. The component in the direction of $\mathbf{e}_e$ is

$$a_e = -\frac{1}{\rho}\frac{\partial p}{\partial e} + \frac{F_e}{\rho}$$

and its component in the direction of $\mathbf{e}_q$ is

$$a_q = -\frac{1}{\rho}\frac{\partial p}{\partial q} + \frac{F_q}{\rho} \tag{A.3}$$

The acceleration component a_q is the acceleration in the q direction. The pressure term on the right side of Eq. (A.3) can be modified by making use of the Tds relation

$$T\frac{\partial s}{\partial q} = \frac{\partial h}{\partial q} - \frac{1}{\rho}\frac{\partial p}{\partial q} \tag{A.4}$$

The directional derivative of the stagnation enthalpy

$$h_0 = h + \frac{1}{2}\left(V_m^2 + V_\theta^2\right) \tag{A.5}$$

in the q direction is

$$\frac{\partial h}{\partial q} = \frac{\partial h_0}{\partial q} - V_m\frac{\partial V_m}{\partial q} - V_\theta\frac{\partial V_\theta}{\partial q} \tag{A.6}$$

Using this in the expression for the pressure gradient in Eq. (A.4) leads to

$$-\frac{1}{\rho}\frac{\partial p}{\partial q} = T\frac{\partial s}{\partial q} - \frac{\partial h_0}{\partial q} + V_m\frac{\partial V_m}{\partial q} + V_\theta\frac{\partial V_\theta}{\partial q} \tag{A.7}$$

Substituting this into Eq. (A.3) gives

$$\sin(\gamma+\phi)V_m\frac{\partial V_m}{\partial m} - \cos(\gamma+\phi)\frac{V_m^2}{R} - V_m\frac{\partial V_m}{\partial q}$$
$$= T\frac{\partial s}{\partial q} - \frac{\partial h_0}{\partial q} + \cos\gamma\frac{V_\theta^2}{r} + V_\theta\frac{\partial V_\theta}{\partial q} + \frac{F_q}{\rho} \tag{A.8}$$

With $q = \dfrac{r}{\cos\gamma}$, terms involving V_θ on the RHS can be combined to yield:

$$\frac{\partial V_m^2}{\partial q} + P(q)V_m^2 = T(q) \tag{A.9}$$

where the functions $P(q)$ and $T(q)$ are defined as

$$P(q) = 2\left[-\frac{\cos(\gamma+\phi)}{R} - \frac{\sin(\gamma+\phi)}{V_m}\frac{\partial V_m}{\partial m}\right] \tag{A.10}$$

$$T(q) = 2\left[\frac{\partial h_0}{\partial q} - T\frac{\partial s}{\partial q} - \frac{1}{2q^2}\frac{\partial}{\partial q}(q^2 V_\theta^2) - \frac{F_q}{\rho}\right] \tag{A.11}$$

In a flow passage of constant height the meridional direction is the z direction and q can be replaced by r. In addition, the streamlines do not bend in the direction of the axis z, so that the radius of curvature R tends to infinity. If the there is no sweep γ is zero, and if the stagnation enthalpy and entropy do not vary across the channel, then, in the absence of a blade force, this equation reduces to

$$\frac{\partial V_z^2}{\partial r} + \frac{V_\theta}{r}\frac{\partial}{\partial r}(rV_\theta) = 0 \tag{A.12}$$

This is the simple radial equilibrium equation used in Chapters 6 and 7.

A.1.2 Formal solution

Since Eq. (A.9) is nonlinear, it must be solved numerically. One way to carry this out is to first find its formal solution. Ignoring the nonlinearity and treating $P(q)$ and $T(q)$ as functions of q, the equation

$$\frac{\partial V_m^2}{\partial q} + P(q)V_m^2 = T(q) \qquad (A.13)$$

becomes a first-order differential equation with variable coefficients. Its solution can then be obtained by using an integrating factor, or by variation of parameters.

To simplify the notation, first the variable V_m^2 is replaced by U and the homogeneous solution of Eq. (A.13) is seen to be

$$U_h = Ce^{-\int P(q)dq} \qquad (A.14)$$

To obtain the particular solution to Eq. (A.13), the solution is assumed to have the form

$$U_p = H(r)e^{-\int P(x)dx}$$

which, when inserted into the equation and after rearrangement, yields

$$\frac{\partial U_p}{\partial q} + P(q)U_p = \frac{\partial H}{\partial q} e^{-\int_{q_0}^{q} P(q')dq'}$$

Comparing this to Eq. (A.13) shows that

$$T = \frac{\partial H}{\partial q} e^{-\int_{q_0}^{q} P(q')dq'}$$

or

$$\frac{\partial H}{\partial q} = T e^{\int_{q_0}^{q} P(q')dq'}$$

Integrating leads to

$$H(q) = \int_{q_0}^{q} T(q') e^{\int_{q_0}^{q'} P(q'')dq''} dq'$$

Therefore, the particular solution is

$$U_p = e^{-\int_{q_0}^{q} P(q')dq'} \int_{q_0}^{q} e^{\int_{q_0}^{q'} P(q'')dq''} T(q')dq' \qquad (A.15)$$

The complete solution in the original variable is

$$V_m^2(q) = V_M^2 e^{-\int_{q_0}^{q} P(q')dq'} + e^{-\int_{q_0}^{q} P(q')dq'} \int_{q_0}^{q} e^{\int_{q_0}^{q'} P(q'')dq''} T(q')dq' \qquad (A.16)$$

If the sweep angle γ is zero, the q direction is the r direction. In such a case integrations in Eq. (A.13) are with respect to r. With the lower limit of integration the middle streamline, the solution becomes

$$V_m^2(r_M) = V_M^2 e^{-\int_{r_M}^{r} P(r')dr'} + e^{-\int_{r_M}^{r} P(r')dr'} \int_{r_M}^{r} e^{\int_{r_M}^{r'} P(r'')dr''} T(r')dr' \qquad (A.17)$$

and the integration constant is identified as the meridional velocity squared on the middle streamline, $V_M^2 = V_m^2(r_M)$. The streamlines are then laid out such that each streamtube has the same mass flow rate, and grid is set up on the qz plane, with grid points on the streamlines a specified distance apart. The solution then proceeds by iteration. The simplest situation is to consider the intergap region so that the blades are not present. The quasiorthogonals q can be spaced and oriented such that the first one is aligned with the trailing edge of a blade and the last one with the leading edge of the next blade row. The curvature is calculated at the grid locations, and the velocity gradients are determined by finite-difference approximations. This fixes the values of P and T. Integration of Eq. (A.17) from the central streamline toward the hub and the casing gives the velocity field. Next, the mass flow rates are calculated for each streamtube. The residual is examined, and the streamlines are adjusted so that the fluid again flows at the same rate through each streamtube. The process is repeated until convergence for the location of the streamlines is obtained. This requires the use of relaxation.

Appendix B
Thermodynamic Tables

B.1 Thermodynamic properties of steam, temperature table

B.2 Thermodynamic properties of steam, pressure table

B.3 Thermodynamic properties of superheated steam

B.4 Thermodynamic properties of air

B.5 Specific heats of common gases

B.6 Molar specific heats of common gases

Table B.1 Thermodynamic Properties of Saturated Steam, Temperature Table

T °C	p bar	$v_f \cdot 10^3$ m³/kg	v_g m³/kg	u_f kJ/kg	u_g kJ/kg	h_f kJ/kg	h_g kJ/kg	s_f kJ/kg K	s_g kJ/kg K
0.01	0.00611	1.0002	206.136	0.00	2375.3	0.01	2501.3	0.0000	9.1562
1	0.00657	1.0002	192.439	4.18	2375.9	4.183	2502.4	0.0153	9.1277
2	0.00706	1.0001	179.762	8.40	2377.3	8.401	2504.2	0.0306	9.1013
3	0.00758	1.0001	168.016	12.61	2378.7	12.61	2506.0	0.0459	9.0752
4	0.00814	1.0001	157.126	16.82	2380.0	16.82	2507.9	0.0611	9.0492
5	0.00873	1.0001	147.024	21.02	2381.4	21.02	2509.7	0.0763	9.0236
6	0.00935	1.0001	137.647	25.22	2382.8	25.22	2511.5	0.0913	8.9981
7	0.01002	1.0001	128.939	29.41	2384.2	29.42	2513.4	0.1063	8.9729
8	0.01073	1.0002	120.847	33.61	2385.6	33.61	2515.2	0.1213	8.9479
9	0.01148	1.0002	113.323	37.80	2386.9	37.80	2517.1	0.1361	8.9232
10	0.01228	1.0003	106.323	41.99	2388.3	41.99	2518.9	0.1510	8.8986
11	0.01313	1.0004	99.808	46.17	2389.7	46.18	2520.7	0.1657	8.8743
12	0.01403	1.0005	93.740	50.36	2391.1	50.36	2522.6	0.1804	8.8502
13	0.01498	1.0006	88.086	54.55	2392.4	54.55	2524.4	0.1951	8.8263
14	0.01599	1.0008	82.814	58.73	2393.8	58.73	2526.2	0.2097	8.8027
15	0.01706	1.0009	77.897	62.92	2395.2	62.92	2528.0	0.2242	8.7792
16	0.01819	1.0011	73.308	67.10	2396.6	67.10	2529.9	0.2387	8.7560
17	0.01938	1.0012	69.023	71.28	2397.9	71.28	2531.7	0.2532	8.7330
18	0.02064	1.0014	65.019	75.47	2399.3	75.47	2533.5	0.2676	8.7101
19	0.02198	1.0016	61.277	79.65	2400.7	79.65	2535.3	0.2819	8.6875
20	0.02339	1.0018	57.778	83.83	2402.0	83.84	2537.2	0.2962	8.6651
21	0.02488	1.0020	54.503	88.02	2403.4	88.02	2539.0	0.3104	8.6428
22	0.02645	1.0023	51.438	92.20	2404.8	92.20	2540.8	0.3246	8.6208
23	0.02810	1.0025	48.568	96.38	2406.1	96.39	2542.6	0.3388	8.5990
24	0.02985	1.0027	45.878	100.57	2407.5	100.57	2544.5	0.3529	8.5773
25	0.03169	1.0030	43.357	104.75	2408.9	104.75	2546.3	0.3670	8.5558
26	0.03363	1.0033	40.992	108.93	2410.2	108.94	2548.1	0.3810	8.5346
27	0.03567	1.0035	38.773	113.12	2411.6	113.12	2549.9	0.3949	8.5135
28	0.03782	1.0038	36.690	117.30	2413.0	117.30	2551.7	0.4088	8.4926
29	0.04008	1.0041	34.734	121.48	2414.3	121.49	2553.5	0.4227	8.4718
30	0.04246	1.0044	32.896	125.67	2415.7	125.67	2555.3	0.4365	8.4513
31	0.04495	1.0047	31.168	129.85	2417.0	129.85	2557.1	0.4503	8.4309
32	0.04758	1.0050	29.543	134.03	2418.4	134.04	2559.0	0.4640	8.4107
33	0.05033	1.0054	28.014	138.22	2419.8	138.22	2560.8	0.4777	8.3906
34	0.05323	1.0057	26.575	142.40	2421.1	142.41	2562.6	0.4914	8.3708
35	0.05627	1.0060	25.220	146.58	2422.5	146.59	2564.4	0.5050	8.3511
40	0.07381	1.0079	19.528	167.50	2429.2	167.50	2573.4	0.5723	8.2550
45	0.09593	1.0099	15.263	188.41	2435.9	188.42	2582.3	0.6385	8.1629
50	0.12345	1.0122	12.037	209.31	2442.6	209.33	2591.2	0.7037	8.0745

Table B.1 (*Continued*)

T °C	p bar	$v_f \cdot 10^3$ m³/kg	v_g m³/kg	u_f kJ/kg	u_g kJ/kg	h_f kJ/kg	h_g kJ/kg	s_f kJ/kg K	s_g kJ/kg K
55	0.1575	1.0146	9.5726	230.22	2449.2	230.24	2600.0	0.7679	7.9896
60	0.1993	1.0171	7.6743	251.13	2455.8	251.15	2608.8	0.8312	7.9080
65	0.2502	1.0199	6.1996	272.05	2462.4	272.08	2617.5	0.8935	7.8295
70	0.3118	1.0228	5.0446	292.98	2468.8	293.01	2626.1	0.9549	7.7540
75	0.3856	1.0258	4.1333	313.92	2475.2	313.96	2634.6	1.0155	7.6813
80	0.4737	1.0290	3.4088	334.88	2481.6	334.93	2643.1	1.0753	7.6112
85	0.5781	1.0324	2.8289	355.86	2487.9	355.92	2651.4	1.1343	7.5436
90	0.7012	1.0359	2.3617	376.86	2494.0	376.93	2659.6	1.1925	7.4784
95	0.8453	1.0396	1.9828	397.89	2500.1	397.98	2667.7	1.2501	7.4154
100	1.013	1.0434	1.6736	418.96	2506.1	419.06	2675.7	1.3069	7.3545
105	1.208	1.0474	1.4200	440.05	2512.1	440.18	2683.6	1.3630	7.2956
110	1.432	1.0515	1.2106	461.19	2517.9	461.34	2691.3	1.4186	7.2386
115	1.690	1.0558	1.0370	482.36	2523.5	482.54	2698.8	1.4735	7.1833
120	1.985	1.0603	0.8922	503.57	2529.1	503.78	2706.2	1.5278	7.1297
125	2.320	1.0649	0.7709	524.82	2534.5	525.07	2713.4	1.5815	7.0777
130	2.700	1.0697	0.6687	546.12	2539.8	546.41	2720.4	1.6346	7.0272
135	3.130	1.0746	0.5824	567.46	2545.0	567.80	2727.2	1.6873	6.9780
140	3.612	1.0797	0.5090	588.85	2550.0	589.24	2733.8	1.7394	6.9302
145	4.153	1.0850	0.4464	610.30	2554.8	610.75	2740.2	1.7910	6.8836
150	4.757	1.0904	0.3929	631.80	2559.5	632.32	2746.4	1.8421	6.8381
160	6.177	1.1019	0.3071	674.97	2568.3	675.65	2758.0	1.9429	6.7503
170	7.915	1.1142	0.2428	718.40	2576.3	719.28	2768.5	2.0421	6.6662
180	10.02	1.1273	0.1940	762.12	2583.4	763.25	2777.8	2.1397	6.5853
190	12.54	1.1414	0.1565	806.17	2589.6	807.60	2785.8	2.2358	6.5071
200	15.55	1.1564	0.1273	850.58	2594.7	852.38	2792.5	2.3308	6.4312
210	19.07	1.1726	0.1044	895.43	2598.7	897.66	2797.7	2.4246	6.3572
220	23.18	1.1900	0.0862	940.75	2601.6	943.51	2801.3	2.5175	6.2847
230	27.95	1.2088	0.0716	986.62	2603.1	990.00	2803.1	2.6097	6.2131
240	33.45	1.2292	0.0597	1033.1	2603.1	1037.2	2803.0	2.7013	6.1423
250	39.74	1.2515	0.0501	1080.4	2601.6	1085.3	2800.7	2.7926	6.0717
260	46.89	1.2758	0.0422	1128.4	2598.4	1134.4	2796.2	2.8838	6.0009
270	55.00	1.3026	0.0356	1177.4	2593.2	1184.6	2789.1	2.9751	5.9293
280	64.13	1.3324	0.0302	1227.5	2585.7	1236.1	2779.2	3.0669	5.8565
290	74.38	1.3658	0.0256	1279.0	2575.7	1289.1	2765.9	3.1595	5.7818
300	85.84	1.4037	0.0217	1332.0	2562.8	1344.1	2748.7	3.2534	5.7042
310	98.61	1.4473	0.0183	1387.0	2546.2	1401.2	2727.0	3.3491	5.6226
320	112.8	1.4984	0.0155	1444.4	2525.2	1461.3	2699.7	3.4476	5.5356
340	145.9	1.6373	0.0108	1569.9	2463.9	1593.8	2621.3	3.6587	5.3345
360	186.6	1.8936	0.0070	1725.6	2352.2	1761.0	2482.0	3.9153	5.0542
374.12	220.9	3.1550	0.0031	2029.6	2029.6	2099.3	2099.3	4.4298	4.4298

Table B.2 Thermodynamic properties of saturated steam, pressure table

p bar	T °C	$v_f \cdot 10^3$ m³/kg	v_g m³/kg	u_f kJ/kg	u_g kJ/kg	h_f kJ/kg	h_g kJ/kg	s_f kJ/kg K	s_g kJ/kg K
0.06	36.17	1.0065	23.737	151.47	2424.0	151.47	2566.5	0.5208	8.3283
0.08	41.49	1.0085	18.128	173.73	2431.0	173.74	2576.0	0.5921	8.2272
0.10	45.79	1.0103	14.693	191.71	2436.8	191.72	2583.7	0.6489	8.1487
0.12	49.40	1.0119	12.377	206.82	2441.6	206.83	2590.1	0.6960	8.0849
0.16	55.30	1.0147	9.4447	231.47	2449.4	231.49	2600.5	0.7718	7.9846
0.20	60.05	1.0171	7.6591	251.32	2455.7	251.34	2608.9	0.8318	7.9072
0.25	64.95	1.0198	6.2120	271.85	2462.1	271.88	2617.4	0.8929	7.8302
0.30	69.09	1.0222	5.2357	289.15	2467.5	289.18	2624.5	0.9438	7.7676
0.40	75.85	1.0263	3.9983	317.48	2476.1	317.52	2636.1	1.0257	7.6692
0.50	81.31	1.0299	3.2442	340.38	2483.1	340.43	2645.3	1.0908	7.5932
0.60	85.92	1.0330	2.7351	359.73	2488.8	359.79	2652.9	1.1451	7.5314
0.70	89.93	1.0359	2.3676	376.56	2493.8	376.64	2659.5	1.1917	7.4793
0.80	93.48	1.0385	2.0895	391.51	2498.1	391.60	2665.3	1.2327	7.4342
0.90	96.69	1.0409	1.8715	405.00	2502.0	405.09	2670.5	1.2693	7.3946
1.00	99.61	1.0431	1.6958	417.30	2505.5	417.40	2675.1	1.3024	7.3592
2.00	120.2	1.0605	0.8865	504.49	2529.2	504.70	2706.5	1.5301	7.1275
2.50	127.4	1.0672	0.7193	535.12	2537.0	535.39	2716.8	1.6073	7.0531
3.00	133.5	1.0731	0.6063	561.19	2543.4	561.51	2725.2	1.6719	6.9923
3.50	138.9	1.0785	0.5246	584.01	2548.8	584.38	2732.4	1.7276	6.9409
4.00	143.6	1.0835	0.4627	604.38	2553.4	604.81	2738.5	1.7768	6.8963
5.00	151.8	1.0925	0.3751	639.74	2561.1	640.29	2748.6	1.8608	6.8216
6.00	158.8	1.1006	0.3158	669.96	2567.2	670.62	2756.7	1.9313	6.7602
7.00	165.0	1.1079	0.2729	696.49	2572.2	697.27	2763.3	1.9923	6.7081
8.00	170.4	1.1147	0.2405	720.25	2576.5	721.14	2768.9	2.0462	6.6627
9.00	175.4	1.1211	0.2150	741.84	2580.1	742.85	2773.6	2.0947	6.6224
10.00	179.9	1.1272	0.1945	761.67	2583.2	762.80	2777.7	2.1387	6.5861
11.00	184.1	1.1329	0.1775	780.06	2585.9	781.31	2781.2	2.1791	6.5531
12.00	188.0	1.1384	0.1633	797.23	2588.3	798.60	2784.3	2.2165	6.5227
13.00	191.6	1.1437	0.1513	813.37	2590.4	814.85	2787.0	2.2514	6.4946
14.00	195.1	1.1488	0.1408	828.60	2592.2	830.21	2789.4	2.2840	6.4684
15.00	198.3	1.1538	0.1318	843.05	2593.9	844.78	2791.5	2.3148	6.4439
16.00	201.4	1.1586	0.1238	856.81	2595.3	858.66	2793.3	2.3439	6.4208
17.00	204.3	1.1633	0.1167	869.95	2596.5	871.93	2795.0	2.3715	6.3990
18.00	207.1	1.1678	0.1104	882.54	2597.7	884.64	2796.4	2.3978	6.3782
19.00	209.8	1.1723	0.1047	894.63	2598.6	896.86	2797.6	2.4230	6.3585
20.00	212.4	1.1766	0.0996	906.27	2599.5	908.62	2798.7	2.4470	6.3397
25.00	224.0	1.1973	0.0800	958.92	2602.3	961.92	2802.2	2.5543	6.2561
30.00	233.9	1.2165	0.0667	1004.59	2603.2	1008.2	2803.3	2.6453	6.1856
35.00	242.6	1.2348	0.0571	1045.26	2602.9	1049.6	2802.6	2.7250	6.1240

Table B.2 (*Continued*)

p bar	T °C	$v_\text{f} \cdot 10^3$ m³/kg	v_g m³/kg	u_f kJ/kg	u_g kJ/kg	h_f kJ/kg	h_g kJ/kg	s_f kJ/kg K	s_g kJ/kg K
40.00	250.4	1.2523	0.0498	1082.18	2601.5	1087.2	2800.6	2.7961	6.0690
45.00	257.5	1.2694	0.0441	1116.14	2599.3	1121.9	2797.6	2.8607	6.0188
50.00	264.0	1.2861	0.0394	1147.74	2596.5	1154.2	2793.7	2.9201	5.9726
55.00	270.0	1.3026	0.0356	1177.39	2593.1	1184.6	2789.1	2.9751	5.9294
60.00	275.6	1.3190	0.0324	1205.42	2589.3	1213.3	2783.9	3.0266	5.8886
65.00	280.9	1.3352	0.0297	1232.06	2584.9	1240.7	2778.1	3.0751	5.8500
70.00	285.9	1.3515	0.0274	1257.52	2580.2	1267.0	2771.8	3.1211	5.8130
75.00	290.6	1.3678	0.0253	1281.96	2575.1	1292.2	2765.0	3.1648	5.7774
80.00	295.0	1.3843	0.0235	1305.51	2569.6	1316.6	2757.8	3.2066	5.7431
85.00	299.3	1.4009	0.0219	1328.27	2563.8	1340.2	2750.1	3.2468	5.7097
90.00	303.4	1.4177	0.0205	1350.36	2557.6	1363.1	2742.0	3.2855	5.6771
95.00	307.3	1.4348	0.0192	1371.84	2551.1	1385.5	2733.4	3.3229	5.6452
100.0	311.0	1.4522	0.0180	1392.79	2544.3	1407.3	2724.5	3.3592	5.6139
105.0	314.6	1.4699	0.0170	1413.27	2537.1	1428.7	2715.1	3.3944	5.5830
110.0	318.1	1.4881	0.0160	1433.34	2529.5	1449.7	2705.4	3.4288	5.5525
115.0	321.5	1.5068	0.0151	1453.06	2521.6	1470.4	2695.1	3.4624	5.5221
120.0	324.7	1.5260	0.0143	1472.47	2513.4	1490.8	2684.5	3.4953	5.4920
125.0	327.9	1.5458	0.0135	1491.61	2504.7	1510.9	2673.4	3.5277	5.4619
130.0	330.9	1.5663	0.0128	1510.55	2495.7	1530.9	2661.8	3.5595	5.4317
135.0	333.8	1.5875	0.0121	1529.31	2486.2	1550.7	2649.7	3.5910	5.4015
140.0	336.7	1.6097	0.0115	1547.94	2476.3	1570.5	2637.1	3.6221	5.3710
145.0	339.5	1.6328	0.0109	1566.49	2465.9	1590.2	2623.9	3.6530	5.3403
150.0	342.2	1.6572	0.0103	1585.01	2455.0	1609.9	2610.0	3.6838	5.3091
155.0	344.8	1.6828	0.0098	1603.55	2443.4	1629.6	2595.5	3.7145	5.2774
160.0	347.4	1.7099	0.0093	1622.17	2431.3	1649.5	2580.2	3.7452	5.2450
165.0	349.9	1.7388	0.0088	1640.92	2418.4	1669.6	2564.1	3.7761	5.2119
170.0	352.3	1.7699	0.0084	1659.89	2404.8	1690.0	2547.1	3.8073	5.1777
175.0	354.7	1.8033	0.0079	1679.18	2390.2	1710.7	2529.0	3.8390	5.1423
180.0	357.0	1.8399	0.0075	1698.88	2374.6	1732.0	2509.7	3.8714	5.1054
185.0	359.3	1.8801	0.0071	1719.17	2357.7	1754.0	2488.9	3.9047	5.0667
190.0	361.5	1.9251	0.0067	1740.22	2339.3	1776.8	2466.3	3.9393	5.0256
195.0	363.7	1.9762	0.0063	1762.34	2319.0	1800.9	2441.4	3.9756	4.9815
200.0	365.8	2.0357	0.0059	1785.94	2296.2	1826.7	2413.7	4.0144	4.9331
205.0	367.9	2.1076	0.0055	1811.76	2269.7	1855.0	2381.6	4.0571	4.8787
210.0	369.9	2.1999	0.0050	1841.25	2237.5	1887.5	2343.0	4.1060	4.8144
215.0	371.8	2.3362	0.0045	1878.57	2193.9	1928.8	2291.0	4.1684	4.7299
220.9	374.1	3.1550	0.0316	2029.60	2029.6	2099.3	2099.3	4.4298	4.4298

Table B.3 Thermodynamics properties of superheated steam

T °C	v m³/kg	u kJ/kg	h kJ/kg	s kJ/kg K	T °C	v m³/kg	u kJ/kg	h kJ/kg	s kJ/kg K
		$p = 0.06$ bar					$p = 0.35$ bar		
36.17	23.739	2424.0	2566.5	8.3283	72.67	4.531	2472.1	2630.7	7.7148
80	27.133	2486.7	2649.5	8.5794	80	4.625	2483.1	2645.0	7.7553
120	30.220	2544.1	2725.5	8.7831	120	5.163	2542.0	2722.7	7.9637
160	33.303	2602.2	2802.0	8.9684	160	5.697	2600.7	2800.1	8.1512
200	36.384	2660.9	2879.2	9.1390	200	6.228	2659.9	2877.9	8.3229
240	39.463	2720.6	2957.4	9.2975	240	6.758	2719.8	2956.3	8.4821
280	42.541	2781.2	3036.4	9.4458	280	7.287	2780.6	3035.6	8.6308
320	45.620	2842.7	3116.4	9.5855	320	7.816	2842.2	3115.8	8.7707
360	48.697	2905.2	3197.4	9.7176	360	8.344	2904.8	3196.9	8.9031
400	51.775	2968.8	3279.5	9.8433	400	8.872	2968.5	3279.0	9.0288
440	54.852	3033.4	3362.6	9.9632	440	9.400	3033.2	3362.2	9.1488
500	59.468	3132.4	3489.2	10.134	500	10.192	3132.2	3488.9	9.3194
		$p = 0.7$ bar					$p = 1.0$ bar		
89.93	2.368	2493.8	2659.5	7.4793	99.61	1.6958	2505.5	2675.1	7.3592
120	2.5709	2539.3	2719.3	7.6370	120	1.7931	2537.0	2716.3	7.4665
160	2.8407	2599.0	2797.8	7.8272	160	1.9838	2597.5	2795.8	7.6591
200	3.1082	2658.7	2876.2	8.0004	200	2.1723	2657.6	2874.8	7.8335
240	3.3744	2718.9	2955.1	8.1603	240	2.3594	2718.1	2954.0	7.9942
280	3.6399	2779.8	3034.6	8.3096	280	2.5458	2779.2	3033.8	8.1438
320	3.9049	2841.6	3115.0	8.4498	320	2.7317	2841.1	3114.3	8.2844
360	4.1697	2904.4	3196.2	8.5824	360	2.9173	2904.0	3195.7	8.4171
400	4.4342	2968.1	3278.5	8.7083	400	3.1027	2967.7	3278.0	8.5432
440	4.6985	3032.8	3361.7	8.8285	440	3.2879	3032.5	3361.3	8.6634
480	4.9627	3098.6	3446.0	8.9434	480	3.4730	3098.3	3445.6	8.7785
520	5.2269	3165.4	3531.3	9.0538	520	3.6581	3165.2	3531.0	8.8889
		$p = 1.5$ bar					$p = 3.0$ bar		
111.37	1.1600	2519.3	2693.3	7.2234	133.55	0.6060	2543.4	2725.2	6.992
160	1.3174	2594.9	2792.5	7.4660	160	0.6506	2586.9	2782.1	7.127
200	1.4443	2655.8	2872.4	7.6425	200	0.7163	2650.2	2865.1	7.310
240	1.5699	2716.7	2952.2	7.8044	240	0.7804	2712.6	2946.7	7.476
280	1.6948	2778.2	3032.4	7.9548	280	0.8438	2775.0	3028.1	7.629
320	1.8192	2840.3	3113.2	8.0958	320	0.9067	2837.8	3109.8	7.771
360	1.9433	2903.3	3194.8	8.2289	360	0.9692	2901.2	3191.9	7.905
400	2.0671	2967.2	3277.2	8.3552	400	1.0315	2965.4	3274.9	8.032
440	2.1908	3032.0	3360.6	8.4756	440	1.0937	3030.5	3358.7	8.153
480	2.3144	3097.9	3445.1	8.5908	480	1.1557	3096.6	3443.4	8.269
520	2.4379	3164.8	3530.5	8.7013	520	1.2177	3163.7	3529.0	8.380
560	2.5613	3232.9	3617.0	8.8077	560	1.2796	3231.9	3615.7	8.486

Table B.3 (*Continued*)

T °C	v m³/kg	u kJ/kg	h kJ/kg	s kJ/kg K	T °C	v m³/kg	u kJ/kg	h kJ/kg	s kJ/kg K
		p = 5 bar					p = 7 bar		
151.86	0.3751	2561.1	2748.6	6.8216	164.97	0.2729	2572.2	2763.3	6.7081
180	0.4045	2609.5	2811.7	6.9652	180	0.2846	2599.6	2798.8	6.7876
220	0.4449	2674.9	2897.4	7.1463	220	0.3146	2668.1	2888.4	6.9771
260	0.4840	2738.9	2980.9	7.3092	260	0.3434	2733.9	2974.2	7.1445
300	0.5225	2802.5	3063.7	7.4591	300	0.3714	2798.6	3058.5	7.2970
340	0.5606	2866.3	3146.6	7.5989	340	0.3989	2863.2	3142.4	7.4385
380	0.5985	2930.7	3229.9	7.7304	380	0.4262	2928.1	3226.4	7.5712
420	0.6361	2995.7	3313.8	7.8550	420	0.4533	2993.6	3310.9	7.6966
460	0.6736	3061.6	3398.4	7.9738	460	0.4802	3059.8	3395.9	7.8160
500	0.7109	3128.5	3483.9	8.0873	500	0.5070	3126.9	3481.8	7.9300
540	0.7482	3196.3	3570.4	8.1964	540	0.5338	3194.9	3568.5	8.0393
560	0.7669	3230.6	3614.0	8.2493	560	0.5741	3229.2	3612.2	8.09243
		p = 10 bar					p = 15 bar		
179.91	0.1945	2583.2	2777.7	6.5861	198.32	0.1318	2593.9	2791.5	6.4439
220	0.2169	2657.5	2874.3	6.7904	220	0.1405	2638.1	2849.0	6.5630
260	0.2378	2726.1	2963.9	6.9652	260	0.1556	2712.6	2945.9	6.7521
300	0.2579	2792.7	3050.6	7.1219	300	0.1696	2782.5	3036.9	6.9168
340	0.2776	2858.5	3136.1	7.2661	340	0.1832	2850.4	3125.2	7.0657
380	0.2970	2924.2	3221.2	7.4006	380	0.1965	2917.6	3212.3	7.2033
420	0.3161	2990.3	3306.5	7.5273	420	0.2095	2984.8	3299.1	7.3322
460	0.3352	3057.0	3392.2	7.6475	460	0.2224	3052.3	3385.9	7.4540
500	0.3541	3124.5	3478.6	7.7622	500	0.2351	3120.4	3473.1	7.5699
540	0.3729	3192.8	3565.7	7.8721	540	0.2478	3189.2	3561.0	7.6806
580	0.3917	3262.0	3653.7	7.9778	580	0.2605	3258.8	3649.6	7.7870
620	0.4105	3332.2	3742.7	8.0797	620	0.2730	3329.4	3739.0	7.8894
		p = 20 bar					p = 30 bar		
212.42	0.0996	2599.5	2798.7	6.3397	233.90	0.06667	2603.2	2803.3	6.1856
240	0.1084	2658.8	2875.6	6.4937	240	0.06818	2618.9	2823.5	6.2251
280	0.1200	2735.6	2975.6	6.6814	280	0.07710	2709.0	2940.3	6.4445
320	0.1308	2807.3	3068.8	6.8441	320	0.08498	2787.6	3042.6	6.6232
360	0.1411	2876.7	3158.9	6.9911	360	0.09232	2861.3	3138.3	6.7794
400	0.1512	2945.1	3247.5	7.1269	400	0.09935	2932.7	3230.7	6.9210
440	0.1611	3013.4	3335.6	7.2539	440	0.10618	3003.0	3321.5	7.0521
480	0.1708	3081.9	3423.6	7.3740	480	0.11287	3073.0	3411.6	7.1750
520	0.1805	3150.9	3511.9	7.4882	520	0.11946	3143.2	3501.6	7.2913
560	0.1901	3220.6	3600.7	7.5975	560	0.12597	3213.8	3591.7	7.4022
600	0.1996	3291.0	3690.2	7.7024	600	0.13243	3285.0	3682.3	7.5084
640	0.2091	3362.4	3780.5	7.8036	640	0.13884	3357.0	3773.5	7.6105

Table B.3 (*Continued*)

T °C	v m³/kg	u kJ/kg	h kJ/kg	s kJ/kg K	T °C	v m³/kg	u kJ/kg	h kJ/kg	s kJ/kg K
		p = 40 bar					p = 60 bar		
250.38	0.04978	2601.5	2800.6	6.0690	276.62	0.03244	2589.3	2783.9	5.888
280	0.05544	2679.0	2900.8	6.2552	280	0.03317	2604.7	2803.7	5.924
320	0.06198	2766.6	3014.5	6.4538	320	0.03874	2719.0	2951.5	6.181
360	0.06787	2845.3	3116.7	6.6207	360	0.04330	2810.6	3070.4	6.377
400	0.07340	2919.8	3213.4	6.7688	400	0.04739	2892.7	3177.0	6.540
440	0.07872	2992.3	3307.2	6.9041	440	0.05121	2970.2	3277.4	6.685
480	0.08388	3064.0	3399.5	7.0301	480	0.05487	3045.3	3374.5	6.817
520	0.08894	3135.4	3491.1	7.1486	520	0.05840	3119.4	3469.8	6.941
560	0.09392	3206.9	3582.6	7.2612	560	0.06186	3193.0	3564.1	7.057
600	0.09884	3278.9	3674.3	7.3687	600	0.06525	3266.6	3658.1	7.167
640	0.10372	3351.5	3766.4	7.4718	640	0.06859	3340.5	3752.1	7.272
680	0.10855	3424.9	3859.1	7.5711	680	0.07189	3414.9	3846.3	7.372
		p = 80 bar					p = 100 bar		
295.04	0.02352	2569.6	2757.8	5.7431	311.04	0.01802	2544.3	2724.5	5.613
320	0.02681	2661.7	2876.2	5.9473	320	0.01925	2588.2	2780.6	5.709
360	0.03088	2771.9	3018.9	6.1805	360	0.02330	2728.0	2961.0	6.004
400	0.03431	2863.5	3138.0	6.3630	400	0.02641	2832.0	3096.1	6.211
440	0.03742	2946.8	3246.2	6.5192	440	0.02911	2922.3	3213.4	6.380
480	0.04034	3026.0	3348.6	6.6589	480	0.03160	3005.8	3321.8	6.528
520	0.04312	3102.9	3447.8	6.7873	520	0.03394	3085.9	3425.3	6.662
560	0.04582	3178.6	3545.2	6.9070	560	0.03619	3164.0	3525.8	6.786
600	0.04845	3254.0	3641.5	7.0200	600	0.03836	3241.1	3624.7	6.902
640	0.05102	3329.3	3737.5	7.1274	640	0.04048	3317.9	3722.7	7.011
680	0.05356	3404.9	3833.4	7.2302	680	0.04256	3394.6	3820.3	7.116
720	0.05607	3480.9	3929.4	7.3289	720	0.04461	3471.6	3917.7	7.216
		p = 120 bar					p = 140 bar		
324.75	0.01426	2513.4	2684.5	5.4920	336.75	0.01148	2476.3	2637.1	5.371
360	0.01810	2677.1	2894.4	5.8341	360	0.01421	2616.0	2815.0	5.657
400	0.02108	2797.8	3050.7	6.0739	400	0.01722	2760.2	3001.3	5.943
440	0.02355	2896.3	3178.9	6.2589	440	0.01955	2868.8	3142.5	6.147
480	0.02576	2984.9	3294.0	6.4161	480	0.02157	2963.1	3265.2	6.315
520	0.02781	3068.4	3402.1	6.5559	520	0.02343	3050.3	3378.3	6.461
560	0.02976	3149.0	3506.1	6.6839	560	0.02517	3133.6	3485.9	6.594
600	0.03163	3228.0	3607.6	6.8029	600	0.02683	3214.7	3590.3	6.716
640	0.03345	3306.3	3707.7	6.9150	640	0.02843	3294.5	3692.5	6.830
680	0.03523	3384.3	3807.0	7.0214	680	0.02999	3373.8	3793.6	6.939
720	0.03697	3462.3	3906.0	7.1231	720	0.03152	3452.8	3894.1	7.042
760	0.03869	3540.6	4004.8	7.2207	760	0.03301	3532.0	3994.2	7.141

Table B.3 (*Continued*)

T °C	v m³/kg	u kJ/kg	h kJ/kg	s kJ/kg K	T °C	v m³/kg	u kJ/kg	h kJ/kg	s kJ/kg K
		p = 160 bar					p = 180 bar		
347.44	0.00931	2431.3	2580.2	5.2450	357.06	0.00750	2374.6	2509.7	5.1054
360	0.01105	2537.5	2714.3	5.4591	360	0.00810	2418.3	2564.1	5.1916
400	0.01427	2718.5	2946.8	5.8162	400	0.01191	2671.7	2886.0	5.6872
440	0.01652	2839.6	3104.0	6.0433	440	0.01415	2808.5	3063.2	5.9432
480	0.01842	2940.5	3235.3	6.2226	480	0.01596	2916.9	3204.2	6.1358
520	0.02013	3031.8	3353.9	6.3761	520	0.01756	3012.7	3328.8	6.2971
560	0.02172	3117.9	3465.4	6.5133	560	0.01903	3101.9	3444.5	6.4394
600	0.02322	3201.1	3572.6	6.6390	600	0.02041	3187.3	3554.8	6.5687
640	0.02466	3282.6	3677.2	6.7561	640	0.02173	3270.5	3661.7	6.6885
680	0.02606	3363.1	3780.1	6.8664	680	0.02301	3352.4	3766.5	6.8008
720	0.02742	3443.3	3882.1	6.9712	720	0.02424	3433.7	3870.0	6.9072
760	0.02876	3523.4	3983.5	7.0714	760	0.02545	3514.7	3972.8	7.0086
		p = 200 bar					p = 240 bar		
365.81	0.00588	2296.2	2413.7	4.9331					
400	0.00995	2617.9	2816.9	5.5521	400	0.00673	2476.0	2637.5	5.2365
440	0.01223	2775.2	3019.8	5.8455	440	0.00929	2700.9	2923.9	5.6511
480	0.01399	2892.3	3172.0	6.0534	480	0.01100	2839.9	3103.9	5.8971
520	0.01551	2993.1	3303.2	6.2232	520	0.01241	2952.1	3250.0	6.0861
560	0.01688	3085.5	3423.2	6.3708	560	0.01366	3051.8	3379.5	6.2456
600	0.01817	3173.3	3536.7	6.5039	600	0.01480	3144.6	3499.8	6.3866
640	0.01939	3258.2	3646.0	6.6264	640	0.01587	3233.3	3614.3	6.5148
680	0.02056	3341.6	3752.8	6.7408	680	0.01690	3319.6	3725.1	6.6336
720	0.02170	3424.0	3857.9	6.8488	720	0.01788	3404.3	3833.4	6.7450
760	0.02280	3505.9	3961.9	6.9515	760	0.01883	3488.2	3940.2	6.8504
800	0.02388	3587.8	4065.4	7.0498	800	0.01976	3571.7	4046.0	6.9508
		p = 280 bar					p = 320 bar		
400	0.00383	2221.7	2328.8	4.7465	400	0.00237	1981.0	2056.8	4.3252
440	0.00712	2613.5	2812.9	5.4497	440	0.00543	2509.0	2682.9	5.2325
480	0.00885	2782.7	3030.5	5.7472	480	0.00722	2720.5	2951.5	5.5998
520	0.01019	2908.9	3194.3	5.9592	520	0.00853	2863.4	3136.2	5.8390
560	0.01135	3016.8	3334.6	6.1319	560	0.00962	2980.6	3288.4	6.0263
600	0.01239	3115.1	3462.1	6.2815	600	0.01059	3084.9	3423.8	6.1851
640	0.01336	3207.9	3582.0	6.4158	640	0.01148	3182.0	3549.4	6.3258
680	0.01428	3297.2	3697.0	6.5390	680	0.01232	3274.6	3668.8	6.4538
720	0.01516	3384.4	3808.8	6.6539	720	0.01312	3364.3	3784.0	6.5722
760	0.01600	3470.3	3918.4	6.7621	760	0.01388	3452.3	3896.4	6.6832
800	0.01682	3555.5	4026.5	6.8647	800	0.01462	3539.2	4006.9	6.7881

Table B.4 Thermodynamic Properties of Air

T	h	s^0	p_r	v_r	T	h	s^0	p_r	v_r
K	kJ/kg	kJ/(kg·, K)			K	kJ/kg	kJ/(kg·K)		
200	199.97	1.29559	0.3363	1707.0	450	451.80	2.11161	5.775	223.6
210	209.97	1.34444	0.3987	1512.0	460	462.02	2.13407	6.245	211.4
220	219.97	1.39105	0.4690	1346.0	470	472.24	2.15604	6.742	200.1
230	230.02	1.43557	0.5477	1205.0	480	482.49	2.17760	7.268	189.5
240	240.02	1.47824	0.6355	1084.0	490	492.74	2.19876	7.824	179.7
250	250.05	1.51917	0.7329	979.0	500	503.02	2.21952	8.411	170.6
260	260.09	1.55848	0.8405	887.8	510	513.32	2.23993	9.031	162.1
270	270.11	1.59634	0.9590	808.0	520	523.63	2.25997	9.684	154.1
280	280.13	1.63279	1.0889	738.0	530	533.98	2.27967	10.37	146.7
285	285.14	1.65055	1.1584	706.1	540	544.35	2.29906	11.10	139.7
290	290.16	1.66802	1.2311	676.1	550	554.74	2.31809	11.86	133.1
295	295.17	1.68515	1.3068	647.9	560	565.17	2.33685	12.66	127.0
300	300.19	1.70203	1.3860	621.2	570	575.59	2.35531	13.50	121.2
305	305.22	1.71865	1.4686	596.0	580	586.04	2.37348	14.38	115.7
310	310.24	1.73498	1.5546	572.3	590	596.52	2.39140	15.31	110.6
315	315.27	1.75106	1.6442	549.8	600	607.02	2.40902	16.28	105.8
320	320.29	1.76690	1.7375	528.6	610	617.53	2.42644	17.30	101.2
325	325.31	1.78249	1.8345	508.4	620	628.63	2.44356	18.36	96.92
330	330.34	1.79783	1.9352	489.4	630	638.63	2.46048	19.84	92.84
340	340.42	1.82790	2.149	454.1	640	649.22	2.47716	20.64	88.99
350	350.49	1.85708	2.379	422.2	650	659.84	2.49364	21.86	85.34
360	360.58	1.88543	2.626	393.4	660	670.47	2.50985	23.13	81.89
370	370.67	1.91313	2.892	367.2	670	681.14	2.52589	24.46	78.61
380	380.77	1.94001	3.176	343.4	680	691.82	2.54175	25.85	75.50
390	390.88	1.96633	3.481	321.5	690	702.52	2.55731	27.29	72.56
400	400.98	1.99194	3.806	301.6	700	713.27	2.57277	28.80	69.76
410	411.12	2.01699	4.153	283.3	710	724.04	2.58810	30.38	67.07
420	421.26	2.40142	4.522	266.6	720	734.82	2.60319	32.02	64.53
430	431.43	2.06533	4.915	251.1	730	745.62	2.61803	33.72	62.13
440	441.61	2.08870	5.332	236.8	740	756.44	2.63280	35.50	59.82

Table B.4 (*Continued*)

T	h	s^0	p_r	v_r	T	h	s^0	p_r	v_r
K	kJ/kg	kJ/(kg · K)			K	kJ/kg	kJ/(kg · K)		
750	769.29	2.64737	37.35	57.63	1300	1395.97	3.27345	330.9	11.275
760	778.18	2.66176	39.07	55.54	1320	1419.76	3.29160	352.5	10.747
770	789.11	2.67595	41.31	53.39	1340	1443.60	3.30959	375.3	10.247
780	800.03	2.69013	43.35	51.64	1360	1467.49	3.32724	399.1	9.780
790	810.99	2.70100	45.55	49.86	1380	1491.44	3.34474	424.2	9.337
800	821.95	2.71787	47.75	48.08	1400	1515.42	3.36200	450.5	8.919
820	843.98	2.74504	52.59	44.84	1420	1539.44	3.37901	478.0	8.526
840	866.08	2.77170	57.60	41.85	1440	1563.51	3.39586	506.9	8.153
860	888.27	2.79783	63.09	39.12	1460	1587.63	3.41217	537.1	7.801
880	910.56	2.82344	68.98	36.61	1480	1611.79	3.42892	568.8	7.468
900	932.93	2.84856	75.29	34.31	1500	1635.97	3.44516	601.9	7.152
920	955.39	2.87324	82.05	32.18	1520	1660.23	3.46120	636.5	6.854
940	977.92	2.89748	89.28	30.22	1540	1684.51	3.47712	672.8	6.569
960	1000.55	2.92128	97.00	28.40	1560	1708.82	3.49276	710.5	6.301
980	1023.25	2.94469	105.2	26.73	1580	1733.17	3.50829	750.0	6.046
1000	1046.04	2.96770	114.0	25.17	1600	1757.57	3.52364	791.2	5.804
1020	1068.89	2.99034	123.4	23.72	1620	1782.00	3.53879	834.1	5.574
1040	1091.85	2.99034	123.4	22.39	1640	1806.46	3.55381	878.9	5.355
1060	1114.86	3.03449	143.9	21.12	1660	1830.96	3.56867	925.3	5.147
1080	1137.89	3.05608	155.2	19.98	1680	1855.50	3.58355	974.2	4.949
1100	1161.07	3.07732	167.1	18.896	1700	1880.1	3.5979	1025	4.761
1120	1184.28	3.09825	179.7	17.886	1750	1941.6	3.6336	1161	4.328
1140	1207.57	3.11883	193.1	16.946	1800	2003.3	3.6681	1310	3.944
1160	1230.92	3.13916	207.2	16.064	1850	2065.3	3.7023	1475	3.601
1180	1254.34	3.15916	222.2	15.241	1900	2127.4	3.7354	1655	3.295
1200	1277.79	3.17888	238.0	14.170	1950	2189.7	3.7677	1852	3.022
1220	1301.31	3.19834	254.7	13.747	2000	2252.1	3.7994	2068	2.776
1240	1324.93	3.21751	272.3	13.069	2050	2314.6	3.8303	2303	2.555
1260	1348.55	3.21751	281.3	12.435	2100	2377.4	3.8605	2995	2.356
1280	1372.24	3.25510	310.4	11.835	2150	2440.3	3.8901	2837	2.175
					2200	2503.2	3.9191	3138	2.012
					2250	2566.4	3.9474	3464	1.864

Table B.5 Specific heats of common gases

T K	CO kJ/kg K	CO_2 kJ/kg K	H_2 kJ/kg K	H_2O kJ/kg K	O_2 kJ/kg K	N_2 kJ/kg K
250	1.0404	0.7957	14.1961	1.8498	0.9094	1.0416
300	1.0400	0.8482	14.3169	1.8629	0.9178	1.0408
350	1.0425	0.8961	14.4000	1.8805	0.9289	1.0428
400	1.0476	0.9398	14.4553	1.9018	0.9421	1.0475
450	1.0548	0.9796	14.4914	1.9263	0.9565	1.0548
500	1.0640	1.0156	14.5160	1.9534	0.9717	1.0647
550	1.0746	1.0482	14.5354	1.9827	0.9871	1.0772
600	1.0863	1.0777	14.5549	2.0136	1.0022	1.0922
650	1.0989	1.1044	14.5787	2.0458	1.0167	1.1097
700	1.1120	1.1285	14.6099	2.0790	1.0304	1.1296
750	1.1253	1.1504	14.6504	2.1128	1.0430	1.1520
800	1.1385	1.1702	14.7011	2.1472	1.0545	1.1769
850	1.1513	1.1884	14.7616	2.1819	1.0648	1.2042
900	1.1634	1.2051	14.8306	2.2167	1.0741	1.2341
950	1.1746	1.2207	14.9054	2.2518	1.0823	1.2666
1000	1.1846	1.2355	14.9825	2.2870	1.0899	1.3016

Table B.6 Molar specific heats of common gases

T K	CO $\dfrac{kJ}{kgmol\ K}$	CO_2 $\dfrac{kJ}{kgmol\ K}$	H_2 $\dfrac{kJ}{kgmol\ K}$	H_2O $\dfrac{kJ}{kgmol\ K}$	O_2 $\dfrac{kJ}{kgmol\ K}$	N_2 $\dfrac{kJ}{kgmol\ K}$
250	29.1421	35.0193	28.6194	33.3337	29.1004	29.1760
300	29.1294	37.3292	28.8628	33.5702	29.3683	29.1533
350	29.1996	39.4394	29.0304	33.8866	29.7249	29.2086
400	29.3419	41.3622	29.1419	34.2710	30.1457	29.3399
450	29.5459	43.1103	29.2148	34.7124	30.6086	29.5453
500	29.8015	44.6961	29.2642	35.2010	31.0944	29.8234
550	30.0989	46.1320	29.3033	35.7277	31.5863	30.1727
600	30.4285	47.4306	29.3426	36.2847	32.0704	30.5923
650	30.7811	48.6044	29.3906	36.8650	32.5355	31.0816
700	31.1478	49.6658	29.4535	37.4628	32.9729	31.6401
750	31.5199	50.6273	29.5352	38.0732	33.3769	32.2677
800	31.8892	51.5014	29.6374	38.6922	33.7443	32.9646
850	32.2475	52.3006	29.7594	39.3170	34.0744	33.7310
900	32.5872	53.0374	29.8984	39.9457	34.3696	34.5679
950	32.9008	53.7242	30.0493	40.5775	34.6347	35.4761
1000	33.1812	54.3736	30.2048	41.2125	34.8772	36.4569

References

1. D. G. Ainley, Performance of axial flow turbines, *Proceedings of the Institute of Mechanical Engineers*, 159, 1948.
2. D. G. Ainley and G. C. R. Mathieson, *A Method for Performance Estimation for Axial-Flow Turbines*, Aeronautical Research Council, Reports and Memoranda, 2974, 1957.
3. G. Agricola, *De Re Metallica*, Translated by Herbert Clark Hoover and Lou Henry Hoover, Dover, New York, 1950.
4. V. Arakeri, Contributions to some cavitation problems in turbomachinery, *Proceedings Acadey of Engineering Sciences*, India, 24:454-483, 1999.
5. O. E. Balje, *Turbo Machines: A Guide to Selection and Theory*, Wiley, New York, 1981.
6. E. A. Baskharone, *Principles of Turbomachiney in Air-Breating Engines*, Cambridge University Press, Cambridge, UK, 2006.
7. W. W. Bathie, *Fundamentals of Gas Turbines*, Wiley, New York, 1984.
8. A. M. Binnie and M. W. Woods, The pressure distribution in a convergent-divergent steam nozzle, *Proceedings of the Institute of Mechanical Engineers*, 138, 1938.
9. C. E. Brennen, *Hydrodynamics of Pumps*, Concepts NREC and Oxford University Press, White River Junction, VT, 1994.
10. C. E. Brennen, *Cavitation and Bubble Dynamics*, Oxford University Press, 1995.
11. H. Brown, *The Challenge of Man's Future*, The Viking Press, New York, 1954.
12. A. F. Burstall, *A History of Mechanical Engineering*, MIT Press, Cambridge, MA, 1965.
13. T. Burton, D. Sharpe, N. Jenkins, and E. Bossanyi, *Wind Energy Handbook*, Wiley, New York, USA, 2001.
14. A. D. S. Carter, *The Low Speed Performance of Related Airfoils in Cascade*, Aeronautical Research Council, CP 29, 1950.

15. H. Cohen, G. F. C. Rogers, and H. I. H. Saravanamuttoo, *Gas Turbine Theory*, 3rd ed., Longman Scientific & Technical, London, 1972.
16. H. R. M. Craig and H. J. A. Cox, Performance estimation of axial-flow turbines, *Proceedings of the Institute of Mechanical Engineers*, 185:407-424, 1970.
17. G. T. Csanady, *Theory of Turbomachines*, McGraw-Hill, New York, 1964.
18. N. A. Cumpsty, *Compressor Aerodynamics*, Longman Scientific & Technical, London, 1989.
19. N. A. Cumpsty, *Jet Propulsion*, Cambridge University Press, Cambridge, UK, 2003.
20. P. de Haller, Das Verhalten von Tragflugelgittern in Axialverdichtern und in Windkanal, *Bernstoff-Wärmer-Kraft* 5, Heft 10, 1953.
21. J. D. Denton, Loss mechanisms in turbomachines, *Journal of Turbomachinery*, 115:621-653, 1993.
22. S. L. Dixon, *Thermodynamics and Fluid Dynamics of Turbomachinery, 5th ed.*, Elsevier, Oxford, 2005.
23. L. F. Drbal, P. G. Boston, and K. L. Westra, *Power Plant Engineering*, Springer, Berlin, 1995.
24. J. Dunham and P. M. Came, *Improvements to the Ainley–Mathiesen Method of Turbine Performance Prediction*, ASME paper 70-GT-2, 1970.
25. D. M. Eggleston and F. S. Stoddard, *Wind Turbine Engineering Design*, Van Nostrand Reinhold, New York, 1987.
26. J. Fullermann, Centrifugal compressors, in *Advances in Petroleum Chemistry and Refining, Engineering*. J. McKetta, ed. Interscience Publishers, New York, 1962.
27. A. J. Glassman, *Computer Program for Design and Analysis of Radial Inflow Turbines*, TN-8164, NASA, 1976.
28. H. Glauert, Airplane propellers, in *Aerodynamic Theory, A General Review of Progress, Vol IV*, W. F. Durand, ed. Dover, New York, 1963.
29. S. Goldstein, On the vortex theory of screw propellers, *Proceedings of the Royal Society*, 123, 929.
30. P. M. Goorjian, An invalid equation in the general momentum theory of actuator disk, *AIAA Journal*, 10, 1972.
31. A. Guha, A unified theory of aerodynamic and condensastion shock waves in vapor-droplet flows with or without a carrier gas, *Physics of Fluids*, 6:1893-1913, 1983.
32. E. Hau, *Wind Turbines, Fundamentals, Technologies, Application, Economics*, Springer, Berlin, 2006.
33. W. R. Hawthorne, *Elements of Turbine and Compressor Theory*, Gas Turbine Laboratory Note, MIT, 1957, (quoted by Horlock, in Axial Flow Turbines [34].)
34. J. H. Horlock, *Axial Flow Compressors*, Butterworth & Co., London, 1958.
35. J. H. Horlock, *Axial Flow Turbines*, Butterworth & Co., London, 1966.
36. A. R. Howell, *The Present Basis of Axial Flow Compressor Design, Part I, Cascade Theory and Performance*, Aeronautical Research Council Reports and Memoranda, No. 2095, 1942.
37. A. R. Howell, Design of axial compressors, *Proceedings of the Institution of Mechanical Engineers*, 153, 1945.
38. A. R. Howell and R. P. Bonham, Overall and stage characteristics of axial-flow compressors, *Proceedings of the Institution of Mechanical Engineers*, 163, 1950.
39. M. K. Hubbert, Energy from fossil fuels, *Science*, 109:103-109, 1945.
40. H. Hugoniot, Propagation des mouvements dans les corps et spécialement dans les gaz parfaits, *Journal de l'Ecole Polytechnique*, 57, 1887.

41. J. C. Hunsaker and B. G. Rightmire, *Engineering Applications of Fluid Mechanics*, McGraw-Hill, New York, 1947.

42. R. A. Huntington, Evaluation of polytropic calculation methods for turbomachinery analysis, *ASME Journal of Engineering for Gas Turbines and Power*, 107:872-879, 1985.

43. *International Energy Outlook 2009*, US-DOE, released May 27, 2009.

44. V. Kadambi and M. Prasad, *An Introduction to Energy Conversion, Vol. III Turbomachinery*, New Age International Publishers, New Delhi, 1977.

45. I. J. Karrasik, J. P. Messina, P. Cooper, and C. C. Heald, *Pump Handbook*, McGraw-Hill, New York, 2007.

46. W. J. Kearton, *Steam Turbine Theory and Practice*, Pitman and Sons, Bath, UK, 1931.

47. J. H. Keenan, *Thermodynamics*, MIT Press, Cambridge USA, 1971.

48. I. M. Khalil, W. Tabakoff, and A. Hamed, Losses in a radial inflow turbine, *Journal of Fluids Engineering*, 98:364-372, 1976.

49. S. A. Korpela, Oil depletion in the world, *Current Science*, 91:1148-1152, 2006.

50. R. I. Lewis, *Turbomachinery Performance Analysis*, Elsevier, Oxford, 1996.

51. S. Lieblein and W. H. Roudebush, *Theoretical Loss Relation for Low-Speed 2D Cascade Flow*, TN 3662, NASA, 1956.

52. E. Logan, Jr., *Turbomachinery; Basic Theory and Applications*, Marcel-Dekker, New York, 1981.

53. A. Lugaresi and A. Massa, Designing Francis turbines: Trends in the last decade, *Water Power & Dam Construction*, 1987.

54. A. Maddison, Contours of world economy and the art of macro-measurement 1500–2001, Rugglers lecture, *IARIW 28th General Conference*, Cork, Ireland, Aug. 2004.

55. M. Mallen and G. Saville, *Polytropic Processes in the Performance Prediction of Centrifugal Compressors*, Institution of Mechanical Engineers, Paper C183/77, 89-96, 1977.

56. L. F. Moody and T. Zowski, Fluid machinery, in *Handbook of Applied Hydraulics*, C. V. Davis and K. W. Sorenson, eds. McGraw-Hill, New-York, 1969.

57. M. Nechleba, *Hydraulic Turbines, Their Design and Equipment*, Artia, Praque, Czechoslovakia, 1957.

58. C. Pfleiderer, *Die Wasserturbinen, mit einem Anhang über Strömungsgetriebe*, Wolfenbütteler Verlag-Anst.,

59. C. Pfleiderer and H. Pertermann, *Strömungsmachinen*, Springer-Verlag, Berlin, 1986. Hannover, Germany, 1947.

60. L. Prandtl, Schraubenpropeller mit geringstem Energieverlust, *Nachrichten Mathematischphyiscalishe Klasse*, Göttingen, Germany, 1919, (Appendix to the article of A. Betz).

61. W. Rankine, On the mechanical principles of propellors, *Transactions of the Institute of Naval Architects*, 6:13-30, 1865.

62. W. Rankine, On the thermodynamic theory of waves of finite longitudinal disturbances. *Philosophical Transactions of the Royal Socciety*, 160:27, 1870.

63. C. Rodgers and R. Gleiser, Performance of a high-efficiency radial/axial turbine. *ASME Journal of Turbomachinery* 109:151-154, 1987.

64. Rolls Royce, *The Jet Engine*, London, 2005.

65. H. Rohlik, Radial-inflow turbines, in *Turbine Design and Applications* A. J. Glassman, ed., SP290, NASA, 1975.

66. S. J. Savonius, The S-rotor and its applications, *Mechanical Engineering*, 53, issue 5, 1931.

67. F. Schweiger and J. Gregory, Developments in the design of Kaplan turbines, *Water Power & Dam Construction*, 1987.

68. D. G. Sheppard, Historical Development of the Windmill, in *Wind Turbine Technology*, D. A. Spera, ed., ASME Press, New York, 1995.

69. J. M. Shultz, The polytropic analysis of centrifugal compressors, *ASME Journal of Engineering for Power*, 69-82, 1962.

70. L. H. Smith, Jr., Casing Boundary Layers in Multistage Compressors, in *Proceedings of the Symposium on Flow Research on Blading*, L. S. DZung, ed., Elsevier, Burlington, MA, 1970.

71. S. F. Smith, A simple correlations of turbine efficiency, *Journal of Royal Aeronautical Society*, 69:467-470, 1965.

72. C. R. Soderberg, *Unpublished Notes of the Gaqs Turbine Laboratory, MIT*, Cambridge, MA, (quoted by Horlock [35]).

73. J. D. Stanitz, Some theoretical aerodynamic investigations of impellers in radial and mixed-flow centrifugal compressors, *ASME Transactions, Series A*, 74:473-497, 1952.

74. A. J. Stepanoff, *Centrifugal and Axial Flow Pumps*, 2nd ed., Wiley, New York, 1957.

75. A. Stodola, *Steam and Gas Turbines*, 6th ed., P. Smith, New York, 1945.

76. W. Traupel, *Thermische Turbomachinen*, 2nd ed. Springer, 1971.

77. J. Volkovitch, Analytical prediction of vortex ring boundaries for helicopters in steep descents, *Journal of American Helicopter Society*, 17(3), 1973.

78. A. Whitfield, The preliminary design of radial inflow turbines, *Journal of Turbomachinery*, 112, 1990.

79. A. Whitfield and N. C. Baines, *Design of Radial Turbomachines*, Longman Scientific and Technical, Essex, UK, 1990.

80. F. J. Wiesner, A review of slip factors for centrifugal compressors, *Journal of Engineering Power*, ASME, 89:558-572, 1967.

81. D. G. Wilson and T. Korakianitis, *Design of High-Efficiency Turbomachinery and Gas Turbines*, 2nd ed., Prentice-Hall, Upper Saddle River, NJ, 1998.

82. R. E. Wilson and P. .B. .E. Lissaman, *Applied Aerodynamics of Wind Power Machines*, Oregon State University , May 1974.

83. A. Wirzenius, *Keskipakopumput (Centrifugal Pumps*, 3rd. ed., (in Finnish), Kustannusyhtymä, Tampere, Finland, 1978.

84. R. H. Zimmerman, *Lecture Notes on Turbomachinery*, The Ohio State University, Columbus, Ohio USA, 1989.

85. O. Zweifel, The Spacing of turbomachinery Blading, Especially with Large Angular Deflection, *Brown Boveri Review*, 32:12, 1945.

Index

actuator disk, 402
advance ratio, 414
Ainley–Mathieson correlation
    profile losses, 205
    secondary losses, 208
air tables, 29
axial compressor
    50% reaction, 227
    blade-loading coefficient, 225
    cascade, 252
    de Haller criterion, 228
    deflection, 228
    design deflection, 231
    diffusion factor, 242
    flow angles, 227
    flow coefficient, 225
    flow deviation, 258
    free vortex defined, 236
    free vortex design, 228, 236
    Lieblein diffusion factor, 242
    multistage reheat factor, 260
    off-design operation, 234
    optimum diffusion, 257
    other losses, 247, 257
    performance characteristics, 234
    polytropic efficiency, 259
    pressure ratio, 221, 224
    radial equilibrium, 235
    reaction, 225
    solidity, 233
    stage efficiency, 250
    stage stagnation temperature rise, 223
    static enthalpy loss coefficients, 250
    typical range for blade-loading coefficient, 222
    typical range of flow coefficient, 222
    work-done factor for multistage compressors, 260
axial turbine
    0% reaction stage, 178
    50% reaction stage, 176
    Ainley–Mathieson correlation for losses, 205
    blade-loading coefficient, 172
    constant mass flux, 188
    fixed nozzle angle, 187
    flow angles, 173
    flow coefficient, 172
    free vortex design, 183
    hub-to-casing ratio, 136
    multistage reheat factor, 215
    off-design operation, 180
    performance characteristics, 199
    polytropic efficiency, 216
    pressure ratio, 193
    radial equilibrium, 181
    reaction, 172
    secondary losses, 208
    Smith chart, 199
    Soderberg loss coefficients, 190
    stage, 167
    stage efficiency, 191
    stage stagnation temperature drop, 181

454 INDEX

stagnation pressure losses for a stage, 193
Zweifel correlation, 204
balance principle defined, 44
blade-loading coefficient
  axial compressor, 225
  axial turbine, 172
  defined, 111
blade element theory for a wind turbine, 409
blade shapes
  axial compressor, 253
  Francis turbine, 373
  single-stage steam turbine, 150
boundary layer
  displacement thickness, 244
  momentum thickness, 244
British gravitational units, 13
buckets, 151
bulb turbine, 361
  specific speed, 363
casing, 166
cavitation
  hydraulic turbines, 380
  pumps, 303
centrifugal compressor
  blade height, 285
  characteristics, 281
  choking of inducer, 280
  diffusion ratio, 284
  history, 12
  illustration of a modern multistage, 12
  inducer, 265
  natural-gas transmission, 5
  optimum inducer angle, 276
  slip, 268
  vaneless diffuser, 287
centrifugal pump
  cavitation, 303
  efficiency, 291, 293
  flow and loading coefficients, 294
  history, 12
  industrial uses, 5
  specific diameter, 294
  specific speed, 294
  vaneless diffuser, 305
  volute design, 306
choking, 65, 67
  compressor, 235
chord and axial chord, 166
Colebrook formula for friction, 84, 368
combustion
  specific heat of gases, 35
  theoretical air, 34
compressible flow
  area change, 61
  choked flow, 65
  converging–diverging nozzle, 67
  converging nozzle, 65
  Fanno flow with area change, 84
  friction in nozzle flow, 75
  Mach waves, 92
  overexpanded, 68

speed of sound, 58
underexpanded, 68
compressor
  characteristics of a radial inflow turbine, 131
  choking, 131
computer software EES, 22
conservation principle defined, 44
Cordier diagram, 294
corrected flow rate, 130
Dalton's model for a mixture of ideal gases, 32
Darcy friction factor, 84
diffusion ratio
  centrifugal compressor, 284
  radial inflow turbine, 350
double-suction pump, 293
efficiency
  axial compressor stage, 225
  axial turbine stage, 191
  centrifugal compressor, 273
  centrifugal pump, 291
  hydraulic turbine, 363
  nozzle, 82
  polytropic, 75
  pressure-compounded steam turbine stage, 149
  radial inflow turbine, 315, 319
  Rankine cycle, 135
  rotor, 139
  steam power plant, 137
  total-to-static, 37, 139
  total-to-total, 36
electricity production, 2
endwalls, 166
energy engineering, 2
energy resources
  biomass, 1
  fossil fuels, 1
  hydraulic, 359
  wind energy, 401
enthalpy
  relative stagnation, 225
  stagnation, 17
Euler equation for turbomachinery, 109
Fanning friction factor, 84
Fanno flow, 84
  stagnation pressure loss, 86
fifty percent (50%) reaction stage, 176
flow angles of absolute and relative velocity, 106
flow coefficient, 111, 125
  axial compressor, 225
  axial turbine, 172
flow function, 63
flow work, 17
fluid coupling
  advantages, 385
  efficiency, 387
  flow rate, 389
  losses, 388
  partially filled, 390
  primary, 385
  secondary, 385
  toroidal shape, 386

torque coefficient, 390
Francis turbine, 362, 370
   specific speed, 363
friction factor
   Colebrook, 84
   Darcy, 84
   Fanning, 84
gas turbine
   electricity generation, 4
   industrial, 11
Gibbs–Dalton model for ideal gases, 33
Helmholtz vortex theorem, 426
hub, 166
hydraulic turbine
   bulb turbine, 361
   capacity factor, 4
   crossflow or Banki–Mitchell turbine, 380
   effective head, 360
   electricity generation, 5
   Francis turbine, 362, 370, 373
   gross head, 360
   history, 8
   Kaplan turbine, 361, 377–378
   mechanical efficiency, 361
   overall efficiency, 360
   Pelton wheel, 362
   pit turbine, 371
   power-specific speed, 362
   synchronous speed, 363
   turgo, 380
   volumetric efficiency, 361
incompressible fluid
   internal energy and irreversibility, 35
   stagnation pressure, 44
induced and forced draft, 39
induction factor, 405
internal heating, 42
jet engine
   gas generator, 221
   history, 11
   spool, 221
Kaplan turbine, 361
   number of blades, 378
   specific speed, 363
kinetic theory of specific heats for gases, 28
Mach number, 59
manometer formula, 127
mass conservation principle, 16
meridional velocity, 106
mixing and pressure change, 53
Mollier diagram, 24
moment of momentum balance, 108
momentum equation, 47
nondimensional groups, 125
normal shock, 68–69
nozzle
   efficiency, 83
   polytropic process, 76
   static enthalpy loss coefficient, 79
   steam, 87
   velocity coefficient, 79

nuclear fuels—uranium and thorium, 1
nucleation—homogeneous and heterogeneous, 90
oblique shock, 68
off-design operation of an axial turbine, 180
overexpanded flow, 68
Pelton wheel, 362
   number of buckets, 363
   specific speed, 363
Pfleiderer correlation for pumps, 293
pitch, 166
polytropic efficiency, 216, 259
positive-displacement machine, 2
power-absorbing machine, 2
power-producing machine, 2
power-specific speed, 362
power coefficient, 127
power ratio of a radial inflow turbine, 315
Prandtl–Meyer theory, 93
pressure compounding, 146
pressure ratio
   axial compressor, 224
   axial turbine, 193
   steam turbine, 136
pressure recovery partial, 51
pressure side of blade, 166
primary energy production
   wind energy, 401
radial equilibrium
   axial compressor, 235
   axial turbine, 181
   constant mass flux, 188
   first power exponent, 241
   fixed nozzle angle, 187
   zero-power exponent, 240
radial inflow turbine, 313
   Balje diagram, 324
   blade height, 351
   efficiency, 319, 345
   minimum exit Mach number, 347
   number of blades, 352
   optimum incidence, 352
   optimum inlet, 339
   radius ratio, 350
   recommended diffusion, 350
   specific diameter, 324
   specific speed, 324
   stator flow, 329
   stator loss coefficients, 333
   typical design parameters, 328
Rankine combined vortex wake, 415
reaction
   axial compressor, 225
   axial turbine, 172
   definition, 116
   in terms of kinetic energies, 116
reheat factor
   axial compressor, 260
   axial turbine, 215
renewable energy, 1
Reynolds number, 125
rothalpy, 114

rotor efficiency
    centrifugal compressor, 270
scale effect, 125
scaling analysis, 124
shape parameter, 128
shock
    normal, 68–69
    Rankine–Hugoniot relations, 73
    shock relations, 72
    strength, 75
similitude, 125
slip, 268
slip stream, 402
sonic state, 62
specific diameter
    centrifugal pump, 294
    radial inflow turbine, 323
specific speed, 128
    centrifugal pump, 294
    hydraulic turbines, 361, 363
    radial inflow turbine, 323
speed of sound, 59
    influence of molecular mass, 59
spouting velocity, 317
stage
    axial compressor, 223
    axial turbine, 167
    normal, 167
stagnation density, 60
stagnation pressure, 60
stagnation pressure loss and entropy, 45
stagnation pressure losses
    axial turbine, 193
    profile losses for axial compressor, 247
stagnation pressure
    low Mach number, 60
stagnation state
    defined, 36
stagnation temperature, 60
static enthalpy loss coefficients, 141
steam tables, 22
steam turbine, 3
    blade shape, 151
    electricity production, 2
    history, 10
    nozzle coefficient, 138
    pressure compounding, 146
    rotor efficiency, 139
    single-stage impulse, 138
    single-stage optimum blade speed, 144
    Soderberg correlation, 160
    types, 136
    velocity compounding, 152
    zero-reaction stage, 158
steam
    computer software EES, 22
    condensation shock, 91
    equation of state, 22
    Mollier diagram, 24
    supersaturation, 24, 90
    undercooling, 91

Wilson line, 24
Zeuner equation, 28
streamline curvature method, 431
subsonic flow defined, 59
suction side of blade, 166
supercritical and ultrasupercritical steam cycle, 3
supersonic flow defined, 59
swirl velocity, 168
thermodynamics
    compressed liquid, 26
    equation of state for air, 29
    equation of state for steam, 22
    first law, 17
    Gibbs equations, 20
    ideal gas, 27
    ideal gas mixutures, 31
    incompressible fluid, 35
    second law, 19
three-dimensional flow
    axial compressor, 235
    axial turbine, 181
tip clearance and leakage flow, 167
torque converter
    efficiency, 393, 396
    torque multiplication, 391
total head, 291
transonic flow defined, 59
trothalpy, 114
turbine characteristics of a radial inflow turbine, 132
turbocharger, 130
turbomachine
    definition of, 2
    history, 7
    household use, 6
    names of components, 2
Tygun formula, 363
underexpanded flow, 68
unloading of a blade, 242
utilization
    definition, 117
    maximum, 119
    relation to reaction, 118
variable specific heats, 41
velocity compounding, 152
velocity triangle, 106
ventilating blower, 39
volute, 306
water wheel history, 7
Wilson line, 24
wind energy
    capacity factor, 401
    Denmark, 401
    Germany, 401
    installed capacity, 5
    United States, 401
wind turbine, 401
    actuator disk, 403
    American windmill, 9
    Betz limit, 406
    blade element theory contributions by
        N. E. Joukovsky, 409

blade element theory development by S. Drzewieci, 402
blade element theory of W. Froude, 402
blade forces for a nonrotating wake, 415
capacity factor, 5
ducted turbine, 408
Glauert theory for an ideal turbine, 424
history, 8
induction factors for an irrotational wake, 415
momentum theory, 403
momentum theory of W. J. M. Rankine, 402
operation as a propeller, 407
power coefficient, 406
Prandtl's tip correction, 426
pressure drop across the actuator disk, 404
Savonius rotor, 9
tip speed ratio, 406, 414
velocity at the actuator disk, 404
wake rotation, 409
windmill, 401
work coefficient, 126
zero percent (0%) reaction, 158, 178
Zeuner's equation, 87
Zweifel correlation, 204

02/1